COMBINATORIAL METHODS IN DISCRETE DISTRIBUTIONS

WILEY SERIES IN PROBABILITY AND STATISTICS

Established by WALTER A. SHEWHART and SAMUEL S. WILKS

Editors: *David J. Balding, Noel A. C. Cressie, Nicholas I. Fisher,
Iain M. Johnstone, J. B. Kadane, Geert Molenberghs, Louise M. Ryan,
David W. Scott, Adrian F. M. Smith, Jozef L. Teugels*
Editors Emeriti: *Vic Barnett, J. Stuart Hunter, David G. Kendall*

A complete list of the titles in this series appears at the end of this volume.

COMBINATORIAL METHODS IN DISCRETE DISTRIBUTIONS

Charalambos A. Charalambides
Department of Mathematics
University of Athens
Athens, Greece

A JOHN WILEY & SONS, INC., PUBLICATION

Copyright © 2005 by John Wiley & Sons, Inc. All rights reserved.

Published by John Wiley & Sons, Inc., Hoboken, New Jersey.
Published simultaneously in Canada.

No part of this publication may be reproduced, stored in a retrieval system, or transmitted in any form or by any means, electronic, mechanical, photocopying, recording, scanning, or otherwise, except as permitted under Section 107 or 108 of the 1976 United States Copyright Act, without either the prior written permission of the Publisher, or authorization through payment of the appropriate per-copy fee to the Copyright Clearance Center, Inc., 222 Rosewood Drive, Danvers, MA 01923, 978-750-8400, fax 978-646-8600, or on the web at www.copyright.com. Requests to the Publisher for permission should be addressed to the Permissions Department, John Wiley & Sons, Inc., 111 River Street, Hoboken, NJ 07030, (201) 748-6011, fax (201) 748-6008.

Limit of Liability/Disclaimer of Warranty: While the publisher and author have used their best efforts in preparing this book, they make no representations or warranties with respect to the accuracy or completeness of the contents of this book and specifically disclaim any implied warranties of merchantability or fitness for a particular purpose. No warranty may be created or extended by sales representatives or written sales materials. The advice and strategies contained herein may not be suitable for your situation. You should consult with a professional where appropriate. Neither the publisher nor author shall be liable for any loss of profit or any other commercial damages, including but not limited to special, incidental, consequential, or other damages.

For general information on our other products and services please contact our Customer Care Department within the U.S. at 877-762-2974, outside the U.S. at 317-572-3993 or fax 317-572-4002.

Wiley also publishes its books in a variety of electronic formats. Some content that appears in print, however, may not be available in electronic format.

Library of Congress Cataloging in Publication Data:

Charalambides, Ch. A.
   Combinatorial methods in discrete distributions / Charalambos A. Charalambides.
     p.   cm.
   Includes bibliographic references and index.
   ISBN 0-471-68027-3 (cloth : acid-free paper)
   1. Combinatorial analysis. 2. Distribution (Probability theory). I. Title

QA164.C52   2004
519.2′4—dc22   2004059097

Printed in the United States of America

10 9 8 7 6 5 4 3 2 1

To Lena

Contents

Preface   xi

1   BASIC COMBINATORICS AND PROBABILITY   1
   1.1   Basic counting principles   1
   1.2   Recurrence relations   9
   1.3   Finite differences   11
   1.4   Discrete probability   16
   1.5   Inclusion and exclusion principle   26
   1.6   Distributions and moments of random variables   40
   1.7   Generating functions   55
   1.8   Reference notes   60
   1.9   Exercises and complements   61

2   STIRLING NUMBERS   73
   2.1   Introduction   73
   2.2   Definitions and generating functions   74
   2.3   Explicit expressions and recurrence relations   85
   2.4   Generalized factorial coefficients   96

	2.5	Enumeration of partitions by subsets and permutations by cycles	108
	2.6	Reference notes	113
	2.7	Exercises and complements	115
3	**GENERALIZED STIRLING AND LAH NUMBERS**	**121**
	3.1	Introduction	121
	3.2	Associated Stirling numbers	122
	3.3	Associated generalized factorial coefficients	129
	3.4	Universal generating functions	134
	3.5	Generalized Stirling numbers	139
	3.6	Generalized Lah numbers	146
	3.7	Reference notes	150
	3.8	Exercises and complements	151
4	**OCCUPANCY DISTRIBUTIONS**	**159**
	4.1	Introduction	159
	4.2	A random occupancy model	160
	4.3	Occupancy distributions	164
	4.4	Particular occupancy distributions	170
		4.4.1 Classical occupancy distribution	170
		4.4.2 Restricted occupancy distribution	173
		4.4.3 Pseudo-contagious occupancy distribution	177
		4.4.4 Restricted Bose-Einstein occupancy distribution	179
	4.5	Statistical applications	182
	4.6	A general random occupancy model	191
	4.7	Reference notes	202
	4.8	Exercises and complements	204
5	**SEQUENTIAL OCCUPANCY DISTRIBUTIONS**	**215**
	5.1	Introduction	215
	5.2	A sequential random occupancy model	216
	5.3	Sequential occupancy distributions	221

5.4	Particular sequential occupancy distributions	225
5.4.1	Sequential classical occupancy distributions	225
5.4.2	Sequential restricted occupancy distributions	230
5.4.3	Sequential pseudo-contagious occupancy distributions	233
5.5	Statistical applications	236
5.6	A reduced sequential occupancy model	241
5.7	Reference notes	244
5.8	Exercises and complements	245

6 CONVOLUTIONS OF TRUNCATED DISTRIBUTIONS 251

6.1	Introduction	251
6.2	Zero truncated discrete distributions	252
6.3	Some particular convolutions	258
6.3.1	Zero truncated Poisson distribution	258
6.3.2	Logarithmic distribution	258
6.3.3	Zero truncated binomial distribution	260
6.3.4	Zero truncated negative binomial distribution	261
6.4	General truncated discrete distributions	262
6.5	Statistical applications	266
6.5.1	Zero truncated power series distribution	266
6.5.2	Left truncated power series distribution	270
6.6	Reference notes	275
6.7	Exercises and complements	276

7 COMPOUND AND MIXTURE DISTRIBUTIONS 281

7.1	Introduction	281
7.2	Compound discrete distributions	282
7.3	Mixture discrete distributions	285
7.4	Particular compounding distributions	287
7.4.1	Poisson compounding distribution	287
7.4.2	Binomial compounding distribution	290
7.4.3	Negative binomial compounding distribution	292
7.4.4	Logarithmic compounding distribution	294

7.5	Compound Poisson distributions	296
	7.5.1 Hermite distribution	297
	7.5.2 Generalized Hermite distribution	300
	7.5.3 Pólya-Aeppli distribution	302
	7.5.4 Poisson distribution of order k	304
7.6	Compound logarithmic distributions	307
	7.6.1 Chebyshev distribution of the first kind	308
	7.6.2 Logarithmic distribution of order k	310
7.7	Compound negative binomial distributions	312
	7.7.1 Chebyshev distribution of the second kind	313
	7.7.2 Gegenbauer distribution	315
	7.7.3 Geometric distribution of order k	317
7.8	Partition polynomials and compound distributions	318
7.9	Reference notes	327
7.10	Exercises and complements	330

Appendix HINTS AND ANSWERS TO EXERCISES   **343**

References   **383**

Author Index   **401**

Subject Index   **405**

Preface

Combinatorial and related discrete mathematical methods occupy a central position in discrete probability theory. Chief among these are combinatorial enumerative methods and basic finite difference calculus methods. Specifically, a considerable number of stochastic phenomena or experiments in discrete probability theory can be described by the stochastic model of distribution (allocation) of balls (objects) into urns (boxes). It is an advantageous approach since urn models can be easily visualized and are very flexible. Further, these models admit many different but equivalent interpretations. The traditional courses in mathematics emphasize infinitesimal calculus ignoring or paying very little attention to finite difference calculus. The lack of knowledge of the basic finite difference methods restricts the treatment of any possible aspect of discrete probability theory by infinitesimal calculus methods and any procedure peculiar to discreteness is presented only as a special device. Of course without differential and integral calculus, discrete probability theory would be very far from its present state of development. Nevertheless, it is admitted that a more systematic treatment using discreteness would enhance the subject. The factorials occupy the same central position in the calculus of finite differences as the powers in the infinitesimal calculus. Thus, the Stirling numbers of the first and second kind, which are the coefficients in the expansion of the factorials into powers and vice versa, constitute a part of the bridge connecting these two calculi.

The present book is devoted to the presentation of combinatorial and related discrete mathematical methods in the theory of certain classes of discrete distributions. As to the contents the following remarks and stressing may be useful. The mathematics prerequisites are modest. They are offered by a basic course in infinitesimal calculus. This should include some power series. The combinatorial, finite differences and probabilistic fundamental background is included in an introductory chapter making the entire text self-contained.

In Chapter 1, the basic counting principles are introduced after a brief presentation of the necessary elements of set theory. The notion of a recurrence (recursive) relation is introduced. Further, finite differences are briefly presented and the connection of a recurrence relation with a finite difference equation is remarked. The probability and the conditional probability on discrete sample spaces are introduced. The inclusion and exclusion principle, which is a basic tool both in combinatorics and discrete probability theory, is extensively treated. The classical Bonferroni inequalities are derived and a limiting form of the general inclusion and exclusion principle is deduced. The last two sections are devoted to an introduction to distributions and moments of random variables and generating functions, respectively.

Chapter 2 presents an elaborate and systematic treatment of the central and noncentral Stirling numbers of the first and second kind. The coefficients of the expansions of the central and noncentral generalized factorials into (usual) factorials, which are connected with the central and noncentral Stirling numbers, are presented. In addition, the enumeration of partitions by subsets and permutations by cycles, a subject furnishing combinatorial interpretations of the central and noncentral Stirling numbers, is discussed.

The discussion of the Stirling numbers and the generalized factorial coefficients, started in Chapter 2, is continued in Chapter 3, moving on to the presentation of the associated Stirling numbers and the associated generalized factorial coefficients. Also, universal generating functions for the numbers of restricted partitions by subsets and permutations by cycles are constructed. Finally, the generalized Stirling and Lah numbers are examined.

Chapter 4 deals with occupancy distributions, which found applications in many fields such as cluster analysis, biology, computer science, and statistical physics. One of the most powerful techniques in the study of occupancy distributions is the reduction of the joint distribution of the occupancy numbers to a joint conditional distribution of independent random variables, given their sum. Exploiting this method, a random occupancy model is specified by the distributions of the independent random occupancy numbers. Then, the distribution of the number of urns each occupied by a specified number of balls is deduced by the aid of the general inclusion and exclusion principle. In the particular case of independent and identically distributed random variables, the occupancy distributions are expressed in terms of finite differences of convolutions of the common distribution of the random variables. Several particular occupancy distributions are presented and their statistical applications are discussed. Finally, a general random occupancy model is introduced and studied.

In Chapter 5, the presentation of the occupancy theory is continued with the discussion of sequential occupancy distributions. Under a sequential random occupancy model, the waiting time and the complementary waiting time distributions are also expressed in terms of finite differences of convolutions of the common distribution of the random variables. Several particular sequential occupancy distributions are examined and their statistical applications are discussed. Finally, a reduced sequential occupancy model is studied.

In Chapter 6, n-fold convolutions of truncated distributions are studied. The sum of the components (random variables) of a random sample of size n from the family of truncated power series distributions with parameter θ (which constitutes a wide class of discrete distributions) is a complete sufficient statistic. Its distribution, which is the n-fold convolution of the parent distribution, facilitates the derivation of minimum variance unbiased estimators for estimable functions of the parameter θ. It is worth noticing that the n-fold convolution of a truncated discrete distribution may be expressed by the n-fold convolution of the corresponding non-truncated discrete distribution, suitably renormed. Further, within the framework of the random occupancy model studied in Chapter 4, the renorming factor, which is a conditional probability, is an occupancy distribution. This connection is utilized in the derivation of the probability function of convolutions of truncated discrete distributions. The n-fold convolution of zero truncated discrete distributions is expressed in terms of finite differences of the convolution of the common distributions of the random variables.

Chapter 7 is concerned with the compound (or generalized) and mixture discrete distributions. Under the random occupancy model, with a random number of urns and the number of balls distributed into any specific urn obeying a discrete probability law, the compound discrete distribution of the total number of balls distributed into the urns is derived. Several particular compound discrete distributions are examined. A class of countable mixture discrete distributions, which is connected to the class of compound discrete distributions, is briefly discussed.

At the end of each chapter, a collection of exercises and complements is provided. Most of these exercises, which are of varying difficulty, aim to consolidate the concepts and results being presented, while others complement, extend or generalize some of the results. So, the working of the exercises must be considered as an integral part of the book. Hints and answers to the exercises are included in an appendix at the end of the book. A clue to difficulty often lies in the elaborateness of the hint. There may be some benefit in first trying an exercise and consulting the hint only when an impasse is reached. Also, the reader may want to consult the hint to the solution of an exercise for an approach that may differ from his/her own.

In order not to interrupt the continuity of the presentation, references through the actual text are kept to a minimum. However, reference notes are included at the end of each chapter to inform on the original contributions. The references are by no means exhaustive; most of them have been included because they were used in one way or another in the text.

This book is intended for graduate students and researchers; it can be used for an advanced course in discrete distributions, requiring some mathematical maturity or exposure to calculus and, perhaps, to probability and statistics. Such a course would include the following sections: 1.3, 1.5, 2.1-2.4, 4.1-4.5, 5.1-5.5, 6.1-6.3, 6.5.1, 7.1-7.5. In principle, the omitted parts of the book do not disturb the main line of the text. Precisely, the omitted sections of Chapter 1 are usually included in a basic probability course. Section 2.5 in Chapter 2 and the entire Chapter 3 treated specialized topics; the associated Stirling numbers and generalized factorial coefficients appear only in the statistical applications of Section 6.5.2, which is also omitted. Sections 4.6 and 5.6 are concerned with extensions of the theory presented in preceded sections, while Sections 7.6 and 7.7 treated very specialized compound distributions.

It is a pleasure to thank my colleague and friend Prof. N. Balakrishnan and the executive editor of John Wiley & Sons, Mr. Steve Quigley, for their support and encouragement throughout the preparation of the book. The detailed and constructive comments of the reviewers of the manuscript were of great help and are gratefully acknowledged. Special thanks are also due to Mrs. Rosa Garderi for the superb job in typesetting the book.

CHARALAMBOS A. CHARALAMBIDES

Athens, January 2005

1
BASIC COMBINATORICS AND PROBABILITY

1.1 BASIC COUNTING PRINCIPLES

The concept of a set is a primitive notion in set theory, like the concepts of a point and a straight line in Euclidean geometry. For the presentation of the basic concepts of combinatorics, only a few elements of set theory are needed and not its axiomatic foundation. In this respect, it is sufficient to consider a *set* as a (well-defined) collection of distinct objects. Sets are usually designated by capital letters of the alphabet with or without subscripts and their elements by lowercase letters. The fact that an element *a belongs to* (or *is a member of*) the set A is expressed by writing $a \in A$. The negation of this statement is expressed by writing $a \notin A$. In describing which objects are contained in a set A, we use the notation

$$A = \{a_1, a_2, \ldots, a_n, \ldots\},$$

which requires that the list of elements of the set A is known, or the notation

$$A = \{a : a \text{ has property } P\},$$

where P is a characteristic property of the elements of the set A.

Two special sets are often of interest: the *universal* set, designated by Ω, which is the set of all objects under consideration, and the *empty* or *null* set, designated by $\emptyset$, which does not contain any of the objects under consideration. A set B is called a *subset* of a set A if and only if for every $b \in B$, we have $b \in A$ (every element of

---

Combinatorial Methods in Discrete Distributions. By Charalambos A. Charalambides
ISBN 0-471-68027-3 Copyright © 2005 John Wiley & Sons, Inc.

B is also an element of A). This is indicated by writing $B \subseteq A$ (and is read as B is a subset of A or B is included in A) or equivalently by writing $A \supseteq B$ (and is read as A is a superset of A or A includes B). If $B \subseteq A$ and there exists $a \in A$ such that $a \notin B$, then B is said to be a *proper subset* of A, and this is indicated by $B \subset A$ or equivalently by $A \supset B$. The fact that $B \subseteq A$ does not exclude $A \subseteq B$; when both relations hold, the sets A and B consist of the same elements and are called *equal* and this is indicated by $A = B$. The set of all subsets of a set A is called the *power set* of A and is denoted by $\mathcal{P}(A)$. In the sequel, all sets under consideration are considered as subsets of a universal set Ω, that is, are all elements of $\mathcal{P}(A)$.

The concept of an ordered pair and generally of an ordered n-tuple is needed for the definition of the Cartesian product of two and generally of n sets. According to the definition of the equality of sets, the pair (two-element set) $\{a, b\}$ is equal to the pair $\{b, a\}$. Further, there are cases where it matters which element is first and which is second, as, for example, in analytic geometry the pair of coordinates (a, b) of a point on the plane designates its abscissa and ordinate, respectively. The necessity of this distinction of the elements of a pair leads to the introduction of the concept of an ordered pair. A pair of elements a and b (not necessarily different) in which a is considered as the first and b as the second is called an *ordered pair* and is denoted by (a, b). According to this definition two pairs (a, b) and (c, d) are equal if and only if $a = c$ and $b = d$. The concept of an ordered n-tuple $(a_1, a_2, \ldots, a_n)$ can be inductively defined as follows. Thus, for $n = 3$, an ordered triple (a_1, a_2, a_3) is defined as

$$(a_1, a_2, a_3) = ((a_1, a_2), a_3),$$

an ordered pair with first element the ordered pair (a_1, a_2) and second the element a_3. Generally, an ordered n-tuple $(a_1, a_2, \ldots, a_n)$ is defined as

$$(a_1, a_2, \ldots, a_{n-1}, a_n) = ((a_1, a_2, \ldots, a_{n-1}), a_n),$$

an ordered pair with first element the ordered $(n-1)$-tuple $(a_1, a_2, \ldots, a_{n-1})$ and second the element a_n. In several cases it is convenient to adopt the vector terminology, where the first element a_1 is called first coordinate, the second element a_2 is called second coordinate, and so on.

The *Cartesian product* of the sets A and B, denoted by $A \times B$, is defined as the set of ordered pairs in which the first coordinate is an element of the set A and the second coordinate is an element of the set B; that is,

$$A \times B = \{(a, b) : a \in A, b \in B\}.$$

This definition can be extended to n sets $A_1, A_2, \ldots, A_n$ as follows:

$$A_1 \times A_2 \times \cdots \times A_n = \{(a_1, a_2, \ldots, a_n) : a_1 \in A_1, a_2 \in A_2, \ldots, a_n \in A_n\}.$$

In particular, if $A_1 = A_2 = \cdots = A_n \equiv A$, the Cartesian product is denoted by A^n.

A subset F of the Cartesian product $A \times B$ is called a *map* (or *function* or *correspondence*) of the set A into the set B if and only if, for every $a \in A$, there exists only one b such that $(a, b) \in F$. Thus, if $(a, b_1) \in F$ and $(a, b_2) \in F$, then $b_1 = b_2$.

In the ordered pair $(a,b) \in F$, the element $a \in A$ is called the *archetype* and the element $b \in B$ is called the *image* of a by F and is usually denoted by $b = F(a)$. Consequently, a map F of the set A into the set B associates to each element $a \in A$ only one element $b \in B$, the image of a by F. If, in addition, for every element $b \in B$ there exists at least one element $a \in A$ such that $(a,b) \in F$, then F is called *surjective* (*onto*). A map F of the set A into the set B is called *injective* (*one-to-one*) if and only if there exists at most one element $a \in A$ such that $(a,b) \in F$. Thus, if $(a_1, b) \in F$ and $(a_2, b) \in F$, then $a_1 = a_2$. Finally, a map F is called *bijective* (*one-to-one* and *onto*) if F is both surjective and injective.

Two sets A and B are called *equivalent* and this is denoted by $A \sim B$ if and only if there exists a bijective map of the set A into the set B.

A set A is called *finite*, with n elements, if and only if it is equivalent to the subset $\{1, 2, \ldots, n\}$ of natural numbers. The empty set $\emptyset$ is considered as finite with 0 elements. A set that is not finite is called *infinite*. A set A is called *countably infinite* if and only if it is equivalent to the set $N = \{1, 2, \ldots, n, \ldots\}$ of natural numbers. Denoting by a_k the element of A that corresponds to the natural number k, for $k = 1, 2, \ldots$, the set A can be represented as

$$A = \{a_1, a_2, \ldots, a_n\},$$

if it is finite with n elements, or as

$$A = \{a_1, a_2, \ldots, a_k, \ldots\},$$

if it is infinitely countable. A set A is called *countable* if it is finite or countably infinite. A set that is not countable is called *uncountable*.

The *union* of two sets A and B is defined as the set that includes the elements of Ω belonging to A or B (the conjunction or being not exclusive) and is denoted by $A \cup B$; that is,

$$A \cup B = \{\omega \in \Omega : \omega \in A \quad \text{or} \quad \omega \in B\}.$$

This definition is extended to a finite or an infinite number of sets:

$$\bigcup_{k=1}^{n} A_k \equiv A_1 \cup A_2 \cup \cdots \cup A_n$$
$$= \{\omega \in \Omega : \omega \in A_k \text{ for at least one subscript } k \in \{1, 2, \ldots, n\}\},$$

$$\bigcup_{k=1}^{\infty} A_k \equiv A_1 \cup A_2 \cup \cdots \cup A_n \cup \cdots$$
$$= \{\omega \in \Omega : \omega \in A_k \text{ for at least one subscript } k \in \{1, 2, \ldots, n, \ldots\}\}.$$

The *intersection* of two sets A and B is defined as the set that includes the common elements of the two sets and is denoted by $A \cap B$; that is,

$$A \cap B = \{\omega \in \Omega : \omega \in A \text{ and } \omega \in B\}.$$

For reasons of economy the notation AB instead of $A \cap B$ is frequently used. This definition is extended to a finite or an infinite number of sets:

$$\bigcap_{k=1}^{n} A_k \equiv A_1 \cap A_2 \cap \cdots \cap A_n$$
$$= \{\omega \in \Omega : \omega \in A_k \text{ for all subscripts } k \in \{1, 2, \ldots, n\}\},$$

$$\bigcap_{k=1}^{\infty} A_k \equiv A_1 \cap A_2 \cap \cdots \cap A_n \cap \cdots$$
$$= \{\omega \in \Omega : \omega \in A_k \text{ for all subscripts } k \in \{1, 2, \ldots, n, \ldots\}\}.$$

The *complement* (with respect to the universal set Ω) of a set A is defined as the set that includes the elements of Ω not belonging to A and is denoted by A' or A^c; that is,

$$A' = \{\omega \in \Omega : \omega \notin A\}.$$

The (set theoretic) *difference* of a set B from a set A is defined as the set that includes the elements of A not belonging to B and is denoted by $A - B$; that is,

$$A - B = \{\omega \in \Omega : \omega \in A, \omega \notin B\}.$$

Note that
$$A' = \Omega - A, \quad A - B = A \cap B'.$$

Two important interrelations between the operations of union, intersection and complementation are known as

De Morgan's formulae: For $A_1, A_2, \ldots, A_n$ subsets of a universal set Ω,

$$(A_1 \cup A_2 \cup \cdots \cup A_n)' = A'_1 \cap A'_2 \cap \cdots \cap A'_n,$$
$$(A_1 \cap A_2 \cap \cdots \cap A_n)' = A'_1 \cup A'_2 \cup \cdots \cup A'_n,$$

and more generally, for a sequence A_k, $k = 1, 2, \ldots$, of subsets of a universal set Ω,

$$\left(\bigcup_{k=1}^{\infty} A_k\right)' = \bigcap_{k=1}^{\infty} A'_k, \quad \left(\bigcap_{k=1}^{\infty} A_k\right)' = \bigcup_{k=1}^{\infty} A'_k.$$

The distinction of two sets according to whether they have or do not have elements in common will be useful in the sequel. In this respect, the next definition is introduced. Two sets A and B are said to be *disjoint* if they do not have elements in common, that is, if $A \cap B = \emptyset$. More generally, the sets $A_1, A_2, \ldots, A_n$ are said to be *pairwise* or *mutually disjoint* if $A_i \cap A_j = \emptyset$, for all pairs of subscripts (i, j), with $i \neq j$, from the set of indices $\{1, 2, \ldots, n\}$. In such a case, the operation of the union is designated by $+$ or Σ instead of $\cup$.

The number of elements of a finite set A is denoted by $N(A)$ or $|A|$ and is called the *cardinal* of it. In the case of a finite universal set Ω its cardinality is taken as

$N(\Omega) \equiv N$. At this point, though clear from the relevant definitions, it is worth noting explicitly the following lemma.

Lemma 1.1 *If A and B are finite and equivalent sets, then*

$$N(A) = N(B).$$

Thus, the cardinal of a finite set A may be deduced by determining a finite set B, equivalent to A, with known cardinality.

The cardinal of the union of finite and pairwise disjoint sets is deduced in the following theorem.

Theorem 1.1 *If $A_1, A_2, \ldots, A_n$ are finite and pairwise disjoint sets, then*

$$N(A_1 + A_2 + \cdots + A_n) = N(A_1) + N(A_2) + \cdots + N(A_n). \quad (1.1)$$

Proof Note first that for A and B finite and disjoint sets, since $A \cap B = \emptyset$, any element of $A + B$ belongs either to A only or to B only and so

$$N(A + B) = N(A) + N(B).$$

Thus, relation (1.1) holds for $n = 2$. Suppose that (1.1) holds for $n - 1$, that is

$$N(A_1 + A_2 + \cdots + A_{n-1}) = N(A_1) + N(A_2) + \cdots + N(A_{n-1}).$$

It will be shown that (1.1) holds also for n. For this reason set

$$A = A_1 + A_2 + \cdots + A_{n-1}, \quad B = A_n,$$

whence

$$A \cap B = (A_1 + A_2 + \cdots + A_{n-1}) \cap A_n$$
$$= A_1 \cap A_n + A_2 \cap A_n + \cdots + A_{n-1} \cap A_n = \emptyset$$

and

$$A + B = (A_1 + A_2 + \cdots + A_{n-1}) + A_n$$
$$= A_1 + A_2 + \cdots + A_{n-1} + A_n.$$

Thus, the sets A and B are finite and disjoint and according to $N(A + B) = N(A) + N(B)$, and the hypothesis that (1.1) holds for $n - 1$, it follows that

$$N(A_1 + A_2 + \cdots + A_n) = N(A_1 + A_2 + \cdots + A_{n-1}) + N(A_n)$$
$$= N(A_1) + N(A_2) + \cdots + N(A_{n-1}) + N(A_n).$$

Hence, according to the principle of mathematical induction, (1.1) holds for every integer $n \geq 2$.

Remark 1.1 Relation (1.1) is often referred to as the *addition principle* and can also be stated as follows: If an element (object) ω_i can be selected in k_i different ways,

$i = 1, 2, \ldots, n$, and the selection of w_i excludes the simultaneous selection of w_j, $i, j = 1, 2, \ldots, n, i \neq j$, then any of the elements (objects) w_1 or w_2 or $\cdots$ or w_n can be selected in $k_1 + k_2 + \cdots + k_n$ ways.

The next theorem is concerned with the cardinality of the Cartesian product of finite sets.

Theorem 1.2 *If A and B are finite sets, then*

$$N(A \times B) = N(A)N(B). \tag{1.2}$$

Proof Let $A = \{a_1, a_2, \ldots, a_k\}$ and $B = \{b_1, b_2, \ldots, b_r\}$. Then the set A may be written in the form

$$A = A_1 + A_2 + \cdots + A_k, A_i = \{a_i\}, i = 1, 2, \ldots, k$$

and the Cartesian product $A \times B$ in the form

$$A \times B = (A_1 + A_2 + \cdots + A_k) \times B$$
$$= (A_1 \times B) + (A_2 \times B) + \cdots + (A_k \times B),$$

where the sets (Cartesian products) $A_1 \times B, A_2 \times B, \ldots, A_k \times B$ are pairwise disjoint. Hence,

$$N(A \times B) = N(A_1 \times B) + N(A_2 \times B) + \cdots + N(A_k \times B).$$

Noting that, for any $i \in \{1, 2, \ldots, k\}$, the Cartesian product $A_i \times B$ is the set of the ordered pairs (a_i, b_j) with first element the only element a_i of the set $A_i = \{a_i\}$ and second element any of the elements b_j, $j = 1, 2, \ldots, r$, of the set B, it follows that

$$N(A_i \times B) = N(B), \quad i = 1, 2, \ldots, k.$$

Thus,

$$N(A \times B) = kN(B) = N(A)N(B)$$

and the proof of the theorem is completed.

A subset S_2 of the Cartesian product Ω^2, of a finite universal set Ω with itself, cannot always be written as a Cartesian product $A \times B$, with $A \subseteq \Omega$ and $B \subseteq \Omega$. Nevertheless, an expression similar to (1.2) may be obtained for the number of elements of S_2 when the number of selections for the first coordinate, and for each of these selections, the number of selections for the second coordinate are known. Specifically, the next corollary, which is readily deduced from Theorems 1.1 and 1.2, is concerned with the number of elements of S_2.

Corollary 1.1 *If $S_2 = (A_1 \times B_1) + (A_2 \times B_2) + \cdots + (A_k \times B_k)$, where $A_1, A_2, \ldots, A_k$ are finite and pairwise disjoint sets and $B_1, B_2, \ldots, B_k$ are finite sets, then*

$$N(S_2) = N(A_1)N(B_1) + N(A_2)N(B_2) + \cdots + N(A_k)N(B_k). \tag{1.3}$$

In particular, if $A_i = \{a_i\}$ and $B_i = \{b_{i,1}, b_{i,2}, \ldots, b_{i,r}\}$, $i = 1, 2, \ldots, k$, whence $S_2 = \{(a_i, b_{i,j}) : i = 1, 2, \ldots, k, \ j = 1, 2, \ldots, r\}$, and introducing $A = \{a_1, a_2, \ldots, a_k\}$, then

$$N(S_2) = N(A)N(B_i) = kr. \tag{1.4}$$

The cardinality of the Cartesian product of more than two finite sets is inductively deduced from Theorem 1.1.

Corollary 1.2 *If $A_1, A_2, \ldots, A_n$ are finite sets, then*

$$N(A_1 \times A_2 \times \cdots \times A_n) = N(A_1)N(A_2)\cdots N(A_n). \tag{1.5}$$

Proof Note first that, according to (1.2), relation (1.5) holds for $n = 2$. Suppose that (1.5) holds for $n - 1$, that is

$$N(A_1 \times A_2 \times \cdots \times A_{n-1}) = N(A_1)N(A_2)\cdots N(A_{n-1}).$$

It will be shown that (1.5) holds also for n. For this reason put

$$A = A_1 \times A_2 \times \cdots \times A_{n-1}, \ B = A_n.$$

Then

$$A \times B = (A_1 \times A_2 \times \cdots \times A_{n-1}) \times A_n = A_1 \times A_2 \times \cdots \times A_{n-1} \times A_n.$$

Thus, according to (1.2), $N(A \times B) = N(A)N(B)$, and the hypothesis that (1.5) holds for $n - 1$, it follows that

$$N(A_1 \times A_2 \times \cdots \times A_n) = N(A_1 \times A_2 \times \cdots \times A_{n-1})N(A_n)$$
$$= N(A_1)N(A_2)\cdots N(A_{n-1})N(A_n)$$

and according to the principle of mathematical induction, (1.5) holds for every integer $n \geq 2$.

An extension of expression (1.4) to a subset S_n of the n-fold Cartesian product Ω^n is given in the following corollary.

Corollary 1.3 *Let S_n be a subset of elements $(\omega_1, \omega_2, \ldots, \omega_n)$ of the n-fold Cartesian product Ω^n, of a finite universal set Ω with itself. Specifically, assume that the first coordinate ω_1 can be selected from a set $A = \{a_1, a_2, \ldots, a_{k_1}\}$ of k_1 elements and, for each selection $\omega_1 = a_{i_1}$, the second coordinate ω_2 can be selected from a set $A_{i_1} = \{a_{i_1,1}, a_{i_1,2}, \ldots, a_{i_1,k_2}\}$ of k_2 elements, $i_1 = 1, 2, \ldots, k_1$, and so on. Finally, assume that for each selection $\omega_1 = a_{i_1}, \omega_2 = a_{i_1,i_2}, \ldots, \omega_{n-1} = a_{i_1,i_2,\ldots,i_{n-1}}$, the last coordinate ω_n can be selected from a set*

$$A_{i_1,i_2,\ldots,i_{n-1}} = \{a_{i_1,i_2,\ldots,i_{n-1},1}, a_{i_1,i_2,\ldots,i_{n-1},2}, \ldots, a_{i_1,i_2,\ldots,i_{n-1},k_n}\}$$

of k_n elements, $i_r = 1, 2, \ldots, k_r$, $r = 1, 2, \ldots, n - 1$. Then

$$N(S_n) = N(A)N(A_{i_1})\cdots N(A_{i_1,i_2,\ldots,i_{n-1}}) = k_1 k_2 \cdots k_n. \tag{1.6}$$

Remark 1.2 Relation (1.6) is often referred to as the *multiplication principle* and can be restated as follows: If an element (object) w_1 can be selected in k_1 ways and for each of these ways an element w_2 can be selected in k_2 ways and so on and for each of these ways an element w_n can be selected in k_n ways, then all the elements w_1 and w_2 and $\cdots$ and w_n can be selected (sequentially) in $k_1 k_2 \cdots k_n$ ways.

In many applications the set of different selections of w_i can not be identified in advance, but only after the selections of $w_1, w_2, \ldots, w_{i-1}$. This does not cause any difficulty in the application of the multiplication principle since the only requirement is the cardinality of the set of selections of w_i. This point is further clarified in the next example.

Example 1.1 *Different kinds of tickets.* Suppose that there are n stations on a railway line. How many different kinds of tickets have to be provided so that booking is possible from any station to any other station?

Let us represent the route from station s_i to station s_j by the pair (s_i, s_j), $i \neq j$. The set of different kinds of tickets that have to be provided is equal to the set S_2 of ordered pairs (a, b) of different stations. Note that the set of different selections of a is $A = \{s_1, s_2, \ldots, s_n\}$, while the set of different selections of b is not known in advance, since b has to be different from a. After the selection $a = s_i$, from the set A, the set of different selections of b is $B_i = A - \{s_i\} = \{s_1, s_2, \ldots, s_{i-1}, s_{i+1}, \ldots, s_n\}$. Irrespective of the possibility of identifying the set B_i its cardinality is in any case equal to $N(B_i) = n - 1$ and hence formula (1.4) can be applied. Therefore,

$$N(S_2) = N(A)N(B_i) = n(n-1),$$

which is the required number of different kinds of tickets.

Example 1.2 *The number of subsets of a finite set.* Consider a finite set $W_n = \{w_1, w_2, \ldots, w_n\}$ and let $\mathcal{A}_n = \mathcal{P}(W_n)$ be the set of subsets of W_n. The number $s_n = N(\mathcal{A}_n)$ may be determined as follows.

Note first that to any subset U of W_n there corresponds an ordered n-tuple $(a_1, a_2, \ldots, a_n)$ such that $a_j = 0$ if $w_j \notin U$ and $a_j = 1$ if $w_j \in U$ for $j = 1, 2, \ldots, n$. In this way, the null set $\emptyset \subseteq W_n$ corresponds to the ordered n-tuple $(0, 0, \ldots, 0)$, with all components zero; the one-element set $\{w_1\} \subseteq W_n$ corresponds to the ordered n-tuple $(1, 0, \ldots, 0)$, with the first component equal to one and the remaining components zero; and the set $W_n \subseteq W_n$ corresponds to the ordered n-tuple $(1, 1, \ldots, 1)$, with all components equal to one. This correspondence is one to one and, according to Lemma 1.1, the number s_n of subsets of W_n is equal to the number of ordered n-tuples $(a_1, a_2, \ldots, a_n)$ with $a_j \in A_j = \{0, 1\}$, which, in turn, is equal to the number of elements of the Cartesian product $A_1 \times A_2 \times \cdots \times A_n$, with $A_j = \{0, 1\}$, $j = 1, 2, \ldots, n$. Then, according to formula (1.5), this number is equal to $s_n = 2^n$.

Example 1.3 *Sum of terms of a geometric progression.* The sum s_n of the first n terms of a geometric progression with the first term equal to 1 and proportion equal to 2 may be derived combinatorially as follows.

Let $\mathcal{E}_n$ be the set of nonempty subsets of a finite set $W_n = \{w_1, w_2, \ldots, w_n\}$ and $\mathcal{C}_k$ the set of subsets of the finite set $W_k = \{w_1, w_2, \ldots, w_k\}$, each of which contains w_k,

$k = 1, 2, \ldots, n$. Then $\mathcal{C}_1, \mathcal{C}_2, \ldots, \mathcal{C}_n$ are pairwise disjoint and $\mathcal{C}_1 + \mathcal{C}_2 + \cdots + \mathcal{C}_n = \mathcal{E}_n$. Note that to each subset of W_k containing the element w_k corresponds a subset of W_{k-1} (without any restriction) from which it is obtained by adding w_k, for $k = 2, 3, \ldots, n$. This correspondence is one to one and hence the number $N(\mathcal{C}_k)$ is equal to the number of subsets of W_{k-1}, which, in turn, according to Example 1.2, is equal to $N(\mathcal{C}_k) = 2^{k-1}$, $k = 2, 3, \ldots, n$. Especially, the number of subsets of $W_1 = \{w_1\}$ containing the element w_1 is equal to $N(\mathcal{C}_1) = 1$. Further, the number of nonempty subsets of W_n is equal to $N(\mathcal{E}_n) = 2^n - 1$. Therefore, according to (1.1),

$$N(\mathcal{C}_1) + N(\mathcal{C}_2) + \cdots + N(\mathcal{C}_n) = N(\mathcal{E}_n),$$

it follows that

$$s_n = 1 + 2 + 2^2 + \cdots + 2^{n-1} = 2^n - 1,$$

which is the required expression.

1.2 RECURRENCE RELATIONS

Consider a sequence of numbers a_n, $n = 0, 1, \ldots$, and let

$$a_{n+r} = F(n, a_n, a_{n+1}, \ldots, a_{n+r-1}), \; n = 0, 1, \ldots.$$

This equation, in which the term a_{n+r} is expressed as a function of the preceding r terms, $a_n, a_{n+1}, \ldots, a_{n+r-1}$, of the sequence, is called a *recurrence relation of order* r. In this book interest is restricted to recurrence relations of the form

$$b_0(n)a_{n+r} + b_1(n)a_{n+r-1} + \cdots + b_r(n)a_n = u(n), \; n = 0, 1, \ldots, \qquad (1.7)$$

where $u(n)$ and the coefficients $b_j(n)$, $j = 0, 1, \ldots, r$, with $b_0(n) \neq 0$ and $b_r(n) \neq 0$, are given functions of n. Recurrence relation (1.7) is particularly called a *linear recurrence relation of order* r. If $u(n) = 0$, $n = 0, 1, \ldots$, (1.7) is called *homogeneous*; otherwise (1.7) is called *complete*. If the coefficients are constant (independent of n): $b_j(n) = b_j$, $j = 0, 1, \ldots, r$, $n = 0, 1, \ldots$, (1.7) is called a *linear recurrence relation of order r with constant coefficients*.

The notion of a recurrence relation is also introduced in the case of a double (two-index) sequence $a_{n,k}$, $n = 0, 1, \ldots$, $k = 0, 1, \ldots$. So, the equation

$$a_{n+r,k+s} = F(n, k, a_{n,k}, a_{n,k+1}, a_{n+1,k}, \ldots, a_{n+r-1,k+s}, a_{n+r,k+s-1}),$$

for $n, k = 0, 1, \ldots$, is called a *recurrence relation of order* (r, s). Note that r is the order of this recurrence relation with respect to the first index (variable) and s is its order with respect to the second index (variable). In this case the term $a_{n+r,k+s}$ is expressed as a function of the $(r+1)(s+1) - 1 = rs + r + s$ terms $a_{n,k}, a_{n,k+1}, a_{n+1,k}, \ldots, a_{n+r-1,k+s}, a_{n+r,k+s-1}$ of the double sequence. In particular, the recurrence relation

$$b_{0,0}(n,k)a_{n+r,k+s} + b_{0,1}(n,k)a_{n+r,k+s-1} + \cdots + b_{r,s}(n,k)a_{n,k} = u(n,k), \quad (1.8)$$

for $n, k = 0, 1, \ldots$, where $u(n, k)$ and the coefficients $b_{i,j}(n, k)$, $i = 0, 1, \ldots, r$, $j = 0, 1, \ldots, s$, with $b_{0,0}(n, k) \neq 0$ and $b_{r,s}(n, k) \neq 0$, are given functions of n and k, is called a *linear recurrence relation of order* (r, s).

Solution of the recurrence relation (1.7) or (1.8) in a set S is called a sequence that makes this equation an identity in S.

Recurrence relations are widely used in combinatorics for the study of many of its problems even in the case where the direct expression of the general term a_n or $a_{n,k}$ in a closed form is possible. From the computational point of view, the tabulation of the number a_n for $n = 0, 1, \ldots$, or the number $a_{n,k}$ for $n = 0, 1, \ldots, k = 0, 1, \ldots$, can be done more easily step by step by using the corresponding recurrence relation. For this purpose, the knowledge of the r *initial conditions* (values) $a_0, a_1, \ldots, a_{r-1}$ in the case of the recurrence relation (1.7) and of the $r + s$ *initial conditions* (sequences) $a_{0,k}, a_{1,k}, \ldots, a_{r-1,k}$ and $a_{n,0}, a_{n,1}, \ldots, a_{n,s-1}$ in the case of the recurrence relation (1.8) is required. The initial conditions guarantee the uniqueness of the solution of the recurrence relation.

Let us consider the *homogeneous linear recurrence relation of order* r, corresponding to (1.7):

$$b_0(n)a_{n+r} + b_1(n)a_{n+r-1} + \cdots + b_r(n)a_n = 0, \quad n = 0, 1, \ldots, \quad (1.9)$$

where the coefficients $b_j(n)$, $j = 0, 1, \ldots, r$, with $b_0(n) \neq 0$ and $b_r(n) \neq 0$, are given functions of n. Note that if $a_1(n)$ and $a_2(n)$ are any solutions of (1.9), then $c_1 a_1(n) + c_2 a_2(n)$, where c_1 and c_2 are arbitrary constants, is also a solution of (1.9). In general, it can be shown that the set of solutions of (1.9) constitutes a linear r dimensional vector space. Note that r solutions $a_1(n), a_2(n), \ldots, a_r(n)$ of (1.9) are linearly independent if and only if their Wronki determinant,

$$W_r(n) = \begin{vmatrix} a_1(n) & a_2(n) & \cdots & a_r(n) \\ a_1(n+1) & a_2(n+1) & \cdots & a_r(n+1) \\ \cdots & \cdots & \cdots & \cdots \\ a_1(n+r-1) & a_2(n+r-1) & \cdots & a_r(n+r-1) \end{vmatrix},$$

is different from zero for some index $n = m$. Consequently, if r solutions $a_1(n)$, $a_2(n), \ldots, a_r(n)$ of the homogeneous linear recurrence relation of order r (1.9) are linearly independent, then they constitute a base for the r dimensional linear vector space of its solutions. Further, every solution of (1.9) is of the form

$$a_n = c_1 a_1(n) + c_2 a_2(n) + \cdots + c_r a_r(n), \quad (1.10)$$

where $c_1, c_2, \ldots, c_r$ are arbitrary constants. The solution (1.10) is called *general solution* of (1.9). In the case where r values $a_m, a_{m+1}, \ldots, a_{m+r-1}$ are given, the system

$$c_1 a_1(m) + c_2 a_2(m) + \cdots + c_r a_r(m) = a_m,$$
$$c_1 a_1(m+1) + c_2 a_2(m+1) + \cdots + c_r a_r(m+1) = a_{m+1},$$
$$\cdots$$
$$c_1 a_1(m+r-1) + c_2 a_2(m+r-1) + \cdots + c_r a_r(m+r-1) = a_{m+r-1},$$

since $W_r(m) \neq 0$, has a unique solution with respect to $c_1, c_2, \ldots, c_r$. Introducing this solution into (1.10), the unique solution of (1.9) is deduced. Further, if $w(n)$ is a *particular solution* of the complete linear recurrence relation of order r (1.7), then according to the preceding analysis it follows directly that

$$a_n = c_1 a_1(n) + c_2 a_2(n) + \cdots + c_r a_r(n) + w(n) \qquad (1.11)$$

is the *general solution* of (1.7).

Example 1.4 *The transfer of the Hanoi tower.* Consider three pegs A, B and C and n cyclic disks of different diameters $\{d_1, d_2, \ldots, d_n\}$, with $d_1 < d_2 < \cdots < d_n$. Initially, the disks are placed on peg A in decreasing order of size from the bottom to the top. Let a_n be the number of movements of disks required for the transfer of the tower from peg A to peg B, using peg C as an auxiliary one, under the restriction that in each movement only one disk is transferred and no disk is placed over a disk of smaller size.

In order to find a recurrence relation for a_n note that the $n - 1$ disks $\{d_1, d_2, \ldots, d_{n-1}\}$ of the tower can be transferred to peg C after a_{n-1} movements. Then, in one movement, the last disk d_n is transferred to peg B. Finally, the tower of the $n - 1$ disks is transferred from peg C to peg B, over the disk d_n, after a_{n-1} movements. Consequently,

$$a_n = 2a_{n-1} + 1, \ n = 2, 3, \ldots, \ a_1 = 1.$$

This is a complete recurrence relation of the first order with constant coefficient. Introducing the sequence $b_n = a_n/2^n$, $n = 1, 2, \ldots$, this recurrence relation is transformed to

$$b_n = b_{n-1} + (1/2)^n, \ n = 2, 3, \ldots, \ b_1 = 1/2.$$

Therefore,

$$b_n = \sum_{r=1}^{n} b_r - \sum_{r=2}^{n} b_{r-1} = \sum_{r=1}^{n} (1/2)^r = 1 - (1/2)^n$$

and

$$a_n = 2^n - 1, \ n = 1, 2, \ldots.$$

1.3 FINITE DIFFERENCES

The first order finite difference, with increment a, of a function $y = f(x)$, denoted by $\Delta_a f(x)$, is defined by

$$\Delta_a f(x) = f(x + a) - f(x).$$

If there is any danger of confusion the difference operator should be denoted by $\Delta_{x,a}$, indicating the independent variable. Recursively, the rth order finite difference of $f(x)$ is defined by

$$\Delta_a^r f(x) = \Delta_a[\Delta_a^{r-1} f(x)], \ r = 2, 3, \ldots.$$

Note that for a function $f(x)$ for which the derivative $Df(x)$ exists, it holds that

$$\lim_{a \to 0} \frac{\Delta_a f(x)}{a} = \lim_{a \to 0} \frac{f(x+a) - f(x)}{a} = Df(x).$$

Thus, symbolically

$$\lim_{a \to 0} \frac{\Delta_a}{a} = D. \tag{1.12}$$

Introducing the shift (displacement), with increment a, of a function $f(x)$, which is defined by

$$E_a f(x) = f(x+a)$$

and recursively by

$$E_a^r f(x) = E_a[E_a^{r-1} f(x)] = f(x + ra), \quad r = 2, 3, \ldots,$$

it follows that

$$\Delta_a f(x) = E_a f(x) - f(x) = (E_a - I) f(x),$$

where $If(x) = f(x)$. Thus, symbolically

$$\Delta_a = E_a - I, \quad E_a = \Delta_a + I.$$

Further, the rth power of the difference operator, $\Delta_a^r = (E_a - I)^r$, on using *Newton's binomial formula*

$$(t+u)^r = \sum_{j=0}^{r} \binom{r}{j} t^j u^{r-j},$$

may be expressed in terms of powers of the shift operator as

$$\Delta_a^r = \sum_{j=0}^{r} (-1)^{r-j} \binom{r}{j} E_a^j, \quad r = 1, 2, \ldots . \tag{1.13}$$

Performing this operator on a function $f(x)$ and since $E_a^j f(x) = f(x + ja)$, we get

$$\Delta_a^r f(x) = \sum_{j=0}^{r} (-1)^{r-j} \binom{r}{j} f(x + ja), \quad r = 1, 2, \ldots .$$

Inversely, the operator relation $E_a^x = (\Delta_a + I)^x$, on using *Newton's general binomial formula*

$$(1+t)^x = \sum_{j=0}^{\infty} \binom{x}{j} t^j, \quad -1 < t < 1, \quad x \in R,$$

may be expanded as

$$E_a^x = \sum_{j=0}^{\infty} \binom{x}{j} \Delta_a^j, \quad x \in R. \tag{1.14}$$

Also, by expanding the operator relation $E_a^x = (I - \Delta_a/E_a)^{-x}$,

$$E_a^x = \sum_{j=0}^{\infty} \binom{x+j-1}{j} \Delta_a^j E_a^{-j}, \quad x \in R. \tag{1.15}$$

It should be noted that when any of the last two operators is performed on a function $f(t)$ the convergence of the resulting series should be secured.

A symbolic formula expressing the difference operator in terms of the derivative operator, and vice versa, may be deduced from the expansion of a function into a Taylor's series. Specifically, the expansion

$$f(x+a) = f(x) + \sum_{r=1}^{\infty} \frac{a^r}{r!} D^r f(x)$$

may be written as

$$\Delta_a f(x) = \sum_{r=1}^{\infty} \frac{a^r}{r!} D^r f(x),$$

implying the symbolic expression

$$\Delta_a = \sum_{r=1}^{\infty} \frac{a^r}{r!} D^r.$$

Therefore,

$$\Delta_a = e^{aD} - 1 \tag{1.16}$$

and

$$aD = \log(1 + \Delta_a). \tag{1.17}$$

In the particular case of increment $a = 1$, the operators Δ_a and E_a are abbreviated to Δ and E, respectively. Also, if the independent variable x should be indicated the notation Δ_x and E_x is used.

A function whose differences are as simple as the derivatives of a power is the factorial. The *factorial of x of order n*, denoted by $(x)_n$, is defined for x a real number and n an integer by

$$(x)_n = x(x-1)\cdots(x-n+1), \; n=1,2,\ldots, \; (x)_0 = 1$$

and for $x \neq -r, r = 1, 2, \ldots, n$, by

$$(x)_{-n} = \frac{1}{(x+n)(x+n-1)\cdots(x+1)}, \; n = 1, 2, \ldots.$$

Remark 1.3 In addition to the notation $(x)_n$ of the factorial of x of order n, adopted in this book, the notation $x^{(n)}$, which indicates the relation of the factorials to the

powers, is also used. The factorial $(x)_n$ is called *descending* (or *falling*) in distinction to the *ascending factorial of x of order n*, which for $n = 1, 2, \ldots$ is defined by the product
$$x(x+1)\cdots(x+n-1).$$
A distinct notation is unnecessary since both factorials can be expressed by the same notation, only the argument being different. Specifically, this product, with the adopted notation, equals
$$(x+n-1)_n = (-1)^n(-x)_n.$$
The *first order difference* of $(x)_n$, $n = 1, 2, \ldots$, is
$$\Delta(x)_n = (x+1)_n - (x)_n = (x+1-x+n-1)(x)_{n-1} = n(x)_{n-1}.$$
Recursively, the rth order difference of $(x)_n$, $n = 1, 2, \ldots$, is obtained as
$$\Delta^r(x)_n = (n)_r(x)_{n-r}, \; r = 1, 2, \ldots, n,$$
$$\Delta^r(x)_n = 0, \; r = n+1, n+2, \ldots .$$
Similarly, for $n = 1, 2, \ldots$,
$$\Delta(x)_{-n} = (x+1)_{-n} - (x)_{-n} = (x+1-x-n-1)(x)_{-n-1}$$
$$= -n(x)_{-n-1}$$
and
$$\Delta^r(x)_{-n} = (-n)_r(x)_{-n-r}, \; r = 1, 2, \ldots .$$
Applying the operator (1.14), with $a = 1$, on the function $f(t) = (t)_n$, $t \in R$,
$$E^x(t)_n = \sum_{r=0}^{\infty} \binom{x}{r} \Delta^r(t)_n, \; t \in R, \; x \in R$$
and since $E^x(t)_n = (t+x)_n$, $\Delta^r(t)_n = (n)_r(t)_{n-r}$, $r = 1, 2, \ldots, n$, $\Delta^r(t)_n = 0$, $r = n+1, n+2, \ldots$, we deduce *Vandermonde's formula*, (N. Vandermonde (1772)),
$$(t+x)_n = \sum_{r=0}^{n} \binom{n}{r}(x)_r(t)_{n-r}, \; t \in R, \; x \in R. \qquad (1.18)$$

The *generalized factorial of x of order n and increment a*, denoted by $(x)_{n,a}$, is defined for x and a real numbers and n integer by
$$(x)_{n,a} = x(x-a)\cdots(x-an+a), \; n = 1, 2, \ldots, \; (x)_{0,a} = 1$$
and for $x \neq -ra$, $r = 1, 2, \ldots, n$, by
$$(x)_{-n,a} = \frac{1}{(x+na)(x+na-a)\cdots(x+a)}, \; n = 1, 2, \ldots .$$

Note that the generalized factorial of x of order n and increment a, $(x)_{n,a}$, may be expressed as a *(generalized) factorial of x of order n* (unit increment) *and scale parameter s*: $(x)_{n,a} = s^{-n}(sx)_n$, $s = 1/a$.

Another equally important function in the calculus of finite differences is the binomial (coefficient). The *binomial (coefficient) of x of order n*, denoted by $\binom{x}{n}$, is defined for every real number x and a nonnegative integer n by

$$\binom{x}{n} = \frac{(x)_n}{n!}, \ n = 1, 2, \ldots, \ \binom{x}{0} = 1.$$

The *difference of the binomial (coefficient) of x of order n*, using Pascal's triangle, is obtained as

$$\Delta \binom{x}{n} = \binom{x+1}{n} - \binom{x}{n} = \binom{x}{n-1}.$$

Recursively,

$$\Delta^r \binom{x}{n} = \binom{x}{n-r}, \ r = 1, 2, \ldots, n,$$

$$\Delta^r \binom{x}{n} = 0, \ r = n+1, n+2, \ldots .$$

Applying the operator (1.14), with $a = 1$, on the function $f(t) = \binom{t}{n}$, $t \in R$,

$$E^x \binom{t}{n} = \sum_{r=0}^{\infty} \binom{x}{r} \Delta^r \binom{t}{n}, \ t \in R, \ x \in R,$$

we readily deduce *Cauchy's formula*, (A. Cauchy (1833)),

$$\binom{x+t}{n} = \sum_{r=0}^{n} \binom{x}{r} \binom{t}{n-r}, \ t \in R, \ x \in R. \quad (1.19)$$

The *generalized binomial (coefficient) of x of order n and increment a*, denoted by $\binom{x}{n}_a$, is defined for x and a real numbers and a nonnegative integer n by

$$\binom{x}{n}_a = \frac{(x)_{n,a}}{n!}, \ n = 1, 2, \ldots, \ \binom{x}{0}_a = 1.$$

An equation of the form

$$\Phi(x, f(x), E_a f(x), \ldots, E_a^r f(x)) = 0, \ x \in R$$

or equivalently of the form

$$\Psi(x, f(x), \Delta_a f(x), \ldots, \Delta_a^r f(x)) = 0, \ x \in R$$

is called a *finite difference equation of order r*. This equation is a general (implicit) form of a difference equation of order r. Under certain conditions it can be written in the following solved form, with respect to $E_a^r(x)$:

$$E_a^r f(x) = F(x, f(x), E_a f(x), \ldots, E_a^{r-1} f(x)), \ x \in R. \quad (1.20)$$

In the particular case where the function $y = f(x)$ is defined only on a countable set of points $\{x_0, x_1, x_2, \ldots, x_n, \ldots\}$, which, in the applications of the calculus of finite differences, are usually equidistant, $x_n = x_0 + na$, $n = 0, 1, \ldots$, the transformation $z_n = (x_n - x_0)/a$, $n = 0, 1, \ldots$ is used. So, the function f is transformed to the function g, with $g(n) = f(x_0 + na)$, which is defined on the set $\{0, 1, \ldots\}$ of nonnegative integers. In this particular case using the sequence

$$y_n = g(n) = f(x_0 + na), \ n = 0, 1, \ldots,$$

and since $E_a^j f(x_n) = f(x_n + ja) = g(n+j) = y_{n+j}$, $j = 1, 2, \ldots$, equation (1.20), with $x = x_n$, may be written as

$$y_{n+r} = F(n, y_n, y_{n+1}, \ldots, y_{n+r-1}), \ n = 0, 1, \ldots, \tag{1.21}$$

which is a recurrence relation of order r. Note that the method of solving the finite difference equation (1.20) is the same as that of solving the recurrence relation (1.21) even when the function $y = f(x)$ is defined for every $x \in R$. In the last case an investigation of the solution is required.

Similarly, considering the partial finite differences of a bivariate function $f(x, y)$: $\Delta_{y,b}^j \Delta_{x,a}^i f(x, y)$, $i, j = 0, 1, \ldots$, where

$$\Delta_{x,a} f(x, y) = f(x + a, y) - f(x, y),$$

$$\Delta_{y,b} f(x, y) = f(x, y + b) - f(x, y),$$

a finite difference equation with partial differences may be defined. Its connection with a recurrence relation of a double (two-index) sequence can also be established.

1.4 DISCRETE PROBABILITY

The theory of probability is concerned with the study of mathematical models, known as *stochastic models*, which are used in explaining *random* or *stochastic phenomena* or *experiments*. The basic characteristic of these experiments is that the conditions under which they are performed and the values of various quantities appearing in them do not predetermine the outcome but do predetermine the set of possible outcomes. The element of randomness lies on the inability of predetermining the outcome of random phenomena or experiments. The set Ω of possible outcomes of a random phenomenon or experiment is called a *sample space* and the elements ω of Ω are called *sample points*. It should be noted that it is possible to define more than one set of possible outcomes for each random phenomenon or experiment and according to the requirements of the specific problem the more appropriate of these is chosen as the sample space. The inappropriate choice of the sample space leads to many paradoxes. The sample space may be finite or countably infinite or uncountable. For the cases of finite or more generally countable sample spaces, on which this presentation concentrates, every subset A of Ω is called an *event*. An event $A = \{\omega\}$ containing only one element of Ω is called an *elementary* or *simple* event.

Definition 1.1 *Let Ω be a discrete sample space. A set function $P(\cdot)$ defined on the set of events $\mathcal{P}(\Omega)$ and assuming real values is called a probability (or a measure of probability) if it satisfies the following properties (axioms):*
(a) *nonnegativity,*
$$P(A) \geq 0 \text{ for every event } A \in \mathcal{P}(\Omega),$$
(b) *normalization,*
$$P(\Omega) = 1,$$
(c) *countable additivity,*
$$P(A_1 + A_2 + \cdots + A_n + \cdots) = P(A_1) + P(A_2) + \cdots + P(A_n) + \cdots,$$
for every sequence of pairwise disjoint events $A_i \in \mathcal{P}(\Omega)$, $i = 1, 2, \ldots$.

Note that a direct consequence of Definition 1.1 is that
$$P(\emptyset) = 0,$$
which follows from axiom (c) by setting $A_i = \emptyset$, $i = 1, 2, \ldots$, and since $P(\emptyset) \geq 0$. Also, the probability is *finitely additive*:
$$P(A_1 + A_2 + \cdots + A_n) = P(A_1) + P(A_2) + \cdots + P(A_n),$$
for every finite sequence of pairwise disjoint events $A_i \in \mathcal{P}(\Omega)$, $i = 1, 2, \ldots, n$. This property follows again from axiom (c) by taking $A_i = \emptyset$, $i = n+1, n+2, \ldots$.

The three axioms are quite general and for the calculation of the probability of any event $A \in \mathcal{P}(\Omega)$ the knowledge of the probabilities $p_i = P(\{\omega_i\})$ of the elementary events $\{\omega_i\}$, $i = 1, 2, \ldots$ is required. Indeed, if $A = \{\omega_{i_1}, \omega_{i_2}, \ldots, \omega_{i_k}, \ldots\}$, then, with $A_{i_r} = \{\omega_{i_r}\}$, $r = 1, 2, \ldots, k, \ldots$, it follows that
$$A = A_{i_1} + A_{i_2} + \cdots + A_{i_k} + \cdots, \quad A_{i_r} \cap A_{i_s} = \emptyset, \ r \neq s,$$
and, according to the axiom of countable additivity, that
$$P(A) = \sum_{r=0}^{\infty} P(\{\omega_{i_r}\}).$$
In the special case of a finite sample space $\Omega = \{\omega_1, \omega_2, \ldots, \omega_N\}$, whose elements (sample points, possibilities) are equally probable,
$$P(\{\omega_1\}) = P(\{\omega_2\}) = \cdots = P(\{\omega_N\}) = \frac{1}{N},$$
it reduces to
$$P(A) = \frac{N(A)}{N},$$
where $N(A)$ is the number of elements of A and $N \equiv N(\Omega)$ is the number of elements of the sample space Ω. This uniform probability measure is known as the

classical probability. The calculation of the probability $P(A)$ of an event A, in a finite sample space Ω whose elements are equally probable, is a purely combinatorial problem of counting the numbers $N(A)$ and N of certain configurations.

Some useful properties of the probability are derived in the following theorem.

Theorem 1.3 (a) *If A' is the complement of an event A, with respect to a sample space Ω, then*

$$P(A') = 1 - P(A). \tag{1.22}$$

(b) *If A and B are any events in Ω, then*

$$P(A - B) = P(A) - P(AB) \tag{1.23}$$

and particularly for $B \subseteq A$,

$$P(A - B) = P(A) - P(B). \tag{1.24}$$

(c) *If A and B are any events in Ω, then*

$$P(A \cup B) = P(A) + P(B) - P(AB). \tag{1.25}$$

Proof (a) The events A and A' are disjoint and $A + A' = \Omega$. Thus, using the finite additivity property and axiom of normalization, we get

$$P(A) + P(A') = P(\Omega) = 1,$$

which implies (1.22).

(b) Clearly, the events $A - B = AB'$ and AB are disjoint and their union is $(A - B) + AB = A$. Hence,

$$P(A) = P(A - B) + P(AB),$$

which implies (1.23). In particular, for $B \subseteq A$ we have $AB = B$ and (1.24) is deduced.

(c) The events $A - B = AB'$ and B are also disjoint and their union equals $(A - B) + B = A \cup B$. Thus, using the finite additivity property, we get

$$P(A \cup B) = P(A - B) + P(B),$$

which by virtue of (1.23) implies (1.25).

Example 1.5 Consider an urn containing n balls numbered $1, 2, \ldots, n$. Suppose that k balls are drawn, at random, with replacement. Let A_r be the event that the largest drawn number equals r, $r = 1, 2, \ldots, n$. Find the probability $P(A_r)$, $r = 1, 2, \ldots, n$.

The derivation of the probability $P(A_r)$ is facilitated by the expression of the event A_r as a difference of two events as follows: Let B_r be the event that the largest drawn number is less than or equal to r, $r = 1, 2, \ldots, n$. Then,

$$A_1 = B_1, \ A_r = B_r - B_{r-1}, \ B_{r-1} \subseteq B_r, \ r = 2, 3, \ldots, n,$$

and according to (1.24),

$$P(A_1) = P(B_1), \ P(A_r) = P(B_r) - P(B_{r-1}), \ r = 2, 3, \ldots, n.$$

The number of elements of the sample space Ω of the stochastic experiment equals $N(\Omega) = n^k$, the number of k-permutations of the set $\{1, 2, \ldots, n\}$ with repetition. Further, the number of elements of the event B_r equals $N(B_r) = r^k$, the number of k-permutations of the set $\{1, 2, \ldots, r\}$ with repetition. Since the sample points are equiprobable it follows that

$$P(B_r) = \frac{r^k}{n^k}, \ r = 1, 2, \ldots, n$$

and so

$$P(A_r) = \frac{r^k - (r-1)^k}{n^k}, \ r = 1, 2, \ldots, n.$$

The need for the introduction of the conditional probability emerges in cases where a partial knowledge about the result of a stochastic experiment lessens the uncertainty by reducing the sample space. Specifically, consider a stochastic experiment with sample space Ω and probability $P(A)$ for every event $A \in \mathcal{P}(\Omega)$. Suppose that at some stage of its performance a certain event A has occurred. Then, as regards the final result of the stochastic experiment, the sample space is reduced to the set A and any event B (with respect to the sample space Ω) is reduced to the event $C = AB$, which is denoted by $B|A$ and read as the event B given the event A. The probability of the event B given the event A, which is denoted by $P(B|A)$, $B \in \mathcal{P}(\Omega)$ and called conditional probability (given the event A), is defined by connecting it with the probabilities $P(A)$ and $P(AB)$. Hence, there is no need to postulate separately the existence of the probability $P(B|A)$, $B \in \mathcal{P}(\Omega)$ and its properties.

Definition 1.2 *Let Ω be a discrete sample space and $A \in \mathcal{P}(\Omega)$ an event with $P(A) > 0$. The conditional probability, given the event A, is a set function $P(\cdot|A)$ defined by*

$$P(B|A) = \frac{P(AB)}{P(A)}, \ B \in \mathcal{P}(\Omega).$$

When $P(A) = 0$, $P(B|A)$ is not defined. For any specific event B, the probability $P(B|A)$ is called the conditional probability of B given A.

The conditional probability can be used in expressing the probability of the intersection of a finite number of events. Specifically, we have the following *multiplicative theorem*.

Theorem 1.4 *Let $A_i \in \mathcal{P}(\Omega)$, $i = 1, 2, \ldots, n$, be events such that $P(A_1 A_2 \cdots A_{n-1}) > 0$. Then*

$$P(A_1 A_2 \cdots A_n) = P(A_1) P(A_2|A_1) \cdots P(A_n|A_1 A_2 \cdots A_{n-1}). \quad (1.26)$$

Proof Note first that

$$A_1 A_2 \cdots A_{n-1} \subseteq A_1 A_2 \cdots A_{n-2} \subseteq \cdots \subseteq A_1 A_2 \subseteq A_1,$$

and so

$$P(A_1 A_2 \cdots A_{n-1}) \leq P(A_1 A_2 \cdots A_{n-2}) \leq \cdots \leq P(A_1 A_2) \leq P(A_1).$$

Further, since $P(A_1 A_2 \cdots A_{n-1}) > 0$, it follows that

$$P(A_1) > 0, \ P(A_1 A_2) > 0, \ldots, \ P(A_1 A_2 \cdots A_{n-2}) > 0.$$

Consequently, the conditional probabilities on the right-hand side of (1.26) are meaningful (defined). Then, according to Definition 1.2, we have

$$P(A_2|A_1) = \frac{P(A_1 A_2)}{P(A_1)}, \ P(A_3|A_1 A_2) = \frac{P(A_1 A_2 A_3)}{P(A_1 A_2)}, \ldots,$$

$$P(A_n|A_1 A_2 \cdots A_{n-1}) = \frac{P(A_1 A_2 \cdots A_{n-1} A_n)}{P(A_1 A_2 \cdots A_{n-1})}$$

and so

$$P(A_1 A_2 \cdots A_n) = P(A_1) \frac{P(A_1 A_2)}{P(A_1)} \frac{P(A_1 A_2 A_3)}{P(A_1 A_2)} \cdots \frac{P(A_1 A_2 \cdots A_{n-1} A_n)}{P(A_1 A_2 \cdots A_{n-1})}$$
$$= P(A_1) P(A_2|A_1) P(A_3|A_1 A_2) \cdots P(A_n|A_1 A_2 \cdots A_{n-1}),$$

which is the required expression.

The probability of any event may be expressed as a sum of probabilities by using conditional probabilities of this event. This expression requires the notion of a partition of the sample space, which is defined as follows. A set of pairwise disjoint events $\{A_1, A_2, \ldots, A_n, \ldots\}$ in a sample space Ω is called a *partition* of Ω if $A_1 + A_2 + \cdots + A_n + \cdots = \Omega$. The next theorem is known as the *total probability theorem*.

Theorem 1.5 *Let* $\{A_1, A_2, \ldots, A_n, \ldots\}$ *be a partition of a sample space* Ω, *with* $P(A_k) > 0$, $k = 1, 2, \ldots$, *and B an event in* Ω. *Then*

$$P(B) = \sum_{k=1}^{\infty} P(A_k) P(B|A_k). \qquad (1.27)$$

Proof Since $\{A_1, A_2, \ldots, A_n, \ldots\}$ is a partition of a sample space Ω, for any event B in Ω, the events $C_k = A_k B$, $k = 1, 2, \ldots$, are pairwise disjoint, $C_i C_j = (A_i A_j) B = \emptyset$, and

$$\sum_{k=1}^{\infty} (A_k B) = \left(\sum_{k=1}^{\infty} A_k \right) B = \Omega B = B.$$

Therefore, according to the additivity property of the probability, it follows that

$$P(B) = \sum_{k=1}^{\infty} P(A_k B)$$

and since $P(A_k) > 0$, $k = 1, 2, \ldots$, on using the multiplicative theorem,

$$P(A_k B) = P(A_k) P(B|A_k),$$

(1.27) is deduced.

An interesting consequence of the conditional probability, in connection with the total probability theorem, which has caused much controversy between statisticians, is *Bayes' theorem*.

Theorem 1.6 *Let* $\{A_1, A_2, \ldots, A_n, \ldots\}$ *be a partition of a sample space* Ω, *with* $P(A_k) > 0$, $k = 1, 2, \ldots$, *and B an event in* Ω *with* $P(B) > 0$. *Then*

$$P(A_r|B) = \frac{P(A_r) P(B|A_r)}{\sum_{k=1}^{\infty} P(A_k) P(B|A_k)}, \quad r = 1, 2, \ldots. \tag{1.28}$$

Proof The expression of the conditional probability $P(A_r|B)$, on using the multiplicative theorem and the total probability theorem, yields

$$P(A_r|B) = \frac{P(A_r B)}{P(B)} = \frac{P(A_r) P(B|A_r)}{\sum_{k=1}^{\infty} P(A_k) P(B|A_k)}, \quad r = 1, 2, \ldots,$$

which is the required expression.

In general, the conditional probability $P(B|A)$ differs from the probability $P(B)$. When $P(B|A) > P(B)$ or $P(B|A) < P(B)$, the knowledge of the occurrence of the event A provides us with positive or negative information increasing or decreasing the probability of the occurrence of the event B, respectively. In the particular case $P(B|A) = P(B)$, the knowledge of the occurrence of the event A has no effect on the probability of the occurrence of the event B. The event B is then called *stochastically independent* of the event A. Since $P(AB) = P(A)P(B|A)$ and $P(AB) = P(A|B)P(B)$, whenever the event B is stochastically independent of the event A, so is A of B and inversely. Further, in this case it follows that $P(AB) = P(A)P(B)$ and this symmetric relation in A and B may be adopted as the definition of independence. Note that this relation holds true even if any of or both the probabilities $P(A)$ and $P(B)$ equal zero. Indeed, since $AB \subseteq A$ and $AB \subseteq B$, whence $P(AB) \leq P(A)$ and $P(AB) \leq P(B)$, it follows that the probability $P(AB)$ equals zero if any of or both the probabilities $P(A)$ and $P(B)$ equal zero. This definition generalizes to more than two events as follows.

Definition 1.3 *The events A_i, $i = 1, 2, \ldots, n$ are called (mutually or completely) stochastically independent if and only if the following relation holds:*

$$P(A_{i_1} A_{i_2} \cdots A_{i_r}) = P(A_{i_1}) P(A_{i_2}) \cdots P(A_{i_r}), \tag{1.29}$$

for every r-combination $\{i_1, i_2, \ldots, i_r\}$ of the n indices $\{1, 2, \ldots, n\}$ and $r = 2, 3, \ldots, n$.

Example 1.6 *Pairwise but not completely independent events.* Suppose that two balls are successively drawn at random and with replacement from an urn containing $2n$ balls numbered $1, 2, \ldots, 2n$. Let A_1 be the event that number drawn at the first trial is even, A_2 be the event that the number drawn at the second trial is even and A_3 be the event that the sum of the numbers drawn at the two trials is even. Examine whether the events A_1, A_2 and A_3 are independent.

The number of elements of the sample space Ω equals $N(\Omega) = (2n)^2$, the number of 2-permutations (i, j) of the set $\{1, 2, \ldots, 2n\}$, with repetition. These sample points are equiprobable and according to the classical definition of probability,

$$P(A_1) = P(A_2) = P(A_3) = \frac{2n^2}{(2n)^2} = \frac{1}{2},$$

$$P(A_1 A_2) = P(A_1 A_3) = P(A_2 A_3) = \frac{n^2}{(2n)^2} = \frac{1}{4},$$

$$P(A_1 A_2 A_3) = \frac{n^2}{(2n)^2} = \frac{1}{4}.$$

Hence,

$$P(A_1 A_2) = P(A_1)P(A_2), \ P(A_1 A_3) = P(A_1)P(A_3),$$
$$P(A_2 A_3) = P(A_2)P(A_3),$$

and

$$P(A_1 A_2 A_3) \neq P(A_1)P(A_2)P(A_3).$$

Consequently, the events A_1, A_2 and A_3 are pairwise independent but not completely independent.

Example 1.7 Consider an urn containing $2n$ balls numbered $\{b_1, b_2, \ldots, b_{2n}\}$. Suppose that all the balls are successively drawn one after the other, at random, without replacement. Let A_i be the event that ball b_{2i-1} is drawn before the drawing of ball b_{2i}, $i = 1, 2, \ldots, n$. Examine whether the events $A_1, A_2, \ldots, A_n$ are independent.

A sample point may be represented by a permutation $(b_{j_1}, b_{j_2}, \ldots, b_{j_{2n}})$ of the set $\{b_1, b_2, \ldots, b_{2n}\}$, where b_{j_r} is the ball drawn at the rth trial, $r = 1, 2, \ldots, 2n$. Thus, the number of elements of the sample space Ω, of the stochastic experiment of the drawings of the $2n$ balls, equals $N(\Omega) = (2n)!$, the number of permutations $(b_{j_1}, b_{j_2}, \ldots, b_{j_{2n}})$ of the set $\{b_1, b_2, \ldots, b_{2n}\}$. Let $\{i_1, i_2, \ldots, i_r\}$ be a selection of r indices out of the n indices $\{1, 2, \ldots, n\}$. The number $N(A_{i_1} A_{i_2} \cdots A_{i_r})$, of permutations $(b_{j_1}, b_{j_2}, \ldots, b_{j_{2n}})$ of the set $\{b_1, b_2, \ldots, b_{2n}\}$, in which in each of

r specific pairs, (b_{2i_k-1}, b_{2i_k}), $k = 1, 2, \ldots, r$, the order of the two elements is specified, equals the product

$$(2n - 2r)! \binom{2n - 2r + 2}{2} \binom{2n - 2r + 4}{2} \cdots \binom{2n}{2},$$

of the number $(2n - 2r)!$ of permutations of the $2n - 2r$ non specified balls, and the number of ways the r specified ordered pairs of balls may be introduced in each such permutation. Since

$$\binom{2n - 2r + 2k}{2} = \frac{(2n - 2r + 2k)(2n - 2r + 2k - 1)}{2}, \quad k = 1, 2, \ldots, r,$$

it follows that

$$N(A_{i_1} A_{i_2} \cdots A_{i_r}) = \frac{(2n)!}{2^r}, \quad r = 1, 2, \ldots, n.$$

Therefore, according to the classical definition of probability, for $r = 1$,

$$P(A_i) = \frac{1}{2}, \quad i = 1, 2, \ldots, n$$

and for $r = 2, 3, \ldots, n$ and every r-combination $\{i_1, i_2, \ldots, i_r\}$ of the n indices $\{1, 2, \ldots, n\}$,

$$P(A_{i_1} A_{i_2} \cdots A_{i_r}) = \frac{1}{2^r}.$$

Thus, all the conditions (1.29) are fulfilled and consequently the events $A_1, A_2, \ldots, A_n$ are independent.

The conditional probability

$$P(C|AB) = \frac{P(ABC)}{P(AB)}, \quad P(AB) > 0,$$

on using the expressions $P(ABC) = P(A)P(BC|A)$ and $P(AB) = P(A)P(B|A)$, may be written as

$$P(C|AB) = \frac{P(BC|A)}{P(B|A)}, \quad P(B|A) > 0.$$

In general, the conditional probability $P(C|AB)$ differs from the conditional probability $P(C|A)$. In the particular case $P(C|AB) = P(C|A)$, given the occurrence of the event A, the additional knowledge of the occurrence of the event B has no effect on the probability of the occurrence of the event C. The event C is then called *conditionally independent* of the event B, given the event A. Since $P(BC|A) = P(B|A)P(C|AB)$ and $P(BC|A) = P(B|AC)P(C|A)$, whenever the event C is conditionally independent of the event B, given the event A, so is B of C and inversely. Further, in this case it follows that $P(BC|A) = P(B|A)P(C|A)$

24 BASIC COMBINATORICS AND PROBABILITY

and this symmetric relation in B and C may be adopted as the definition of conditional independence given A. This definition generalizes to more than two events as follows.

Definition 1.4 *The events B_i, $i = 1, 2, \ldots, n$ are called (mutually or completely) conditionally independent, given the event A, if and only if the following relation holds:*

$$P(B_{i_1} B_{i_2} \cdots B_{i_r} | A) = P(B_{i_1} | A) P(B_{i_2} | A) \cdots P(B_{i_r} | A), \qquad (1.30)$$

for every r-combination $\{i_1, i_2, \ldots, i_r\}$ of the n indices $\{1, 2, \ldots, n\}$ and $r = 2, 3, \ldots, n$.

Example 1.8 *Independent but not conditionally independent events.* Consider a lottery-urn containing ten balls bearing the digits $0, 1, \ldots, 9$. Suppose that three balls are successively drawn at random and with replacement. Let A be the event that at the first drawing the digit 0 is drawn, B the event that at the second drawing the digit 0 is drawn and C the event that in exactly two drawings the same digit is drawn. Show that the events B and C are independent, while the same events are not conditionally independent, given the event A.

A sample point may be represented by an ordered triple (j_1, j_2, j_3), where j_r is the digit drawn at the rth drawing, $r = 1, 2, 3$. The number of elements of the sample space Ω equals $N(\Omega) = 10^3$, the number of 3-permutations (j_1, j_2, j_3) of the set $\{0, 1, \ldots, 9\}$, with repetition. These sample points are equiprobable and according to the classical definition of probability,

$$P(B) = \frac{10^2}{10^3} = \frac{1}{10}, \; P(C) = \frac{3 \cdot 10 \cdot 9}{10^3} = \frac{27}{100}, \; P(BC) = \frac{3 \cdot 9}{10^3} = \frac{27}{1000}.$$

Thus,
$$P(BC) = P(B)P(C)$$

and so the events B and C are independent. Further,

$$P(B|A) = \frac{10}{10^2} = \frac{1}{10}, \; P(C|A) = \frac{3 \cdot 9}{10^2} = \frac{27}{100}, \; P(BC|A) = \frac{9}{10^2} = \frac{9}{100}.$$

Therefore,
$$P(BC|A) \neq P(B|A)P(C|A)$$

and so the events B and C are not conditionally independent, given the event A.

Example 1.9 *Dependent but conditionally independent events.* Suppose that a ball is randomly drawn from an urn containing 31 balls numbered $1, 2, \ldots, 31$. Let A, B and C be the events that the number drawn is divisible by the primes 5, 3 and 2, respectively. Show that the events B and C are not independent, while the same events are conditionally independent, given the event A.

Clearly,
$$P(B) = \frac{10}{31}, \; P(C) = \frac{15}{31}, \; P(BC) = \frac{5}{31}.$$

Thus
$$P(BC) \neq P(B)P(C)$$
and so the events B and C are not independent. Further,
$$P(B|A) = \frac{2}{6} = \frac{1}{3}, \ P(C|A) = \frac{3}{6} = \frac{1}{2}, \ P(BC|A) = \frac{1}{6}.$$
Therefore,
$$P(BC|A) = P(B|A)P(C|A)$$
and so the events B and C are conditionally independent, given the event A.

The notion of independent trials of a random experiment is a basic element in most of the models studied in probability theory. For the introduction of this notion let us consider a sequence of n random experiments, with Ω_i, $i = 1, 2, \ldots, n$, the corresponding sequence of sample spaces. The successive (or simultaneous) execution of the first, second and so on up to the nth random experiment defines an n-dimensional (compound) random experiment. An appropriate sample space for the study of this random experiment is the Cartesian product
$$W = \Omega_1 \times \Omega_2 \times \cdots \times \Omega_n$$
$$= \{w = (\omega_1, \omega_2, \ldots, \omega_n) : \omega_i \in \Omega_i, \ i = 1, 2, \ldots, n\}.$$

An n-dimensional (compound) random experiment, which is composed of the successive execution of a random experiment with sample space Ω, is particularly called a *sequence of n trials* of this experiment. In this particular case, in which $\Omega_i = \Omega$, $i = 1, 2, \ldots, n$, the sample space is the n-fold Cartesian product of Ω with itself,
$$W = \Omega^n = \{w = (\omega_1, \omega_2, \ldots, \omega_n) : \omega_i \in \Omega, \ i = 1, 2, \ldots, n\}.$$

If the sample spaces Ω_i, $i = 1, 2, \ldots, n$, are discrete, then any event $B \in \mathcal{P}(W)$ may be expressed as $B = A_1 \times A_2 \times \cdots \times A_n$, where $A_i \in \mathcal{P}(\Omega_i)$, $i = 1, 2, \ldots, n$. Further, any event $A_i \in \mathcal{P}(\Omega_i)$ may be expressed as an event $B_i \in \mathcal{P}(W)$, $i = 1, 2, \ldots, n$, with $B_1 = A_1 \times \Omega_2 \times \cdots \times \Omega_n$, $B_2 = \Omega_1 \times A_2 \times \cdots \times \Omega_n, \ldots,$ $B_n = \Omega_1 \times \cdots \times \Omega_{n-1} \times A_n$. Such an event $B_i \in \mathcal{P}(W)$ is called an *event depending on the ith experiment*, $i = 1, 2, \ldots, n$. The notion of independent events is carried over to independent random experiments and, particularly, to independent trials of a random experiment. Specifically, the following definition is adopted.

Definition 1.5 *The random experiments with sample spaces Ω_i, $i = 1, 2, \ldots, n$, are called stochastically independent if and only if the following relation holds:*
$$P(B_1 B_2 \cdots B_n) = P(B_1)P(B_2) \cdots P(B_n), \tag{1.31}$$
for every events $B_1 = A_1 \times \Omega_2 \times \cdots \times \Omega_n$, $B_2 = \Omega_1 \times A_2 \times \cdots \times \Omega_n, \ldots,$ $B_n = \Omega_1 \times \cdots \times \Omega_{n-1} \times A_n$.

The importance of independent random experiments and particularly of independent trials of a random experiment primarily lies on their usage in constructing useful stochastic models. In this respect the probabilities $P_i(A_i)$ for every event $A_i \in \mathcal{P}(\Omega_i)$, $i = 1, 2, \ldots, n$, are defined and assuming the independence of the random experiments or the trials of a random experiment, the probability $P(B)$ for every event $B \in \mathcal{P}(W)$ is defined so that (1.31) is satisfied. Specifically, the probabilities of the elementary events $\{w\} = \{(\omega_1, \omega_2, \ldots, \omega_n)\}$ of the sample space $W = \Omega_1 \times \Omega_2 \times \cdots \times \Omega_n$ are defined by

$$P(\{w\}) \equiv P(\{(\omega_1, \omega_2, \ldots, \omega_n)\}) = P_1(\{\omega_1\}) P_2(\{\omega_2\}) \cdots P_n(\{\omega_n\}).$$

and so

$$P(A_1 \times A_2 \times \cdots \times A_n) = P_1(A_1) P_2(A_2) \cdots P_n(A_n).$$

Then, the probability $P(B)$, for every event $B \in \mathcal{P}(W)$, is defined by

$$P(B) = \sum_{w \in B} P(\{w\}).$$

Clearly, the intersection of $B_1 = A_1 \times \Omega_2 \times \cdots \times \Omega_n$, $B_2 = \Omega_1 \times A_2 \times \cdots \times \Omega_n, \ldots$, $B_n = \Omega_1 \times \cdots \times \Omega_{n-1} \times A_n$ may be expressed as

$$B_1 B_2 \cdots B_n = A_1 \times A_2 \times \cdots \times A_n$$

and since $P(B_i) = P_i(A_i)$, it follows that (1.31) is satisfied.

1.5 INCLUSION AND EXCLUSION PRINCIPLE

A generalization of (1.25) to n events $A_1, A_2, \ldots, A_n$ in a sample space Ω, known as the *inclusion and exclusion principle* or *Poincaré formula*, is derived in the next theorem. This expression holds also for the *cardinality measure* $N(\cdot)$ instead of probability measure $P(\cdot)$.

Theorem 1.7 (Inclusion and exclusion principle) *The probability $Q_{n,1} = P(A_1 \cup A_2 \cup \cdots \cup A_n)$, that at least one among n events $A_1, A_2, \ldots, A_n$ occurs, is given by*

$$Q_{n,1} = S_{n,1} - S_{n,2} + \cdots + (-1)^{n-1} S_{n,n}, \tag{1.32}$$

with

$$S_{n,r} = \sum P(A_{i_1} A_{i_2} \cdots A_{i_r}), \ r = 1, 2, \ldots, n, \tag{1.33}$$

where the summation is extended over all r-combinations $\{i_1, i_2, \ldots, i_r\}$ of the n indices $\{1, 2, \ldots, n\}$.

Proof Note first that (1.32), according to (1.25), holds for $n = 2$. Then, supposing that (1.32) holds for $n - 1$, it should be shown that it also holds for n. From (1.25), with $A = A_1 \cup A_2 \cup \cdots \cup A_{n-1}$, $B = A_n$ and since

$$AB = (A_1 A_n) \cup (A_2 A_n) \cup \cdots \cup (A_{n-1} A_n),$$

it follows that the probability $Q_{n,1} = P(A_1 \cup A_2 \cup \cdots \cup A_n)$ may be expressed as

$$Q_{n,1} = P(A_1 \cup A_2 \cup \cdots \cup A_{n-1}) + P(A_n) - P(B_1 \cup B_2 \cup \cdots \cup B_{n-1}),$$

where $B_1 = A_1 A_n$, $B_2 = A_2 A_n, \ldots, B_{n-1} = A_{n-1} A_n$. Thus, by the induction hypothesis,

$$Q_{n,1} = \sum_{r=1}^{n-1} (-1)^{r-1} S_{n-1,r} + P(A_n) - \sum_{r=1}^{n-1} (-1)^{r-1} W_{n-1,r},$$

with

$$S_{n-1,r} = \sum P(A_{i_1} A_{i_2} \cdots A_{i_r}),$$

$$W_{n-1,r} = \sum P(B_{i_1} B_{i_2} \cdots B_{i_r}) = \sum P(A_{i_1} A_{i_2} \cdots A_{i_r} A_n),$$

where the summation is extended over all r-combinations $\{i_1, i_2, \ldots, i_r\}$ of the $n-1$ indices $\{1, 2, \ldots, n-1\}$. Note that

$$S_{n-1,1} + P(A_n) = \sum_{i=1}^{n-1} P(A_i) + P(A_n) = \sum_{i=1}^{n} P(A_i) = S_{n,1}.$$

Further, for $r = 2, 3, \ldots, n-1$,

$$S_{n-1,r} + W_{n-1,r-1} = \sum P(A_{i_1} A_{i_2} \cdots A_{i_r}) + \sum P(A_{i_1} A_{i_2} \cdots A_{i_{r-1}} A_n)$$

is the sum of the terms $P(A_{i_1} A_{i_2} \cdots A_{i_r})$, with the summation extended over the

$$\binom{n-1}{r}$$

r-combinations $\{i_1, i_2, \ldots, i_r\}$ of the indices $\{1, 2, \ldots, n-1, n\}$ that do not include the index n and over the

$$\binom{n-1}{r-1}$$

r-combinations $\{i_1, i_2, \ldots, i_{r-1}, n\}$ of the n indices $\{1, 2, \ldots, n-1, n\}$ that include the index n. Thus,

$$S_{n-1,r} + W_{n-1,r-1} = \sum P(A_{i_1} A_{i_2} \cdots A_{i_r}) = S_{n,r}, \quad r = 2, 3, \ldots, n-1,$$

where the summation is extended over the

$$\binom{n-1}{r} + \binom{n-1}{r-1} = \binom{n}{r}$$

r-combinations $\{i_1, i_2, \ldots, i_r\}$ of the n indices $\{1, 2, \ldots, n\}$. Also,

$$W_{n-1,n-1} = P(B_1 B_2 \cdots B_{n-1}) = P(A_1 A_2 \cdots A_{n-1} A_n) = S_{n,n}.$$

Consequently,
$$Q_{n,1} = S_{n,1} - S_{n,2} + \cdots + (-1)^{n-1} S_{n,n}$$
and so it is shown that (1.32) also holds for n. Therefore, according to the principle of mathematical induction, (1.32) holds for every integer $n \geq 2$.

Corollary 1.4 *The probability $P_{n,0} = P(A'_1 A'_2 \cdots A'_n)$, that none among the n events $A_1, A_2, \ldots, A_n$ occurs, is given by*

$$P_{n,0} = S_{n,0} - S_{n,1} + \cdots + (-1)^n S_{n,n}, \tag{1.34}$$

where $S_{n,0} = 1$ and $S_{n,r}$, $r = 1, 2, \ldots, n$, is given by (1.33).

Proof Since by De Morgan's formula $(A_1 \cup A_2 \cup \cdots \cup A_n)' = A'_1 A'_2 \cdots A'_n$, applying (1.22), with $A = A_1 \cup A_2 \cup \cdots \cup A_n$, it follows that

$$P(A'_1 A'_2 \cdots A'_n) = 1 - P(A_1 \cup A_2 \cup \cdots \cup A_n).$$

Introducing into it expression (1.32), formula (1.34) is deduced.

The probability $P_{n,k}$ that exactly k among n events $A_1, A_2, \ldots, A_n$ occur in the particular case $k = 0$ has been expressed, by the aid of the inclusion and exclusion principle, in the form of an alternating sum of the sums of probabilities $S_{n,r} = \sum P(A_{i_2} A_{i_2} \cdots A_{i_r})$, $r = 1, 2, \ldots, n$. An analogous expression in the general case, with $0 \leq k \leq n$, extending the inclusion and exclusion principle, is given in the next theorem.

Theorem 1.8 (General inclusion and exclusion principle) *The probability $P_{n,k}$ that exactly k among n events $A_1, A_2, \ldots, A_n$ occur is given by*

$$P_{n,k} = \sum_{r=k}^{n} (-1)^{r-k} \binom{r}{k} S_{n,r}, \quad k = 0, 1, \ldots, n, \tag{1.35}$$

where $S_{n,0} = 1$ and $S_{n,r}$, $r = 1, 2, \ldots, n$, is given by (1.33).

Proof Clearly,

$$P_{n,k} = \sum_{i_1,\ldots,i_k} P(A_{i_1} A_{i_2} \cdots A_{i_k} A'_{j_1} A'_{j_2} \cdots A'_{j_{n-k}}), \tag{1.36}$$

where the summation is extended over all k-combinations $\{i_1, i_2, \ldots, i_k\}$ of the n indices $\{1, 2, \ldots, n\}$ and $\{j_1, j_2, \ldots, j_{n-k}\} = \{1, 2, \ldots, n\} - \{i_1, i_2, \ldots, i_k\}$.

Let us first calculate the probability $P(A_{i_1} A_{i_2} \cdots A_{i_k} A'_{j_1} A'_{j_2} \cdots A'_{j_{n-k}})$, that k specified events, among the n events $A_1, A_2, \ldots, A_n$, occur. Applying (1.23) to the events $A = A_{i_1} A_{i_2} \cdots A_{i_k}$ and $B = A_{j_1} \cup A_{j_2} \cup \cdots \cup A_{j_{n-k}}$, whence $B' = A'_{j_1} A'_{j_2} \cdots A'_{j_{n-k}}$, and putting $C_{j_s} = A_{i_1} A_{i_2} \cdots A_{i_k} A_{j_s}$, $s = 1, 2, \ldots, n-k$, we deduce the relation

$$P(A_{i_1} A_{i_2} \cdots A_{i_k} A'_{j_1} A'_{j_2} \cdots A'_{j_{n-k}})$$
$$= P(A_{i_1} A_{i_2} \cdots A_{i_k}) - P(C_{j_1} \cup C_{j_2} \cup \cdots \cup C_{j_{n-k}}),$$

from which, on using (1.32), we conclude the expression

$$P(A_{i_1}A_{i_2}\cdots A_{i_k}A'_{j_1}A'_{j_2}\ldots A'_{j_{n-k}}) = \sum_{r=k}^{n}(-1)^{r-k}S_{n,k,r}(i_1,i_2,\ldots,i_k),$$

with

$$S_{n,k,r}(i_1,i_2,\ldots,i_k) = \sum_{h_1,\ldots,h_{r-k}} P(A_{i_1}A_{i_2}\cdots A_{i_k}A_{h_1}A_{h_2}\cdots A_{h_{r-k}}),$$

for $r = k, k+1, \ldots, n$, where the summation is extended over all $(r-k)$-combinations $\{h_1, h_2, \ldots, h_{r-k}\}$ of the $n - k$ indices $\{j_1, j_2, \ldots, j_{n-k}\}$. Introducing this expression into expression (1.36) of the probability $P_{n,k}$, we get

$$P_{n,k} = \sum_{i_1,\ldots,i_k}\sum_{r=k}^{n}(-1)^{r-k}S_{n,k,r}(i_1,i_2,\ldots,i_k) = \sum_{r=k}^{n}(-1)^{r-k}S_{n,k,r}, \quad (1.37)$$

where

$$S_{n,k,r} = \sum_{i_1,\ldots,i_k} S_{n,k,r}(i_1,i_2,\ldots,i_k)$$
$$= \sum_{i_1,\ldots,i_k}\sum_{h_1,\ldots,h_{r-k}} P(A_{i_1}A_{i_2}\cdots A_{i_k}A_{h_1}A_{h_2}\cdots A_{h_{r-k}}).$$

Note that this sum includes

$$\binom{n}{k}\binom{n-k}{r-k}$$

terms of the form $P(A_{m_1}A_{m_2}\cdots A_{m_r})$, $\{m_1, m_2, \ldots, m_r\} \subseteq \{1, 2, \ldots, n\}$. Further, among these terms, the distinct ones are equal to

$$\binom{n}{r},$$

the number of r-combinations $\{m_1, m_2, \ldots, m_r\}$ of the n indices $\{1, 2, \ldots, n\}$, and each such term is included in the sum as many times as the number

$$\binom{r}{k}$$

of ways of selecting k indices $\{i_1, i_2, \ldots, i_k\}$ from the set $\{m_1, m_2, \ldots, m_r\}$, so that

$$\binom{n}{r}\binom{r}{k} = \binom{n}{k}\binom{n-k}{r-k},$$

which is a simple combinatorial identity. Consequently,

$$S_{n,k,r} = \binom{r}{k}S_{n,r}, \quad S_{n,r} = \sum_{m_1,\ldots,m_r} P(A_{m_1}A_{m_2}\cdots A_{m_r}),$$

where the summation is extended over all r-combinations $\{m_1, m_2, \ldots, m_r\}$ of the n indices $\{1, 2, \ldots, n\}$. Introducing it into expression (1.37), we deduce (1.35).

An interesting corollary may be deduced from Theorem 1.8 in the case of the existence of some kind of symmetry with respect to the events. Such symmetry, appearing in several applications, constitutes the case where the sets are exchangeable in the following sense.

Definition 1.6 *The events $A_1, A_2, \ldots, A_n$ are called exchangeable if, for every collection of r indices $\{i_1, i_2, \ldots, i_r\}$ from the set of n indices $\{1, 2, \ldots, n\}$, the probability*

$$P(A_{i_1} A_{i_2} \cdots A_{i_r}) = p_r, \; r = 1, 2, \ldots, n, \tag{1.38}$$

depends only on r and not on the specific collection of the indices.

Corollary 1.5 *Let $A_1, A_2, \ldots, A_n$ be exchangeable events. Then the probability $P_{n,k}$ that exactly k among these n events occur is given by*

$$P_{n,k} = \binom{n}{k} \sum_{j=0}^{n-k} (-1)^j \binom{n-k}{j} p_{k+j}, \; k = 0, 1, \ldots, n, \tag{1.39}$$

or equivalently by

$$\begin{aligned} P_{n,k} &= \binom{n}{k} \sum_{j=0}^{n-k} (-1)^{n-k-j} \binom{n-k}{j} p_{n-j} \\ &= \binom{n}{k} [\Delta_u^{n-k} p_{n-u}]_{u=0}, \; k = 0, 1, \ldots, n, \end{aligned} \tag{1.40}$$

where $p_0 = 1$ and $p_r = P(A_{i_1} A_{i_2} \cdots A_{i_r})$, $r = 1, 2, \ldots, n$.

Proof The sum

$$S_{n,r} = \sum P(A_{i_1} A_{i_2} \cdots A_{i_r}), \; r = k, k+1, \ldots, n,$$

where the summation is extended over all r-combinations $\{i_1, i_2, \ldots, i_r\}$ of the n indices $\{1, 2, \ldots, n\}$, under the assumption $P(A_{i_1} A_{i_2} \cdots A_{i_r}) = p_r$, of the exchangeability of the n events $A_1, A_2, \ldots, A_n$, is given by

$$S_{n,r} = \binom{n}{r} p_r, \; r = k, k+1, \ldots, n.$$

Introducing it into expression (1.35) of the probability $P_{n,k}$ and using the identity

$$\binom{r}{k}\binom{n}{r} = \binom{n}{k}\binom{n-k}{r-k},$$

we get

$$P_{n,k} = \binom{n}{k} \sum_{r=k}^{n} (-1)^{r-k} \binom{n-k}{r-k} p_r, \quad k = 0, 1, \ldots, n.$$

Replacing the variable r by $j = r - k$, expression (1.39) is established. Also, changing the variable r into $j = n - r$ we get the first part of (1.40). Further, performing the $(n-k)$th order difference operator Δ_u^{n-k}, with unit increment, on the function $f(u) = p_{n-u}$ and using (1.13), we find

$$\left[\Delta_u^{n-k} p_{n-u}\right]_{u=0} = \sum_{j=0}^{n-k} (-1)^{n-k-j} \binom{n-k}{j} \left[E_u^j p_{n-u}\right]_{u=0}$$

$$= \sum_{j=0}^{n-k} (-1)^{n-k-j} \binom{n-k}{j} p_{n-j},$$

and so the second part of expression (1.40) is deduced.

It is worth noticing that the sum $S_{n,r} = \sum P(A_{i_2} A_{i_2} \cdots A_{i_r})$ represents the rth binomial moment of the sequence of probabilities $P_{n,k}$, $k = 0, 1, \ldots, n$,

$$b_{(r)} = \sum_{k=r}^{n} \binom{k}{r} P_{n,k}, \quad r = 1, 2, \ldots, n. \tag{1.41}$$

Specifically, inverting (1.35), we deduce the following corollary of Theorem 1.8.

Corollary 1.6 *Consider the sequence of probabilities $P_{n,k}$, $k = 0, 1, \ldots, n$, that exactly k among n events $A_1, A_2, \ldots, A_n$ occur. Its rth binomial moment $b_{(r)}$ is given by*

$$b_{(r)} = S_{n,r}, \quad r = 1, 2, \ldots, n, \tag{1.42}$$

where $S_{n,r}$, $r = 1, 2, \ldots, n$, is given by (1.33). In particular, if the events $A_1, A_2, \ldots, A_n$ are exchangeable then

$$b_{(r)} = \binom{n}{r} p_r, \quad r = 1, 2, \ldots, n, \tag{1.43}$$

with $p_r = P(A_{i_1} A_{i_2} \cdots A_{i_r})$, $r = 1, 2, \ldots, n$.

Proof The rth binomial moment (1.41) of the sequence of probabilities $P_{n,k}$, $k = 0, 1, \ldots, n$, on using expression (1.35), may be written as

$$b_{(r)} = \sum_{k=r}^{n} \sum_{j=k}^{n} (-1)^{j-k} \binom{k}{r} \binom{j}{k} S_{n,j} = \sum_{j=r}^{n} \left\{ \sum_{k=r}^{j} (-1)^{j-k} \binom{k}{r} \binom{j}{k} \right\} S_{n,j},$$

where the inner sum

$$c_{j,r} = \sum_{k=r}^{j} (-1)^{j-k} \binom{k}{r} \binom{j}{k}$$

is evaluated by the aid of the combinatorial identity

$$\binom{k}{r}\binom{j}{k} = \binom{j}{r}\binom{j-r}{k-r}$$

and is equal to

$$c_{j,r} = \binom{j}{r}\sum_{k=r}^{j}(-1)^{j-k}\binom{j-r}{k-r} = \binom{j}{r}(1-1)^{j-r} = \delta_{j,r},$$

the Kronecker delta: $\delta_{r,r} = 1$, $\delta_{j,r} = 0$, $j \neq r$. Consequently,

$$b_{(r)} = \sum_{j=r}^{n}\delta_{j,r}S_{n,j} = S_{n,r}, \quad r = 1, 2, \ldots, n.$$

Under the assumption, $P(A_{i_1}A_{i_2}\cdots A_{i_r}) = p_r$, of the exchangeability of the n events $A_1, A_2, \ldots, A_n$,

$$S_{n,r} = \binom{n}{r}p_r, \quad r = 1, 2, \ldots, n,$$

yielding (1.43).

The probability $Q_{n,k}$ that at least k among n events $A_1, A_2, \ldots, A_n$ occur, which is connected with the probability $P_{n,k}$, is deduced from Theorem 1.8 in the following corollary.

Corollary 1.7 *The probability $Q_{n,k}$ that at least k among n events $A_1, A_2, \ldots, A_n$ occur is given by*

$$Q_{n,k} = \sum_{r=k}^{n}(-1)^{r-k}\binom{r-1}{k-1}S_{n,r}, \quad k = 1, 2, \ldots, n, \qquad (1.44)$$

where $S_{n,r}$, $r = 1, 2, \ldots, n$, is given by (1.33). In particular, if the events $A_1, A_2, \ldots, A_n$ are exchangeable, then

$$Q_{n,k} = n\binom{n-1}{k-1}\sum_{j=0}^{n-k}(-1)^j\binom{n-k}{j}\frac{p_{k+j}}{k+j}, \quad k = 1, 2, \ldots, n, \qquad (1.45)$$

or equivalently

$$Q_{n,k} = n\binom{n-1}{k-1}\sum_{j=0}^{n-k}(-1)^{n-k-j}\binom{n-k}{j}\frac{p_{n-j}}{n-j}$$

$$= n\binom{n-1}{k-1}\left[\Delta_u^{n-k}\frac{p_{n-u}}{n-u}\right]_{u=0}, \quad k = 1, 2, \ldots, n, \qquad (1.46)$$

where $p_r = P(A_{i_1}A_{i_2}\cdots A_{i_r})$, $r = 1, 2, \ldots, n$.

Proof The probability $Q_{n,k}$, on using (1.35), may be expressed as

$$Q_{n,k} = \sum_{j=k}^{n} P_{n,j} = \sum_{j=k}^{n} \sum_{r=j}^{n} (-1)^{r-j} \binom{r}{j} S_{n,r}.$$

Interchanging the order of summation and using Pascal's triangle,

$$\binom{r}{j} = \binom{r-1}{j} + \binom{r-1}{j-1},$$

we get

$$Q_{n,k} = \sum_{r=k}^{n} \left\{ \sum_{j=k}^{r} (-1)^{r-j} \binom{r-1}{j} + \sum_{j=k}^{r} (-1)^{r-j} \binom{r-1}{j-1} \right\} S_{n,r}$$

$$= \sum_{r=k}^{n} (-1)^{r-k} \binom{r-1}{k-1} S_{n,r}.$$

Under the assumption $P(A_{i_1} A_{i_2} \cdots A_{i_r}) = p_r$, of the exchangeability of the n events $A_1, A_2, \ldots, A_n$, we have

$$S_{n,r} = \binom{n}{r} p_r, \ r = k, k+1, \ldots, n.$$

Introducing it into (1.44) and successively using the combinatorial identities

$$\binom{n}{r} = \frac{n}{r} \binom{n-1}{r-1}$$

and

$$\binom{r-1}{k-1} \binom{n-1}{r-1} = \binom{n-1}{k-1} \binom{n-k}{r-k},$$

we get

$$Q_{n,k} = n \binom{n-1}{k-1} \sum_{r=k}^{n} (-1)^{r-k} \binom{n-k}{r-k} \frac{p_r}{r}, \ k = 1, 2, \ldots, n.$$

Replacing the variable r by $j = r-k$, expression (1.45) is established. Also, changing the variable r into $j = n - r$ the first part of (1.46) is deduced. Further, since

$$\left[\Delta_u^{n-k} \frac{p_{n-u}}{n-u} \right]_{u=0} = \sum_{j=0}^{n-k} (-1)^{n-k-j} \binom{n-k}{j} \left[E_u^j \frac{p_{n-u}}{n-u} \right]_{u=0}$$

$$= \sum_{j=0}^{n-k} (-1)^{n-k-j} \binom{n-k}{j} \frac{p_{n-j}}{n-j},$$

the second part of the expression (1.46) is deduced.

The probability $R_{k,n}$ that n trials (of a random experiment) are required until the kth occurrence of an event A, which is of great interest in waiting-time distribution theory, is deduced from Corollary 1.7.

Corollary 1.8 *Consider a sequence of random experiments (or trials of a random experiment) and let A be an event that may occur at the ith experiment (trial), $i = 1, 2, \ldots$. Also, let A_i be the event that A occurs at the ith experiment (trial), $i = 1, 2, \ldots$. The probability $R_{k,n}$ that n experiments (trials) are required until the kth occurrence of the event A is given by*

$$R_{k,n} = \sum_{r=k}^{n} (-1)^{r-k} \binom{r-1}{k-1} W_{n-1, r-1}, \quad n = k, k+1, \ldots, \tag{1.47}$$

with

$$W_{n-1, r-1} = \sum P(A_{i_1} A_{i_2} \cdots A_{i_{r-1}} A_n), \tag{1.48}$$

where the summation is extended over all $(r-1)$-combinations $\{i_1, i_2, \ldots, i_{r-1}\}$ of the $n-1$ indices $\{1, 2, \ldots, n-1\}$. In particular, if the events $A_1, A_2, \ldots, A_n$ are exchangeable, then

$$R_{k,n} = \binom{n-1}{k-1} \sum_{j=0}^{n-k} (-1)^j \binom{n-k}{j} p_{k+j}, \quad n = k, k+1, \ldots \tag{1.49}$$

or equivalently

$$R_{k,n} = \binom{n-1}{k-1} \sum_{j=0}^{n-k} (-1)^{n-k-j} \binom{n-k}{j} p_{n-j}$$

$$= \binom{n-1}{k-1} [\Delta_u^{n-k} p_{n-u}]_{u=0}, \quad n = k, k+1, \ldots, \tag{1.50}$$

where $p_r = P(A_{i_1} A_{i_1} \cdots A_{i_r})$, $r = 1, 2, \ldots, n$.

Proof The probability $R_{k,n}$ that n experiments (trials) are required until the kth occurrence of the event A may be expressed, in terms of the probability $Q_{n,k}$ that at least k among the n events $A_1, A_2, \ldots, A_n$ occur, as

$$R_{k,n} = Q_{n,k} - Q_{n-1,k}.$$

Thus, upon using expression (1.44), and since

$$S_{n,r} - S_{n-1,r} = W_{n-1,r-1}, \quad r = 1, 2, \ldots, n-1, \quad S_{n,n} = W_{n-1,n-1},$$

(1.47) is deduced. Under the assumption $P(A_{i_1} A_{i_2} \cdots A_{i_r}) = p_r$, of the exchangeability of the n events $A_1, A_2, \ldots, A_n$, we have

$$W_{n-1, r-1} = \binom{n-1}{r-1} p_r, \quad r = k, k+1, \ldots, n.$$

Introducing it into (1.47) and using the combinatorial identity

$$\binom{r-1}{k-1}\binom{n-1}{r-1} = \binom{n-1}{k-1}\binom{n-k}{r-k},$$

we get

$$R_{k,n} = \binom{n-1}{k-1}\sum_{r=k}^{n}(-1)^{r-k}\binom{n-k}{r-k}p_r, \quad n = k, k+1, \ldots.$$

Replacing the (bound) variable r by $j = r - k$, expression (1.49) is deduced. Also, changing the variable r into $j = n - r$, the first part of (1.50) is obtained. Further, since

$$[\Delta_u^{n-k} p_{n-u}]_{u=0} = \sum_{j=0}^{n-k}(-1)^{n-k-j}\binom{n-k}{j}[E_u^j p_{n-u}]_{u=0}$$

$$= \sum_{j=0}^{n-k}(-1)^{n-k-j}\binom{n-k}{j}p_{n-j},$$

the second part of (1.50) is deduced.

Example 1.10 *The classical problem of coincidences (rencontres).* Consider an urn containing n balls numbered $1, 2, \ldots, n$. Suppose that all the balls are successively drawn one after the other, at random, without replacement. The drawing of the ball bearing the number j at the jth trial, $j = 1, 2, \ldots, n$, is called a *coincidence (rencontre)*. Calculate (a) the probability that exactly k coincidences occur in the n trials, (b) the probability that exactly k coincidences occur in r specified trials and (c) the probability that the kth coincidence occurs at the rth trial.

(a) Note first that each drawing of the n balls corresponds to a permutation $(j_1, j_2, \ldots, j_n)$ of the set $\{1, 2, \ldots, n\}$, where j_r is the number drawn at the rth trial, $r = 1, 2, \ldots, n$. Thus the number of elements of the sample space Ω, of the stochastic experiment of the drawings of the n balls, equals $N(\Omega) = n!$, the number of the permutations $(j_1, j_2, \ldots, j_n)$ of the set $\{1, 2, \ldots, n\}$. Let A_i be the event that a coincidence occurs at the ith trial, $i = 1, 2, \ldots, n$ and consider a selection of r indices $\{i_1, i_2, \ldots, i_r\}$ out of the n indices $\{1, 2, \ldots, n\}$. Then

$$p_r = P(A_{i_1} A_{i_2} \cdots A_{i_r}) = \frac{(n-r)!}{n!}, \quad r = 1, 2, \ldots, n,$$

since at the $n-r$ non specified trials any of the numbers $\{1, 2, \ldots, n\} - \{i_1, i_2, \ldots, i_r\}$ may be drawn in $(n-r)!$ ways. Therefore, the events $A_1, A_2, \ldots, A_n$ are exchangeable and, according to Corollary 1.5, the probability $P_{n,k}$ that exactly k coincidences occur in the n trials is given by

$$P_{n,k} = \frac{1}{k!}\sum_{r=0}^{n-k}\frac{(-1)^r}{r!}, \quad k = 0, 1, \ldots, n.$$

Notice that for large n and since

$$\sum_{r=0}^{\infty} \frac{(-1)^r}{r!} = e^{-1},$$

this probability can be approximated by

$$P_{n,k} \cong \frac{e^{-1}}{k!}, \quad k = 0, 1, \ldots.$$

(b) Let $\{s_1, s_2, \ldots, s_r\} \subseteq \{1, 2, \ldots, n\}$ be the set of the r specified trials and consider the events $A_{s_1}, A_{s_2}, \ldots, A_{s_r}$. Then, as in the preceding case, for any selection of m indices $\{i_1, i_2, \ldots, i_m\}$ out of the r indices $\{s_1, s_2, \ldots, s_r\}$,

$$p_m = P(A_{i_1} A_{i_2} \cdots A_{i_m}) = \frac{(n-m)!}{n!}, \quad m = 1, 2, \ldots, r.$$

Consequently, on using (1.39) with r instead of n, the probability $P_{r,k;n}$ that exactly k coincidences occur in r specified trials is deduced as

$$P_{r,k;n} = \binom{r}{k} \sum_{j=0}^{r-k} (-1)^j \binom{r-k}{j} \frac{(n-k-j)!}{n!}, \quad k = 0, 1, \ldots, r.$$

Note that this probability depends only on the number r of the specified trials and not on the specific trials.

(c) The probability $R_{k,r;n}$ that the kth coincidence occurs at the rth trial, on using (1.49) with r instead of n, may be expressed as

$$R_{k,r;n} = \binom{r-1}{k-1} \sum_{j=0}^{r-k} (-1)^j \binom{r-k}{j} \frac{(n-k-j)!}{n!}, \quad r = k, k+1, \ldots, n.$$

Also, using (1.50) with r instead of n,

$$R_{k,r;n} = \binom{r-1}{k-1} \sum_{j=0}^{r-k} (-1)^{r-k-j} \binom{r-k}{j} \frac{(n-r+j)!}{n!}$$

$$= \binom{r-1}{k-1} \frac{[\Delta_u^{r-k}(n-r+u)!]_{u=0}}{n!},$$

for $r = k, k+1, \ldots, n$.

A series of alternating inequalities for the probability $P_{n,k}$ that exactly k among n events occur, and the probability $Q_{n,k}$ that at least k among n events occur, may be deduced from relations (1.35) and (1.44) and their inverses, respectively.

Theorem 1.9 (Bonferroni's inequalities) *Let $P_{n,k}$ and $Q_{n,k}$ be the probabilities that exactly k and at least k among n events $A_1, A_2, \ldots, A_n$ occur, respectively. Then, for $s = k, k+1, \ldots$, the following alternating inequalities hold true:*

$$(-1)^{s-k+1} \left\{ P_{n,k} - \sum_{r=k}^{s} (-1)^{r-k} \binom{r}{k} S_{n,r} \right\} \geq 0, \tag{1.51}$$

INCLUSION AND EXCLUSION PRINCIPLE 37

for $k = 0, 1, \ldots, n$, and

$$(-1)^{s-k+1}\left\{Q_{n,k} - \sum_{r=k}^{s}(-1)^{r-k}\binom{r-1}{k-1}S_{n,r}\right\} \geq 0, \quad (1.52)$$

for $k = 1, 2, \ldots, n$, where $S_{n,0} = 1$ and $S_{n,r}$, $r = 1, 2, \ldots, n$, is given by (1.33).

Proof Consider, for fixed n and $s = k, k+1, \ldots, k = 0, 1, \ldots, n$, the sequence

$$H_{n,s,k} = (-1)^{s-k+1}\left\{P_{n,k} - \sum_{r=k}^{s}(-1)^{r-k}\binom{r}{k}S_{n,r}\right\},$$

which, by virtue of (1.35) vanishes for $s = n$ and any $k = 0, 1, \ldots, n$, while for $s = k, k+1, \ldots, n-1$ is given by

$$H_{n,s,k} = \sum_{r=s+1}^{n}(-1)^{r-s-1}\binom{r}{k}S_{n,r}, \quad k = 0, 1, \ldots, n-1.$$

The inverse of relation (1.35) is the relation (cf. Corollary 1.6)

$$S_{n,r} = \sum_{j=r}^{n}\binom{j}{r}P_{n,j}.$$

The sequence $H_{n,s,k}$, $k = 0, 1, \ldots, n$, upon introducing into it the last expression, may be written as a linear combination of the probabilities $P_{n,j}, j = s+1, s+2, \ldots, n$:

$$H_{n,s,k} = \sum_{r=s+1}^{n}\sum_{j=r}^{n}(-1)^{r-s-1}\binom{r}{k}\binom{j}{r}P_{n,j}$$

$$= \sum_{j=s+1}^{n}\left\{\sum_{r=s+1}^{j}(-1)^{r-s-1}\binom{r}{k}\binom{j}{r}\right\}P_{n,j}.$$

The coefficient of the general term of this expression,

$$c_{j,s} = \sum_{r=s+1}^{j}(-1)^{r-s-1}\binom{r}{k}\binom{j}{r} = \binom{j}{k}\sum_{r=s+1}^{j}(-1)^{r-s-1}\binom{j-k}{r-k},$$

for $j = s+1, s+2, \ldots, n$, on using Pascal's triangle

$$\binom{j-k}{r-k} = \binom{j-k-1}{r-k} + \binom{j-k-1}{r-k-1},$$

reduces to

$$c_{j,s} = \binom{j}{k}\sum_{r=s+1}^{j-1}(-1)^{r-s-1}\binom{j-k-1}{r-k}$$

$$+ \binom{j}{k}\sum_{r=s+1}^{j}(-1)^{r-s-1}\binom{j-k-1}{r-k-1} = \binom{j}{k}\binom{j-k-1}{s-k},$$

for $j = s+1, s+2, \ldots, n$, that is, is positive, which implies (1.51). The inequalities (1.52) may be similarly deduced from (1.44) and its inverse.

The inequalities (1.51) and (1.52) in the particular cases $s = k + 2i + 1$ and $s = k + 2j$ yield the classical lower and upper bounds for the probabilities $P_{n,k}$ and $Q_{n,k}$, respectively.

Corollary 1.9 (Classical Bonferroni's inequalities) *Let $P_{n,k}$ and $Q_{n,k}$ be the probabilities that exactly k and at least k among n events $A_1, A_2, \ldots, A_n$ occur, respectively. Then, for nonnegative integers i and j,*

$$\sum_{r=k}^{k+2i+1} (-1)^{r-k} \binom{r}{k} S_{n,r} \leq P_{n,k} \leq \sum_{r=k}^{k+2j} (-1)^{r-k} \binom{r}{k} S_{n,r}, \qquad (1.53)$$

for $k = 0, 1, \ldots, n$, and

$$\sum_{r=k}^{k+2i+1} (-1)^{r-k} \binom{r-1}{k-1} S_{n,r} \leq Q_{n,k} \leq \sum_{r=k}^{k+2j} (-1)^{r-k} \binom{r-1}{k-1} S_{n,r}, \qquad (1.54)$$

for $k = 1, 2, \ldots, n$, where $S_{n,0} = 1$ and $S_{n,r}$, $r = 0, 1, \ldots, n$, is given by (1.33).

A limiting form of the general inclusion and exclusion theorem may be deduced by the aid of the classical Bonferroni's inequalities (1.53).

Theorem 1.10 *Let $P_{n,k}$ be the probability that exactly k among n events $A_1, A_2, \ldots, A_n$ occur, $k = 0, 1, \ldots, n$, and $b_{(r)} = S_{n,r}$, $r = 0, 1, \ldots, n$, its binomial moments, where $S_{n,0} = 1$ and $S_{n,r}$, $r = 1, 2, \ldots, n$, is given by (1.33). If the limit*

$$\lim_{n \to \infty} S_{n,r} = S_r \qquad (1.55)$$

exists for all $r \geq k$, and if the limit

$$\lim_{n \to \infty} \sum_{r=k}^{n} (-1)^{r-k} \binom{r}{k} S_r = \sum_{r=k}^{\infty} (-1)^{r-k} \binom{r}{k} S_r \equiv P_k \qquad (1.56)$$

exists for all $k \geq 0$, then

$$\lim_{n \to \infty} P_{n,k} = P_k. \qquad (1.57)$$

Proof The classical Bonferroni's inequalities (1.53), by letting $n \to \infty$ and using the assumption (1.55), entail for all nonnegative integers i and j the inequalities

$$\sum_{r=k}^{k+2i+1} (-1)^{r-k} \binom{r}{k} S_r \leq \liminf_{n \to \infty} P_{n,k}$$

$$\leq \limsup_{n \to \infty} P_{n,k} \leq \sum_{r=k}^{k+2j} (-1)^{r-k} \binom{r}{k} S_r.$$

INCLUSION AND EXCLUSION PRINCIPLE

Further, letting $i \to \infty$ and $j \to \infty$, and using the assumption (1.56), the extreme members of these inequalities tend to P_k. Consequently, it follows that

$$\liminf_{n\to\infty} P_{n,k} = \limsup_{n\to\infty} P_{n,k} = P_k,$$

which implies (1.57).

Example 1.11 *Distribution of lottery numbers not drawn in a given number of drawings.* Consider a lottery-urn containing n balls numbered from 1 to n. Assume that, in a drawing, s numbers are randomly drawn from the urn and returned to it before the next drawing. Let $p_k(n, s, r)$ be the probability that k among the n numbers are not drawn in a series of r (consecutive) lottery-drawings. (a) Evaluate the probability $p_k(n, s, r)$ and (b) assuming that $r = (n/s) \log n - (n/s) \log \lambda + o(n)$, with $0 < \lambda < \infty$, show that

$$\lim_{n\to\infty} p_k(n, s, r) = e^{-\lambda} \frac{\lambda^k}{k!}, \quad k = 0, 1, \ldots.$$

(a) Let Ω be the sample space of the random experiment of r (consecutive) lottery-drawings in each of which s numbers are randomly drawn from an urn containing the numbers $\{1, 2, \ldots, n\}$. Further, let $A_i \subseteq \Omega$ be the event that number i is not drawn, $i = 1, 2, \ldots, n$, and consider a selection of j indices $\{i_1, i_2, \ldots, i_j\}$ out of the n indices $\{1, 2, \ldots, n\}$. Then

$$p_j = P(A_{i_1} A_{i_2} \cdots A_{i_j}) = \binom{n-j}{s}^r \bigg/ \binom{n}{s}^r, \quad j = 1, 2, \ldots, n.$$

Thus, the events $A_1, A_2, \ldots, A_n$ are exchangeable and, by Corollary 1.5,

$$p_k(n, s, r) = \binom{n}{k} \sum_{j=0}^{n-k} (-1)^j \binom{n-k}{j} \binom{n-k-j}{s}^r \bigg/ \binom{n}{s}^r$$

$$= \binom{n}{k} \left[\Delta_u^{n-k} \binom{u}{s}^r \right]_{u=0} \bigg/ \binom{n}{s}^r, \quad k = 0, 1, \ldots, n.$$

(b) Further,

$$S_{n,j} = \binom{n}{j} \binom{n-j}{s}^r \bigg/ \binom{n}{s}^r.$$

Since

$$S_{n,j} \cong \frac{n^j}{j!} \left[\left(1 - \frac{j}{n}\right) \left(1 - \frac{j}{n-1}\right) \cdots \left(1 - \frac{j}{n-s+1}\right) \right]^r$$

$$= \frac{n^j}{j!} \exp\left\{ r \sum_{i=n-s+1}^{n} \log\left(1 - \frac{j}{i}\right) \right\},$$

on using the expansion $-\log(1-t) = \sum_{k=1}^{\infty} t^k/k = t + o(t^2)$, we get the expression

$$S_{n,j} = \frac{n^j}{j!} \exp\left\{-rj\left(\sum_{i=n-s+1}^{n} \frac{1}{i} + o(n^{-2})\right)\right\}$$

$$= \frac{n^j}{j!} \exp\left\{-rj\left(\frac{1}{n}\sum_{i=1}^{s}\left(1 - \frac{s-i}{n}\right)^{-1} + o(n^{-2})\right)\right\}$$

$$= \frac{n^j}{j!} \exp\left\{-rj\left(\frac{s}{n} + o(n^{-2})\right)\right\}.$$

Letting $r = (n/s)\log n - (n/s)\log\lambda + o(n)$, with $0 < \lambda < \infty$, and taking the limit as $n \to \infty$, it follows that

$$\lim_{n \to \infty} S_{n,j} = S_j = \frac{\lambda^j}{j!},$$

and

$$\lim_{n \to \infty} \sum_{j=k}^{n} (-1)^{j-k} \binom{j}{k} S_j = \frac{\lambda^k}{k!} \sum_{j=k}^{\infty} \frac{(-\lambda)^{j-k}}{(j-k)!} = e^{-\lambda} \frac{\lambda^k}{k!}.$$

Applying Theorem 1.10, the required asymptotic expression of $p_k(n, s, r)$ is deduced.

1.6 DISTRIBUTIONS AND MOMENTS OF RANDOM VARIABLES

Consider a stochastic experiment with sample space Ω and probability $P(A)$ for every event A in Ω. The probabilistic study of one or more quantitative or qualitative characteristics of the sample points is facilitated by assigning to each $\omega \in \Omega$ a number x or a vector of numbers $(x_1, x_2, \ldots, x_k)$, respectively.

Definition 1.7 (a) *A function X defined on a sample space Ω and assuming real values is called a random variable if and only if the set*

$$\{\omega \in \Omega : X(\omega) \leq x\}$$

is an event in Ω. The function F_X defined by

$$F_X(x) = P(X \leq x) = P(\{\omega \in \Omega : X(\omega) \leq x\}), \quad x \in R$$

is called the (cumulative) distribution function of the random variable X.

(b) *A pair of functions (X, Y) defined on a sample space Ω and assuming real values is called a bivariate random variable if and only if the set*

$$\{\omega \in \Omega : X(\omega) \leq x, \; Y(\omega) \leq y\}$$

is an event in Ω. The function $F_{X,Y}$ defined by

$$F_{X,Y}(x, y) = P(X \leq x, Y \leq y)$$
$$= P(\{\omega \in \Omega : X(\omega) \leq x, Y(\omega) \leq y\}), \quad x, y \in R$$

is called the *distribution function of the bivariate random variable* (X, Y) or the *joint distribution function of the random variables X and Y*.

Clearly, the distribution functions $F_X(x)$, $x \in R$ and $F_{X,Y}(x, y)$, $x, y \in R$ take on values in the interval $[0, 1]$. Further, it can easily be shown that F_X is *nondecreasing*, $F_X(x_1) \leq F_X(x_2)$, $x_1 < x_2$, *right continuous*, $F_X(x+) \equiv \lim_{n \to \infty} F_X(x_n) = F_X(x)$, $x_n \downarrow x$ and satisfies $F_X(-\infty) \equiv \lim_{x \to -\infty} F_X(x) = 0$ and $F_X(+\infty) \equiv \lim_{x \to +\infty} F_X(x) = 1$. Note that a nondecreasing function such as F_X may have discontinuities that can only be jumps. The set of discontinuities of a distribution function is at most countable. A jump of F_X at a point x equals $F_X(x) - F_X(x-) = P(X = x)$, where $F_X(x-) \equiv \lim_{n \to \infty} F_X(x_n)$, $x_n \uparrow x$. If F_X is continuous, then $F_X(x) = F_X(x-)$ and $P(X = x) = 0$ for all $x \in R$. The bivariate distribution function $F_{X,Y}$ is *nondecreasing* and *right continuous*, with respect to each of the variables. Also, $F_{X,Y}(-\infty, y) \equiv \lim_{s \to -\infty} F_{X,Y}(x, y) = 0$, $F_{X,Y}(x, -\infty) \equiv \lim_{y \to -\infty} F_{X,Y}(x, y) = 0$ and $F_{X,Y}(+\infty, +\infty) \equiv \lim_{x \to +\infty, y \to +\infty} F_{X,Y}(x, y) = 1$. In addition, the bivariate function $F_{X,Y}$ possesses the following limiting properties:

$$F_{X,Y}(x, +\infty) \equiv \lim_{y \to +\infty} F_{X,Y}(x, y) = F_X(x),$$

$$F_{X,Y}(+\infty, y) \equiv \lim_{x \to +\infty} F_{X,Y}(x, y) = F_Y(y).$$

The distribution function F_X of X, considered in the framework of the probability distribution of the bivariate random variable (X, Y), is called the *marginal distribution function* of the random variable X. Similarly, F_Y is called the *marginal distribution function* of the random variable Y.

Definition 1.8 *Let X and Y be random variables with joint distribution function $F_{X,Y}(x, y)$, $x, y \in R$ and marginal distribution functions $F_X(x)$, $x \in R$ and $F_Y(y)$, $y \in R$. The random variables X and Y are called independent if and only if*

$$F_{X,Y}(x, y) = F_X(x) F_Y(y), \text{ for all } x, y \in R.$$

In the sequel, study is restricted to random variables with discontinuous distribution functions.

Definition 1.9 (a) *A random variable X is said to be discrete if it assumes, with probability 1, a countable (finite or infinite) set of values $\{x_0, x_1, \ldots, x_n, \ldots\}$,*

$$\sum_{i=0}^{\infty} P(X = x_i) = \sum_{i=0}^{\infty} P(\{\omega \in \Omega : X(\omega) = x_i\}) = 1.$$

The function

$$f_X(x_i) = P(X = x_i), \quad i = 0, 1, \ldots,$$

is called the probability mass function or simply the probability function of the random variable X.

(b) *A bivariate random variable (X, Y) is said to be discrete if it assumes, with probability 1, a countable (finite or infinite) set of values, $\{(x_0, y_0), (x_0, y_1), (x_1, y_0), \ldots, (x_n, y_m), \ldots\}$,*

$$\sum_{j=0}^{\infty} \sum_{j=0}^{\infty} P(X = x_i, Y = y_j)$$

$$= \sum_{j=0}^{\infty} \sum_{i=0}^{\infty} P(\{\omega \in \Omega : X(\omega) = x_i, Y(\omega) = y_j\}) = 1.$$

The function

$$f_{X,Y}(x_i, y_j) = P(X = x_i, Y = y_j), \quad i, j = 0, 1, \ldots,$$

is called the probability function of the bivariate random variable (X, Y) or the joint probability function of the random variables X and Y.

Clearly, if $x_0 < x_1 < \cdots < x_n < \cdots$, then

$$F_X(x) = \sum_{i=0}^{r} f_X(x_i), \quad x_r \leq x < x_{r+1}, \quad r = 0, 1, \ldots$$

and

$$f_X(x_0) = F_X(x_0), \quad f_X(x_i) = F_X(x_i) - F_X(x_{i-1}), \quad i = 1, 2, \ldots.$$

The bivariate distribution and probability functions are similarly connected. Specifically, if $x_0 < x_1 < \cdots < x_n < \cdots$ and $y_0 < y_1 < \cdots < y_m < \cdots$, then

$$F_{X,Y}(x, y) = \sum_{j=0}^{s} \sum_{i=0}^{r} f_{X,Y}(x_i, y_j), \quad x_r \leq x < x_{r+1}, \quad y_s \leq y < y_{s+1},$$

for $r, s = 0, 1, \ldots$, and

$$f_{X,Y}(x_i, y_j) = F_{X,Y}(x_i, y_j) - F_{X,Y}(x_{i-1}, y_j) \\ - F_{X,Y}(x_i, y_{j-1}) + F_{X,Y}(x_{i-1}, y_{j-1}),$$

for $i, j = 0, 1, \ldots$, with $x_{-1} = y_{-1} = -\infty$.

Let (X, Y) be a discrete bivariate random variable with probability function $f_{X,Y}(x_i, y_j) = P(X = x_i, Y = y_j)$, $i, j = 0, 1, \ldots$. The *marginal probability functions* $f_X(x_i) = P(X = x_i)$, $i = 0, 1, \ldots$, and $f_Y(y_j) = P(Y = y_j)$, $j = 0, 1, \ldots$, are deduced from the joint probability function as

$$f_X(x_i) = \sum_{j=0}^{\infty} f_{X,Y}(x_i, y_j), \quad f_Y(y_j) = \sum_{i=0}^{\infty} f_{X,Y}(x_i, y_j).$$

The function
$$f_{X|Y}(x_i|y_j) = \frac{f_{X,Y}(x_i, y_j)}{f_Y(y_j)}, \quad i = 0, 1, \ldots, \quad (j = 0, 1, \ldots),$$
considered as a function of x_i, with y_j a given value of the random variable Y for which $f_Y(y_j) > 0$, is a probability function and since
$$f_{X|Y}(x_i|y_j) = P(X = x_i|Y = y_j), \quad i = 0, 1, \ldots, \quad (j = 0, 1, \ldots,),$$
it is called the *conditional probability function* of X given that $Y = y_j$. If the discrete random variables X and Y are independent, $F_{X,Y}(x,y) = F_X(x)F_Y(y)$, $x, y \in R$, then
$$f_{X,Y}(x_i, y_j) = f_X(x_i)f_Y(y_j), \quad i, j = 0, 1, \ldots$$
and inversely. Equivalently, the discrete random variables X and Y are independent if and only if
$$f_{X|Y}(x_i|y_j) = f_X(x_i), \quad i = 0, 1, \ldots, \quad j = 0, 1, \ldots$$
or
$$f_{Y|X}(y_j|x_i) = f_Y(y_j), \quad j = 0, 1, \ldots, \quad i = 0, 1, \ldots.$$

Let X be a discrete random variable with probability (mass) function $f_X(x_i) = P(X = x_i)$, $i = 0, 1, \ldots$, and consider a function $y = g(x)$. If $g(x_i) = y_r$ for every $x_i \in \{x_{0,r}, x_{1,r}, \ldots, x_{k,r}, \ldots\}$, $r = 0, 1, \ldots$, then $Y = g(X)$ is a discrete random variable and its probability function $f_Y(y_r) = P(Y = y_r)$, $r = 0, 1, \ldots$, is given by
$$f_Y(y_r) = \sum_{k=0}^{\infty} f_X(x_{k,r}), \quad r = 0, 1, \ldots. \tag{1.58}$$
Indeed,
$$P(Y = y_r) = P[g(X) = y_r] = \sum_{k=0}^{\infty} P(X = x_{k,r}).$$

Let (X, Y) be a discrete bivariate random variable with probability function $f_{X,Y}(x_i, y_j) = P(X = x_i, Y = y_j)$, $i, j = 0, 1, \ldots$, and consider a function $z = g(x, y)$. If $g(x_i, y_j) = z_r$ for every $(x_i, y_j) \in \{(x_{0,r}, y_{0,r}), (x_{0,r}, y_{1,r}), (x_{1,r}, y_{0,r}), \ldots, (x_{k,r}, y_{s,r}), \ldots\}$, $r = 0, 1, \ldots$, then $Z = g(X, Y)$ is a discrete random variable and its probability function $f_Z(z_r) = P(Z = z_r)$, $r = 0, 1, \ldots$, is given by
$$f_Z(z_r) = \sum_{s=0}^{\infty}\sum_{k=0}^{\infty} f_{X,Y}(x_{k,r}, y_{s,r}), \quad r = 0, 1, \ldots. \tag{1.59}$$
Indeed,
$$P(Z = z_r) = P[g(X,Y) = z_r] = \sum_{s=0}^{\infty}\sum_{k=0}^{\infty} P(X = x_{k,r}, Y = y_{s,r}).$$

The probability distribution of a random variable may be equivalently expressed either by the distribution function or by the probability (density) function. A brief description of the probabilistic behavior of a random variable may be given by some basic parameters of its distribution. Such parameters are the moments (of the distribution) of a random variable. The most widely used moments are the expected value (mean) and the variance. An important moment (of the distribution) of a bivariate random variable is the covariance. The expected value of a random variable is introduced first in the following definition.

Definition 1.10 (a) *Let X be a discrete random variable with probability function $f_X(x_i)$ $i = 0, 1, \ldots$. Then its expected value or mean, denoted by $E(X)$ or by μ_X or simply by μ, is defined by*

$$\mu = E(X) = \sum_{i=0}^{\infty} x_i f_X(x_i),$$

provided that the series converges absolutely.

(b) *Let (X, Y) be a discrete bivariate random variable and $f_{X|Y}(x_i|y_j)$, $i = 0, 1, \ldots$, $(j = 0, 1, \ldots)$ the conditional probability function of X given that $Y = y_j$. Then the conditional expected value or mean of X given that $Y = y_j$, denoted by $E(X|y_j)$ or by $m_{X|Y}(y_j)$ or simply by $m(y_j)$, is defined by*

$$m_{X|Y}(y_j) = E(X|y_j) = \sum_{i=0}^{\infty} x_i f_{X|Y}(x_i|y_j),$$

provided that the series converges absolutely.

The calculation of the expected value of a function $Y = g(X)$, according to Definition 1.10 (a), may be carried out by first deriving the probability function of the random variable Y. This is not necessary to be done in every particular case. Specifically, the expected value of Y,

$$E(Y) = E[g(X)] = \sum_{r=0}^{\infty} y_r f_Y(y_r),$$

on using (1.58), may be expressed as

$$E[g(X)] = \sum_{r=0}^{\infty} y_r \sum_{k=0}^{\infty} f_X(x_{k,r}) = \sum_{r=0}^{\infty} \sum_{k=0}^{\infty} y_r f_X(x_{k,r})$$

$$= \sum_{r=0}^{\infty} \sum_{k=0}^{\infty} g(x_{k,r}) f_X(x_{k,r})$$

and so

$$E[g(X)] = \sum_{i=0}^{\infty} g(x_i) f_X(x_i). \tag{1.60}$$

Similarly,

$$E[g(X,Y)|y_j] = \sum_{i=0}^{\infty} g(x_i, y_j) f_{X|Y}(x_i|y_j) \quad (1.61)$$

and

$$E[g(X,Y)] = \sum_{j=0}^{\infty}\sum_{i=0}^{\infty} g(x_i, y_j) f_{X,Y}(x_i, y_j). \quad (1.62)$$

Clearly, for $g_i(X)$, $i = 1, 2, \ldots, n$, functions of a random variable X and a_i, $i = 1, 2, \ldots, n$ constants,

$$E\left[\sum_{i=1}^{n} a_i g_i(X)\right] = \sum_{i=1}^{n} a_i E[g_i(X)]. \quad (1.63)$$

Also, for $g_i(X_i), i = 1, 2, \ldots, n$ functions of the random variables $X_i, i = 1, 2, \ldots, n$, respectively, and $a_i, i = 1, 2, \ldots, n$ constants,

$$E\left[\sum_{i=1}^{n} a_i g_i(X_i)\right] = \sum_{i=1}^{n} a_i E[g_i(X_i)]. \quad (1.64)$$

Further, if X and Y are independent random variables, then for any functions $g(X)$ and $h(Y)$,

$$E[g(X)h(Y)] = E[g(X)]E[h(Y)]. \quad (1.65)$$

In particular,

$$E(XY) = E(X)E(Y). \quad (1.66)$$

The conditional expected value, according to Definition 1.10 (b), is a real valued function $m(y) = E(X|y)$, which is defined for every $y = y_j$, $j = 0, 1, \ldots$, and assigns the real number $w_j = m(y_j) = E(X|y_j)$ to the point $y = y_j, j = 0, 1, \ldots$. Consequently, $W = m(Y) = E(X|Y)$ is a discrete random variable assuming the values $w_j = m(y_j) = E(X|y_j)$, $j = 0, 1, \ldots$. More generally, $W = h(Y) = E[g(X,Y)|Y]$ is a discrete random variable taking on the values $w_j = h(y_j) = E[g(X,Y)|y_j]$, $j = 0, 1, \ldots$. The expected value of this random variable is connected to the expected value of the function $g(X,Y)$. Specifically, the expression

$$E\{E[g(X,Y)|Y]\} = \sum_{j=0}^{\infty} E[g(X,Y)|y_j] f_Y(y_j),$$

on using (1.61) and the relation $f_{X,Y}(x_i, y_j) = f_{X|Y}(x_i|y_j) f_Y(y_j), i, j = 0, 1, \ldots$, is transformed to

$$E\{E[g(X,Y)|Y]\} = \sum_{j=0}^{\infty}\sum_{i=0}^{\infty} g(x_i, y_j) f_{X|Y}(x_i|y_j) f_Y(y_j)$$

$$= \sum_{j=0}^{\infty}\sum_{i=0}^{\infty} g(x_i, y_j) f_{X,Y}(x_i, y_j),$$

and so by (1.62),

$$E\{E[g(X,Y)|Y]\} = E[g(X,Y)]. \tag{1.67}$$

The variance of a random variable and the covariance of two random variables are defined as expected values of certain functions of the random variables and so for their calculation expressions (1.60), (1.61) and (1.62) may be used. In the following definitions the series involved are assumed to be absolutely convergent.

Definition 1.11 (a) Let X be a discrete random variable with probability function $f_X(x_i)$, $i = 0, 1, \ldots$. Then its variance, denoted by $V(X)$ or by σ_X^2 or simply by σ^2, is defined by

$$\sigma^2 = V(X) = E[(X - \mu)^2].$$

The positive square root of the variance

$$\sigma = \sqrt{V(X)}$$

is called the standard deviation of the random variable X.

(b) Let (X, Y) be a discrete bivariate random variable and $f_{X|Y}(x_i|y_j)$, $i = 0, 1, \ldots$, $(j = 0, 1, \ldots)$ the conditional probability function of X given that $Y = y_j$. Then the conditional variance of X given that $Y = y_j$, denoted by $V(X|y_j)$ or by $\sigma_{X|Y}^2(y_j)$, is defined by

$$\sigma_{X|Y}^2(y_j) = V(X|y_j) = E\{[X - m_{X|Y}(y_j)]^2|y_j\}.$$

The positive square root of the conditional variance

$$\sigma_{X|Y}(y_j) = \sqrt{V(X|y_j)}$$

is called the conditional standard deviation of the random variable X given that $Y = y_j$.

(c) Let (X, Y) be a discrete bivariate random variable with probability function $f_{X,Y}(x_i, y_j)$, $i, j = 0, 1, \ldots$. Then the covariance of X and Y, denoted by $C(X, Y)$ or by $\sigma_{X,Y}$, is defined by

$$\sigma_{X,Y} = C(X, Y) = E[(X - \mu_X)(Y - \mu_Y)].$$

The correlation coefficient of X and Y, denoted by $\rho(X, Y)$ or by $\rho_{X,Y}$ or simply by ρ, is defined by

$$\rho = \rho(X, Y) = \frac{C(X, Y)}{\sqrt{V(X)}\sqrt{V(Y)}} = \frac{\sigma_{X,Y}}{\sigma_X \sigma_Y}.$$

The expected value and the variance of a probability distribution constitute the probabilistic analogue of the center of gravity and the moment of inertia of a mass distribution. Generally, on the analogy of the moments of a mass distribution, the moments of a probability distribution are introduced. They are defined as expected

DISTRIBUTIONS AND MOMENTS OF RANDOM VARIABLES 47

values of certain functions of the random variable and so for their calculation expression (1.60) may be used. In the following definitions the series involved are assumed to be absolutely convergent.

Definition 1.12 *Let X be a discrete random variable with probability function $f(x_i)$, $i = 0, 1, \ldots$, and c any constant. The expected value of the function $g(X) = (X-c)^r$,*

$$\mu_r(c) = E[(X - c)^r], \quad r = 1, 2, \ldots,$$

is called the rth order (power) moment of the random variable X about the point c. In particular,

$$\mu'_r = E(X^r), \quad r = 1, 2, \ldots,$$

is called the rth order (power) moment of the random variable X, while

$$\mu_r = E[(X - \mu)^r], \quad \mu = E(X), \quad r = 1, 2, \ldots,$$

is called the rth order central moment of the random variable X.

The central moments μ_r, $r = 0, 1, \ldots$, with $\mu_0 = 1$, may be expressed in terms of the moments μ'_r, $r = 0, 1, \ldots$, with $\mu'_0 = 1$, and vice versa. Indeed, on using Newton's binomial formula, $(t + u)^r = \sum_{k=0}^{r} \binom{r}{k} u^k t^{r-k}$, along with (1.63),

$$\mu_r = \sum_{k=0}^{r}(-1)^k \binom{r}{k}\mu^k \mu'_{r-k}, \quad \mu'_r = \sum_{k=0}^{r} \binom{r}{k}\mu^k \mu_{r-k}.$$

In the particular case of a nonnegative integer valued random variable the factorial or equivalently the binomial moments can be more easily calculated. These moments are introduced in the following definition.

Definition 1.13 *Let X be a nonnegative integer valued random variable with probability function $f(x) = P(X = x)$, $x = 0, 1, \ldots$, and c any constant. The expected values*

$$\mu_{(r)}(c) = E[(X - c)_r], \quad r = 1, 2, \ldots,$$

and

$$b_{(r)}(c) = E\left[\binom{X-c}{r}\right], \quad r = 1, 2, \ldots,$$

are called the rth order factorial and binomial moments, respectively, of the random variable X about the point c. In particular,

$$\mu_{(r)} = E[(X)_r], \quad r = 1, 2, \ldots,$$

and

$$b_{(r)} = E\left[\binom{X}{r}\right], \quad r = 1, 2, \ldots,$$

are called the rth order factorial and binomial moments, respectively, of the random variable X.

In certain cases of nonnegative integer valued random variables the ascending factorial or binomial moments can be more easily calculated. A random variable expressing the waiting time until the occurrence of an event in a sequence of trials constitutes such a case.

Definition 1.14 *Let X be a nonnegative integer valued random variable with probability function $f(x) = P(X = x)$, $x = 0, 1, \ldots$, and c any constant. The expected values*

$$\mu_{[r]}(c) = E[(X - c + r - 1)_r], \quad r = 1, 2, \ldots,$$

and

$$b_{[r]}(c) = E\left[\binom{X - c + r - 1}{r}\right], \quad r = 1, 2, \ldots,$$

are called the rth order ascending factorial and binomial moments, respectively, of the random variable X about the point c. In particular,

$$\mu_{[r]} = E[(X + r - 1)_r], \quad r = 1, 2, \ldots,$$

and

$$b_{[r]} = E\left[\binom{X + r - 1}{r}\right], \quad r = 1, 2, \ldots$$

are called the rth order ascending factorial and binomial moments, respectively, of the random variable X.

Notice that the rth binomial moment is simply equal to the corresponding rth factorial moment divided by $r!$. The reason for introducing the binomial moments, in addition to the introduction of the factorial moments, is that in the class of probability functions derived by the general inclusion and exclusion principle emerge more naturally. The binomial moments in such a case are deduced in Corollary 1.6. Further, the binomial or factorial moments of a random variable, as expected values of functions of the random variable, are calculated by using (1.60). Thus, the rth binomial moment is given by

$$b_{(r)} = E\left[\binom{X}{r}\right] = \sum_{x=r}^{\infty} \binom{x}{r} f(x), \quad r = 1, 2, \ldots.$$

Also, on using the expression $f(x) = [1 - F(x-1)] - [1 - F(x)]$, $x = 1, 2, \ldots$, and the recurrence relation

$$\binom{x}{r} = \binom{x-1}{r} + \binom{x-1}{r-1}, \quad x = 1, 2, \ldots, \quad r = 1, 2, \ldots,$$

$b_{(r)}$ may be expressed in terms of the distribution function as

$$b_{(r)} = E\left[\binom{X}{r}\right] = \sum_{x=r-1}^{\infty} \binom{x}{r-1}[1 - F(x)], \quad r = 1, 2, \ldots.$$

Similarly, using the recurrence relation

$$\binom{x+r-1}{r} = \binom{x+r-2}{r} + \binom{x+r-2}{r-1}, \quad x=1,2,\ldots, \quad r=1,2,\ldots,$$

$b_{[r]}$ may be expressed in terms of the distribution function as

$$b_{[r]} = E\left[\binom{X+r-1}{r}\right] = \sum_{x=0}^{\infty} \binom{x+r-1}{r-1}[1-F(x)], \quad r=1,2,\ldots.$$

The probability function of a nonnegative integer valued random variable X may be expressed in terms of its binomial moments. Indeed, multiplying the expression

$$b_{(r)} = \sum_{k=r}^{\infty} \binom{k}{r} f(k)$$

by $(-1)^{r-x}\binom{r}{x}$ and summing for $r = x, x+1, \ldots$, we get

$$\sum_{r=x}^{\infty}(-1)^{r-x}\binom{r}{x}b_{(r)} = \sum_{r=x}^{\infty}(-1)^{r-x}\binom{r}{x}\sum_{k=r}^{\infty}\binom{k}{r}f(k)$$

$$= \sum_{k=x}^{\infty}\left\{\sum_{r=x}^{k}(-1)^{r-x}\binom{r}{x}\binom{k}{r}\right\}f(k)$$

and since

$$\sum_{r=x}^{k}(-1)^{r-x}\binom{r}{x}\binom{k}{r} = \binom{k}{x}\sum_{r=x}^{k}(-1)^{r-x}\binom{k-x}{r-x} = \binom{k}{x}\delta_{k,x},$$

we conclude that

$$f(x) = \sum_{r=s}^{\infty}(-1)^{r-x}\binom{r}{x}b_{(r)}, \tag{1.68}$$

provided that the series is absolutely convergent.

The ascending binomial moments about a point c_1, $b_{[r]}(c_1)$, $r = 1, 2, \ldots$, may be expressed in terms of the descending binomial moments about a point c_2, $b_{(r)}(c_2)$, $r = 1, 2, \ldots$, and vice versa. Such an expression facilitates the derivation of the binomial moments of certain occupancy distributions.

Theorem 1.11 *Let*

$$b_{[r]}(c_1) = E\left[\binom{X-c_1+r-1}{r}\right], \quad b_{(r)}(c_2) = E\left[\binom{X-c_2}{r}\right],$$

for $r = 1, 2, \ldots$, with c_1 and c_2 constants. Then

$$b_{[r]}(c_1) = \sum_{k=0}^{r} \binom{r + c_2 - c_1 - 1}{r - k} b_{(k)}(c_2)$$

$$= \sum_{k=0}^{r} (-1)^{r-k} \binom{c_1 - c_2 - k}{r - k} b_{(k)}(c_2) \qquad (1.69)$$

and

$$b_{(r)}(c_2) = \sum_{k=0}^{r} (-1)^{r-k} \binom{r + c_2 - c_1 - 1}{r - k} b_{[k]}(c_1)$$

$$= \sum_{k=0}^{r} \binom{c_1 - c_2 - k}{r - k} b_{[k]}(c_1), \qquad (1.70)$$

where, by convention, $b_{(0)}(c_2) = b_{[0]}(c_1) = 1$.

Proof Cauchy's formula

$$\binom{u + v}{r} = \sum_{k=0}^{r} \binom{u}{k} \binom{v}{r - k},$$

with $u = X - c_2$ and $v = r + c_2 - c_1 - 1$, may be rewritten as

$$\binom{X - c_1 + r - 1}{r} = \sum_{k=0}^{r} \binom{r + c_2 - c_1 - 1}{r - k} \binom{X - c_2}{k}$$

$$= \sum_{k=0}^{r} (-1)^{r-k} \binom{c_1 - c_2 - k}{r - k} \binom{X - c_2}{k}.$$

Taking expectations on both sides of these expressions and using (1.63) we deduce (1.69). Again, Cauchy's formula with $u = c_1 - X$ and $v = r + c_2 - c_1 - 1$, and since

$$\binom{-X + c_2 + r - 1}{r} = (-1)^r \binom{X - c_2}{r},$$

$$\binom{-X + c_1}{k} = (-1)^k \binom{X - c_1 + k - 1}{k},$$

yields the expressions

$$\binom{X - c_2}{r} = \sum_{k=0}^{r} (-1)^{r-k} \binom{r + c_2 - c_1 - 1}{r - k} \binom{X - c_1 + k - 1}{k}$$

$$= \sum_{k=0}^{r} \binom{c_1 - c_2 - k}{r - k} \binom{X - c_1 + k - 1}{k}$$

from which we deduce (1.70).

Example 1.12 *Indicator random variables.* Consider a stochastic experiment with sample space Ω and let $A_1, A_2, \ldots, A_n$ be any events in Ω. Further, consider the indicator random variables I_j, $j = 1, 2, \ldots, n$, defined by

$$I_j(\omega) = \begin{cases} 1, & \text{if } \omega \in A_j \\ 0, & \text{if } \omega \notin A_j \end{cases}, \quad j = 1, 2, \ldots, n.$$

Clearly, for any k-combination $\{j_1, j_2, \ldots, j_k\}$ of the set of n indices $\{1, 2, \ldots, n\}$,

$$E(I_{j_1} I_{j_2} \cdots I_{j_k}) = P(A_{j_1} A_{j_2} \cdots A_{j_k}).$$

(a) Express the binomial moments of the random variable $K = n - \sum_{j=1}^{n} I_j$ in terms of the probabilities $P(A_{j_1} A_{j_2} \cdots A_{j_k})$, $\{j_1, j_2, \ldots, j_k\} \subseteq \{1, 2, \ldots, n\}$.

(b) Assuming that the events $A_1, A_2, \ldots, A_n$ are exchangeable, express the probability function $p_{n,k} = P(K = k)$, $k = 0, 1, \ldots, n$, as a finite difference of the probability $p_r = P(A_{j_1} A_{j_2} \cdots A_{j_r})$.

(a) Let $x = i_1 + i_2 + \cdots + i_n$ and consider the multivariate Cauchy's formula

$$\binom{x}{k} = \sum \binom{i_1}{k_1}\binom{i_2}{k_2} \cdots \binom{i_n}{k_n},$$

where the summation is extended over all $k_j \geq 0$, $j = 1, 2, \ldots, n$, such that $k_1 + k_2 + \cdots + k_n = k$. For $i_j = 0, 1$ and since $\binom{i_j}{k_j} = 0$ if $k_j \geq 2$, $j = 1, 2, \ldots, n$, this formula reduces to

$$\binom{x}{k} = \sum i_{j_1} i_{j_2} \cdots i_{j_k},$$

where the summation is extended over all k-combinations $\{j_1, j_2, \ldots, j_k\}$ of the n indices $\{1, 2, \ldots, n\}$. Therefore the kth binomial moment of $X = \sum_{j=1}^{n} I_j$ is given by

$$a_{(k)} = E\left[\binom{X}{k}\right] = S_{n,k},$$

with

$$S_{n,k} = \sum P(A_{j_1} A_{j_2} \cdots A_{j_k}),$$

where the summation is extended over all k-combinations $\{j_1, j_2, \ldots, j_k\}$ of the n indices $\{1, 2, \ldots, n\}$. The rth binomial moment of $K = n - X$ may be written as

$$b_{(r)} = E\left[\binom{K}{r}\right] = E\left[\binom{n-X}{r}\right] = (-1)^r E\left[\binom{X-n+r-1}{r}\right],$$

and since by Theorem 1.11,

$$E\left[\binom{X-n+r-1}{r}\right] = \sum_{k=0}^{r}(-1)^{r-k}\binom{n-k}{r-k}E\left[\binom{X}{k}\right],$$

it follows that

$$b_{(r)} = E\left[\binom{K}{r}\right] = \sum_{k=0}^{r}(-1)^k \binom{n-k}{r-k} S_{n,k}.$$

(b) If the events $A_1, A_2, \ldots, A_n$ are exchangeable, then $S_{n,k} = \binom{n}{k} p_k$, with $p_k = P(A_{j_1} A_{j_2} \cdots A_{j_k})$ and so

$$a_{(k)} = E\left[\binom{X}{k}\right] = \binom{n}{k} p_k.$$

Hence,

$$b_{(r)} = E\left[\binom{K}{r}\right] = \sum_{k=0}^{r} (-1)^k \binom{n-k}{r-k} \binom{n}{k} p_k$$

$$= \binom{n}{r} \sum_{k=0}^{r} (-1)^k \binom{r}{k} p_k.$$

Since

$$[\Delta_u^r p_{r-u}]_{u=0} = \sum_{k=0}^{r} (-1)^{r-k} \binom{r}{k} [E_u^k p_{r-u}]_{u=0}$$

$$= \sum_{k=0}^{r} (-1)^{r-k} \binom{r}{k} p_{r-k},$$

the rth binomial moment of K may be written in the form

$$b_{(r)} = E\left[\binom{K}{r}\right] = \binom{n}{r} [\Delta_u^r p_{r-u}]_{u=0}.$$

The probability function $P(K = k) = P(X = n - k)$, $k = 0, 1, \ldots, n$, on using Corollary 1.5, with $n - k$ instead of k, is given by

$$P(K = k) = \binom{n}{k} [\Delta_u^k p_{n-u}]_{u=0}, \quad k = 0, 1, \ldots, n.$$

Example 1.13 Consider a supply of m balls successively distributed into n distinguishable urns. Suppose that, at any trial, the probability of allocating a ball in the jth urn is p_j, $j = 1, 2, \ldots, n$, with $p_1 + p_2 + \cdots + p_n = 1$. Let K be the number of occupied urns (by at least one ball each). Find the probability function $p_k(m, n) = P(K = k)$, $k = 1, 2, \ldots, n$, the binomial moments $b_{(r)}(m, n) = E[\binom{K}{r}]$, $r = 1, 2, \ldots$, and the probabilities $P_k(m, n) = P(K \leq k)$, $k = 1, 2, \ldots, n$.

Let A_j be the event that the jth urn remains empty, $j = 1, 2, \ldots, n$. Then $p_k(m, n) = P(K = k)$ is the probability that $n - k$ among the n events $A_1, A_2, \ldots, A_n$ occur. Further, for any r indices $\{i_1, i_2, \ldots, i_r\}$ out of the n indices $\{1, 2, \ldots, n\}$,

$$P(A_{i_1} A_{i_2} \cdots A_{i_r}) = (1 - p_{i_1} - p_{i_2} - \cdots - p_{i_r})^m$$
$$= (p_{j_1} + p_{j_2} + \cdots + p_{j_{n-r}})^m,$$

with $\{j_1, j_2, \ldots, j_{n-r}\} = \{1, 2, \ldots, n\} - \{i_1, i_2, \ldots, i_r\}$ and

$$S_{n,r} = \sum (p_{j_1} + p_{j_2} + \cdots + p_{j_{n-r}})^m,$$

where the summation is extended over all $(n-r)$-combinations $\{j_1, j_2, \ldots, j_{n-r}\}$ of the n indices $\{1, 2, \ldots, n\}$. Thus, according to the general inclusion and exclusion principle (1.35),

$$p_k(m,n) = \sum_{r=0}^{k} (-1)^{k-r} \binom{n-r}{k-r} S_{n,n-r}, \quad k = 1, 2, \ldots, n.$$

The kth binomial moment $E\left[\binom{n-K}{k}\right]$ of the number $K_0 = n - K$ of empty urns, according to Corollary 1.6, is given by

$$E\left[\binom{n-K}{k}\right] = S_{n,k}, \quad k = 1, 2, \ldots, n.$$

Thus,

$$E\left[\binom{K-n+k-1}{k}\right] = (-1)^k E\left[\binom{n-K}{k}\right] = (-1)^k S_{n,k},$$

and using Theorem 1.11, the rth binomial moment of the number K of occupied urns is obtained as

$$b_{(r)}(m,n) = E\left[\binom{K}{r}\right] = \sum_{k=0}^{r} \binom{n-k}{r-k} E\left[\binom{K-n+k-1}{k}\right]$$

$$= \sum_{k=0}^{r} (-1)^k \binom{n-k}{r-k} S_{n,k}.$$

The probability $P_k(m,n) = P(K \leq k)$ equals the probability $P(K_0 \geq n-k)$ that at least $n-k$ among the n events $A_1, A_2, \ldots, A_n$ occur, and according to Corollary 1.7, is deduced as

$$P_k(m,n) = \sum_{r=0}^{k} (-1)^{k-r} \binom{n-r-1}{n-k-1} S_{n,n-r}, \quad k = 1, 2, \ldots, n.$$

Example 1.14 Consider a supply of m like balls randomly distributed into n distinguishable urns. Assume that the jth urn is divided into s_j distinguishable cells (compartments), $j = 1, 2, \ldots, n$, each of capacity limited to one ball and set $s = s_1 + s_2 + \cdots + s_n$. Let K be the number of occupied urns (by at least one ball each). Find the probability function $p_k(m,n;s) = P(K = k)$, $k = 1, 2, \ldots, n$, the binomial moments $b_{(r)}(m,n;s) = E\left[\binom{K}{r}\right]$, $r = 1, 2, \ldots$, and the probabilities $P_k(m,n;s) = P(K \leq k)$ $k = 1, 2, \ldots, n$.

Let A_j be the event that the jth urn remains empty, $j = 1, 2, \ldots, n$. Then $p_k(m, n; s) = P(K = k)$ is the probability that $n-k$ among the n events $A_1, A_2, \ldots, A_n$ occur. Further, for any r indices $\{i_1, i_2, \ldots, i_r\}$ out of the n indices $\{1, 2, \ldots, n\}$,

$$P(A_{i_1} A_{i_2} \cdots A_{i_r}) = \frac{(s - s_{i_1} - s_{i_2} - \cdots - s_{i_r})_m}{(s)_m}$$

$$= \frac{(s_{j_1} + s_{j_2} + \cdots + s_{j_{n-r}})_m}{(s)_m},$$

with $\{j_1, j_2, \ldots, j_{n-r}\} = \{1, 2, \ldots, n\} - \{i_1, i_2, \ldots, i_r\}$ and

$$S_{n,r} = \sum \frac{(s_{j_1} + s_{j_2} + \cdots + s_{j_{n-r}})_m}{(s)_m},$$

where the summation is extended over all $(n - r)$- combinations $\{j_1, j_2, \ldots, j_{n-r}\}$ of the n indices $\{1, 2, \ldots, n\}$. Thus, according to the general inclusion and exclusion principle (1.35),

$$p_k(m, n; s) = \sum_{r=0}^{k}(-1)^{k-r}\binom{n-r}{k-r}S_{n,n-r}, \quad k = 1, 2, \ldots, n.$$

The kth binomial moment $E\left[\binom{n-K}{k}\right]$ of the number $K_0 = n - K$ of empty urns, according to Corollary 1.6, is given by

$$E\left[\binom{n-K}{k}\right] = S_{n,k}, \quad k = 1, 2, \ldots, n.$$

Hence,

$$E\left[\binom{K-n+k-1}{k}\right] = (-1)^k E\left[\binom{n-K}{k}\right] = (-1)^k S_{n,k},$$

and using Theorem 1.11, the rth binomial moment of the number K of occupied urns is obtained as

$$b_{(r)}(m, n; s) = E\left[\binom{K}{r}\right] = \sum_{k=0}^{r}\binom{n-k}{r-k}E\left[\binom{K-n+k-1}{k}\right]$$

$$= \sum_{k=0}^{r}(-1)^k\binom{n-k}{r-k}S_{n,k}.$$

The probability $P_k(m, n; s) = P(K \leq k) = P(K_0 \geq n - k)$ that at least $n - k$ among the n events $A_1, A_2, \ldots, A_n$ occur, according to Corollary 1.7, is deduced as

$$P_k(m, n; s) = \sum_{r=0}^{k}(-1)^{k-r}\binom{n-r-1}{n-k-1}S_{n,n-r}, \quad k = 1, 2, \ldots, n.$$

1.7 GENERATING FUNCTIONS

The study of the distributions of random variables is facilitated by the use of generating functions. This powerful tool has been introduced in probability theory by De Moivre and Laplace. In general, the generating function $G(t)$, of a sequence of real numbers a_r, $r = 0, 1, \ldots$, with respect to the sequence of linearly independent real functions $g_r(t)$, $t \in R$, $r = 0, 1, \ldots$, is defined by

$$G(t) = \sum_{r=0}^{\infty} a_r g_r(t), \quad |t| < \rho,$$

provided that the series is absolutely convergent. The assumption of the linear independence of the functions $g_r(t)$, $t \in R$, $r = 0, 1, \ldots$, secures the uniqueness of the generating function. The bivariate generating function $G(t, u)$, of a double (two index) sequence of real numbers $a_{r,s}$, $r = 0, 1, \ldots$, $s = 0, 1, \ldots$, with respect to the sequences of linearly independent real functions $g_r(t)$, $r = 0, 1, \ldots$ and $h_s(u)$, $u \in R$, $s = 0, 1, \ldots$, is defined by

$$G(t, u) = \sum_{s=0}^{\infty} \sum_{r=0}^{\infty} a_{r,s} g_r(t) h_s(u), \quad |t| < \rho_1, \ |u| < \rho_2,$$

provided that the double series is absolutely convergent. The multivariate generating function $G(t_1, t_2, \ldots, t_k)$, of a multiple (multi-index) sequence of real numbers $a_{r_1, r_2, \ldots, r_k}$, $r_j = 0, 1, \ldots$, $j = 1, 2, \ldots, k$, is correspondingly defined. The most frequently used sequences of linearly independent functions in combinatorics and probability theory are

$$g_r(t) = t^r, \quad r = 0, 1, \ldots, \quad h_s(u) = u^s, \quad s = 0, 1, \ldots$$

and

$$g_r(t) = t^r/r!, \quad r = 0, 1, \ldots, \quad h_s(u) = u^s/s!, \quad s = 0, 1, \ldots .$$

In probability theory the sequences a_r, $r = 0, 1, \ldots$, and $a_{r,s}$, $r = 0, 1, \ldots$, $s = 0, 1, \ldots$, are either sequences of probabilities or sequences of moments.

The most important discrete random variables are those assuming nonnegative integer values. The study of their distributions is facilitated by the use of the probability generating functions.

Definition 1.15 (a) *Let X be a nonnegative integer valued random variable with probability function $f(x) = P(X = x)$, $x = 0, 1, \ldots, n, \ldots$. Then the probability generating function of X, denoted by $P(t)$ or by $P_X(t)$, is defined by*

$$P(t) = \sum_{x=0}^{\infty} f(x) t^x. \tag{1.71}$$

(b) *Let X and Y be nonnegative integer valued random variables with joint probability function $f(x, y) = P(X = x, Y = y)$, $x = 0, 1, \ldots, y = 0, 1, \ldots$. Then the* joint probability generating function of X and Y, denoted by $P(t, u)$ or by $P_{X,Y}(t, u)$, is defined by

$$P(t, u) = \sum_{y=0}^{\infty} \sum_{x=0}^{\infty} f(x, y) t^x u^y. \tag{1.72}$$

Note that both series (1.71) and (1.72) are absolutely convergent at least for $-1 \leq t \leq 1$ and $-1 \leq u \leq 1$. Also, the factorial or binomial moment generating functions

$$G(t) = \sum_{r=0}^{\infty} \mu_{(r)} \frac{t^r}{r!} = \sum_{r=0}^{\infty} b_{(r)} t^r,$$

and

$$G(t, u) = \sum_{s=0}^{\infty} \sum_{r=0}^{\infty} \mu_{(r,s)} \frac{t^r}{r!} \cdot \frac{u^s}{s!} = \sum_{s=0}^{\infty} \sum_{r=0}^{\infty} b_{(r,s)} t^r u^s,$$

where

$$\mu_{(r)} = E[(X)_r], \quad b_{(r)} = E\left[\binom{X}{r}\right], \quad r = 0, 1, \ldots,$$

and

$$\mu_{(r,s)} = E[(X)_r (Y)_s], \quad b_{(r,s)} = E\left[\binom{X}{r}\binom{Y}{s}\right], \quad r = 0, 1, \ldots, \quad s = 0, 1, \ldots,$$

may be expressed in terms of the corresponding probability generating functions as

$$G(t) = P(t+1), \quad G(t, u) = P(t+1, u+1)$$

and inversely

$$P(t) = G(t-1), \quad P(t, u) = G(t-1, u-1).$$

Indeed,

$$G(t) = \sum_{r=0}^{\infty} \left\{ \sum_{x=r}^{\infty} \binom{x}{r} f(x) \right\} t^r = \sum_{x=0}^{\infty} f(x) \left\{ \sum_{r=0}^{x} \binom{x}{r} t^r \right\}$$

$$= \sum_{x=0}^{\infty} f(x)(t+1)^x = P(t+1)$$

and

$$G(t, u) = \sum_{s=0}^{\infty} \sum_{r=0}^{\infty} \left\{ \sum_{y=s}^{\infty} \sum_{x=r}^{\infty} \binom{x}{r}\binom{y}{s} f(x, y) \right\} t^r u^s$$

$$= \sum_{y=0}^{\infty} \sum_{x=0}^{\infty} f(x, y) \left\{ \sum_{r=0}^{x} \binom{x}{r} t^r \right\} \left\{ \sum_{s=0}^{y} \binom{y}{s} u^s \right\}$$

$$= \sum_{y=0}^{\infty} \sum_{x=0}^{\infty} f(x, y)(t+1)^x (u+1)^s = P(t+1, u+1).$$

Further, consider the distribution function $F(x) = P(X \leq x)$ and the reliability (survival) function $R(x) = P(X > x)$ both at the jump (discontinuity) points $x = 0, 1, \ldots$. The generating functions

$$H(t) = \sum_{x=0}^{\infty} F(x)t^x, \quad Q(t) = \sum_{x=0}^{\infty} R(x)t^x,$$

on using the expressions

$$F(x) = \sum_{r=0}^{x} f(r), \quad R(x) = \sum_{r=x+1}^{\infty} f(r), \quad x = 0, 1, \ldots,$$

may also be expressed in terms of the corresponding probability generating functions as

$$H(t) = \frac{P(t)}{1-t}, \quad Q(t) = \frac{1-P(t)}{1-t}.$$

Moreover, the distribution of the sum of independent nonnegative integer valued random variables may be derived by the aid of probability generating functions. Specifically, we have the following theorem.

Theorem 1.12 (a) *Let X_j, $j = 1, 2, \ldots, n$, be nonnegative integer valued and independent random variables with probability generating functions $P_{X_j}(t)$, $j = 1, 2, \ldots, n$. Then the probability generating function $P_{S_n}(t)$ of the sum $S_n = X_1 + X_2 + \cdots + X_n$ is given by the product*

$$P_{S_n}(t) = P_{X_1}(t)P_{X_2}(t) \cdots P_{X_n}(t). \tag{1.73}$$

If, in addition, the random variables X_j, $j = 1, 2, \ldots, n$ are identically distributed with common probability generating function $P_X(t)$, then

$$P_{S_n}(t) = [P_X(t)]^n. \tag{1.74}$$

(b) *Let X_j, $j = 1, 2, \ldots$ be a sequence of nonnegative integer valued independent and identically distributed random variables with common probability generating functions $P_X(t)$. Further, let N be a nonnegative integer valued random variable, independent of the random variables X_j, $j = 1, 2, \ldots$, with probability generating function $P_N(t)$. Then the probability generating function $P_{S_N}(t)$ of the sum $S_N = X_1 + X_2 + \cdots + X_N$ is given by the composite function*

$$P_{S_N}(t) = P_N(P_X(t)). \tag{1.75}$$

Proof (a) The assumption of the independence of the random variables X_j, $j = 1, 2, \ldots, n$, permits the application of (1.65), yielding the expression

$$E(t^{S_n}) = E(t^{X_1+X_2+\cdots+X_n}) = E(t^{X_1}t^{X_2}\cdots t^{X_n})$$
$$= E(t^{X_1})E(t^{X_2})\cdots E(t^{X_n}),$$

from which (1.73) is obtained. Further, since $P_{X_j}(t) = P_X(t)$, $j = 0, 1, \ldots, n$, (1.74) is readily deduced.

(b) Applying the conditional expectation technique (1.67), it follows that

$$P_{S_N}(t) = E(t^{S_N}) = E[E(t^{S_N}|N=n)] = \sum_{n=0}^{\infty} E(t^{S_N}|N=n)P(N=n).$$

Since the random variables S_n and N are independent,

$$E(t^{S_N}|N=n) = E(t^{S_n}|N=n) = E(t^{S_n}) = P_{S_n}(t).$$

Further, since the random variables X_j, $j = 1, 2, \ldots, n$ are identically distributed, $P_{S_n}(t) = [P_X(t)]^n$ and so

$$P_{S_N}(t) = \sum_{n=0}^{\infty} [P_X(t)]^n P(N=n) = P_N(P_X(t)).$$

The proof of the theorem is thus completed.

Example 1.15 *Montmort-Moivre problem.* Consider an urn containing n balls bearing the numbers $0, 1, \ldots, n-1$. Suppose that k balls are successively drawn, at random, with replacement. Let X_j be the number on the ball drawn at the jth drawing, $j = 1, 2, \ldots, k$. Find the probability and the distribution function of the sum $S_k = X_1 + X_2 + \cdots + X_k$.

The assumption that the drawings are made with replacement implies that the random variables X_j, $j = 1, 2, \ldots, k$ are independent and identically distributed with common distribution being the discrete uniform,

$$f_X(x) = \frac{1}{n}, \quad x = 0, 1, \ldots, n-1,$$

and so

$$P_X(t) = \frac{1}{n} \sum_{x=0}^{n-1} t^x = \frac{1-t^n}{n(1-t)}.$$

The probability generating function of the sum $S_k = X_1 + X_2 + \cdots + X_k$, according to Theorem 1.12, is given by

$$P_{S_k}(t) = \frac{(1-t^n)^k}{n^k(1-t)^k}.$$

Expanding it into powers of t we get

$$P_{S_k}(t) = \frac{1}{n^k} \left[\sum_{r=0}^{k} (-1)^r \binom{k}{r} t^{nr} \right] \cdot \left[\sum_{i=0}^{\infty} \binom{k+i-1}{k-1} t^i \right]$$

$$= \sum_{s=0}^{\infty} \left\{ \frac{1}{n^k} \sum_{r=0}^{[s/n]} (-1)^r \binom{k}{r} \binom{k+s-nr-1}{k-1} \right\} t^s$$

and so the probability function $f_{S_k}(s) = P(S_k = s)$, $s = 0, 1, \ldots, (n-1)k$, is deduced as

$$f_{S_k}(s) = \frac{1}{n^k} \sum_{r=0}^{[s/n]} (-1)^r \binom{k}{r} \binom{k+s-nr-1}{k-1}.$$

Further, the generating function of the distribution function $F_{S_k}(s) = P(S_k \leq s)$, $s = 0, 1, \ldots, (n-1)k$,

$$H_{S_k}(t) = \sum_{s=0}^{\infty} F_{S_k}(s) t^s,$$

is given by

$$H_{S_k}(t) = \frac{P_{S_k}(t)}{1-t} = \frac{(1-t^n)^k}{n^k (1-t)^{k+1}}.$$

Expanding it into powers of t, we get

$$H_{S_k}(t) = \frac{1}{n^k} \left[\sum_{r=0}^{k} (-1)^r \binom{k}{r} t^{nr} \right] \cdot \left[\sum_{i=0}^{\infty} \binom{k+i}{k} t^i \right]$$

$$= \sum_{s=0}^{\infty} \left\{ \frac{1}{n^k} \sum_{r=0}^{[s/n]} (-1)^r \binom{k}{r} \binom{k+s-nr}{k} \right\} t^s$$

and so

$$F_{S_k}(s) = \frac{1}{n^k} \sum_{r=0}^{[s/n]} (-1)^r \binom{k}{r} \binom{k+s-nr}{k}.$$

Example 1.16 *Compound Poisson distribution.* Consider a sequence of nonnegative integer valued, independent and identically distributed random variables X_j, $j = 0, 1, \ldots$, with common probability generating function $P_X(t) = \sum_{x=0}^{\infty} f_X(x) t^x$. Let N be a nonnegative integer valued random variable, independent of the random variables X_j, $j = 0, 1, \ldots$, with Poisson probability function

$$f_N(n) = P(N = n) = e^{-\lambda} \frac{\lambda^n}{n!}, \quad n = 0, 1, \ldots, \quad 0 < \lambda < \infty.$$

The probability generating function of the random variable N is

$$P_N(t) = e^{-\lambda(1-t)}$$

and, according to Theorem 1.12, the probability generating function of the sum $S_N = X_1 + X_2 + \cdots + X_N$ is given by

$$P_{S_N}(t) = \exp\{-\lambda[1 - P_X(t)]\}.$$

The distribution of the random variable S_N is called *compound Poisson distribution*.

A representation of the compound Poisson random variable S_N as a linear combination of Poisson random variables is worth noticing. Letting $\lambda_k = \lambda f_X(k)$, $k = 0, 1, \ldots$, the generating function $P_{S_N}(t)$ can be factored as

$$P_{S_N}(t) = \prod_{k=1}^{\infty} e^{-\lambda_k(1-t^k)}.$$

Note that if Y_k is a Poisson random variable, with

$$f_{Y_k}(y) = P(Y_k = y) = e^{-\lambda_k} \frac{\lambda_k^y}{y!}, \quad y = 0, 1, \ldots,$$

then

$$P_{Y_k}(t) = e^{-\lambda_k(1-t)}$$

and the random variable $Z_k = kY_k$ has probability generating function

$$P_{Z_k}(t) = e^{-\lambda_k(1-t^k)}.$$

Therefore,

$$S_N = \sum_{k=1}^{\infty} kY_k,$$

where $Y_k, k = 1, 2, \ldots$ are independent Poisson random variables with parameters λ_k, $k = 1, 2, \ldots$. Note that Y_k is the number of those variables among $X_1, X_2, \ldots, X_N$ that are equal to k, $k = 1, 2, \ldots$.

1.8 REFERENCE NOTES

Basic combinatorics and probability were presented in this chapter. Besides the classical books of P. A. MacMahon (1915, 1916), J. Riordan (1958) and L. Comtet (1974), the recent book of Ch. A. Charalambides (2002) may be used for further reading on enumerative combinatorics. The classical book of Ch. Jordan (1939a) on the calculus of finite differences remains a rich source the interested reader may consult.

Considerable emphasis was given to the presentation of the inclusion and exclusion principle, which is a basic tool both in combinatorics and discrete probability theory. Theorem 1.7 is attributed to H. Poincaré (1896). However, it seems that this formula was known to J. Sylvester (1883) and to L. Euler (1760), when he evaluated $\phi(n)$, now known as Euler's function. The inclusion and exclusion formula was also known to A. De Moivre (1718) in the case of exchangeable events; he was able to generalize results that had been previously used by P. R. Montmort (1708). After De Moivre, C. Jordan (1867) derived, for the cardinality measure, expressions analogous to those in Theorem 1.8 and its Corollary 1.7 for the probability measure. Ch. Jordan (1926, 1927a,b, 1934, 1939a,b) obtained Theorem 1.8 and its Corollaries 1.5 - 1.8 and presented several applications. More general formulations of the inclusion and exclusion principle were given by M. Fréchet (1940), K. L. Chung (1941, 1943b,c) and L. Takács (1967), who furnished a very extensive reference list. Further, L. Takács (1991) derived Theorem 1.10. The work of C. E. Bonferroni (1936) is considered the foundation of the inequalities in Theorem 1.9 and its Corollary 1.9.

A multivariate extension of the inclusion and exclusion formula was established by M. Fréchet (1943) and K. L. Chung (1943a); a bivariate version of this formula is given in Exercises 1.33 - 1.35. Further, a bivariate inclusion and exclusion formula for pairs of disjoint events was deduced by Ch. A. Charalambides (2005a); Exercises 1.36 - 1.40 are based on this paper.

Theorem 1.11, in which the ascending binomial moments are expressed in terms of descending binomial moments and vice-versa, was formulated and shown by Ch. A. Charalambides (1986a) in connection with the derivation of the binomial moments of certain occupancy distributions.

1.9 EXERCISES AND COMPLEMENTS

1.1 (a) Show that the number of distributions of k distinguishable balls into n distinguishable urns is equal to n^k, if the order in which the balls are placed in the urns does not count, while it is equal to $(n+k-1)_k$, if the order of balls in the urns counts.

(b) Show that the number of distributions of k indistinguishable balls into n distinguishable urns is equal to
$$\binom{n+k-1}{k},$$
if the urns are of unlimited capacity, while it is equal to
$$\binom{n}{k},$$
if the capacity of each urn is limited to one ball.

1.2 Consider a collection of m balls composed of $r_i \geq 0$ distinguishable kinds of balls, each including i like balls, $i = 1, 2, \ldots, m$, so that $r_1 + 2r_2 + \cdots + mr_m = m$. Show that the number of distributions of this collection of balls into n distinguishable urns is equal to
$$\prod_{i=1}^{m} \binom{n+i-1}{i}^{r_i},$$
if the urns are of unlimited capacity, while it is equal to
$$\prod_{i=1}^{m} \binom{n}{i}^{r_i},$$
if the capacity of each urn is limited to one ball from each kind.

1.3 Consider an urn containing $2n$ balls bearing the numbers $1, 2, \ldots, 2n$. Assume that $2k+1$ balls are successively drawn, at random, with replacement. Let A be the event that the sum of the drawn numbers is greater than $(2n+1)k+n$. Show that the event A and its complementary event A' are equivalent sets and conclude that $P(A) = 1/2$.

1.4 Suppose that m balls are successively drawn, without replacement, from an urn that contains r white and s black balls and placed in an empty urn. Then n balls are successively drawn without replacement. Show that the probability that k balls, among the n balls drawn from the second urn, are white is given by

$$p_{n,k} = \binom{r}{k}\binom{s}{n-k} \bigg/ \binom{r+s}{n}, \quad k = 0, 1, \ldots, n,$$

which is independent of the number m of balls drawn from the first urn.

1.5 Consider an urn containing a total of n white and black balls. Let A_k be the event that the urn contains k white balls and $P(A_k) = p_k$, $k = 0, 1, \ldots, n$. Suppose that m balls are successively drawn without replacement and let B_j be the event that a white ball is drawn at the jth drawing, $j = 1, 2, \ldots, m$. Show that the events $B_1, B_2, \ldots, B_m$ are exchangeable.

1.6 *(Continuation).* If $p_k = 1/(n+1)$, $k = 0, 1, \ldots, n$, find the conditional probability that there are no black balls in the urn given that all m drawn balls are white.

1.7 *A total conditional probability formula.* Let A and B be events in a sample space Ω, with $P(A) > 0$. Further, let $\{A_1, A_2, \ldots, A_n, \ldots\}$ be a set of pairwise disjoint events in Ω, with $A \subseteq A_1 + A_2 + \cdots + A_n + \cdots$ and $P(A_k|A) > 0$, $k = 1, 2, \ldots, n, \ldots$. Prove that

$$P(B|A) = \sum_{k=1}^{\infty} P(A_k|A) P(B|AA_k).$$

In the particular case $A = A_1 + A_2 + \cdots + A_n + \cdots$, conclude the expression

$$P(B|A) = \frac{\sum_{k=1}^{\infty} P(A_k) P(B|A_k)}{\sum_{k=1}^{\infty} P(A_k)}.$$

Notice that both formulae for $A = \Omega$ reduce to the classical total probability formula.

1.8 Consider an urn containing r white and s black balls. Assume that balls are successively drawn one after the other without replacement. Let A be the event that at least k white balls are drawn in the first n trials and B the event that a white ball is drawn at the $(n+1)$st trial. Show that

$$P(B|A) = \frac{r}{r+s} \frac{H_{n,k}(r-1, s)}{H_{n,k}(r, s)}, \quad n = k, k+1, \ldots, \quad k = 0, 1, \ldots,$$

where
$$H_{n,k}(r,s) = \sum_{j=k}^{n} \binom{r}{j}\binom{s}{n-j} \Big/ \binom{r+s}{n}$$
is the tail probability function of a hypergeometric distribution. In particular, for $k=0$, whence $P(B|A) = P(B)$, it follows that $P(B) = r/(r+s)$.

1.9 *Updating the prior probabilities.* Let $\{A_1, A_2, \ldots, A_k, \ldots\}$ be a partition of a sample space Ω, with $P(A_k) > 0$, $k = 1, 2, \ldots$, and $B_1, B_2, \ldots, B_m$ any events in Ω, with $P(B_1 B_2 \cdots B_m) > 0$. The posterior probabilities $P(A_r|B_1)$, $r = 1, 2, \ldots$, by Bayes' formula, are expressed in terms of the prior probabilities $P(A_k)$, $k = 1, 2, \ldots$, as
$$P(A_r|B_1) = \frac{P(A_r)P(B_1|A_r)}{\sum_{k=1}^{\infty} P(A_k)P(B_1|A_k)}.$$
Similarly, by Bayes' formula, the posterior probabilities $P(A_r|B_1 B_2)$, $r = 1, 2, \ldots$, are expressed in terms of the prior probabilities $P(A_k)$, $k = 1, 2, \ldots$, as
$$P(A_r|B_1 B_2) = \frac{P(A_r)P(B_1 B_2|A_r)}{\sum_{k=1}^{\infty} P(A_k)P(B_1 B_2|A_k)}.$$
Show that the posterior probabilities $P(A_r|B_1 B_2)$, $r = 1, 2, \ldots$, may be expressed, equivalently, in terms of the *updated prior probabilities* $P(A_k|B_1)$, $k = 1, 2, \ldots$, as
$$P(A_r|B_1 B_2) = \frac{P(A_r|B_1)P(B_2|B_1 A_r)}{\sum_{k=1}^{\infty} P(A_k|B_1)P(B_2|B_1 A_k)}$$
and conclude that
$$P(A_r|B_1 B_2 \cdots B_m) = \frac{P(A_r|B_1 B_2 \cdots B_{m-1})P(B_m|B_1 B_2 \cdots B_{m-1} A_r)}{\sum_{k=1}^{\infty} P(A_k|B_1 B_2 \cdots B_{m-1})P(B_m|B_1 B_2 \cdots B_{m-1} A_k)}.$$

1.10 *(Continuation).* If, in addition, the events $B_1, B_2, \ldots, B_m$ are conditionally independent given the event A_k, for all $k = 1, 2, \ldots$, show that
$$P(A_r|B_1 B_2 \cdots B_m) = \frac{P(A_r)P(B_1|A_r)P(B_2|A_r)\cdots P(B_m|A_r)}{\sum_{k=1}^{\infty} P(A_k)P(B_1|A_k)P(B_2|A_k)\cdots P(B_m|A_k)}$$
and
$$P(A_r|B_1 B_2 \cdots B_m) = \frac{P(A_r|B_1 B_2 \cdots B_{m-1})P(B_m|A_r)}{\sum_{k=1}^{\infty} P(A_k|B_1 B_2 \cdots B_{m-1})P(B_m|A_k)}.$$

1.11 *Conditional independence and exchangeability.* Let $\{A_1, A_2, \ldots, A_k, \ldots\}$ be a partition of a sample space Ω, with $P(A_k) > 0$, $k = 1, 2, \ldots$. If the events $B_1, B_2, \ldots, B_n$ in Ω are conditionally independent, given the event A_k, and $P(B_1|A_k) = P(B_2|A_k) = \cdots = P(B_n|A_k) \equiv a_k$, for all $k = 1, 2, \ldots$, then show that they are exchangeable.

1.12 Let $A_1, A_2, \ldots, A_n$ be exchangeable events and $B_{n,k}$ the event that exactly k among these n events occur. Show that

$$P(A_{i_1} A_{i_2} \cdots A_{i_r} | B_{n,k}) = \frac{(k)_r}{(n)_r}, \quad r = 0, 1, \ldots, k, \ \ k = 0, 1, \ldots, n.$$

Further, let $A_{j_1}, A_{j_2}, \ldots, A_{j_m}$ be m events from the n events $A_1, A_2, \ldots, A_n$ and consider the event $B_{m,r}$ that exactly r among these m events occur. Show that

$$P(B_{m,r}) = \sum_{k=r}^{n-m+r} \frac{\binom{k}{r}\binom{n-k}{m-r}}{\binom{n}{m}} P(B_{n,k}), \quad r = 0, 1, \ldots, m.$$

1.13 Prove that the events $A_1, A_2, \ldots, A_n$ are exchangeable if and only if for every collection of r indices $\{i_1, i_2, \ldots, i_r\}$ from the n indices $\{1, 2, \ldots, n\}$, the probability

$$P(A_{i_1} A_{i_2} \cdots A_{i_r} A'_{i_{r+1}} A'_{i_{r+2}} \cdots A'_{i_n}) = P_r, \quad r = 1, 2, \ldots, n,$$

where $\{i_{r+1}, i_{r+2}, \ldots, i_n\} = \{1, 2, \ldots, n\} - \{i_1, i_2, \ldots, i_r\}$, depends only on r and not on the specific collection of the indices.

1.14 Consider n distinguishable urns each of which contains Np white and Nq black balls, with $q = 1 - p$ and $0 < p < 1$. Assume that s_j balls are successively drawn, with replacement, from the jth urn and let B_j be the event that k_j balls, among the s_j balls drawn, are white, $j = 1, 2, \ldots, n$. Clearly, the events $B_1, B_2, \ldots, B_n$ are independent. Further, let A be the event that a total of $k = k_1 + k_2 + \cdots + k_n$ balls, among the $s = s_1 + s_2 + \cdots + s_n$ balls drawn from the n urns, are white. Show that the events $B_1, B_2, \ldots, B_n$ are not conditionally independent, given the event A.

1.15 Consider an urn containing Np white and Nq black balls, with $q = 1 - p$ and $0 < p < 1$. Assume that balls are successively drawn, with replacement, and let B_j be the event that k_j white balls are required to be drawn after the drawing of the m_{j-1}th ball and until the drawing of the m_jth black ball, where $m_j = \sum_{i=1}^{j} s_i$, $j = 1, 2, \ldots, n$, with $m_0 = 0$. Also, let A be the event that a total of $k = k_1 + k_2 + \cdots + k_n$ white balls are required to be drawn until the drawing of sth black ball, with $s = s_1 + s_2 + \cdots + s_n$. Show that the events $B_1, B_2, \ldots, B_n$ are not conditionally independent, given the event A.

1.16 *Independence and conditional probabilities* (P. T. Strait (1971)). Let A, B and C be events with $P(A)$, $P(B)$ and $P(C) > 0$. Show that if the events A and B are independent, then

$$P(C|A) = P(B)P(C|AB) + P(B')P(C|AB').$$

Conversely, if this relation holds and the events B and C are not conditionally independent given A, then the events A and B are independent.

1.17 Show that the events A_i, $i = 1, 2, \ldots, n$, are (mutually or completely) stochastically independent, if and only if the following relation holds:

$$P(A_{i_1} A_{i_2} \cdots A_{i_r} A'_{i_{r+1}} A'_{i_{r+2}} \cdots A'_{i_n})$$
$$= P(A_{i_1}) P(A_{i_2}) \cdots P(A_{i_r}) P(A'_{i_{r+1}}) P(A'_{i_{r+2}}) \cdots P(A'_{i_n}),$$

for every r-combination $\{i_1, i_2, \ldots, i_r\}$ of the n indices $\{1, 2, \ldots, n\}$ and $r = 0, 1, \ldots, n$, with $\{i_{r+1}, i_{r+2}, \ldots, i_n\} = \{1, 2, \ldots, n\} - \{i_1, i_2, \ldots, i_r\}$.

Similarly, show that the events B_i, $i = 1, 2, \ldots, n$, are (mutually or completely) conditionally independent, given the event A, if and only if the following relation holds:

$$P(B_{i_1} B_{i_2} \cdots B_{i_r} B'_{i_{r+1}} \cdots B'_{i_n} | A)$$
$$= P(B_{i_1}|A) P(B_{i_2}|A) \cdots P(B_{i_r}|A) P(B'_{i_{r+1}}|A) \cdots P(B'_{i_n}|A),$$

for every r-combination $\{i_1, i_2, \ldots, i_r\}$ of the n indices $\{1, 2, \ldots, n\}$ and $r = 0, 1, \ldots, n$, with $\{i_{r+1}, i_{r+2}, \ldots, i_n\} = \{1, 2, \ldots, n\} - \{i_1, i_2, \ldots, i_r\}$.

1.18 Consider an urn containing n balls numbered $\{1, 2, \ldots, n\}$. Suppose that all the balls are successively drawn one after the other, at random, without replacement. Let A_r be the event that at the rth drawing a number greater than the numbers drawn in the $r - 1$ previous drawings, $r = 1, 2, \ldots, n$, is drawn. (a) Show that $P(A_r) = 1/r$, $r = 1, 2, \ldots, n$ and (b) examine whether the events $A_1, A_2, \ldots, A_n$ are independent.

1.19 Consider a collection of m balls composed of $r_i \geq 0$ distinguishable kinds of balls, each including i like balls, $i = 1, 2, \ldots, m$, so that $r_1 + 2r_2 + \cdots + mr_m = m$. Assume that the balls are distributed into n distinguishable urns at random and that all the distributions are equally probable. Show that the probability $p_{n,k}$ that k urns are occupied is given by

$$p_{n,k} = \binom{n}{k} \sum_{j=0}^{k} (-1)^{k-j} \binom{k}{j} \prod_{i=1}^{m} \binom{j+i-1}{i}^{r_i} \bigg/ \prod_{i=1}^{m} \binom{n+i-1}{i}^{r_i}$$

$$= \binom{n}{k} \left[\Delta_u^k \prod_{i=1}^{m} \binom{u+i-1}{i}^{r_i} \right]_{u=0} \bigg/ \prod_{i=1}^{m} \binom{n+i-1}{i}^{r_i}$$

if the urns are of unlimited capacity, while it is given by

$$p_{n,k} = \binom{n}{k} \sum_{j=0}^{k} (-1)^{k-j} \binom{k}{j} \prod_{i=1}^{m} \binom{j}{i}^{r_i} \bigg/ \prod_{i=1}^{m} \binom{n}{i}^{r_i}$$

$$= \binom{n}{k} \left[\Delta_u^k \prod_{i=1}^{m} \binom{u}{i}^{r_i} \right]_{u=0} \bigg/ \prod_{i=1}^{m} \binom{n}{i}^{r_i}$$

if the capacity of each urn is limited to one ball from each kind.

1.20 (*Continuation*). Show that the rth binomial moment $b_{(r)}$ of the number of occupied urns is given by

$$b_{(r)} = \binom{n}{r} \sum_{j=0}^{r} (-1)^{r-j} \binom{r}{j} \prod_{i=1}^{m} \binom{n-r+j+i-1}{i}^{r_i} \bigg/ \prod_{i=1}^{m} \binom{n+i-1}{i}^{r_i}$$

$$= \binom{n}{r} \left[\Delta_u^r \prod_{i=1}^{m} \binom{u+i-1}{i}^{r_i} \right]_{u=n-r} \bigg/ \prod_{i=1}^{m} \binom{n+i-1}{i}^{r_i}$$

if the urns are of unlimited capacity, while it is given by

$$b_{(r)} = \binom{n}{r} \sum_{j=0}^{r} (-1)^{r-j} \binom{r}{j} \prod_{i=1}^{m} \binom{n-r+j}{i}^{r_i} \bigg/ \prod_{i=1}^{m} \binom{n}{i}^{r_i}$$

$$= \binom{n}{r} \left[\Delta_u^r \prod_{i=1}^{m} \binom{u}{i}^{r_i} \right]_{u=n-r} \bigg/ \prod_{i=1}^{m} \binom{n}{i}^{r_i}$$

if the capacity of each urn is limited to one ball from each kind.

1.21 *A generalization of the problem of coincidences*. Consider an urn containing s like series of balls with n balls, numbered from 1 to n, in each series. Assume that n balls are randomly drawn one after the other without replacement. Show that the probability $p_{n,k;s}$ that k coincidences occur is given by

$$p_{n,k;s} = \binom{n}{k} \sum_{j=0}^{n-k} (-1)^j \binom{n-k}{j} \frac{s^{k+j}}{(sn)_{k+j}}$$

$$= \binom{n}{k} \frac{[\Delta_u^{n-k} s^{n-u}(sn-n+u)_u]_{u=0}}{(sn)_n}, \quad k=0,1,\ldots,n.$$

1.22 (*Continuation*). Show that

$$\lim_{n\to\infty} p_{n,k;s} = e^{-1} \frac{1}{k!}, \quad k=0,1,\ldots,$$

which is independent of s.

1.23 (*Continuation*). Assume that the successive drawing of balls is terminated with the appearance of the first coincidence. Show that the probability $q_{n,k;s}$ that the first coincidence appears at the kth drawing is given by

$$q_{n,k;s} = \sum_{j=0}^{k-1} (-1)^j \binom{k-1}{j} \frac{s^{j+1}}{(sn)_{j+1}}$$

$$= \frac{[\Delta_u^{k-1} s^{k-u}(sn-k+u)_{n-k+u}]_{u=0}}{(sn)_n}, \quad k=1,2,\ldots,n.$$

1.24 *Gambler's ruin.* A gambler plays a series of games against an adversary. The probability of the gambler to win one euro on a given game is p and that of losing one euro is $q = 1 - p$. Assume that initially the gambler possesses k euros and the adversary possesses $n - k$ euros. Let p_k be the probability of the gambler's ruin. Derive the recurrence relation

$$p_k = p p_{k+1} + q p_{k-1}, \quad k = 1, 2, \ldots, n-1,$$

with initial conditions $p_0 = 1$ and $p_n = 0$. Deduce that for $p \neq 1/2$,

$$p_k = \frac{(q/p)^k - (q/p)^n}{1 - (q/p)^n}, \quad k = 0, 1, \ldots, n,$$

while for $p = 1/2$,

$$p_k = \frac{n - k}{n}, \quad k = 0, 1, \ldots, n.$$

1.25 *Distribution of shares.* Consider two players a and b contending in a series of games in which the winner is the one who first wins n games. Assume that the probability of a to win in a game is p and so that of b is $q = 1 - p$. Further, suppose that for some reason the series of games is interrupted when a has won $n - k$ games and b has won $n - r$ games, with $k, r < n$. Let $p_{k,r}$ be the probability of a to win the series of games, when k wins by a are required before r wins by b. Derive the recurrence relation

$$p_{k,r} = p p_{k-1,r} + q p_{k,r-1}, \quad k = 1, 2, \ldots, \quad r = 1, 2, \ldots,$$

with initial conditions $p_{0,r} = 1, r = 1, 2, \ldots, p_{k,0} = 0, k = 0, 1, \ldots$. Show that the generating function

$$P(t, u) = \sum_{k=1}^{\infty} \sum_{r=1}^{\infty} p_{k,r} t^r u^k$$

is given by

$$P(t, u) = \frac{ptu}{(1-t)(1-qt-pu)}$$

and, expanding it into powers of t and u, deduce the probability $p_{k,r}$, $k = 1, 2, \ldots$, $r = 1, 2, \ldots$.

1.26 *Multiple coincidences.* Consider r urns each containing n balls numbered from 1 to n. Assume that a ball is drawn from each urn and, without replacement, this multiple drawing is repeated until all the balls are drawn. The drawing of the jth ball from each of the urns at the jth trial constitutes an r-tuple coincidence. Show the probability $p_k(n, r)$ of occurrence of k r-tuple coincidences in n trials is given by

$$p_k(n, r) = \binom{n}{k} \sum_{j=0}^{n-k} (-1)^{n-k-j} \binom{n-k}{j} \frac{(j!)^r}{(n!)^r}$$

$$= \binom{n}{k} \frac{\left[\Delta_u^{n-k}(u!)^r\right]_{u=0}}{(n!)^r}.$$

1.27 Consider an urn containing m distinguishable balls bearing the numbers $\{s, s+1, \ldots, s+m-1\}$. Assume that n balls are randomly drawn with replacement so that all sample points are equally probable. Show that (a) the probability $p_k(n, m, s)$ that the sum of the numbers drawn is k, with $ns \leq k \leq n(s+m-1)$, is given by

$$p_k(n, m, s) = \frac{1}{m^n} \sum_{r=0}^{n} (-1)^r \binom{n}{r} \binom{k - n(s-1) - rm - 1}{n - 1}$$

and (b) the probability $P_k(n, m, s)$ that the sum of the numbers drawn is less than or equal to k, with $ns \leq k \leq n(s+m-1)$, is given by

$$P_k(n, m, s) = \frac{1}{m^n} \sum_{r=0}^{n} (-1)^r \binom{n}{r} \binom{k - n(s-1) - rm}{n}.$$

1.28 Let $U_1, U_2, \ldots, U_n$ be independent and identically distributed random variables with a common uniform distribution over the interval $[0, 1]$. Using a generalized inclusion and exclusion principle (L. Takács (1967)), show that the distribution function of the sum $X_n = \sum_{i=1}^{n} U_i$, $F_n(x) = P(X_n \leq x)$, is given by

$$F_n(x) = \frac{1}{n!} \sum_{r=0}^{[x]} (-1)^r \binom{n}{r} (x - r)^n, \quad 0 \leq x < n,$$

where $[x]$ denotes the integral part of x, and $F_n(x) = 0$, $-\infty < x < 0$, $F_n(x) = 1$, $n \leq x < \infty$.

1.29 Consider a supply of m like balls randomly distributed into n distinguishable urns. Assume that the jth urn is divided into s_j distinguishable cells (compartments), $j = 1, 2, \ldots, n$, each of unlimited capacity and set $s = s_1 + s_2 + \cdots + s_n$. Let K be the number of occupied urns (by at least one ball each). Find the probability function $p_k(m, n; s) = P(K = k)$, $k = 1, 2, \ldots, n$, the binomial moments $b_{(r)}(m, n; s) = E[\binom{K}{r}]$, $r = 1, 2, \ldots$, and the probabilities $P_k(m, n; s) = P(K \leq k)$, $k = 1, 2, \ldots, n$.

1.30 *General ballot formula* (L. Takács (1962)). Consider an urn containing n balls marked with nonnegative integers $\{k_1, k_2, \ldots, k_n\}$, where $k_1 + k_2 + \cdots + k_n = k \leq n$. Suppose that all n balls are randomly drawn one after the other without replacement. Let $A_{n,k}$ be the event that the sum of the numbers drawn up to the rth drawing is less than r for all $r = 1, 2, \ldots, n$. Show that $P(A_{n,k}) = (n - k)/n$.

1.31 *Generalized binomial distribution* (S. D. Poisson (1837)). Consider a sequence of n independent Bernoulli trials and assume that the probability of success at the ith trial equals p_i, $i = 1, 2, \ldots, n$. Using the inclusion and exclusion principle, show that the probability $p_k(n)$ of the occurrence of k successes in the n trials is given by

$$p_k(n) = \sum_{r=k}^{n} (-1)^{r-k} \binom{r}{k} S_{n,r}, \quad k = 0, 1, \ldots, n,$$

with

$$S_{n,r} = \sum p_{i_1} p_{i_2} \cdots p_{i_r},$$

where the summation is extended over all r-combinations $\{i_1, i_2, \cdots, i_r\}$ of the n indices $\{1, 2, \ldots, n\}$.

1.32 *(Continuation). Generalized negative binomial distribution.* Consider a sequence of independent Bernoulli trials and assume that the probability of success at the ith trial equals p_i, $i = 1, 2, \ldots$. Show that the probability $q_n(k)$ that n trials are required until the occurrence of the kth success is given by

$$q_n(k) = p_n \sum_{r=k}^{n} (-1)^{r-k} \binom{r-1}{k-1} S_{n-1, r-1}, \quad n = k, k+1, \ldots.$$

1.33 *Bivariate inclusion and exclusion formula.* Let $A_1, A_2, \ldots, A_m$ and $B_1, B_2, \ldots, B_n$ be events in a sample space Ω. Show that the probability $P_{m,n;k,r}$ that exactly k among the m events $A_1, A_2, \ldots, A_m$ and exactly r among the n events $B_1, B_2, \ldots, B_n$ occur, for $k = 0, 1, \ldots, m$, $r = 0, 1, \ldots, n$, is given by

$$P_{m,n;k,r} = \sum_{j=r}^{n} \sum_{i=k}^{m} (-1)^{i+j-k-r} \binom{i}{k} \binom{j}{r} S_{m,n;i,j},$$

with

$$S_{m,n;i,j} = \sum P(A_{u_1} A_{u_2} \cdots A_{u_i} B_{v_1} B_{v_2} \cdots B_{v_j}),$$

where the summation is extended over all i-combinations $\{u_1, u_2, \ldots, u_i\}$ of the m indices $\{1, 2, \ldots, m\}$ and all j-combinations $\{v_1, v_2, \ldots, v_j\}$ of the n indices $\{1, 2, \ldots, n\}$. In particular, if the events $A_1, A_2, \ldots, A_m$ and $B_1, B_2, \ldots, B_n$ are exchangeable in the sense that, for every subset $\{u_1, u_2, \ldots, u_i\}$ of $\{1, 2, \ldots, m\}$ and every subset $\{v_1, v_2, \ldots, v_j\}$ of $\{1, 2, \ldots, n\}$, the probability

$$P(A_{u_1} A_{u_2} \cdots A_{u_i} B_{v_1} B_{v_2} \cdots B_{v_j}) = p_{i,j},$$

for $i = 0, 1, \ldots, m$, $j = 0, 1, \ldots, n$, $i + j \neq 0$, depends only on the cardinalities i and j and not on the specific subsets of indices, deduce that

$$P_{m,n;k,r} = \binom{m}{k} \binom{n}{r} \sum_{j=r}^{n} \sum_{i=k}^{m} (-1)^{i+j-k-r} \binom{m-k}{i-k} \binom{n-r}{j-r} p_{i,j},$$

where $p_{0,0} = 1$.

1.34 *(Continuation).* Show that the binomial moment

$$b_{(i,j)} = \sum_{r=j}^{n} \sum_{k=i}^{m} \binom{k}{i} \binom{r}{j} P_{m,n;k,r}$$

is given by
$$b_{(i,j)} = S_{m,n;i,j}, \quad i = 0, 1, \ldots, m, \quad j = 0, 1, \ldots, n.$$

In particular, if the events $A_1, A_2, \ldots, A_m$ and $B_1, B_2, \ldots, B_n$ are exchangeable, deduce that
$$b_{(i,j)} = \binom{m}{i}\binom{n}{j} p_{i,j}, \quad i = 0, 1, \ldots, m, \quad j = 0, 1, \ldots, n.$$

1.35 (*Continuation*). Show that the probability $Q_{m,n;k,r}$ that at least k among the m events $A_1, A_2, \ldots, A_m$ and at least r among the n events $B_1, B_2, \ldots, B_n$ occur, for $r = 1, 2, \ldots, m$, $k = 1, 2, \ldots, n$, is given by
$$Q_{m,n;k,r} = \sum_{j=r}^{n}\sum_{i=k}^{m} (-1)^{i+j-k-r} \binom{i-1}{k-1}\binom{j-1}{r-1} S_{m,n;i,j}.$$

In particular, if the events $A_1, A_2, \ldots, A_m$ and $B_1, B_2, \ldots, B_n$ are exchangeable, deduce that
$$Q_{m,n;k,r} = \binom{m-1}{k-1}\binom{n-1}{r-1} \sum_{j=r}^{n}\sum_{i=k}^{m} (-1)^{i+j-k-r} \frac{m}{i}\binom{m-k}{i-k}\frac{n}{j}\binom{n-r}{j-r} p_{i,j}.$$

1.36 *Bivariate inclusion and exclusion formula for pairs of disjoint events.* Let (A_h, B_h), $h = 1, 2, \ldots, n$ be pairs of disjoint events in a sample space Ω. Show that the probability $P_{n;k,r}$ that exactly k among the events $A_1, A_2, \ldots, A_n$ and exactly r among the events $B_1, B_2, \ldots, B_n$ occur, for $k = 0, 1, \ldots, n-r$, $r = 0, 1, \ldots, n$, is given by
$$P_{n;k,r} = \sum_{j=r}^{n-k}\sum_{i=k}^{n-j} (-1)^{i+j-k-r} \binom{i}{k}\binom{j}{r} S_{n;i,j},$$

with $S_{n;0,0} = 1$ and
$$S_{n;i,j} = \sum P(A_{u_1} A_{u_2} \cdots A_{u_i} B_{v_1} B_{v_2} \cdots B_{v_j}),$$

for $i = 0, 1, \ldots, n-j$, $j = 0, 1, \ldots, n$, $i + j \neq 0$, where the summation is extended over all pairs of disjoint subsets $\{u_1, u_2, \ldots, u_i\}$ and $\{v_1, v_2, \ldots, v_j\}$ of the n indices $\{1, 2, \ldots, n\}$. In particular, if the pairs (A_h, B_h), $h = 1, 2, \ldots, n$ of disjoint events are exchangeable in the sense that, for every pair of disjoint subsets $\{u_1, u_2, \ldots, u_i\}$ and $\{v_1, v_2, \ldots, v_j\}$ of the n indices $\{1, 2, \ldots, n\}$, the probability
$$P(A_{u_1} A_{u_2} \cdots A_{u_i} B_{v_1} B_{v_2} \cdots B_{v_j}) = p_{i,j},$$

for $i = 0, 1, \ldots, n-j$, $j = 0, 1, \ldots, n$, $i + j \neq 0$, depends only the cardinalities i and j and not on the specific subsets of indices, deduce that
$$P_{n;k,r} = \binom{n}{k,r} \sum_{j=0}^{n-k-r}\sum_{i=0}^{n-k-r-j} (-1)^{i+j} \binom{n-k-r}{i,j} p_{k+i,r+j},$$

where
$$\binom{n}{k,r} = \frac{n!}{k!r!(n-k-r)!}.$$

1.37 (*Continuation*). Show that the binomial moment
$$b_{(i,j)} = \sum_{r=j}^{n}\sum_{k=i}^{n-r}\binom{k}{i}\binom{r}{j}P_{n;k,r}$$
is given by
$$b_{(i,j)} = S_{n;i,j}, \ i = 0, 1, \ldots, n-j, \ j = 0, 1, \ldots, n.$$

In particular, if the pairs of disjoint events (A_h, B_h), $h = 1, 2, \ldots, n$ are exchangeable, deduce that
$$b_{(i,j)} = \binom{n}{i,j}p_{i,j}, \ i = 0, 1, \ldots, n-j, \ j = 0, 1, \ldots, n.$$

1.38 (*Continuation*). Show that the probability $Q_{n;k,r}$ that at least k among the events $A_1, A_2, \ldots, A_n$ and at least r among the events $B_1, B_2, \ldots, B_n$ occur, for $k = 1, 2, \ldots, n-r$, $r = 1, 2, \ldots, n-1$, is given by
$$Q_{n;k,r} = \sum_{j=r}^{n-k}\sum_{i=k}^{n-j}(-1)^{i+j-k-r}\binom{i-1}{k-1}\binom{j-1}{r-1}S_{n;i,j}.$$

In particular, if the disjoint pairs of events (A_h, B_h), $h = 1, 2, \ldots, n$ are exchangeable, deduce that
$$Q_{n;k,r} = \binom{n-2}{k-1,r-1}\sum_{j=0}^{n-k-r}\sum_{i=0}^{n-k-r-j}(-1)^{i+j}\binom{n-k-r}{i,j}\frac{n(n-1)p_{k+i,r+j}}{(k+i)(r+j)}.$$

1.39 *Generalized trinomial distribution.* Consider a sequence of independent trials of a random experiment with sample space Ω. Let A and B be two disjoint events in Ω and $C = A'B'$. Assume that at the ith trial the probability of occurrence of A is p_i and the probability of occurrence of B is q_i so that the probability of occurrence of C is $1 - p_i - q_i$, $i = 1, 2, \ldots, n$. Show that the probability $p_{k,r}(n)$ of exactly k occurrences of A and r occurrences of B in n trials is given by
$$p_{k,r}(n) = \sum_{j=r}^{n-k}\sum_{i=k}^{n-j}(-1)^{i+j-k-r}\binom{i}{k}\binom{j}{r}S_{n;i,j},$$
for $k = 0, 1, \ldots, n-r$, $r = 0, 1, \ldots, n$, with
$$S_{n;i,j} = \sum p_{u_1}p_{u_2}\cdots p_{u_i}q_{v_1}q_{v_2}\cdots q_{v_j},$$

where the summation is extended over all pairs of disjoint subsets $\{u_1, u_2, \ldots, u_i\}$ and $\{v_1, v_2, \ldots, v_j\}$ of the set $\{1, 2, \ldots, n\}$. In particular, for $p_i = p$, $q_i = q$, $i = 1, 2, \ldots, n$, deduce the probability function of the trinomial distribution as

$$p_{k,r}(n) = \binom{n}{k,r} p^k q^r (1 - p - q)^{n-k-r},$$

for $k = 0, 1, \ldots, n - r$, $r = 0, 1, \ldots, n$.

1.40 (*Continuation*). *Generalized negative trinomial distribution.* Show that the probability $q_{k,r}(s)$ of exactly k occurrences of A and r occurrences of B until the sth occurrence of the event C is given by

$$q_{k,r}(s) = (1 - p_{s+k+r} - q_{s+k+r}) \sum_{j=r}^{s+r-1} \sum_{i=k}^{s+k+r-j-1} (-1)^{i+j-k-r} \binom{i}{k}\binom{j}{r} S_{n;i,j},$$

for $k = 0, 1, \ldots$, $r = 0, 1, \ldots$. In particular, for $p_i = p$, $q_i = q$, $i = 1, 2, \ldots$, deduce the probability function of the negative trinomial distribution as

$$q_{k,r}(s) = \binom{s+k+r-1}{k,r} p^k q^r (1 - p - q)^s,$$

for $k = 0, 1, \ldots$, $r = 0, 1, \ldots$.

2
STIRLING NUMBERS

2.1 INTRODUCTION

The Stirling numbers of the first and second kind, which are the coefficients of the expansions of factorials into powers and of powers into factorials, respectively, were introduced by James Stirling in his *Methodus Differentialis* (1730). Equivalently, the Stirling numbers of the first kind are derivatives of factorials at zero, while the Stirling numbers of the second kind are finite differences of powers at zero. Since the factorials occupy the same central position in the calculus of finite differences as the powers in the infinitesimal calculus, the Stirling numbers constitute a part of the bridge connecting these two calculi. In the classical occupancy problem, the number of ways of distributing n distinguishable balls into k distinguishable urns, so that no urn remains empty, was expressed by Abraham De Moivre (1718) in the form of a simple sum of elementary terms with alternating sign; it is essentially the Stirling number of the second kind multiplied by $k!$. The number of different results of a tossing of n distinguishable dice in which each of the $k = 6$ faces appears (at least once), derived by P. R. Montmort (1708), probably inspired De Moivre's more general result.

The Stirling numbers under different names attracted the attention of several other well-known mathematicians of the 18th and 19th centuries. The classical book of Ch. Jordan (1939a) on the calculus of finite differences revived the interest in these numbers. A variety of applications of the Stirling numbers in combinatorics and in probability theory was provided. The coefficients of the expansion of generalized

---

Combinatorial Methods in Discrete Distributions. By Charalambos A. Charalambides
ISBN 0-471-68027-3 Copyright © 2005 John Wiley & Sons, Inc.

74 STIRLING NUMBERS

factorials into factorials are connected with the Stirling numbers and have applications in combinatorics, in occupancy problems, and in probability theory. The finite differences of powers at an arbitrary point, which were treated in N. Nielsen (1906), are closely connected with the Stirling numbers of the second kind; the name noncentral Stirling numbers of the second kind, given to them recently, expresses this connection. These numbers were studied together the corresponding noncentral Stirling numbers of the first kind.

In this chapter, the central and noncentral Stirling numbers and generalized factorial coefficients are thoroughly examined. Further, the enumeration of partitions by subsets and permutations by cycles, a subject furnishing combinatorial interpretations of the Stirling numbers, is discussed.

2.2 DEFINITIONS AND GENERATING FUNCTIONS

Consider the factorial of t of order n:

$$(t)_n = t(t-1)\cdots(t-n+1), \ n = 1, 2\ldots, \ (t)_0 = 1. \tag{2.1}$$

Clearly, this is a polynomial of t of degree n. Executing the multiplications and arranging the terms in ascending order of powers of t we get

$$(t)_n = \sum_{k=0}^{n} s(n,k) t^k, \ n = 0, 1, \ldots. \tag{2.2}$$

Inversely, the nth power of t may be expressed in the form of a polynomial of the factorials of t of degree n. Specifically, using (2.1), we get successively the expressions

$$t^0 = (t)_0 = 1, \ t^1 = (t)_1, \ t^2 = t[1 + (t-1)] = (t)_1 + (t)_2,$$

$$t^3 = (t)_1 t + (t)_2 t = (t)_1[1 + (t-1)] + (t)_2[2 + (t-2)] = (t)_1 + 3(t)_2 + (t)_3$$

and generally

$$t^n = \sum_{k=0}^{n} S(n,k)(t)_k, \ n = 0, 1, \ldots. \tag{2.3}$$

Clearly, $s(0,0) = S(0,0) = 1$, $s(n,0) = S(n,0) = 0$, for all $n > 0$ and $s(n,k) = S(n,k) = 0$, for $k > n$. Further, replacing in (2.2) t by $-t$, and since $(t+n-1)_n = (-1)^n(-t)_n$, we deduce the expression

$$(t+n-1)_n = \sum_{k=0}^{n} |s(n,k)| t^k, \ n = 0, 1, \ldots, \tag{2.4}$$

where

$$|s(n,k)| = (-1)^{n-k} s(n,k). \tag{2.5}$$

Note that $|s(n,k)|$, according to (2.4), as a sum of products of $n-k$ positive integers from the set $\{1, 2, \ldots, n-1\}$, is a positive integer. Based on expansions (2.2), (2.3) and (2.4), the following definition is introduced.

Definition 2.1 *The coefficients $s(n,k)$ and $S(n,k)$ in the expansions of factorials into powers and of powers into factorials are called Stirling numbers of the first and second kind, respectively. The coefficient $|s(n,k)|$ in the expansion of ascending factorials into powers is the signless or absolute Stirling number of the first kind.*

An interesting and useful extension of the Stirling numbers, in combinatorics and discrete probability, is provided by the coefficients of the expansions of noncentral factorials into powers and of powers into noncentral factorials. Specifically, let

$$(t-r)_n = \sum_{k=0}^{n} s(n,k;r) t^k, \ n = 0, 1, \ldots \quad (2.6)$$

and

$$t^n = \sum_{k=0}^{n} S(n,k;r)(t-r)_k, \ n = 0, 1, \ldots . \quad (2.7)$$

Clearly, $s(0,0;r) = S(0,0;r) = 1$, $s(n,0;r) = (-r)_n$, $S(n,0;r) = r^n$, for $n > 0$, and $s(n,k;r) = S(n,k;r) = 0$, for $k > n$. Further, replacing t by $-t$ in (2.6) and since $(t+r+n-1)_n = (-1)^n(-t-r)_n$, we deduce the expression

$$(t+r+n-1)_n = \sum_{k=0}^{n} |s(n,k;r)| t^k, \ n = 0, 1, \ldots, \quad (2.8)$$

where the coefficient

$$|s(n,k;r)| = (-1)^{n-k} s(n,k;r), \ k = 0, 1, \ldots, n, \ n = 0, 1, \ldots, \quad (2.9)$$

for $r > 0$, as a sum of products of positive numbers, is positive. Then, the following definition is introduced.

Definition 2.2 *The coefficients $s(n,k;r)$ and $S(n,k;r)$ in the expansions of noncentral factorials into powers and of powers into noncentral factorials are called noncentral Stirling numbers of the first and second kind, respectively. The coefficient $|s(n,k;r)|$, for $r > 0$, in the expansion of noncentral ascending factorials into powers is the noncentral signless or absolute Stirling number of the first kind.*

Expansions (2.2), (2.3), (2.6) and (2.7) readily imply that the central and noncentral Stirling numbers of the first kind are derivatives of factorials and the central and noncentral Stirling numbers of the second kind are finite differences of powers. Specifically, a function $f(t)$ for which the derivatives at zero, $[D^k f(t)]_{t=0}$, $k = 0, 1, \ldots$, exist, may be expanded, according to the Maclaurin formula, into powers of t as

$$f(t) = \sum_{k=0}^{\infty} \left[\frac{1}{k!} D^k f(t)\right]_{t=0} \cdot t^k.$$

In the case of the nth order noncentral factorial of t, we have $D^k(t-r)_n = 0$, $k > n$, and so

$$(t-r)_n = \sum_{k=0}^{n} \left[\frac{1}{k!} D^k(t-r)_n\right]_{t=0} \cdot t^k, \quad n = 0, 1, \ldots.$$

Hence,

$$s(n, k; r) = \left[\frac{1}{k!} D^k(t-r)_n\right]_{t=0}, \qquad (2.10)$$

for $k = 0, 1, \ldots, n$, $n = 0, 1, \ldots$, and

$$|s(n, k; r)| = \left[\frac{1}{k!} D^k(t+n-1)_n\right]_{t=r}, \qquad (2.11)$$

for $k = 0, 1, \ldots, n$, $n = 0, 1, \ldots$. Further, a function $f(t)$ for which the differences at zero, $[\Delta^k f(t)]_{t=0}$, $k = 0, 1, \ldots$, exist, may be expanded, according to Newton's formula, into noncentral factorials of t as

$$f(t) = \sum_{k=0}^{\infty} \left[\frac{1}{k!} \Delta^k f(t)\right]_{t=r} \cdot (t-r)_k.$$

In the case of the nth power t, we have $\Delta^k t^k = 0$, $k > n$, and so

$$t^n = \sum_{k=0}^{n} \left[\frac{1}{k!} \Delta^k t^n\right]_{t=r} \cdot (t-r)_k, \quad n = 0, 1, \ldots.$$

Therefore,

$$S(n, k; r) = \left[\frac{1}{k!} \Delta^k t^n\right]_{t=r}, \qquad (2.12)$$

for $k = 0, 1, \ldots, n$, $n = 0, 1, \ldots$.

Notice that, for $r = 0$, the noncentral Stirling numbers reduce to the corresponding (central) Stirling numbers. For $r \neq 0$, these numbers may be expressed in terms of the corresponding central Stirling numbers. Specifically, expanding the noncentral ascending factorial of t of order n, $(t+r+n-1)_n$, into powers of $u = t+r$, using (2.4), and then expanding the powers of $u = t+r$ into powers of t, using Newton's binomial formula, we deduce the expansion

$$(t+r+n-1)_n = \sum_{j=0}^{n} |s(n,j)|(t+r)^j = \sum_{j=0}^{n} |s(n,j)| \sum_{k=0}^{j} \binom{j}{k} t^k r^{j-k}$$

$$= \sum_{k=0}^{n} \left\{\sum_{j=k}^{n} \binom{j}{k} r^{j-k} |s(n,j)|\right\} t^k,$$

which, compared to (2.8), yields the expression

$$|s(n,k;r)| = \sum_{j=k}^{n} \binom{j}{k} r^{j-k} |s(n,j)|. \qquad (2.13)$$

Also, expanding the noncentral ascending factorial of t of order n, $(t+r+n-1)_n$, into ascending factorials of t, using Vandermonde's formula, and then expanding the ascending factorials of t into powers of t, using (2.4), we get the expansion

$$(t+r+n-1)_n = \sum_{j=0}^{n} \binom{n}{j} (t+j-1)_j (r+n-j-1)_{n-j}$$

$$= \sum_{j=0}^{n} \binom{n}{j} (r+n-j-1)_{n-j} \sum_{k=0}^{j} |s(j,k)| t^k$$

$$= \sum_{k=0}^{n} \left\{ \sum_{j=k}^{n} \binom{n}{j} (r+n-j-1)_{n-j} |s(j,k)| \right\} t^k,$$

which, compared to (2.8), implies the expression

$$|s(n,k;r)| = \sum_{j=k}^{n} \binom{n}{j} (r+n-j-1)_{n-j} |s(j,k)|. \qquad (2.14)$$

Similarly,

$$S(n,k;r) = \sum_{j=k}^{n} \binom{j}{k} (r)_{j-k} S(n,j) \qquad (2.15)$$

and

$$S(n,k;r) = \sum_{j=k}^{n} \binom{n}{j} r^{n-j} S(j,k). \qquad (2.16)$$

The noncentral Stirling numbers retain almost all the properties of the (central) Stirling numbers. In the sequel, their common properties are stated in terms of the noncentral Stirling numbers; the corresponding properties of the (central) Stirling numbers, whenever needed, are deduced by setting the noncentrality parameter equal to zero. Further, there are very few cases in which the expression of a property of the noncentral Stirling numbers is not as simple and elegant as the expression of the same property of the (central) Stirling numbers. Such a property is stated only for the (central) Stirling numbers; the corresponding property of the noncentral Stirling numbers is deduced by using a suitable expression among (2.13), (2.14), (2.15) and (2.16).

As has been noted, the signless Stirling numbers of the first kind and the noncentral signless Stirling numbers of the first kind, with positive noncentrality parameter, are positive numbers since both are expressed as sums of positive numbers. These expressions are derived in the following theorem.

Theorem 2.1 *The noncentral signless Stirling number of the first kind $|s(n,k;r)|$, for $r > 0$, is given by the sum*

$$|s(n,k;r)| = \sum (i_1 + r)(i_2 + r) \cdots (i_{n-k} + r), \qquad (2.17)$$

where the summation is extended over all $(n-k)$-combinations $\{i_1, i_2, \ldots, i_{n-k}\}$ of the n nonnegative integers $\{0, 1, \ldots, n-1\}$.

Further, the signless Stirling number of the first kind $|s(n,k)|$ is given by the sum

$$|s(n,k)| = \sum i_1 i_2 \cdots i_{n-k}, \qquad (2.18)$$

where the summation is extended over all $(n-k)$-combinations $\{i_1, i_2, \ldots, i_{n-k}\}$ of the $n-1$ positive integers $\{1, 2, \ldots, n-1\}$.

Proof According to definition (2.8) of $|s(n,k;r)|$, we have for $n = 1, 2, \ldots$,

$$(t+r)(t+r+1) \cdots (t+r+n-1) = \sum_{k=0}^{n} |s(n,k;r)| t^k.$$

Note that the ith factor of the product of the left-hand side,

$$p_i(t) = t + r + i, \quad i = 0, 1, \ldots, n-1,$$

is a monomial with constant term $r + i$. Executing the multiplications, the kth order power of t is formed by multiplying the constant terms $\{i_1 + r, i_2 + r, \ldots, i_{n-k} + r\}$ of any $n-k$ factors $\{i_1, i_2, \ldots, i_{n-k}\}$, out of the n factors $\{0, 1, \ldots, n-1\}$, together with the first order terms t of the remaining k factors. Thus, by the multiplication principle, (2.17) is deduced.

Further, starting from definition (2.4) of $|s(n,k)|$ and first cancelling one t from both sides of it, the derivation of (2.18) can be similarly carried out.

Remark 2.1 Expression (2.17), of the noncentral signless Stirling number of the first kind $|s(n,k;r)|$, for $n > 0$, may be transformed as

$$|s(n,k;r)| = (r+n-1)_n \sum \frac{1}{(j_1+r)(j_2+r) \cdots (j_k+r)}, \qquad (2.19)$$

where the summation is extended over all k-combinations $\{j_1, j_2, \ldots, j_k\}$ of the n nonnegative integers $\{0, 1, \ldots, n-1\}$. Indeed, to each $(n-k)$-combination $\{i_1, i_2, \ldots, i_{n-k}\}$ of $\{0, 1, \ldots, n-1\}$ there uniquely corresponds the k-combination $\{j_1, j_2, \ldots, j_k\} = \{0, 1, \ldots, n-1\} - \{i_1, i_2, \ldots, i_{n-k}\}$ and vice versa. Therefore,

$$(i_1+r)(i_2+r) \cdots (i_{n-k}+r) = \frac{(r+n-1)_n}{(j_1+r)(j_2+r) \cdots (j_k+r)}.$$

DEFINITIONS AND GENERATING FUNCTIONS 79

Introducing this transformation into expression (2.17), (2.19) is deduced. Similarly, expression (2.18), of the signless Stirling number of the first kind $|s(n,k)|$, may be transformed to

$$|s(n,k)| = (n-1)! \sum \frac{1}{j_1 j_2 \cdots j_{k-1}},$$

where the summation is extended over all $(k-1)$-combinations $\{j_1, j_2, \ldots, j_{k-1}\}$ of the $n-1$ positive integers $\{1, 2, \ldots, n-1\}$.

The noncentral Stirling numbers of the first and second kind constitute a pair of orthogonal bivariate sequences. This is shown in the next theorem.

Theorem 2.2 *The noncentral Stirling numbers of the first and second kind satisfy the following orthogonality relations:*

$$\sum_{j=k}^{n} s(n,j;r)S(j,k;r) = \delta_{n,k}, \quad \sum_{j=k}^{n} S(n,j;r)s(j,k;r) = \delta_{n,k}, \qquad (2.20)$$

where $\delta_{n,k} = 1$, if $k = n$ and $\delta_{n,k} = 0$, if $k \neq n$ is the Kronecker delta.

Proof Expanding the nth order noncentral factorial of t into powers of t, by using (2.6), and in the resulting expression expanding the powers of t into noncentral factorials of t, by using (2.7), we get the relation

$$(t-r)_n = \sum_{j=0}^{n} s(n,j;r)t^j = \sum_{j=0}^{n} s(n,j;r) \sum_{k=0}^{j} S(j,k;r)(t-r)_k$$

$$= \sum_{k=0}^{n} \left\{ \sum_{j=k}^{n} s(n,j;r)S(j,k;r) \right\} (t-r)_k,$$

which implies the first of (2.20). Similarly, expanding the nth power of t into noncentral factorials of t and in the resulting expression expanding the noncentral factorials of t into powers of t, we deduce the second of (2.20).

Note that, for fixed n, the (power) generating function of the sequence of noncentral Stirling numbers of the first kind $s(n,k;r)$, $k = 0, 1, \ldots, n$, according to (2.6), is given by

$$s_n(t;r) = \sum_{k=0}^{n} s(n,k;r)t^k = (t-r)_n, \quad n = 0, 1, \ldots,$$

while the factorial generating function of the sequence of noncentral Stirling numbers of the second kind $S(n,k;r)$, $k = 0, 1, \ldots, n$, according to (2.7), is given by

$$S_n(t;r) = \sum_{k=0}^{n} S(n,k;r)(t)_k = (t+r)^n, \quad n = 0, 1, \ldots.$$

Further, from (2.6), on using Newton's general binomial formula,

$$(1+u)^x = \sum_{n=0}^{\infty} \binom{x}{n} u^n,$$

we deduce the double generating function

$$g(t,u;r) = \sum_{n=0}^{\infty} \sum_{k=0}^{n} s(n,k;r) t^k \frac{u^n}{n!} = (1+u)^{t-r}, \qquad (2.21)$$

while from (2.7) we get

$$f(t,u;r) = \sum_{n=0}^{\infty} \sum_{k=0}^{n} S(n,k;r)(t)_k \frac{u^n}{n!} = e^{(t+r)u}. \qquad (2.22)$$

The exponential generating functions of the sequences of noncentral Stirling numbers $s(n,k;r)$, $n = k, k+1, \ldots$, and $S(n,k;r)$, $n = k, k+1, \ldots$, for fixed k, are derived in the next theorem.

Theorem 2.3 (a) *The exponential generating function of the noncentral Stirling numbers of the first kind $s(n,k;r)$, $n = k, k+1, \ldots$, for fixed k, is given by*

$$g_k(u;r) = \sum_{n=k}^{\infty} s(n,k;r) \frac{u^n}{n!} = (1+u)^{-r} \frac{[\log(1+u)]^k}{k!}, \quad k = 0, 1, \ldots. \qquad (2.23)$$

(b) *The exponential generating function of the noncentral Stirling numbers of the second kind $S(n,k;r)$, $n = k, k+1, \ldots$, for fixed k, is given by*

$$f_k(u;r) = \sum_{n=k}^{\infty} S(n,k;r) \frac{u^n}{n!} = e^{ru} \frac{(e^u - 1)^k}{k!}, \quad k = 0, 1, \ldots. \qquad (2.24)$$

Proof (a) Interchanging the order of summation in (2.21), we get

$$g(t,u;r) = \sum_{k=0}^{\infty} \sum_{n=k}^{\infty} s(n,k;r) \frac{u^n}{n!} t^k = \sum_{k=0}^{\infty} g_k(u;r) t^k.$$

Further, writing (2.21) as $g(t,u;r) = (1+u)^{-r} \exp\{t \log(1+u)\}$ and expanding it into powers of t, we find

$$g(t,u;r) = (1+u)^{-r} \sum_{k=0}^{\infty} \frac{[\log(1+u)]^k}{k!} t^k.$$

Comparing these two expansions, we deduce (2.23).

(b) Similarly, interchanging the order of summation in (2.22), we get

$$f(t,u;r) = \sum_{k=0}^{\infty} \sum_{n=k}^{\infty} S(n,k;r) \frac{u^n}{n!} (t)_k = \sum_{k=0}^{\infty} f_k(u;r)(t)_k.$$

Also, writing (2.22) as $f(t,u;r) = e^{ru}[1+(e^u-1)]^t$ and expanding it into factorials of t, we find

$$f(t,u;r) = e^{ru}\sum_{k=0}^{\infty}\binom{t}{k}(e^u-1)^k = e^{ru}\sum_{k=0}^{\infty}\frac{(e^u-1)^k}{k!}(t)_k.$$

A comparison of these two expansions entails (2.24).

The generating functions of the noncentral signless Stirling numbers of the first kind may be obtained from the corresponding generating functions of the numbers $s(n,k;r)$ by using (2.9). Thus, from the first part of Theorem 2.3, we deduce the following corollary.

Corollary 2.1 *The exponential generating function of the noncentral signless Stirling numbers of the first kind $|s(n,k;r)|$, $n = k, k+1, \ldots$, for fixed k, is given by*

$$h_k(u;r) = \sum_{n=k}^{\infty}|s(n,k;r)|\frac{u^n}{n!} = (1-u)^{-r}\frac{[-\log(1-u)]^k}{k!}, \quad k=0,1,\ldots. \quad (2.25)$$

In the following example, a probabilistic application of the noncentral signless Stirling numbers of the first kind is given.

Example 2.1 *Bernoulli trials with varying success probability.* Suppose that balls are successively drawn one after the other from an urn containing w white and b black balls, according to the following scheme. After each trial the drawn ball is placed back in the urn along with s black balls. Determine (a) the probability $p(k;n)$ of drawing k white balls in n trials and (b) the probability $q(n;k)$ that n trials are required until the kth white ball is drawn.

(a) Let A_j be the event of drawing a white ball at the jth trial, $j = 1, 2, \ldots, n$. Then, setting $\theta = w/s$ and $r = b/s$, we have

$$p_j = P(A_j) = \frac{\theta}{\theta + r + j - 1},$$

$$q_j = P(A'_j) = 1 - P(A_j) = \frac{r+j-1}{\theta+r+j-1}, \quad j=1,2,\ldots,n$$

and

$$P(A_{j_1}A_{j_2}\cdots A_{j_k}A'_{j_{k+1}}\cdots A'_{j_n})$$
$$= P(A_{j_1})P(A_{j_2})\cdots P(A_{j_k})P(A'_{j_{k+1}})\cdots P(A'_{j_n})$$
$$= \frac{\theta^k}{(\theta+r+n-1)_n}(r+j_{k+1}-1)(r+j_{k+2}-1)\cdots(r+j_n-1).$$

Thus, the probability $p(k;n)$ of drawing k white balls in n trials is given by the sum

$$p(k;n) = \frac{\theta^k}{(\theta+r+n-1)_n}\sum(r+j_{k+1}-1)(r+j_{k+2}-1)\cdots(r+j_n-1),$$

where the summation is extended over all $(n-k)$-combinations $\{j_{k+1}, j_{k+2}, \ldots, j_n\}$ of the n positive integers $\{1, 2, \ldots, n\}$. Putting $i_m = j_{k+m} - 1$, $m = 1, 2, \ldots, n-k$, and using (2.17), we get

$$p(k; n) = \frac{|s(n, k; r)|\theta^k}{(\theta + r + n - 1)_n}, \quad k = 0, 1, \ldots, n.$$

(b) The probability $q(n; k)$ that n trials are required until the kth white ball is drawn equals the probability $p(k-1; n-1)$ of drawing $k-1$ white balls in $n-1$ trials multiplied by the probability $p_n = \theta/(\theta + r + n - 1)$ of drawing a white ball at the nth trial. Therefore,

$$q(n; k) = \frac{|s(n-1, k-1; r)|\theta^k}{(\theta + r + n - 1)_n}, \quad n = k, k+1, \ldots .$$

An interesting application of the orthogonality property of the noncentral Stirling numbers is presented in the next example.

Example 2.2 *Inverse relations.* Consider a sequence x_k, $k = 0, 1, \ldots$, of real numbers, and let

$$y_j = \sum_{k=0}^{j} a_{j,k} x_k, \quad j = 0, 1, \ldots,$$

where $a_{j,k}$, $k = 0, 1, \ldots, j$, $j = 0, 1, \ldots$, are given coefficients. If there exist coefficients $b_{n,j}$, $j = 0, 1, \ldots, n$, $n = 0, 1, \ldots$, orthogonal to the coefficients $a_{j,k}$, $k = 0, 1, \ldots, j$, $j = 0, 1, \ldots$,

$$\sum_{j=k}^{n} b_{n,j} a_{j,k} = \delta_{n,k},$$

then

$$\sum_{j=k}^{n} b_{n,j} y_j = \sum_{j=k}^{n} b_{n,j} \sum_{k=0}^{j} a_{j,k} x_k = \sum_{k=0}^{n} \left\{ \sum_{j=k}^{n} b_{n,j} a_{j,k} \right\} x_k$$

and so

$$x_n = \sum_{j=0}^{n} b_{n,j} y_j, \quad n = 0, 1, \ldots .$$

The expressions of y_j, $j = 0, 1, \ldots$, in terms of x_k, $k = 0, 1, \ldots, j$, and inversely of x_n, $n = 0, 1, \ldots$, in terms of y_j, $j = 0, 1, \ldots, n$, constitute a pair of inverse relations. Similarly, if

$$u_j = \sum_{n=j}^{\infty} a_{n,j} w_n, \quad j = 0, 1, \ldots$$

and

$$\sum_{j=k}^{n} a_{n,j} b_{j,k} = \delta_{n,k},$$

then
$$w_k = \sum_{j=k}^{\infty} b_{j,k} u_j, \quad k = 0, 1, \ldots.$$

In conclusion, every pair of orthogonal sequences entails a pair of inverse relations. According to Theorem 2.2, the noncentral Stirling numbers of the first and second kind constitute a pair of orthogonal sequences. Consequently, if

$$y_j = \sum_{k=0}^{j} s(j, k; r) x_k, \quad j = 0, 1, \ldots,$$

then

$$x_n = \sum_{j=0}^{n} S(n, j; r) y_j, \quad n = 0, 1, \ldots$$

and inversely. Also, if

$$u_j = \sum_{n=j}^{\infty} s(n, j; r) w_n, \quad j = 0, 1, \ldots,$$

then

$$w_k = \sum_{j=k}^{\infty} S(j, k; r) u_j, \quad k = 0, 1, \ldots$$

and inversely.

The following examples provide additional probabilistic and mathematical applications of the Stirling numbers.

Example 2.3 *Connection of ordinary and factorial moments.* Consider a sequence a_x, $x = 0, 1, \ldots$, (of numbers with combinatorial interpretation or probabilities or masses). The jth order (power) moment μ'_j and the jth order factorial moment $\mu_{(j)}$ of this sequence are defined by

$$\mu'_j = \sum_{x=0}^{\infty} x^j a_x, \quad \mu_{(j)} = \sum_{x=0}^{\infty} (x)_j a_x, \quad j = 0, 1, \ldots.$$

Using (2.2), the jth order factorial moment $\mu_{(j)}$ may be expressed as

$$\mu_{(j)} = \sum_{x=0}^{\infty} \sum_{k=0}^{j} s(j, k) x^k a_x = \sum_{k=0}^{j} s(j, k) \sum_{x=0}^{\infty} x^k a_x$$

and so

$$\mu_{(j)} = \sum_{k=0}^{j} s(j, k) \mu'_k, \quad j = 0, 1, \ldots.$$

84 STIRLING NUMBERS

In the same way, using (2.3), we conclude that

$$\mu'_n = \sum_{j=0}^{n} S(n,j)\mu_{(j)}, \quad n=0,1,\ldots.$$

More generally, if

$$\mu'_j(a) = \sum_{x=0}^{\infty}(x-a)^j a_x, \quad \mu_{(j)}(b) = \sum_{x=0}^{\infty}(x-b)_j a_x, \quad j=0,1,\ldots,$$

then, using (2.6) and (2.7), it can be similarly shown that

$$\mu_{(j)}(b) = \sum_{k=0}^{j} s(j,k;r)\mu'_k(a), \quad j=0,1,\ldots,$$

and

$$\mu'_n(a) = \sum_{j=0}^{n} S(n,j;r)\mu_{(j)}(b), \quad n=0,1,\ldots,$$

with $r = b-a$. Notice that the relations connecting the power to the factorial moments and vice versa constitute pairs of inverse relations.

Example 2.4 *The operator* $\Theta = tD$. In the infinitesimal calculus, besides the usual derivative operator D, the operator $\Theta = tD$ is frequently used. Express the nth power of the operator Θ in terms of powers of the operator D.

Clearly,
$$\Theta f(t) = tDf(t),$$
$$\Theta^2 f(t) = \Theta(\Theta f(t)) = tD(tDf(t)) = tDf(t) + t^2 D^2 f(t),$$

$$\Theta^3 f(t) = \Theta(\Theta^2 f(t)) = tD(tDf(t) + t^2 D^2 f(t))$$
$$= tDf(t) + 3t^2 D^2 f(t) + t^3 D^3 f(t)$$

and generally

$$\Theta^n f(t) = \sum_{k=1}^{n} C_{n,k} t^k D^k f(t),$$

where the coefficients $C_{n,k}$, $k = 1, 2, \ldots, n$ are independent of the function $f(t)$. Therefore, for their determination the most convenient function can be chosen. Let $f(t) = t^u$, whence

$$\Theta^n t^u = u^n t^u, \quad D^k t^u = (u)_k t^{u-k}.$$

Consequently,

$$u^n = \sum_{k=1}^{n} C_{n,k} (u)_k$$

and, according to (2.3), $C_{n,k} = S(n,k)$, $k = 1, 2, \ldots, n$. Hence,

$$\Theta^n f(t) = \sum_{k=1}^{n} S(n,k) t^k D^k f(t).$$

Example 2.5 *The operator $\Psi = t\Delta$.* In the calculus of finite differences, besides the usual difference operator Δ, the operator $\Psi = t\Delta$ is frequently used. Express the nth power of the operator Ψ in terms of powers of the operator Δ.

Clearly,
$$\Psi f(t) = t\Delta f(t),$$
$$\Psi^2 f(t) = \Psi(\Psi f(t)) = t\Delta(t\Delta f(t)) = t\Delta f(t) + (t+1)_2 \Delta^2 f(t),$$

$$\Psi^3 f(t) = \Psi(\Psi^2 f(t)) = t\Delta(t\Delta f(t) + (t+1)_2 \Delta^2 f(t))$$
$$= t\Delta f(t) + 3(t+1)_2 \Delta^2 f(t) + (t+2)_3 \Delta^3 f(t)$$

and generally

$$\Psi^n f(t) = \sum_{k=1}^{n} B_{n,k} (t+k-1)_k \Delta^k f(t),$$

where the coefficients $B_{n,k}$, $k = 1, 2, \ldots, n$ are independent of the function $f(t)$. Consequently, for their determination the most convenient function can be chosen. Let $f(t) = (t+u-1)_u$, whence

$$\Psi^n (t+u-1)_u = u^n (t+u-1)_u, \quad \Delta^k (t+u-1)_u = (u)_k (t+u-1)_{u-k}.$$

Therefore,

$$u^n = \sum_{k=1}^{n} B_{n,k} (u)_k$$

and, according to (2.3), $B_{n,k} = S(n,k)$, $k = 1, 2, \ldots, n$. Hence,

$$\Psi^n f(t) = \sum_{k=1}^{n} S(n,k)(t+k-1)_k \Delta^k f(t).$$

2.3 EXPLICIT EXPRESSIONS AND RECURRENCE RELATIONS

An explicit expression of the noncentral Stirling numbers of the second kind is deduced in the following theorem.

Theorem 2.4 *The noncentral Stirling number of the second kind $S(n,k;r)$, $k = 0, 1, \ldots, n$, $n = 0, 1, \ldots$, is given by the sum*

$$S(n,k;r) = \frac{1}{k!} \sum_{j=0}^{k} (-1)^{k-j} \binom{k}{j} (j+r)^n. \tag{2.26}$$

Proof Expanding the generating function (2.24) into powers of u,

$$\sum_{n=k}^{\infty} S(n,k;r) \frac{u^n}{n!} = \frac{1}{k!} \sum_{j=0}^{k} (-1)^{k-j} \binom{k}{j} e^{(j+r)u}$$

$$= \frac{1}{k!} \sum_{j=0}^{k} (-1)^{k-j} \binom{k}{j} \sum_{n=0}^{\infty} (j+r)^n \frac{u^n}{n!},$$

we get the expression

$$\sum_{n=k}^{\infty} S(n,k;r) \frac{u^n}{n!} = \sum_{n=0}^{\infty} \left\{ \frac{1}{k!} \sum_{j=0}^{k} (-1)^{k-j} \binom{k}{j} (j+r)^n \right\} \frac{u^n}{n!},$$

which implies (2.26).

Remark 2.2 The explicit expression of the noncentral Stirling numbers of the second kind, (2.26), may be deduced from (2.12) by using the expression of the kth power of the difference operator in terms of the shift operator. Indeed,

$$S(n,k;r) = \left[\frac{1}{k!} \Delta^k t^n \right]_{t=r} = \frac{1}{k!} \sum_{j=0}^{k} (-1)^{k-j} \binom{k}{j} \left[E^j t^n \right]_{t=r}$$

and, since $\left[E^j t^n \right]_{t=r} = (j+r)^n$, (2.26) is deduced.

The Stirling numbers of the second kind $S(n,k) = S(n,k;0)$, according to Theorem 2.4, are expressed in the form of a single summation of elementary terms, which are products and quotients of factorials and powers. There does not exist an analogous expression for the Stirling numbers of the first kind. An expression of the Stirling numbers of the first kind $s(n,k) = s(n,k;0)$ in the form of a single summation of terms that are products of binomial coefficients and Stirling numbers of the second kind, due to O. Schlömilch (1852), is given in Exercise 2.2. For the Stirling numbers of the first kind it leads to an expression in the form of a double summation of elementary terms.

Expressions of the Stirling numbers $s(n,k)$ and $S(n,k)$ as multiple sums over all compositions as well as over all partitions of n into k parts are derived in the following theorem.

Theorem 2.5 *The Stirling numbers of the first and second kind $s(n,k)$ and $S(n,k)$ are given by*

$$|s(n,k)| = \frac{n!}{k!} \sum \frac{1}{j_1 \cdot j_2 \cdots j_k} \qquad (2.27)$$

and

$$S(n,k) = \frac{n!}{k!} \sum \frac{1}{j_1! j_2! \cdots j_k!}, \qquad (2.28)$$

respectively, where the summation in both sums is extended over all compositions of n into k parts, that is, over all positive integer solutions of the equation

$$j_1 + j_2 + \cdots + j_k = n.$$

Alternatively,

$$|s(n,k)| = \sum \frac{n!}{k_1! k_2! \cdots k_n!} \left(\frac{1}{1}\right)^{k_1} \left(\frac{1}{2}\right)^{k_2} \cdots \left(\frac{1}{n}\right)^{k_n} \tag{2.29}$$

and

$$S(n,k) = \sum \frac{n!}{k_1! k_2! \cdots k_n!} \left(\frac{1}{1!}\right)^{k_1} \left(\frac{1}{2!}\right)^{k_2} \cdots \left(\frac{1}{n!}\right)^{k_n}, \tag{2.30}$$

respectively, where the summation in both sums is extended over all partitions of n into k parts, that is, over all nonnegative integer solutions of the equations

$$k_1 + 2k_2 + \cdots + nk_n = n, \quad k_1 + k_2 + \cdots + k_n = k.$$

Proof The exponential generating function of the signless Stirling numbers of the first kind, (2.25) with $r = 0$, on using the expansion $-\log(1-u) = \sum_{j=1}^{\infty} u^j/j$ and then Cauchy's rule of multiplication of series, may be written as

$$\sum_{n=k}^{\infty} |s(n,k)| \frac{u^n}{n!} = \frac{1}{k!} \left(\sum_{j=1}^{\infty} \frac{u^j}{j}\right)^k = \frac{1}{k!} \prod_{i=1}^{k} \left(\sum_{j_i=1}^{\infty} \frac{u^{j_i}}{j_i}\right)$$

$$= \sum_{n=k}^{\infty} \left\{\frac{n!}{k!} \sum \frac{1}{j_1 \cdot j_2 \cdots j_k}\right\} \frac{u^n}{n!},$$

yielding (2.27). Similarly, the exponential generating function of the Stirling numbers of the second kind, (2.24) with $r = 0$, may be expressed as

$$\sum_{n=k}^{\infty} S(n,k) \frac{u^n}{n!} = \frac{1}{k!} \left(\sum_{j=1}^{\infty} \frac{u^j}{j!}\right)^k = \frac{1}{k!} \prod_{i=1}^{k} \left(\sum_{j_i=1}^{\infty} \frac{u^{j_i}}{j_i!}\right)$$

$$= \sum_{n=k}^{\infty} \left\{\frac{n!}{k!} \sum \frac{1}{j_1! j_2! \cdots j_k!}\right\} \frac{u^n}{n!},$$

yielding (2.28). Using the multinomial theorem instead of Cauchy's rule, the generating functions (2.25) and (2.24), with $r = 0$, may be expressed as

$$h_k(u) = \sum_{n=k}^{\infty} |s(n,k)| \frac{u^n}{n!} = \frac{1}{k!} \left(u + \frac{u^2}{2} + \cdots + \frac{u^n}{n} + \cdots\right)^k$$

$$= \sum_{n=k}^{\infty} \left\{\sum \frac{n!}{k_1! k_2! \cdots k_n!} \left(\frac{1}{1}\right)^{k_1} \left(\frac{1}{2}\right)^{k_2} \cdots \left(\frac{1}{n}\right)^{k_n}\right\} \frac{u^n}{n!},$$

and

$$f_k(u) = \sum_{n=k}^{\infty} S(n,k) \frac{u^n}{n!} = \frac{1}{k!} \left(u + \frac{u^2}{2!} + \cdots + \frac{u^n}{n!} + \cdots\right)^k$$

$$= \sum_{n=k}^{\infty} \left\{ \sum \frac{n!}{k_1! k_2! \cdots k_n!} \left(\frac{1}{1!}\right)^{k_1} \left(\frac{1}{2!}\right)^{k_2} \cdots \left(\frac{1}{n!}\right)^{k_n} \right\} \frac{u^n}{n!},$$

yielding (2.29) and (2.30), respectively.

Remark 2.3 *Bell numbers.* In the framework of the second part of Theorem 2.5, it is worth mentioning a useful sequence of numbers connected with the Stirling numbers of the second kind. The numbers

$$B_n = \sum_{k=0}^{n} S(n,k), \quad n = 0, 1, \ldots, \tag{2.31}$$

are called *Bell numbers*. These numbers, using (2.30), may be expressed as

$$B_n = \sum \frac{n!}{k_1! k_2! \cdots k_n!} \left(\frac{1}{1!}\right)^{k_1} \left(\frac{1}{2!}\right)^{k_2} \cdots \left(\frac{1}{n!}\right)^{k_n}, \tag{2.32}$$

where the summation is extended over all partitions of n, that is, over all nonnegative integer solutions of the equation

$$k_1 + 2k_2 + \cdots + nk_n = n.$$

In the next theorem, triangular recurrence relations of the noncentral Stirling numbers of the first and second kind, useful for tabulation purposes, are derived.

Theorem 2.6 (a) *The noncentral Stirling numbers of the first kind* $s(n,k;r)$, $k = 0, 1, \ldots, n$, $n = 0, 1, \ldots$, *satisfy the triangular recurrence relation*

$$s(n+1,k;r) = s(n,k-1;r) - (n+r)s(n,k;r), \tag{2.33}$$

for $k = 1, 2, \ldots, n+1$, $n = 0, 1, \ldots$, *with initial conditions*

$$s(0,0;r) = 1, \quad s(n,0;r) = (-r)_n, \quad n > 0, \quad s(n,k;r) = 0, \quad k > n.$$

(b) *The noncentral Stirling numbers of the second kind* $S(n,k;r)$, $k = 0, 1, \ldots, n$, $n = 0, 1, \ldots$, *satisfy the triangular recurrence relation*

$$S(n+1,k;r) = S(n,k-1;r) + (k+r)S(n,k;r), \tag{2.34}$$

for $k = 1, 2, \ldots, n+1$, $n = 0, 1, \ldots$, *with initial conditions*

$$S(0,0;r) = 1, \quad S(n,0;r) = r^n, \quad n > 0, \quad S(n,k;r) = 0, \quad k > n.$$

Proof (a) Expanding both members of the recurrence relation

$$(t-r)_{n+1} = (t-r-n)(t-r)_n$$

into powers of t, according to (2.6), we get the relation

$$\sum_{k=0}^{n+1} s(n+1,k;r)t^k = \sum_{k=1}^{n+1} s(n,k-1;r)t^k - \sum_{k=0}^{n}(n+r)s(n,k;r)t^k,$$

which implies (2.33). The initial conditions follow directly from (2.6).

(b) Expanding both members of the recurrence relation

$$(t+r)^{n+1} = (t+r)(t+r)^n$$

into factorials of t, according to (2.7), we have

$$\sum_{k=0}^{n+1} S(n+1,k;r)(t)_k = (t+r)\sum_{j=0}^{n} S(n,j;r)(t)_j.$$

Since $(t)_{j+1} = (t-j)(t)_j$, whence $(t+r)(t)_j = (t)_{j+1} + (j+r)(t)_j$, we deduce the relation

$$\sum_{k=0}^{n+1} S(n+1,k;r)(t)_k = \sum_{j=0}^{n} S(n,j;r)(t)_{j+1} + \sum_{j=0}^{n}(j+r)S(n,j;r)(t)_j$$

$$= \sum_{k=1}^{n+1} S(n,k-1;r)(t)_k + \sum_{k=0}^{n}(k+r)S(n,k;r)(t)_k,$$

which implies (2.34). The initial conditions follow directly from (2.7).

Multiplying the triangular recurrence relation (2.33) by $(-1)^{n-k+1}$ and using (2.9), we deduce for the noncentral signless Stirling numbers of the first kind the following corollary.

Corollary 2.2 *The noncentral signless Stirling numbers of the first kind* $|s(n,k;r)|$, $k = 0, 1, \ldots, n$, $n = 0, 1, \ldots$, *satisfy the triangular recurrence relation*

$$|s(n+1,k;r)| = |s(n,k-1;r)| + (n+r)|s(n,k;r)|, \qquad (2.35)$$

for $k = 1, 2, \ldots, n+1$, $n = 0, 1, \ldots$, *with initial conditions*

$$|s(0,0;r)| = 1, \ |s(n,0;r)| = (r+n-1)_n, \ n > 0, \ |s(n,k;r)| = 0, \ k > n.$$

The signless Stirling numbers of the first kind $|s(n,k)|$ can be tabulated by using the recurrence relation (2.35) for $r = 0$ and its initial conditions. Table 2.1 gives the numbers $|s(n,k)|$, $k = 1, 2, \ldots, n$, $n = 1, 2, \ldots, 9$. Similarly, the Stirling numbers of the second kind $S(n,k)$ can be tabulated by using the recurrence relation (2.34) for $r = 0$ and its initial conditions. Table 2.2 gives the numbers $S(n,k)$, $k = 1, 2, \ldots, n$, $n = 1, 2, \ldots, 10$. The Bell numbers can be tabulated by using (2.31) and Table 2.2. Table 2.3 gives the numbers B_n for $n = 1, 2, \ldots, 10$.

Table 2.1 Signless Stirling numbers of the first kind $|s(n,k)|$

n \ k	1	2	3	4	5	6	7	8	9
1	1								
2	1	1							
3	2	3	1						
4	6	11	6	1					
5	24	50	35	10	1				
6	120	274	225	85	15	1			
7	720	1764	1624	735	175	21	1		
8	5040	13068	13132	6769	1960	322	28	1	
9	40320	109584	118124	67284	22449	4536	546	36	1

Table 2.2 Stirling numbers of the second kind $S(n,k)$

n \ k	1	2	3	4	5	6	7	8	9	10
1	1									
2	1	1								
3	1	3	1							
4	1	7	6	1						
5	1	15	25	10	1					
6	1	31	90	65	15	1				
7	1	63	301	350	140	21	1			
8	1	127	966	1701	1050	266	28	1		
9	1	255	3025	7770	6951	2646	462	36	1	
10	1	511	9330	34105	42525	22827	5880	750	45	1

Table 2.3 Bell numbers B_n

n	1	2	3	4	5	6	7	8	9	10
B_n	1	2	5	15	52	203	877	4140	21147	115975

Vertical and horizontal recurrence relations for the signless (absolute) Stirling numbers of the first kind and the Stirling numbers of the second kind are derived in the following theorems.

Theorem 2.7 *The signless (absolute) Stirling numbers of the first kind* $|s(n,k)|$, $k = 0, 1, \ldots, n$, $n = 0, 1, \ldots$, *with* $|s(0,0)| = 1$, *satisfy* (a) *the vertical recurrence relation*

$$|s(n+1, k+1)| = \sum_{j=k}^{n} (n)_{n-j} |s(j,k)| \qquad (2.36)$$

and (b) *the horizontal recurrence relation*

$$|s(n+1, k+1)| = \sum_{j=k}^{n} \binom{j}{k} |s(n,j)|. \qquad (2.37)$$

Proof Note first that according to (2.4), for $n = 0, 1, \ldots$ and since $|s(n+1, 0)| = 0$, we have

$$\sum_{k=0}^{n} |s(n+1, k+1)| t^k = (t+1)(t+2) \cdots (t+n) = (t+n)_n.$$

(a) Expanding the ascending factorial of $t + 1$ of order n,

$$(t+n)_n = (-1)^n(-t-1)_n,$$

into ascending factorials of t, using Vandermonde's formula,

$$(t+n)_n = (-1)^n \sum_{j=0}^{n} \binom{n}{j} (-1)_{n-j}(-t)_j = \sum_{j=0}^{n} (n)_{n-j}(t+j-1)_j,$$

we find

$$\sum_{k=0}^{n} |s(n+1, k+1)| t^k = \sum_{j=0}^{n} (n)_{n-j}(t+j-1)_j.$$

Further, expanding the ascending factorials on the right-hand side into powers of t, using (2.4), we get the relation

$$\sum_{k=0}^{n} |s(n+1, k+1)| t^k = \sum_{j=0}^{n} (n)_{n-j} \sum_{k=0}^{j} |s(j,k)| t^k$$

$$= \sum_{k=0}^{n} \left\{ \sum_{j=k}^{n} (n)_{n-j} |s(j,k)| \right\} t^k,$$

from which (2.36) is deduced.

(b) Expanding now the ascending factorial of $t+1$ of order n, $(t+n)_n$, into powers of $t + 1$, using (2.4), and then expanding the powers of $t + 1$ into powers of t, using Newton's binomial formula, we get the relation

$$\sum_{k=0}^{n} |s(n+1, k+1)| t^k = \sum_{j=0}^{n} |s(n,j)|(t+1)^j = \sum_{j=0}^{n} |s(n,j)| \sum_{k=0}^{j} \binom{j}{k} t^k$$

$$= \sum_{k=0}^{n} \left\{ \sum_{j=k}^{n} \binom{j}{k} |s(n,j)| \right\} t^k,$$

from which (2.37) is readily deduced.

Theorem 2.8 *The Stirling numbers of the second kind $S(n,k)$, $k = 0, 1, \ldots, n$, $n = 0, 1, \ldots$, with $S(0,0) = 1$, satisfy* (a) *the vertical recurrence relation*

$$S(n+1, k+1) = \sum_{j=k}^{n} \binom{n}{j} S(j,k) \tag{2.38}$$

and (b) *the horizontal recurrence relation*

$$S(n,k) = \sum_{j=k}^{n} (-1)^{j-k} (j)_{j-k} S(n+1, j+1). \tag{2.39}$$

Proof (a) According to (2.3), for $n = 0, 1, \ldots$ and since $S(n+1, 0) = 0$, we have

$$\sum_{k=0}^{n} S(n+1, k+1)(t+1)_{k+1} = (t+1)^{n+1}.$$

Further, since $(t+1)_{k+1} = (t+1)(t)_k$, we conclude that

$$\sum_{k=0}^{n} S(n+1, k+1)(t)_k = (t+1)^n.$$

Expanding the right-hand side into powers of t, by Newton's binomial formula, and in the resulting expression expanding the powers into factorials, using (2.3), we get

$$\sum_{k=0}^{n} S(n+1, k+1)(t)_k = \sum_{j=0}^{n} \binom{n}{j} \sum_{k=0}^{j} S(j,k)(t)_k = \sum_{k=0}^{n} \left\{ \sum_{j=k}^{n} \binom{n}{j} S(j,k) \right\} (t)_k,$$

yielding (2.38).

(b) Multiplying the recurrence relation

$$S(n+1, j+1) = S(n,j) + (j+1)S(n, j+1)$$

by $(-1)^{j-k}(j)_{j-k}$ and summing for $j = k, k+1, \ldots, n$, since $S(n, n+1) = 0$, we get

$$\sum_{j=k}^{n} (-1)^{j-k} (j)_{j-k} S(n+1, j+1) = \sum_{j=k}^{n} (-1)^{j-k} (j)_{j-k} S(n,j)$$

$$+ \sum_{j=k}^{n-1} (-1)^{j-k} (j+1)_{j+1-k} S(n, j+1).$$

The right-hand side of this expression, upon introducing the bound variable $i = j+1$ in the second sum, equals

$$\sum_{j=k}^{n} (-1)^{j-k} (j)_{j-k} S(n,j) - \sum_{i=k+1}^{n} (-1)^{i-k} (i)_{i-k} S(n,i) = S(n,k)$$

and so (2.39) is established.

The triangular recurrence relation (2.34) can be used for the determination of the (power) generating function of the sequence of the noncentral Stirling numbers of the second kind $S(n, k; r)$, $n = k, k+1, \ldots$, for fixed k. This generating function (a) suitably expanded leads to an interesting expression of the numbers $S(n, k; r)$ and (b) transformed yields the expansion of the reciprocal noncentral factorials into reciprocal powers. Specifically, we have the following theorems.

Theorem 2.9 *The (power) generating function of the noncentral Stirling numbers of the second kind $S(n, k; r)$, $n = k, k+1, \ldots$, for fixed k, is given by*

$$\phi_k(u; r) = \sum_{n=k}^{\infty} S(n, k; r) u^n = u^k \prod_{j=0}^{k} (1 - ru - ju)^{-1}, \qquad (2.40)$$

for $k = 1, 2, \ldots$ and $|u| < 1/(r+k)$.

Proof Multiplying (2.34) by u^{n+1} and summing for $n = k-1, k, \ldots$, since $S(k-1, k; r) = 0$, we get

$$\sum_{n=k-1}^{\infty} S(n+1, k; r) u^{n+1} = u \sum_{n=k-1}^{\infty} S(n, k-1; r) u^n$$
$$+ (k+r) u \sum_{n=k}^{\infty} S(n, k; r) u^n,$$

for $k = 1, 2, \ldots$. Consequently,

$$\phi_k(u; r) = u \phi_{k-1}(u; r) + (k+r) u \phi_k(u; r), \quad k = 1, 2, \ldots$$

and so

$$\phi_k(u; r) = u(1 - ru - ku)^{-1} \phi_{k-1}(u), \quad k = 1, 2, \ldots.$$

Applying this recurrence relation repeatedly, we find

$$\phi_k(u; r) = \phi_0(u; r) u^k \prod_{j=1}^{k} (1 - ru - ju)^{-1}$$

and since, according to the initial conditions of (2.34), $\phi_0(u; r) = (1 - ru)^{-1}$, we deduce (2.40).

Theorem 2.10 *The noncentral Stirling number of the second kind is given by the sum*

$$S(n, k; r) = \sum (i_1 + r)(i_2 + r) \cdots (i_{n-k} + r), \qquad (2.41)$$

where the summation is extended over all $(n - k)$-combinations $\{i_1, i_2, \ldots, i_{n-k}\}$ with repetition of the $k+1$ nonnegative integers $\{0, 1, \ldots, k\}$.

In particular, the Stirling number of the second kind is given by the sum

$$S(n,k) = \sum i_1 i_2 \cdots i_{n-k}, \tag{2.42}$$

where the summation is extended over all $(n-k)$-combinations $\{i_1, i_2, \ldots, i_{n-k}\}$ with repetition of the k positive integers $\{1, 2, \ldots, k\}$.

Proof Expanding each factor in (2.40), using the geometric series, we find

$$\phi_k(u;r) = \sum_{n=k}^{\infty} S(n,k;r) u^n = u^k \prod_{i=0}^{k} \left(\sum_{j_i=0}^{\infty} (r+i)^{j_i} u^{j_i} \right)$$

$$= \sum_{n=k}^{\infty} \left\{ \sum r^{j_0}(r+1)^{j_1} \cdots (r+k)^{j_k} \right\} u^n.$$

This expansion implies the expression

$$S(n,k;r) = \sum r^{j_0}(r+1)^{j_1} \cdots (r+k)^{j_k},$$

where the summation is extended over all integers $j_i \geq 0$, $i = 0, 1, \ldots, k$, with $j_0 + j_1 + \cdots + j_k = n - k$. Clearly, this expression is equivalent to (2.41). Also for $r = 0$, $j_0 = 0$ and (2.42) is readily deduced.

Theorem 2.11 (a) *The reciprocal noncentral factorial $1/(t-r)_k$, $k = 1, 2, \ldots$, $t > r + k - 1$, is expanded into a series of reciprocal powers $1/t^n$, $n = k, k+1, \ldots$, as*

$$\frac{1}{(t-r)_k} = \sum_{n=k}^{\infty} S(n-1, k-1; r) \frac{1}{t^n}, \quad k = 1, 2, \ldots. \tag{2.43}$$

(b) *The reciprocal power $1/t^k$, $k = 1, 2, \ldots$, is expanded into a series of reciprocal noncentral ascending factorials $1/(t+r+n-1)_n$, $n = k, k+1, \ldots$, $t > -r$, as*

$$\frac{1}{t^k} = \sum_{n=k}^{\infty} |s(n-1, k-1; r)| \frac{1}{(t+r+n-1)_n}, \quad k = 1, 2, \ldots. \tag{2.44}$$

Proof (a) In (2.40), replacing n by $n-1$ and k by $k-1$ and then setting $u = 1/t$, we get the expression

$$\sum_{n=k}^{\infty} S(n-1, k-1; r) \frac{1}{t^{n-1}} = \frac{t}{(t-r)_k},$$

which implies (2.43).

(b) Note that the expansion of the reciprocal noncentral ascending factorial into reciprocal powers may be deduced from (2.43), by replacing t by $-t$, as

$$\frac{1}{(t+r+k-1)_k} = \sum_{n=k}^{\infty} (-1)^{n-k} S(n-1, k-1; r) \frac{1}{t^n}.$$

In this expansion, replacing the fixed number k by n and the bound variable n by m and then multiplying the resulting expression by

$$|s(n-1, k-1; r)| = (-1)^{n-k} s(n-1, k-1; r)$$

and summing for $n = k, k+1, \ldots$, we find

$$\sum_{n=k}^{\infty} |s(n-1, k-1; r)| \frac{1}{(t+r+n-1)_n}$$

$$= \sum_{n=k}^{\infty} \sum_{m=n}^{\infty} (-1)^{m-k} S(m-1, n-1; r) s(n-1, k-1; r) \frac{1}{t^m}$$

$$= \sum_{m=k}^{\infty} (-1)^{m-k} \left\{ \sum_{n=k}^{m} S(m-1, n-1; r) s(n-1, k-1; r) \right\} \frac{1}{t^m}.$$

By the second of the orthogonality relations (2.20), it holds that

$$\sum_{n=k}^{m} S(m-1, n-1; r) s(n-1, k-1; r) = \delta_{m-1, k-1}$$

and so

$$\sum_{n=k}^{\infty} |s(n-1, k-1; r)| \frac{1}{(t+r+n-1)_n} = \sum_{m=k}^{\infty} (-1)^{m-k} \delta_{m-1, k-1} \frac{1}{t^m},$$

which implies (2.44).

Example 2.6 *A Markov chain.* Suppose that balls are successively drawn one after the other from an urn, initially containing m white and r black balls, according to the following scheme. After each trial if the drawn ball is white, a black ball is placed in the urn, while if the drawn ball is black, it is placed back in the urn. Determine (a) the probability $p(k; n, r)$ of drawing k white balls in n trials, with $n \leq m$, and (b) the probability $q(n; k, r)$ that n trials are required until the kth white ball is drawn, with $k \leq m$.

(a) The probability of drawing a white ball at any trial, given that j white balls are drawn in the previous trials, is $p_j = (m-j)/(m+r)$, $j = 0, 1, \ldots, n-1$. Then, the probability $p(k; n, r)$ of drawing k white balls in n trials satisfies the triangular recurrence relation

$$p(k; n, r) = \frac{k+r}{m+r} p(k; n-1, r) + \frac{m-k+1}{m+r} p(k-1; n-1, r),$$

for $k = 1, 2, \ldots, n$, $n = 1, 2, \ldots$, with initial conditions

$$p(0; 0, r) = 1, \ p(0; n, r) = r^n/(m+r)^n, \ n > 0, \ p(k; n, r) = 0, \ k > n.$$

Further, since $p(1; n, r) = m/(m+r)^n$ and $p(n; n) = (m)_n/(m+r)^n$, we may set

$$p(k; n, r) = c_{n,k} \frac{(m)_k}{(m+r)^n}, \quad k = 0, 1, \ldots, n.$$

Then, the coefficient $c_{n,k}$ satisfies the triangular recurrence relation

$$c_{n,k} = (k+r)c_{n-1,k} + c_{n-1,k-1}, \quad k = 1, 2, \ldots, n, \ n = 1, 2, \ldots,$$

with initial conditions

$$c_{0,0} = 1, \ c_{n,0} = r^n, \ n > 0, \ c_{n,k} = 0, \ k > n.$$

Comparing this recurrence with the triangular recurrence relation of the noncentral Stirling numbers of the second kind, (2.34), we get $c_{n,k} = S(n, k; r)$ and so

$$p(k; n, r) = \frac{S(n, k; r)(m)_k}{(m+r)^n}, \quad k = 1, 2, \ldots, n.$$

(b) The probability $q(n; k, r)$ that n trials are required until the kth white ball is drawn equals the probability $p(k-1; n-1, r)$ of drawing $k-1$ white balls in $n-1$ trials multiplied by the probability $p_{k-1} = (m-k+1)/(m+r)$ of drawing a white ball at the nth trial, given that $k-1$ white balls are drawn in the previous trials. Thus,

$$q(n; k, r) = \frac{S(n-1, k-1; r)(m)_k}{(m+r)^n}, \quad n = k, k+1, \ldots.$$

2.4 GENERALIZED FACTORIAL COEFFICIENTS

Consider the generalized factorial of t of order n and scale parameter s,

$$(st)_n = st(st-1) \cdots (st-n+1), \ n = 1, 2, \ldots, \ (st)_0 = 1, \tag{2.45}$$

with s a real number. It can be expressed as a polynomial of factorials of t of degree n. Specifically, we successively get the expressions

$$(st)_0 = (t)_0 = 1, \ (st)_1 = s(t)_1,$$

$$(st)_2 = st[s(t-1) + (s-1)] = s^2(t)_2 + (s)_2(t)_1,$$

$$(st)_3 = s^2(t)_2[s(t-2) + 2(s-1)] + (s)_2 t[s(t-1) + (s-2)]$$
$$= s^3(t)_3 + 3s(s)_2(t)_2 + (s)_3(t)_1$$

and generally

$$(st)_n = \sum_{k=0}^{n} C(n, k; s)(t)_k, \ n = 0, 1, \ldots. \tag{2.46}$$

In particular, for $s = -1$ and introducing the coefficient $L(n, k) = C(n, k; -1)$, we deduce the expression

$$(-t)_n = \sum_{k=0}^{n} L(n, k)(t)_k, \quad n = 0, 1, \ldots . \tag{2.47}$$

Further, since $(-t)_n = (-1)^n(t+n-1)_n$ and setting $|L(n,k)| = (-1)^n L(n,k)$, we get

$$(t+n-1)_n = \sum_{k=0}^{n} |L(n,k)|(t)_k, \quad n = 0, 1, \ldots . \tag{2.48}$$

Based on expansions (2.46), (2.47) and (2.48), the following definition is introduced.

Definition 2.3 *The coefficient $C(n, k; s)$ of the kth order factorial of t in the expansion of the nth order generalized factorial of t, with scale parameter s, is called the generalized factorial coefficient. In particular, the coefficients $L(n, k) = C(n, k; -1)$ and $|L(n, k)| = (-1)^n L(n, k)$ are called Lah and signless or absolute Lah numbers, respectively.*

A useful extension of the generalized factorial coefficients is provided by the coefficients of the expansion of the noncentral generalized factorials into (usual) factorials. Specifically, let

$$(st+r)_n = \sum_{k=0}^{n} C(n, k; s, r)(t)_k, \quad n = 0, 1, \ldots . \tag{2.49}$$

In particular, for $s = -1$ and introducing $L(n, k; r) = C(n, k; -1, r)$, we get

$$(-t+r)_n = \sum_{k=0}^{n} L(n, k; r)(t)_k, \quad n = 0, 1, \ldots . \tag{2.50}$$

Since $(-t+r)_n = (-1)^n(t-r+n-1)_n$ this expression may be written as

$$(t-r+n-1)_n = \sum_{k=0}^{n} |L(n, k; r)|(t)_k, \tag{2.51}$$

where $|L(n, k; r)| = (-1)^n L(n, k; r)$. Then, the following definition is introduced.

Definition 2.4 *The coefficient $C(n, k; s, r)$ of the kth order factorial of t in the expansion of the nth order noncentral generalized factorial of t, with scale parameter s and noncentrality parameter r, is called the noncentral generalized factorial coefficient. In particular, the coefficients $L(n, k; r) = C(n, k; -1, r)$ and $|L(n, k; r)| = (-1)^n L(n, k; r)$ are called noncentral Lah and noncentral signless or absolute Lah numbers, respectively.*

Clearly, this definition implies

$$C(0, 0; s, r) = 1, \; C(n, 0; s, r) = (r)_n, \; n > 0, \; C(n, k; s, r) = 0, \; k > n.$$

Further, expansion (2.49) entails that the noncentral generalized factorial coefficients are differences of noncentral generalized factorials. Specifically,

$$C(n,k;s,r) = \left[\frac{1}{k!}\Delta^k (st+r)_n\right]_{t=0}, \qquad (2.52)$$

for $k = 0, 1, \ldots, n$, $n = 0, 1, \ldots$. Also, using Vandermonde's formula, the noncentral ascending factorial of t of order n and noncentrality parameter r,

$$(t-r+n-1)_n = (-1)^n(-t+r)_n,$$

may be expanded into ascending factorials of t as

$$(t-r+n-1)_n = \sum_{k=0}^{n}\binom{n}{k}(n-r-1)_{n-k}(t)_k = \sum_{k=0}^{n}\frac{n!}{k!}\binom{n-r-1}{n-k}(t)_k$$

and so, by (2.51), we get for the noncentral Lah numbers the expression

$$|L(n,k;r)| = (-1)^n L(n,k;r) = \frac{n!}{k!}\binom{n-r-1}{n-k}. \qquad (2.53)$$

In particular,

$$|L(n,k)| = (-1)^n L(n,k) = \frac{n!}{k!}\binom{n-1}{k-1}. \qquad (2.54)$$

Finally, note that, for $r = 0$, the noncentral generalized factorial coefficients reduce to the central generalized factorial coefficients. For $r \neq 0$, these coefficients may be expressed in terms of the central generalized factorial coefficients. Specifically, expanding the noncentral generalized factorial $(st+s\rho)_n = (s(t+\rho))_n$ into factorials of $u = t+\rho$, using (2.46), and then expanding the factorials of $u = t+\rho$ into factorials of t, using Vandermonde's formula, we get the expansion

$$(st+s\rho)_n = \sum_{j=0}^{n} C(n,j;s)(t+\rho)_j = \sum_{j=0}^{n} C(n,j;s) \sum_{k=0}^{j} \binom{j}{k}(t)_k(\rho)_{j-k}$$

$$= \sum_{k=0}^{n}\left\{\sum_{j=k}^{n} \binom{j}{k}(\rho)_{j-k}C(n,j;s)\right\}(t)_k,$$

which, compared to (2.49), yields the expression

$$C(n,k;s,\rho s) = \sum_{j=k}^{n}\binom{j}{k}(\rho)_{j-k}C(n,j;s). \qquad (2.55)$$

Also, expanding the noncentral generalized factorials $(st+r)_n$ into generalized factorials $(st)_j$, $j = 0, 1, \ldots, n$, using Vandermonde's formula, and then expanding the generalized factorials $(st)_j$, $j = 0, 1, \ldots, n$, into factorials of t, using (2.46), we deduce the expansion

$$(st+r)_n = \sum_{j=0}^{n} \binom{n}{j}(st)_j(r)_{n-j} = \sum_{j=0}^{n} \binom{n}{j}(r)_{n-j} \sum_{k=0}^{j} C(j,k;s)(t)_k$$

$$= \sum_{k=0}^{n} \left\{ \sum_{j=k}^{n} \binom{n}{j}(r)_{n-j} C(j,k;s) \right\} (t)_k,$$

which, compared to (2.49), implies the expression

$$C(n,k;s,r) = \sum_{j=k}^{n} \binom{n}{j}(r)_{n-j} C(j,k;s). \qquad (2.56)$$

The noncentral generalized factorial coefficient $C(n,k;s,\rho s - r)$ is a polynomial in s of degree n. Specifically, the following theorem is derived.

Theorem 2.12 *The noncentral generalized factorial coefficient $C(n,k;s,\rho s - r)$ is a polynomial in s of degree n, the coefficient of the general term of which being a product of the noncentral Stirling numbers of the first and second kind:*

$$C(n,k;s,\rho s - r) = \sum_{j=k}^{n} s(n,j;r) S(j,k;\rho) s^j. \qquad (2.57)$$

Proof Expanding the noncentral generalized factorial of $u = t + \rho$ of order n, scale parameter s and noncentrality parameter r into powers of $su = s(t+\rho)$, using (2.6), and in the resulting expression expanding the powers of $t + \rho$ into factorials of t, using (2.7), we deduce the expression

$$(s(t+\rho) - r)_n = \sum_{j=0}^{n} s(n,j;r) s^j (t+\rho)^j$$

$$= \sum_{j=0}^{n} s(n,j;r) s^j \sum_{k=0}^{j} S(j,k;\rho)(t)_k$$

$$= \sum_{k=0}^{n} \left\{ \sum_{j=k}^{n} s(n,j;r) S(j,k;\rho) s^j \right\} (t)_k.$$

Also, according to (2.49),

$$(st + \rho s - r)_n = \sum_{k=0}^{n} C(n,k;s,\rho s - r)(t)_k.$$

Then, a comparison of these two expressions entails (2.57).

Remark 2.4 In the particular case $s = 1$ and $\rho = r$, and since $C(n, k; 1, 0) = \delta_{n,k}$, it follows as a corollary of (2.57) the first of (2.20). Also, in the case $s = -1$, we have $C(n, k; -1, -\rho - r) = L(n, k; -\rho - r)$ and so from (2.57), using (2.9) and (2.53), we deduce the relation

$$\sum_{j=k}^{n} |s(n, j; r)| S(j, k; \rho) = \frac{n!}{k!} \binom{n + r + \rho - 1}{n - k},$$

for $k = 1, 2, \ldots, n$, $n = 1, 2, \ldots$.

Note that, for fixed n, the factorial generating function of the sequence of noncentral generalized factorial coefficients $C(n, k; s, r)$, $k = 0, 1, \ldots, n$, according to (2.49), is given by

$$C_n(t; s, r) = \sum_{k=0}^{n} C(n, k; s, r)(t)_k = (st + r)_n, \quad n = 0, 1, \ldots.$$

Further, from (2.49), on using Newton's general binomial theorem, we deduce the double generating function

$$g(t, u; s, r) = \sum_{n=0}^{\infty} \sum_{k=0}^{n} C(n, k; s, r)(t)_k \frac{u^n}{n!} = (1 + u)^{st+r}. \tag{2.58}$$

The exponential generating function of the sequence of noncentral generalized factorial coefficients $C(n, k; s, r)$, $n = k, k+1, \ldots$, for fixed k, is derived in the next theorem.

Theorem 2.13 *The exponential generating function of the noncentral generalized factorial coefficients $C(n, k; s, r)$, $n = k, k+1, \ldots$, for fixed k, is given by*

$$f_k(u; s, r) = \sum_{n=k}^{\infty} C(n, k; s, r) \frac{u^n}{n!} = (1 + u)^r \frac{[(1 + u)^s - 1]^k}{k!}, \tag{2.59}$$

for $k = 0, 1, \ldots$.

Proof Interchanging the order of summation in (2.58), we get

$$g(t, u; s, r) = \sum_{k=0}^{\infty} \sum_{n=k}^{\infty} C(n, k; s, r) \frac{u^n}{n!} (t)_k = \sum_{k=0}^{\infty} f_k(u; s, r)(t)_k.$$

Also, writing (2.58) as $g(t, u; s, r) = (1 + u)^r (1 + [(1 + u)^s - 1])^t$ and expanding it into factorials of t, we find

$$g(t, u; s, r) = (1 + u)^r \sum_{k=0}^{\infty} \frac{[(1 + u)^s - 1]^k}{k!} (t)_k.$$

Comparing these two expansions, we deduce (2.59).

An explicit expression of the noncentral generalized factorial coefficient is derived in the following theorem.

Theorem 2.14 *The noncentral generalized factorial coefficient $C(n, k; s, r)$, $k = 0, 1, \ldots, n$, $n = 0, 1, \ldots$, is given by the sum*

$$C(n, k; s, r) = \frac{1}{k!} \sum_{j=0}^{k} (-1)^{k-j} \binom{k}{j} (sj + r)_n. \tag{2.60}$$

Proof Expanding the generating function (2.59) into powers of u, we get the expression

$$\sum_{n=k}^{\infty} C(n, k; s, r) \frac{u^n}{n!} = \frac{1}{k!} \sum_{j=0}^{k} (-1)^{k-j} \binom{k}{j} (1+u)^{sj+r}$$

$$= \frac{1}{k!} \sum_{j=0}^{k} (-1)^{k-j} \binom{k}{j} \sum_{n=0}^{\infty} \binom{sj+r}{n} u^n$$

$$= \sum_{n=0}^{\infty} \left\{ \frac{1}{k!} \sum_{j=0}^{k} (-1)^{k-j} \binom{k}{j} (sj+r)_n \right\} \frac{u^n}{n!},$$

which implies (2.60).

Remark 2.5 The explicit expression of the noncentral generalized factorial coefficient, (2.60), may be deduced by using the expression of the kth power of the difference operator in terms of the shift operator. Indeed,

$$C(n, k; s, r) = \left[\frac{1}{k!} \Delta^k (st + r)_n \right]_{t=0} = \frac{1}{k!} \sum_{j=0}^{k} (-1)^{k-j} \binom{k}{j} \left[E^j (st + r)_n \right]_{t=0}$$

and, since $\left[E^j (st + r)_n \right]_{t=0} = (sj + r)_n$, (2.60) is deduced.

Expressions of the generalized factorial coefficient $C(n, k; s)$ as multiple sums over all compositions as well as over all partitions of n into k parts are given in the following theorem.

Theorem 2.15 *The generalized factorial coefficient $C(n, k; s)$, $k = 0, 1, \ldots, n$, $n = 0, 1, \ldots$, is given by*

$$C(n, k; s) = \frac{n!}{k!} \sum \binom{s}{j_1} \binom{s}{j_2} \cdots \binom{s}{j_k}, \tag{2.61}$$

where the summation is extended over all compositions of n into k parts, that is over all positive integer solutions of the equation $j_1 + j_2 + \cdots + j_k = n$. Alternatively,

$$C(n, k; s) = \sum \frac{n!}{k_1! k_2! \cdots k_n!} \binom{s}{1}^{k_1} \binom{s}{2}^{k_2} \cdots \binom{s}{n}^{k_n}, \tag{2.62}$$

where the summation is extended over all partitions of n into k parts, that is over all nonnegative integer solutions of the equations

$$k_1 + 2k_2 + \cdots + nk_n = n, \quad k_1 + k_2 + \cdots + k_n = k.$$

Proof Expanding $(1+u)^s - 1$ into powers of u and then, using the Cauchy rule of multiplication of series, the exponential generating function (2.59), with $r = 0$, may be expressed as

$$\sum_{n=k}^{\infty} C(n,k;s)\frac{u^n}{n!} = \frac{1}{k!}\left[\sum_{j=1}^{\infty}\binom{s}{j}u^j\right]^k = \frac{1}{k!}\prod_{i=1}^{k}\left[\sum_{j_i=1}^{\infty}\binom{s}{j_i}u^{j_i}\right]$$

$$= \sum_{n=k}^{\infty}\left\{\frac{n!}{k!}\sum\binom{s}{j_1}\binom{s}{j_2}\cdots\binom{s}{j_k}\right\}\frac{u^n}{n!},$$

yielding (2.61). Using the multinomial theorem instead of the Cauchy rule, we get the expansion

$$\sum_{n=k}^{\infty} C(n,k;s)\frac{u^n}{n!} = \frac{1}{k!}\left[\binom{s}{1}u + \binom{s}{2}u^2 + \cdots + \binom{s}{n}u^n + \cdots\right]^k$$

$$= \sum_{n=k}^{\infty}\left\{\sum \frac{n!}{k_1!k_2!\cdots k_n!}\binom{s}{1}^{k_1}\binom{s}{2}^{k_2}\cdots\binom{s}{n}^{k_n}\right\}\frac{u^n}{n!},$$

which implies (2.62).

Limiting expressions as $s \to 0$ and as $s \to \infty$ and an orthogonality relation for the noncentral generalized factorial coefficient are deduced in the following theorems.

Theorem 2.16 *Let $C(n,k;s,r)$ be the noncentral generalized factorial coefficient. Then*

$$\lim_{s \to 0} s^{-k}C(n,k;s,r) = s(n,k;-r), \tag{2.63}$$

and

$$\lim_{s \to \infty} s^{-n}C(n,k;s,s\rho) = S(n,k;\rho), \tag{2.64}$$

where $s(n,k;r)$ and $S(n,k;\rho)$ are the noncentral Stirling numbers of the first and second kind, respectively.

Proof By using $\lim_{s \to 0} s^{-1}[(1+u)^s - 1] = \log(1+u)$, we deduce for the generating function (2.59) the limiting expression

$$\lim_{s \to 0} s^{-k} f_k(u;s,r) = \sum_{n=k}^{\infty}[\lim_{s \to 0} s^{-k}C(n,k;s,r)]\frac{u^n}{n!}$$

$$= (1+u)^r \frac{\lim_{s \to 0} s^{-k}[(1+u)^s - 1]^k}{k!} = (1+u)^r \frac{[\log(1+u)]^k}{k!},$$

which, compared with the generating function (2.23) of the noncentral Stirling numbers of the first kind, implies (2.63). Similarly, and since $\lim_{s\to\infty}(1+u/s)^s = e^u$, we deduce the limiting expression

$$\lim_{s\to\infty} f_k(u/s;s,s\rho) = \sum_{n=k}^{\infty} [\lim_{s\to\infty} s^{-n}C(n,k;s,s\rho)] \frac{u^n}{n!}$$

$$= \frac{\lim_{s\to\infty}(1+u/s)^{s\rho}[(1+u/s)^s - 1]^k}{k!} = e^{\rho u}\frac{(e^u - 1)^k}{k!},$$

which, compared with the generating function (2.24) of the noncentral Stirling numbers of the second kind, implies (2.64).

Theorem 2.17 *The noncentral generalized factorial coefficients $C(n,k;s,r)$, $k = 0, 1, \ldots, n$, $n = 0, 1, \ldots$, satisfy the relation*

$$\sum_{j=k}^{n} C(n,j;s_1,r_1)C(j,k;s_2,r_2) = C(n,k;s_1s_2, s_1r_2 + r_1). \qquad (2.65)$$

In particular, they satisfy the orthogonality relation

$$\sum_{j=k}^{n} C(n,j;s,r)C(j,k;s^{-1}, -rs^{-1}) = \delta_{n,k}, \qquad (2.66)$$

where $\delta_{n,k} = 1$, if $k = n$ and $\delta_{n,k} = 0$, if $k \neq n$ is the Kronecker delta.

Proof Expanding the nth order factorial of $s_1(s_2 t + r_2) + r_1$ into factorials of $s_2 t + r_2$, using (2.49), and in the resulting expression expanding the factorials of $s_2 t + r_2$ into factorials of t, we get the relation

$$(s_1(s_2 t + r_2) + r_1)_n = \sum_{j=0}^{n} C(n,j;s_1,r_1)(s_2 t + r_2)_j$$

$$= \sum_{j=0}^{n} C(n,j;s_1,r_1) \sum_{k=0}^{j} C(j,k;s_2,r_2)(t)_k$$

$$= \sum_{k=0}^{n} \left\{ \sum_{j=k}^{n} C(n,j;s_1,r_1)C(j,k;s_2,r_2) \right\}(t)_k.$$

Further, according to (2.49), we have

$$(s_1 s_2 t + s_1 r_2 + r_1)_n = \sum_{k=0}^{n} C(n,k;s_1 s_2, s_1 r_2 + r_1)(t)_k$$

and so

$$\sum_{k=0}^{n} \left\{ \sum_{j=k}^{n} C(n,j;s_1,r_1)C(j,k;s_2,r_2) \right\}(t)_k = \sum_{k=0}^{n} C(n,k;s_1 s_2, s_1 r_2 + r_1)(t)_k.$$

The last relation implies (2.65). In the particular case $s_1 = s, r_1 = r, s_2 = s^{-1}$ and $r_2 = -rs^{-1}$, since $C(n, k; 1, 0) = \delta_{n,k}$, (2.66) is deduced.

A triangular recurrence relation for the noncentral generalized factorial coefficients is derived in the following theorem.

Theorem 2.18 *The noncentral generalized factorial coefficients $C(n, k; s, r)$, $k = 0, 1, \ldots, n$, $n = 0, 1, \ldots$, satisfy the triangular recurrence relation*

$$C(n+1, k; s, r) = (sk + r - n)C(n, k; s, r) + sC(n, k-1; s, r), \qquad (2.67)$$

for $k = 1, 2, \ldots, n+1$, $n = 0, 1, \ldots$, with initial conditions

$$C(0, 0; s, r) = 1, \ C(n, 0; s, r) = (r)_n, \ n > 0, \ C(n, k; s, r) = 0, \ k > n.$$

Proof Expanding both members of the recurrence relation

$$(st + r)_{n+1} = (st + r - n)(st + r)_n$$

into factorials of t, according to (2.49), we have

$$\sum_{k=0}^{n+1} C(n+1, k; s, r)(t)_k = st \sum_{j=0}^{n} C(n, j; s, r)(t)_j + (r - n) \sum_{k=0}^{n} C(n, k; s, r)(t)_k.$$

Since $(t)_{j+1} = (t - j)(t)_j$, whence $t(t)_j = (t)_{j+1} + j(t)_j$, the right-hand side of this relation may be written as

$$s \sum_{j=0}^{n} C(n, j; s, r)(t)_{j+1} + s \sum_{j=0}^{n} jC(n, j; s, r)(t)_j + (r - n) \sum_{k=0}^{n} C(n, k; s, r)(t)_k.$$

Therefore,

$$\sum_{k=0}^{n+1} C(n+1, k; s, r)(t)_k = \sum_{k=0}^{n} (sk + r - n)C(n, k; s, r)(t)_k$$
$$+ \sum_{k=1}^{n+1} sC(n, k-1; s, r)(t)_k.$$

Equating the coefficients of $(t)_k$ on both sides of this relation, we conclude (2.67). The initial conditions follow directly from (2.49).

Remark 2.6 (a) If the scale parameter s is a positive integer and the noncentrality parameter r is a nonnegative integer, then the numbers $C(n, k; s, r)$, $n = k, k+1, \ldots, sk+r$, $k = 0, 1, \ldots$, are positive integers. Indeed, from the triangular recurrence relation (2.67) and its initial conditions it follows that these numbers result from successive summation of positive integers. Further, in addition to $C(n, k; s, r) = 0$, $k > n$, it follows from (2.60) that $C(n, k; s, r) = 0$, $sk + r < n$.

(b) Also, if s is a positive integer and r is a nonnegative integer, then the numbers

$$|C(n, k; -s, -r)| = (-1)^n C(n, k; -s, -r), \tag{2.68}$$

for $n = k, k+1, \ldots$, $k = 0, 1, \ldots$, are positive integers. Indeed (2.67) implies for the numbers $|C(n, k; -s, -r)|$ the triangular recurrence relation

$$\begin{aligned}|C(n+1, k; -s, -r)| &= (sk + r + n)|C(n, k; -s, -r)| \\ &+ s|C(n, k-1; -s, -r)|,\end{aligned} \tag{2.69}$$

for $k = 1, 2, \ldots, n+1$, $n = 0, 1, \ldots$, with

$$|C(0, 0; -s, -r)| = 1, \ |C(n, 0; -s, -r)| = (r+n-1)_n, \ n > 0,$$
$$|C(n, k; -s, -r)| = 0, \ k > n,$$

which means that these numbers result from successive summation of positive integers. Note that

$$|C(n, k; -s, -r)| = \left[\frac{1}{k!}\Delta^k(st + r + n - 1)_n\right]_{t=0}, \tag{2.70}$$

for $k = 0, 1, \ldots, n$, $n = 0, 1, \ldots$. The number $|C(n, k; -s, -r)|$, for s a positive integer and r a nonnegative integer, is referred to as the *noncentral absolute generalized factorial coefficient*.

In the following theorem a vertical recurrence relation for the generalized factorial coefficients is deduced.

Theorem 2.19 *The generalized factorial coefficients $C(n, k; s)$, $k = 0, 1, \ldots, n$, $n = 0, 1, \ldots$, with $C(0, 0; s) = 1$, satisfy the vertical recurrence relation*

$$C(n+1, k+1; s) = \sum_{j=k}^{n} \binom{n}{j}(s)_{n-j+1} C(j, k; s). \tag{2.71}$$

Proof According to (2.46) and since $C(n+1, 0; s) = 0$, $n = 0, 1, \ldots$, we have

$$\sum_{k=0}^{n} C(n+1, k+1; s)(t+1)_{k+1} = (st+s)_{n+1}.$$

Further, since $(t+1)_{k+1} = (t+1)(t)_k$, we conclude that

$$\sum_{k=0}^{n} C(n+1, k+1; s)(t)_k = s(st+s-1)_n.$$

Expanding the right-hand side into factorials of st, by Vandermonde's formula, and in the resulting expression expanding the factorials of st into factorials of t, by using

Table 2.4 Generalized factorial coefficients $C(n, k; s)$

n \ k	1	2	3	4	5
1	$(s)_1$				
2	$(s)_2$	s^2			
3	$(s)_3$	$3(s)_2 s$	s^3		
4	$(s)_4$	$7(s)_3 s + 3(s)_2 s$	$6(s)_2 s^2$	s^4	
5	$(s)_5$	$15(s)_4 s + 20(s)_3 s$	$25(s)_3 s^2 + 15(s)_2 s^2$	$10(s)_2 s^3$	s^5

(2.46), we successively get

$$\sum_{k=0}^{n} C(n+1, k+1; s)(t)_k = \sum_{j=0}^{n} \binom{n}{j} (s)_{n-j+1} \sum_{k=0}^{j} C(j, k; s)(t)_k$$
$$= \sum_{k=0}^{n} \left\{ \sum_{j=k}^{n} \binom{n}{j} (s)_{n-j+1} C(j, k; s) \right\} (t)_k,$$

yielding (2.71).

The generalized factorial coefficients $C(n, k; s) = C(n, k; s, 0)$ can be tabulated by using recurrence relation (2.67), for $r = 0$, and its initial conditions. Table 2.4 gives the numbers $C(n, k; s)$, $k = 1, 2, \ldots, n$, $n = 1, 2, \ldots, 5$.

The numbers $C(n, k; s, r)$, $k = 0, 1, \ldots, n$, $n = 0, 1, \ldots$, are defined as the coefficients of the factorials in the expansion of the noncentral generalized factorial, with scale parameter s and noncentrality parameter r. These numbers emerge also in the expansion of the reciprocal noncentral generalized factorials into reciprocal factorials. Specifically, we have the following theorem.

Theorem 2.20 *The reciprocal noncentral generalized factorial* $1/((t - r)/s)_k$, $k = 1, 2, \ldots$, *is expanded into a series of (central) reciprocal factorials* $1/(t)_n$, $n = k, k+1, \ldots$, *as*

$$\frac{1}{((t-r)/s)_k} = \sum_{n=k}^{\infty} sC(n-1, k-1; s, r) \frac{1}{(t)_n}, \quad k = 1, 2, \ldots. \quad (2.72)$$

Proof Multiplying both members of the recurrence relation

$$C(n, k-1; s, r) = [s(k-1) + r - n + 1] C(n-1, k-1; s, r) + sC(n-1, k-2; s, r)$$

by $1/(t)_n$ and using the recurrence relation $1/(t)_n = (t-n)/(t)_{n+1}$, we find

$$tC(n, k-1; s, r)\frac{1}{(t)_{n+1}} - nC(n, k-1; s, r)\frac{1}{(t)_{n+1}}$$
$$= [s(k-1) + r]C(n-1, k-1; s, r)\frac{1}{(t)_n}$$
$$- (n-1)C(n-1, k-1; s, r)\frac{1}{(t)_n} + sC(n-1, k-2; s, r)\frac{1}{(t)_n}.$$

Summing the last relation for $n = k-1, k, \ldots$, we derive for the sum

$$c_k(t; s, r) = \sum_{n=k}^{\infty} C(n-1, k-1; s, r)\frac{1}{(t)_n}, \quad k = 0, 1, \ldots,$$

the recurrence relation

$$c_k(t; s, r) = \frac{s}{t - r - s(k-1)} c_{k-1}(t; s, r), \quad k = 2, 3, \ldots.$$

The initial value $c_1(t; s, r) = \sum_{n=1}^{\infty} (r)_{n-1}/(t)_n$, on using the combinatorial identity

$$\sum_{j=0}^{\infty} \binom{m+j-1}{j} \frac{(x)_j}{(y-m)_j} = \frac{(y)_m}{(y-x)_m},$$

with $m = 1$, $x = r$ and $y = t$, equals $c_1(t; s, r) = 1/(t-r)$. Iterating the recurrence relation, we conclude that

$$c_k(t; s, r) = \frac{1}{s((t-r)/s)_k}$$

and so (2.72) is established.

Remark 2.7 Replacing t by $st + r$, expansion (2.72) may be written as

$$\frac{1}{(t)_{k+1}} = \sum_{n=k}^{\infty} sC(n, k; s)\frac{1}{(st)_{n+1}}, \quad k = 0, 1, \ldots,$$

an expression useful in waiting-time probability problems.

Example 2.7 *Coupon collector's problem.* Consider an urn containing s identical series of coupons, each consisting of m coupons bearing the numbers $1, 2, \ldots, m$ and r additional coupons all bearing the number $m+1$. Suppose that coupons are drawn one after the other, at random, without replacement. Calculate (a) the probability $p(k; n)$ of drawing exactly k of the numbers $\{1, 2, \ldots, m\}$ in n drawings, with $n \leq m$, and (b) the probability $q(n; k)$ that n drawings are required until the kth different number, among the m numbers $\{1, 2, \ldots, m\}$, is drawn, with $k \leq m$.

(a) Let A_i be the event that number i is not drawn in n drawings, $i = 1, 2, \ldots, m$. Then, the probability $p(k; n)$ equals the probability that exactly $m - k$ among the m

events $A_1, A_2, \ldots, A_m$ occur. Further, for any j indices $\{i_1, i_2, \ldots, i_j\}$ out of the m indices $\{1, 2, \ldots, m\}$,

$$p_j = P(A_{i_1} A_{i_2} \cdots A_{i_j}) = \frac{(s(m-j)+r)_n}{(sm+r)_n}, \; j = 1, 2, \ldots, m.$$

Therefore, the events $A_1, A_2, \ldots, A_m$ are exchangeable and according to Corollary 1.5, the probability $p(k; n)$ is given by

$$p(k; n) = \binom{m}{n} \sum_{j=0}^{k} (-1)^{k-j} \binom{k}{j} \frac{(sj+r)_n}{(sm+r)_n}$$

and so, by virtue of (2.60),

$$p(k; n) = \frac{C(n, k; s, r)(m)_k}{(sm+r)_n}, \; k = 0, 1, \ldots, n.$$

(b) The probability $q(n; k)$ that n drawings are required until the kth different number, among the m numbers $\{1, 2, \ldots, m\}$, is drawn equals the probability $p(k-1; n-1)$ of drawing $k-1$ of the m numbers in $n-1$ drawings multiplied by the probability $s(m-k+1)/(sm+r-n+1)$ that one of the $m-k+1$ not already drawn numbers is drawn at the nth drawing. Hence,

$$q(n; k) = \frac{C(n-1, k-1; s, r)s(m)_k}{(sm+r)_n}, \; n = k, k+1, \ldots, sm+r.$$

2.5 ENUMERATION OF PARTITIONS BY SUBSETS AND PERMUTATIONS BY CYCLES

A *partition of a set W into k subsets* is a (non ordered) set of k non empty subsets of W, $\{A_1, A_2, \ldots, A_k\}$, which are pairwise disjoint sets and their union is W. A partition of W into an unspecified number of subsets is simply referred to as a *partition of W*. It is natural to consider, along with partitions of a set the corresponding ordered collections, which are called divisions. A *division of a set W into k subsets* is an ordered collection of k subsets of W, $(A_1, A_2, \ldots, A_k)$, which are pairwise disjoint sets and their union is W. A division of W into an unspecified number of subsets is simply referred to as a *division of W*. Note that in a partition, as opposed to a division of a set, no empty sets are included and the sets constituting it are not ordered. The number of divisions of a finite set into a given number of subsets with specified cardinalities is derived in the following theorem.

Theorem 2.21 *The number of divisions $(A_1, A_2, \ldots, A_k)$ of a finite set W_n, with $N(W_n) = n$, into k (ordered) subsets, with $N(A_i) = r_i \geq 0$, $i = 1, 2, \ldots, k$, equals*

$$C(n, k; r_1, r_2, \ldots, r_{k-1}) = \frac{n!}{r_1! r_2! \cdots r_{k-1}! r_k!}, \tag{2.73}$$

where $r_k = n - r_1 - r_2 - \cdots - r_{k-1}$.

Proof In a division $(A_1, A_2, \ldots, A_k)$ of a finite set W_n, with $N(W_n) = n$, the r_1 elements of the first set A_1 can be chosen from the set W_n, of n elements, in $\binom{n}{r_1}$ ways. After the selection of the r_1 elements of the first set A_1, the r_2 elements of the second set A_2 can be chosen from the set $W_n - A_1$, of $n - r_1$ elements, in $\binom{n-r_1}{r_2}$ ways. Continuing in this manner, after the selection of the $r_1, r_2, \ldots, r_{k-2}$ elements of the sets $A_1, A_2, \ldots, A_{k-2}$, respectively, the r_{k-1} elements of the set A_{k-1} can be chosen from the set $W_n - A_1 - A_2 - \cdots - A_{k-2}$, of $n - r_1 - r_2 - \cdots - r_{k-2}$ elements, in $\binom{n-r_1-r_2-\cdots-r_{k-2}}{r_{k-1}}$ ways. Finally, the selection of the $r_1, r_2, \ldots, r_{k-1}$ elements of the sets $A_1, A_2, \ldots, A_{k-1}$, respectively, determines the r_k elements of the last set A_k, since $W_n - A_1 - A_2 - \cdots - A_{k-1}$ is a set of $n - r_1 - r_2 - \cdots - r_{k-1} = r_k$ elements. Thus, by the multiplication principle, the number $C(n, k; r_1, r_2, \ldots, r_{k-1})$ is given by

$$\binom{n}{r_1}\binom{n-r_1}{r_2}\cdots\binom{n-r_1-r_2-\cdots-r_{k-2}}{r_{k-1}}$$
$$= \frac{n!}{r_1!(n-r_1)!} \cdot \frac{(n-r_1)!}{r_2!(n-r_1-r_2)!} \cdots \frac{(n-r_1-r_2-\cdots-r_{k-2})!}{r_{k-1}!(n-r_1-r_2-\cdots-r_{k-1})!}$$

and so, after some simplifications, (2.73) is deduced.

The next theorem is concerned with the enumeration of the partitions of a finite set by their subsets.

Theorem 2.22 *The number of partitions* $\{A_1, A_2, \ldots, A_k\}$ *of a finite set W_n, with $N(W_n) = n$, into k (unordered) subsets, among which $k_i \geq 0$ include i elements, $i = 1, 2, \ldots, n$, equals*

$$S(n, k; k_1, k_2, \ldots, k_n) = \frac{n!}{k_1!(1!)^{k_1} k_2!(2!)^{k_2} \cdots k_n!(n)^{k_n}}, \quad (2.74)$$

where $k_1 + 2k_2 + \cdots + nk_n = n$ and $k_1 + k_2 + \cdots + k_n = k$.

Proof Consider a partition $\{A_1, A_2, \ldots, A_k\}$ of a finite set W_n, with $N(W_n) = n$, into $k_i \geq 0$ subsets including i elements each, $i = 1, 2, \ldots, n$, with $k_1 + 2k_2 + \cdots + nk_n = n$ and $k_1 + k_2 + \cdots + k_n = k$. To it there correspond $k_1! k_2! \cdots k_n!$ divisions of W_n into k subsets, with $k_i \geq 0$ subsets including i elements each, $i = 1, 2, \ldots, n$; these divisions are formed by internal permutation of the $k_1 \geq 0$ subsets with one element, the $k_2 \geq 0$ subsets with two elements, and finally the $k_n \geq 0$ subsets with n elements in all possible ways. Thus,

$$C(n, k; 1, \ldots, 1, 2, \ldots, 2, \ldots) = k_1! k_2! \cdots k_n! S(n, k; k_1, k_2, \ldots, k_n)$$

and also by (2.73),

$$C(n, k; 1, \ldots, 1, 2, \ldots, 2, \ldots) = \frac{n!}{(1!)^{k_1}(2!)^{k_2}\cdots(n!)^{k_n}}.$$

From these two relations, (2.74) is deduced.

The (total) number of partitions of a finite set of n elements into k subsets may be obtained by summing the numbers (2.74) over all partitions of n into k parts, that is over all nonnegative integer solutions of the equations $k_1 + 2k_2 + \cdots + nk_n = n$ and $k_1 + k_2 + \cdots + k_n = k$. Then, by virtue of (2.30) and (2.31), we deduce the following corollary.

Corollary 2.3 *The number of partitions of a finite set of n elements into k (unordered) subsets equals*
$$S(n, k),$$
the Stirling number of the second kind, and the (total) number of partitions of a finite set of n elements equals
$$B_n = \sum_{k=0}^{n} S(n, k),$$
the Bell number.

A *permutation* of a finite set $W_n = \{w_1, w_2, \ldots, w_n\}$, of n elements, is an ordered n-tuple $(w_{i_1}, w_{i_2}, \ldots, w_{i_n})$ with $w_{i_r} \in W_n$, $r = 1, 2, \ldots, n$ and $w_{i_r} \neq w_{i_s}$ for $r \neq s$. Such a permutation can be equivalently considered as a rearrangement of a fixed ordering. The replacement of the fixed ordering $(w_1, w_2, \ldots, w_n)$ by $(w_{i_1}, w_{i_2}, \ldots, w_{i_n})$ is denoted by

$$\begin{pmatrix} w_1 & w_2 & \ldots & w_n \\ w_{i_1} & w_{i_2} & \ldots & w_{i_n} \end{pmatrix}, \tag{2.75}$$

which defines the rule (mapping) $\sigma(w_r) = w_{i_r}$ of the replacement of the point w_r by the point w_{i_r}, $r = 1, 2, \ldots, n$. A point w_r for which $w_{i_r} = w_r$ is called a *fixed point* of this permutation. The columns in (2.75) corresponding to the fixed points may be omitted. Consider a subset $W_k = \{w_{i_1}, w_{i_2}, \ldots, w_{i_k}\}$, of the set W_n, $k \leq n$. The permutation

$$\begin{pmatrix} w_{i_1} & w_{i_2} & \ldots & w_{i_{k-1}} & w_{i_k} \\ w_{i_2} & w_{i_3} & \ldots & w_{i_k} & w_{i_1} \end{pmatrix}, \tag{2.76}$$

in which the replacements, $\sigma(w_{i_1}) = w_{i_2}$, $\sigma(w_{i_2}) = w_{i_3}, \ldots, \sigma(w_{i_{k-1}}) = w_{i_k}$, $\sigma(w_{i_k}) = w_{i_1}$, close a cycle, is called a *cycle*. The notation (2.76) of a cycle may be abbreviated by writing in parentheses the rule of succession of the permuted (non fixed) points,

$$(w_{i_1}, w_{i_2}, \ldots, w_{i_{k-1}}, w_{i_k}). \tag{2.77}$$

The number k, of the elements of the cycle (2.77), is called the *length* of it. The representation (2.77) of a cycle is not unique. The k permutations

$$(w_{i_1}, w_{i_2}, \ldots, w_{i_{k-1}}, w_{i_k}), (w_{i_2}, \ldots, w_{i_k}, w_{i_1}), \ldots, (w_{i_k}, w_{i_1}, \ldots, w_{i_{k-1}}),$$

which are formed by putting as first any of the k elements and keeping the same rule of succession, correspond to (represent the) same cycle. It is conventional to write a cycle with its smallest element in the first position, making the representation (2.77) unique. Evidently, there are $(n)_k/k$ cycles of length k. A cycle of length n, which includes all the elements of the set W_n, is particularly called a *cyclic permutation*. Thus, there are $(n)_n/n = (n-1)!$ cyclic permutations of the set W_n. Clearly, any permutation of W_n either is a cycle or can be decomposed into disjoint cycles. This decomposition is unique up to a rearrangement of the cycles. The next theorem is concerned with the enumeration of the permutations of a finite set by their decomposition into cycles.

Theorem 2.23 *The number of permutations of a finite set W_n, with $N(W_n) = n$, that are decomposed into k cycles, among which $k_i \geq 0$ are of length i, $i = 1, 2, \ldots, n$, equals*

$$c(n, k; k_1, k_2, \ldots, k_n) = \frac{n!}{k_1! 1^{k_1} k_2! 2^{k_2} \cdots k_n! n^{k_n}}, \qquad (2.78)$$

where $k_1 + 2k_2 + \cdots + nk_n = n$ and $k_1 + k_2 + \cdots + k_n = k$.

Proof Consider a partition $\{A_1, A_2, \ldots, A_k\}$ of a finite set W_n, with $N(W_n) = n$, into $k_i \geq 0$ subsets including i elements each, $i = 1, 2, \ldots, n$, with $k_1 + 2k_2 + \cdots + nk_n = n$ and $k_1 + k_2 + \cdots + k_n = n$. To it there correspond

$$[(1-1)!]^{k_1}[(2-1)!]^{k_2} \cdots [(n-1)!]^{k_n}$$

permutations of the set W_n that are decomposed into k cycles, among which $k_i \geq 0$ are of length i, $i = 1, 2, \ldots, n$; these permutations are formed by permuting the $i - 1$ elements of each of the $k_i \geq 0$ subsets with i elements (excluding the smallest element, which, by convention, occupies the first position) in $[(i-1)!]^{k_i}$ ways, for $i = 1, 2, \ldots, n$. Hence,

$$c(n, k; k_1, k_2, \ldots, k_n)$$
$$= [(1-1)!]^{k_1}[(2-1)!]^{k_2} \cdots [(n-1)!]^{k_n} S(n, k; k_1, k_2, \ldots, k_n)$$

and by (2.74), expression (2.78) is deduced.

The number of permutations of a finite set of n elements that are decomposed into k cycles may be obtained by summing the numbers (2.78) over all partitions of n into k parts, that is over all nonnegative integer solutions of the equations $k_1 + 2k_2 + \cdots + nk_n = n$ and $k_1 + k_2 + \cdots + k_n = k$. Then, by virtue of (2.29), we deduce the following corollary of Theorem 2.23.

Corollary 2.4 *The number of permutations of a finite set of n elements that are decomposed into k cycles equals*

$$|s(n,k)|,$$

the signless Stirling number of the first kind.

The combinatorial interpretations of the Stirling numbers of the first and second kind, furnished by Corollaries 2.3 and 2.4, can be used as definitions of these numbers.

112 STIRLING NUMBERS

In the following examples, similar combinatorial interpretations are furnished for the noncentral Stirling numbers of the first and second kind.

Example 2.8 *Enumeration of restricted partitions by subsets.* Consider a finite set $W_{n+r} = \{w_1, w_2, \ldots, w_{n+r}\}$, with $w_1 < w_2 < \cdots < w_{n+r}$, and its subset $W_r = \{w_1, w_2, \ldots, w_r\}$. Determine the number of partitions of the set W_{n+r} into $k+r$ subsets, such that the r elements of the set W_r belong in r distinct subsets.

The k subsets of such a partition may contain a total of j elements from the set $W_{n+r} - W_r = \{w_{r+1}, w_{r+2}, \ldots, w_{n+r}\}$, of n elements, for $j = k, k+1, \ldots, n$. The j elements can be chosen in $\binom{n}{j}$ different ways. Further, the number of partitions of a set of j elements into k subsets, according to Corollary 2.3, equals the Stirling number of the second kind $S(j, k)$, while the remaining $n - j$ elements of the set $W_{n+r} - W_r$ can be distributed into the r distinct subsets in r^{n-j} different ways. Consequently, the number of partitions of the set W_{n+r} into $k+r$ subsets, such that the r elements of the set W_r belong in r distinct subsets, equals

$$S(n, k; r) = \sum_{j=k}^{n} \binom{n}{j} r^{n-j} S(j, k),$$

which, by (2.16), is the noncentral Stirling number of the second kind.

Example 2.9 *Enumeration of restricted permutations by cycles.* As in Example 2.8, consider a finite set $W_{n+r} = \{w_1, w_2, \ldots, w_{n+r}\}$, with $w_1 < w_2 < \cdots < w_{n+r}$, and its subset $W_r = \{w_1, w_2, \ldots, w_r\}$. Determine the number of permutations of the set W_{n+r} that are decomposed into $k+r$ cycles, such that the r elements of the set W_r belong in r distinct cycles.

The k cycles of such a permutation may contain a total of j elements from the set $W_{n+r} - W_r = \{w_{r+1}, w_{r+2}, \ldots, w_{n+r}\}$, of n elements, for $j = k, k+1, \ldots, n$. The j elements can be chosen in $\binom{n}{j}$ different ways. Further, the number of permutations of a set of j elements that are decomposed into k cycles, according to Corollary 2.4, equals the signless Stirling number of the first kind $|s(j, k)|$. Also, the remaining $n - j$ elements of the set $W_{n+r} - W_r$ can be placed into the r distinct cycles in $(r + n - j - 1)_{n-j} = r(r+1)\cdots(r+n-j-1)$ different ways. Indeed, the first element can be placed in any of the r places between the r elements $w_1, w_2, \ldots, w_r$ or after the last element. After the first element is placed, the second element can be placed in any of the $r + 1$ places between the $r + 1$ elements or after the last element. Finally, after the $n - j - 1$ elements are placed, the last element can be placed in any of the $r + n - j - 1$ places between the $r + n - j - 1$ elements or after the last element. Consequently, the number of permutations of the set W_{n+r} that are decomposed into $k+r$ subsets, such that the r elements of the set W_r belong in r distinct cycles, equals

$$|s(n, k; r)| = \sum_{j=k}^{n} \binom{n}{j} (r + n - j - 1)_{n-j} |s(j, k)|,$$

which, by (2.14), is the noncentral signless Stirling number of the first kind.

2.6 REFERENCE NOTES

The Stirling numbers were so named by N. Nielsen (1906) in honor of James Stirling, who introduced them in his *Methodus Differentialis* (1730), without using any notation for them. The notation adopted in this book is due to J. Riordan (1958). Recurrence relations and certain number theoretic properties of the Stirling numbers of the first kind were derived by L. Lagrange (1770). P. S. Laplace (1812) and A. Cayley (1887) provided several approximations of the Stirling numbers of the second kind. J. A. Grunert (1822, 1843), A. Cauchy (1833), O. Schlömilch (1852, 1895), G. Boole (1860), L. Schläfli (1867), J. Blissard (1867) and J. Worpitzky (1883) explored further the Stirling numbers of both kinds.

The work of N. E. Nörlund (1924) inspired several publications on certain generalizations of the Stirling numbers and their connection with the Bernoulli and generalized Bernoulli numbers. Also, the books by N. Nielsen (1906, 1923), E. Netto (1927), J. F. Steffensen (1927) and L. M. Milne-Thomson (1933) are worth mentioning.

A thorough presentation of the Stirling numbers and their most important properties provided by Ch. Jordan (1933) in an excellent paper, which was included as Chapter 4, in his classical book on the calculus of finite differences (1939a), revived interest in the Stirling numbers. Since then a large number of publications on these numbers have appeared in the literature. The more recent book of L. Comtet (1974) devotes a chapter to these numbers and provides a very rich bibliography. The expression of the nth power of the operator $\Theta = tD$ in terms of powers of the operator D was obtained by J. A. Grunert (1843). The derivation of this expression in Example 2.4 is due to Ch. Jordan (1933, 1939a), who also derived the expression of the nth power of the operator $\Psi = t\Delta$ in terms of powers of the operator Δ, given in Example 2.5. H. W. Gould (1964) expressed the operator $(a^t D_t)^n$ in terms of powers of the operator D_t with the Stirling numbers of the first kind as coefficients. L. Comtet (1973) expressed the more general operator $(\lambda(t) D_t)^n$ in terms of powers of the operator D_t with Bell partition polynomials as coefficients. A generalization to another direction was discussed by L. Carlitz (1932). Y. Taga and K. Isii (1959) proposed a stochastic model for the pattern of communication and diffusion of news in a social group and deduced a recurrence relation for the conditional probability function of the number of people hearing the news, given that n people hear the news. D. J. Bartholomew (1982) derived this probability function in the form of the probability function of Example 2.1(a), with $r = 0$ and θ the ratio of the intensities of source and between individuals transmission of news. The same distribution emerged as the distribution of the number genetic types in a biological population generated by a parent population of n genetic types, which was derived by S. Karlin and J. McGregor (1972); the parameter θ is the rate of mutation of genetic types to new genetic types. The distributions of the record numbers and record times, which are also of the form of the distributions of Example 2.1, were derived by A. Rényi (1962), F. N. David and D. E. Barton (1962), and R. W. Shorrock (1972); Exercise 2.6 is based on these papers.

Besides the classical triangular, vertical and horizontal recurrence relations for the Stirling numbers of both kinds, there are several other recurrence relations derived by L. Lagrange (1770), D. S. Mitrinovic and D. Djokovic (1960), Chr. Karanicoloff

(1961) and M. Glaymann and D. Djokovic (1962). O. Schlömilch (1852) used the same symbol C_k^n to denote the Stirling numbers of both kinds, calling them *factorial coefficients*. In this unified notation $C_k^n = |s(n,k)|$ and $C_k^{-n} = S(k,n)$. N. E. Nörlund (1924) introduced the generalized Bernoulli numbers $B_n^{(r)}$, $n = 0, 1, \ldots$, with generating function the rth power of the generating function of the usual Bernoulli numbers. Then $s(n,k) = \binom{n-1}{k-1} B_{n-k}^{(n)}$ and $S(n,k) = \binom{n}{k} B_{n-k}^{(-k)}$ (see Exercise 2.12). Properties of the generalized Bernoulli numbers were studied by L. Carlitz (1960). F. N. David and D. E. Barton (1962) devoted a chapter to these numbers. The Bell numbers, given in Remark 2.3, were introduced by E. T. Bell (1934a, b) and so named by J. Riordan (1958). H. Yamato (1990) and J. G. Leite and C. A. de B. Pereira (1994) provided probabilistic derivations of several classical properties of the Stirling numbers of the first and second kind, respectively.

The first short table of the Stirling numbers of the second kind, up to $n = 9$, was published by J. Stirling (1730). Extensive tables of the Stirling numbers of both kinds were constructed by H. Gupta (1950), R. A. Fisher and F. Yates (1953), F. N. David, M. G. Kendall and D. E. Barton (1966) and M. Abramowitz and I. A. Stegun (1965). A variety of asymptotic expressions for the Stirling numbers exist in the literature. The approximate expressions given in Exercise 2.10 are due to Ch. Jordan (1933, 1939a). Several references on other approximations are given in the review paper by Ch. A. Charalambides and J. Singh (1988).

The coefficients $L(n,k)$, $k = 0, 1, \ldots, n$, $n = 0, 1, \ldots$, of the expansion of the rising factorials into falling factorials, introduced by I. Lah (1955), were called *Lah numbers* by J. Riordan (1958). Extending these numbers, Ch. A. Charalambides (1976, 1977a, 1979, 1984a) systematically studied the coefficients $C(n,k;s)$ of the generalized factorials into falling factorials. These numbers were noted before by Ch. Jordan (1933) and appeared in E. T. Bell (1934a), as coefficients of a generalized Hermite polynomial, and in several other forms in H. W. Gould (1958), D. E. Barton and F. N. David (1959a,c), R. Shumway and J. Gurland (1960a,b), L. Bernstein (1965), L. Carlitz (1965) and W. Feller (1968). The connection constants in the expression of the operator $(t^{b+1} D_t)^n$ in terms of powers of the derivative operator D_t, examined in L. Comtet (1973), are merely generalized factorial coefficients (see Exercise 2.19). Later, L. Carlitz (1979) studied these numbers under the name degenerate Stirling numbers.

The noncentral Stirling numbers of the second kind appeared in N. Nielsen (1906) as differences of the powers at an arbitrary point. J. Riordan (1937) used them as connection constants of power moments about an arbitrary point and factorial moments (cf. Example 2.3). Also, they appeared as coefficients in a modification of the classical occupancy problem discussed by D. E. Barton and F. N. David (1959a). Recently, these numbers were studied by L. Carlitz (1980a,b) as weighted Stirling numbers, by M. Koutras (1982) as noncentral Stirling numbers, by A. Z. Broder (1984) as r-Stirling numbers and also by R. Shanmugan (1984). The differences of the generalized factorials at an arbitrary point were studied by Ch. A. Charalambides and M. Koutras (1983). A rich bibliography of the Stirling and noncentral Stirling numbers is included in the review paper by Ch. A. Charalambides and J. Singh (1988).

2.7 EXERCISES AND COMPLEMENTS

2.1 *Additional recurrence relations for the Stirling numbers.* Show that the Stirling numbers of the first and second kinds satisfy the recurrence relations

$$s(n,k) = \sum_{r=k}^{n} n^{r-k} s(n+1, r+1), \quad k = 0, 1, \ldots, n, \ n = 0, 1, \ldots,$$

and

$$S(n,k) = \sum_{r=k}^{n} k^{n-r} S(r-1, k-1), \quad k = 1, 2, \ldots, n, \ n = 1, 2, \ldots.$$

2.2 *Schlömilch's formula.* Consider the power series $u = \phi(t) = e^t - 1$, with $\phi(0) = 0$, and its inverse power series $t = \phi^{-1}(u) = \log(1+u)$. Applying the Lagrange inversion formula

$$\frac{1}{k!}\left[\frac{d^n}{du^n}(\phi^{-1}(u))^k\right]_{u=0} = \binom{n-1}{k-1}\left[\frac{d^{n-k}}{dt^{n-k}}\left(\frac{\phi(t)}{t}\right)^{-n}\right]_{t=0},$$

show that

$$s(n,k) = \sum_{r=0}^{n-k}(-1)^r \binom{n+r-1}{k-1}\binom{2n-k}{n-k-r} S(n-k+r, r)$$

and deduce the explicit expression of the Stirling number of the first kind:

$$s(n,k) = \sum_{r=0}^{n-k}\sum_{j=0}^{r}(-1)^j \binom{r}{j}\binom{n+r-1}{k-1}\binom{2n-k}{n-k-r}\frac{j^{n-k+r}}{r!}.$$

2.3 *(Continuation).* Show that

$$S(n, n-k) = \sum_{r=0}^{k}\binom{k-n}{k+r}\binom{k+n}{k-r}|s(k+r,r)|$$

and conclude its inverse relation

$$|s(n, n-k)| = \sum_{r=0}^{k}\binom{k-n}{k+r}\binom{k+n}{k-r} S(k+r, r).$$

2.4 Show that

$$\binom{k+r}{k} s(n, k+r) = \sum_{j=k}^{n-r}\binom{n}{j} s(j,k) s(n-j, r)$$

and conclude its inverse relation

$$\binom{n}{i}s(n-i,r) = \sum_{k=i}^{n-r}\binom{k+r}{k}S(k,i)s(n,k+r).$$

2.5 Show that

$$\binom{k+r}{k}S(n,k+r) = \sum_{j=k}^{n-r}\binom{n}{j}S(j,k)S(n-j,r)$$

and conclude its inverse relation

$$\binom{n}{i}S(n-i,r) = \sum_{k=i}^{n-r}\binom{k+r}{k}s(k,i)S(n,k+r).$$

2.6 *Record numbers and record times.* Consider a sequence $X_1, X_2, \ldots$ of independent and identically distributed random variables with a common continuous distribution function $F(x) = P(X \leq x)$, $x \in R$. The observation X_j is a *record* if $X_j > X_i$ for all $i = 1, 2, \ldots, j-1$. By convention X_1 is considered as a record. Let N_n be the number of records up to time (index) n and L_k be the time of the kth record. Show that

$$P(N_n = k) = \frac{|s(n,k)|}{n!}, \quad k = 1, 2, \ldots, n$$

and

$$P(L_k = n) = \frac{|s(n-1, k-1)|}{n!}, \quad n = k, k+1, \ldots,$$

where $|s(n,k)|$ is the signless Stirling number of the first kind.

2.7 Let $C(n, k; s)$ be the generalized factorial coefficient. Show that

$$(-1)^k C(n, n-k; s^{-1}) = \sum_{r=0}^{k}\binom{k-n}{k+r}\binom{k+n}{k-r}C(k+r, r; s).$$

2.8 *(Continuation).* Show that

$$\binom{k+r}{k}C(n, k+r; s) = \sum_{j=k}^{n-r}\binom{n}{j}C(j, k; s)C(n-j, r; s)$$

and

$$\binom{n}{i}C(n-i, r; s) = \sum_{k=i}^{n-r}\binom{k+r}{k}C(k, i; s^{-1})C(n, k+r; s).$$

2.9 Show that the signless Stirling number of the first kind is given by

$$|s(n+1,k+1)| = \frac{n!}{k!} \sum \frac{(-1)^{k+r_1+r_2+\cdots+r_k} k!}{1^{r_1} r_1! 2^{r_2} r_2! \cdots k^{r_k} r_k!} \left(\zeta_n(1)\right)^{r_1} \left(\zeta_n(2)\right)^{r_2} \cdots \left(\zeta_n(k)\right)^{r_k},$$

where $\zeta_n(s) = \sum_{j=1}^{n} 1/j^s$, and the summation is extended over all nonnegative integer solutions of the equation $r_1 + 2r_2 + \cdots + kr_k = k$.

2.10 *Asymptotic expressions for the Stirling numbers.* For fixed k and $n \to \infty$ show that

$$|s(n+1, k+1)| \cong n![\log n + C]^k / k!$$

and

$$S(n,k) \cong k^n/k!,$$

where $C = 0.57721$ is Euler's constant. Also, for fixed k, s and $n \to \infty$ derive for the generalized factorial coefficients $C(n,k;s)$ the asymptotic expression

$$C(n,k;s) \cong (sk)_n/k!.$$

2.11 *Bernoulli numbers.* The sequence of the Bernoulli numbers $B_n, n = 0, 1, \ldots,$ has generating function

$$g(t) = \sum_{n=0}^{\infty} B_n \frac{t^n}{n!} = \frac{t}{e^t - 1}.$$

Show that

$$B_n = \sum_{k=0}^{n} \binom{n}{k} B_k, \quad n = 1, 2, \ldots, \quad B_0 = 1$$

and

$$B_n = \sum_{k=0}^{n} \frac{(-1)^k k!}{k+1} S(n,k), \quad \sum_{n=0}^{r} s(r,n) B_n = \frac{(-1)^r r!}{r+1},$$

where $s(n,k)$ and $S(r,n)$ are the Stirling numbers of the first and second kind, respectively.

2.12 *Nörlund-Bernoulli numbers.* The sequence of numbers $B_n^{(r)}, n = 0, 1, \ldots,$ where r is a real (or generally a complex) number, with generating function

$$g_r(t) = \sum_{n=0}^{\infty} B_n^{(r)} \frac{t^n}{n!} = \left(\frac{t}{e^t - 1}\right)^r,$$

has been defined by Nörlund and in the particular case $r = 1$ reduces to the sequence of Bernoulli numbers. Show that

$$B_n^{(r+1)} = \left(1 - \frac{n}{r}\right) B_n^{(r)} - n B_{n-1}^{(r)}, \quad n = 1, 2, \ldots, \quad B_0^{(r)} = 1$$

and conclude that

$$B_n^{(r)} = s(r, r-n)/\binom{r-1}{n}, \quad n = 0, 1, \ldots, r-1, \ r = 1, 2, \ldots,$$

$$B_n^{(-r)} = S(n+r, r)/\binom{n+r}{n}, \quad n = 0, 1, \ldots, \ r = 1, 2, \ldots,$$

where $s(n, k)$ and $S(n, k)$ are the Stirling numbers of the first and second kind, respectively.

2.13 Show that the noncentral signless Stirling numbers of the first kind $|s(n, k; r)|$, $k = 0, 1, \ldots, n$, $n = 0, 1, \ldots$, satisfy the relation

$$|s(n, k; r)| = \sum_{j=k}^{n} \binom{n}{j} (m+n-j-1)_{n-j} |s(j, k; r-m)|$$

and conclude that

$$|s(n, k; r)| = \sum_{n=k}^{n} (n)_{n-j} |s(j, k; r-1)|$$

and

$$|s(n, k; r)| = \sum_{n=k}^{n} \binom{n}{j} (r+n-j-1)_{n-j} |s(j, k)|.$$

2.14 Show that the noncentral Stirling numbers of the second kind $S(n, k; r)$, $k = 0, 1, \ldots, n$, $n = 0, 1, \ldots$, satisfy the relation

$$S(n, k; r) = \sum_{j=k}^{n} \binom{n}{j} m^{n-j} S(j, k; r-m)$$

and conclude that

$$S(n, k; r) = \sum_{j=k}^{n} \binom{n}{j} S(j, k; r-1)$$

and

$$S(n, k; r) = \sum_{j=k}^{n} \binom{n}{j} r^{n-j} S(j, k).$$

2.15 *(Continuation).* Show that

$$\sum_{j=k}^{n} s(n, j; r_1) S(j, k; r_2) = \binom{n}{k} (r_2 - r_1)_{n-k},$$

and

$$\sum_{j=k}^{n} S(n, j; r_1) s(j, k; r_2) = \binom{n}{k} (r_1 - r_2)^{n-k},$$

and conclude that

$$\sum_{j=k}^{n} s(n,j;r)S(j,k;r) = \delta_{n,k}, \quad \sum_{j=k}^{n} S(n,j;r)s(j,k;r) = \delta_{n,k}.$$

2.16 (*Continuation*). Show that

$$\binom{k+m}{k} s(n,k+m;r_1+r_2) = \sum_{j=k}^{n-m} \binom{n}{j} s(j,k;r_1)s(n-j,m;r_2)$$

and conclude its inverse relation

$$\binom{n}{i} s(n-i,m;r_2) = \sum_{k=i}^{n-m} \binom{k+m}{k} S(k,i;r_1)s(n,k+m;r_1+r_2).$$

2.17 (*Continuation*). Show that

$$\binom{k+m}{k} S(n,k+m;r_1+r_2) = \sum_{j=k}^{n-m} \binom{n}{j} S(j,k;r_1)S(n-j,m;r_2)$$

and conclude its inverse relation

$$\binom{n}{i} S(n-i,m;r_2) = \sum_{k=i}^{n-m} \binom{k+m}{k} s(k,i;r_1)S(n,k+m;r_1+r_2).$$

2.18 (*Continuation*). Show that

$$\binom{k+m}{k} C(n,k+m;s,r_1+r_2) = \sum_{j=k}^{n-m} \binom{n}{j} C(j,k;s,r_1)C(n-j,m;s,r_2)$$

and conclude its inverse relation

$$\binom{n}{i} C(n-i,m;s,r_2) =$$
$$\sum_{k=i}^{n-m} \binom{k+m}{k} C(k,i;s^{-1},-r_1 s^{-1})C(n,k+m;s,r_1+r_2).$$

2.19 (*Continuation*). Show that

$$(t^{b+1}D)^n f(t) = \sum_{k=0}^{n} (-1)^{n-k} b^n a^{-k} C(n,k;s,r) t^{bn-ak} (Dt^{a+1})^k f(t),$$

where $s = a/b$, $r = (a+1)/b$, and conclude that

$$(t^{b+1}D)^n f(t) = \sum_{k=0}^{n} (-1)^n b^n C(n,k;s) t^{bn+k} D^k f(t), \quad s = -1/b$$

and

$$(tD)^n f(t) = \sum_{k=0}^{n} S(n,k) t^k D^k f(t).$$

2.20 Consider an urn containing s identical series of coupons, each consisting of m coupons bearing the numbers $1, 2, \ldots, m$ and r additional coupons all bearing the number $m+1$. Suppose that coupons are drawn one after the other, at random, by returning into the urn, after each drawing, the chosen coupon together with another coupon bearing the same number. Show that (a) the probability $p(k;n)$ of drawing exactly k of the m numbers $\{1, 2, \ldots, m\}$ in n drawings is given by

$$p(k,n) = \frac{|C(n,k;-s,-r)|(m+k-1)_k}{(sm+r+n-1)_n}, \quad k = 0, 1, \ldots, n,$$

where $|C(n,k;-s,-r)| = (-1)^n C(n,k;-s,-r)$. Further, show that (b) the probability $q(n;k)$ that n drawings are required until the kth different number, among the m numbers $\{1, 2, \ldots, m\}$, is drawn is given by

$$q(n;k) = \frac{|C(n-1,k-1;-s,-r)|s(m+k-1)_k}{(sm+r+n-1)_n}, \quad n = k, k+1, \ldots.$$

3
GENERALIZED STIRLING AND LAH NUMBERS

3.1 INTRODUCTION

The Stirling numbers of the first and second kind, discussed in Chapter 2, have been extended and generalized in several directions. In this chapter some of these numbers, mainly of combinatorial and statistical interest, are presented. Specifically, the associated Stirling numbers, which are the coefficients of the representation of the Stirling numbers $s(n, n - k)$ and $S(n, n - k)$ as sums of binomials of n, are first examined. The presentation is extended to the r-associated Stirling numbers of the first and second kind, which emerge in the enumeration of permutations and partitions of a finite set into a given number of cycles and subsets, respectively. The discussion of the enumeration of partitions by subsets and permutations by cycles started in Chapter 2 is continued, moving on to the derivation of universal generating functions. Other applications of the associated Stirling numbers, in occupancy and convolutions of truncated distributions, are given in subsequent chapters.

The generalized Stirling numbers of the first and second kind, which are the coefficients of the expansions of generalized factorials, with increments $(a_0, a_1, a_2, \ldots)$, into powers and of powers into generalized factorials, with increments $(a_0, a_1, a_2, \ldots)$, respectively, are then discussed. Further, the generalized Lah numbers, which are the coefficients of the expansion of generalized factorials into generalized factorials with different increments, are examined. Finally, the q-Stirling numbers of the first and second kind, the q-Lah numbers and the generalized q-factorial coefficients are briefly presented as particular cases of the generalized Stirling and Lah numbers.

Combinatorial Methods in Discrete Distributions. By Charalambos A. Charalambides
ISBN 0-471-68027-3 Copyright © 2005 John Wiley & Sons, Inc.

3.2 ASSOCIATED STIRLING NUMBERS

Consider the triangular recurrence relation for the Stirling numbers of the first kind:
$$s(n+1, k) = s(n, k-1) - ns(n, k), \tag{3.1}$$
for $k = 1, 2, \ldots, n+1$, $n = 0, 1, \ldots$, with initial conditions
$$s(0,0) = 1, \ s(n,0) = 0, \ n > 0, \ s(n,k) = 0, \ k > n.$$
For $k = n+1$ it reduces to $s(n+1, n+1) = s(n, n)$, $n = 0, 1, \ldots$, with $s(0,0) = 1$, which implies $s(n, n) = 1$, $n = 0, 1, \ldots$. Further, (3.1) for $k = n$ reduces to
$$s(n+1, n) = s(n, n-1) - n, \ n = 1, 2, \ldots.$$
Thus,
$$s(n, n-1) = \sum_{r=1}^{n-1}\{s(r+1, r) - s(r, r-1)\} = -\sum_{r=1}^{n-1} r$$
and so
$$s(n, n-1) = -\binom{n}{2}, \ n = 2, 3, \ldots.$$
Also, for $k = n-1$, (3.1) reduces to
$$s(n+1, n-1) = s(n, n-2) + n\binom{n}{2}, \ n = 2, 3, \ldots.$$
Consequently,
$$s(n, n-2) = \sum_{r=2}^{n-1}\{s(r+1, r-1) - s(r, r-2)\}$$
$$= \sum_{r=2}^{n-1} r\binom{r}{2} = 2\sum_{r=2}^{n-1}\binom{r}{2} + 3\sum_{r=3}^{n-1}\binom{r}{3},$$
implying
$$s(n, n-2) = 2\binom{n}{3} + 3\binom{n}{4}, \ n = 3, 4, \ldots.$$
In general, $s(n, n-k)$ may be expressed as a sum of binomials of n as
$$s(n, n-k) = \sum_{j=0}^{k} s_2(k+j, j)\binom{n}{k+j}, \ n = k+1, k+2, \ldots. \tag{3.2}$$
A generating function of the coefficients $s_2(n, k)$, $n = 2k, 2k+1, \ldots$, for fixed k, is determined as follows. Expression (3.2) may be written, equivalently, as
$$s(n, k) = \sum_{j=0}^{k} s_2(n-j, k-j)\binom{n}{j}, \ s_2(n, k) = 0, \ k > n/2.$$

Multiplying it by t^k and summing for $k = 0, 1, \ldots, n$, we get

$$(t)_n = \sum_{k=0}^{n} s(n,k) t^k = \sum_{k=0}^{n} \sum_{j=0}^{k} s_2(n-j, k-j) \binom{n}{j} t^k$$

$$= \sum_{j=0}^{n} \binom{n}{j} t^j \sum_{k=j}^{n} s_2(n-j, k-j) t^{k-j},$$

and setting $a_n(t) = \sum_{k=0}^{[n/2]} s_2(n,k) t^k$, we deduce the expression

$$(t)_n = \sum_{j=0}^{n} \binom{n}{j} t^j a_{n-j}(t).$$

Multiplying the last expression by $u^n/n!$ and summing for $n = 0, 1, \ldots$, we find

$$(1+u)^t = \sum_{n=0}^{\infty} \sum_{j=0}^{n} \binom{n}{j} t^j a_{n-j}(t) \frac{u^n}{n!}$$

$$= \sum_{j=0}^{\infty} \frac{(ut)^j}{j!} \sum_{n=j}^{\infty} a_{n-j}(t) \frac{u^{n-j}}{(n-j)!} = e^{ut} \sum_{n=0}^{\infty} a_n(t) \frac{u^n}{n!}.$$

Hence,

$$g_2(t,u) = \sum_{k=0}^{\infty} \sum_{n=2k}^{\infty} s_2(n,k) \frac{u^n}{n!} t^k = \sum_{n=0}^{\infty} \sum_{k=0}^{[n/2]} s_2(n,k) t^k \frac{u^n}{n!} = (1+u)^t e^{-ut}.$$

Further, since

$$(1+u)^t e^{-ut} = e^{t[\log(1+u)-u]} = \sum_{k=0}^{\infty} \frac{[\log(1+u)-u]^k}{k!} t^k,$$

it follows that

$$g_{k,2}(u) = \sum_{n=2k}^{\infty} s_2(n,k) \frac{u^n}{n!} = \frac{[\log(1+u)-u]^k}{k!}, \quad k = 0, 1, \ldots. \qquad (3.3)$$

Similarly, from the triangular recurrence relation for the Stirling numbers of the second kind

$$S(n+1, k) = S(n, k-1) + kS(n, k), \qquad (3.4)$$

for $k = 1, 2, \ldots, n+1$, $n = 0, 1, \ldots$, with initial conditions

$$S(0,0) = 1, \quad S(n,0) = 0, \; n > 0, \quad S(n,k) = 0, \; k > n,$$

we successively deduce $S(n,n) = 1, n = 0, 1, \ldots, S(n, n-1) = \binom{n}{2}, n = 2, 3, \ldots$, and

$$S(n, n-2) = \binom{n}{3} + 3\binom{n}{4}, \quad n = 3, 4, \ldots.$$

In general, $S(n, n-k)$ may be expressed as a sum of binomials of n as

$$S(n, n-k) = \sum_{j=0}^{k} S_2(k+j, j) \binom{n}{k+j}, \quad n = k+1, k+2, \ldots, \quad (3.5)$$

or equivalently,

$$S(n, k) = \sum_{j=0}^{k} S_2(n-j, k-j) \binom{n}{j}, \quad S_2(n, k) = 0, \; k > n/2.$$

Proceeding as in the previous case, a generating function of the coefficients $S_2(n, k)$, $n = 2k, 2k+1, \ldots$, for fixed k, is derived. Thus, multiplying the last expression by $t^k u^n / n!$ and summing for $k = 0, 1, \ldots, n$, $n = 0, 1, \ldots$, we get the bivariate generating function

$$f_2(t, u) = \sum_{k=0}^{\infty} \sum_{n=2k}^{\infty} S_2(n, k) \frac{u^n}{n!} t^k = \sum_{n=0}^{\infty} \sum_{k=0}^{[n/2]} S_2(n, k) t^k \frac{u^n}{n!} = e^{t(e^u - 1 - u)},$$

which implies the required generating function:

$$f_{k,2}(u) = \sum_{n=2k}^{\infty} S_2(n, k) \frac{u^n}{n!} = \frac{(e^u - 1 - u)^k}{k!}, \quad k = 0, 1, \ldots . \quad (3.6)$$

The coefficients $s_2(n, k)$ and $S_2(n, k)$ are called *associated Stirling numbers of the first and second kind*, respectively. The numbers $|s_2(n, k)| = (-1)^{n-k} s_2(n, k)$, with generating function

$$h_{k,2}(u) = \sum_{n=2k}^{\infty} |s_2(n, k)| \frac{u^n}{n!} = \frac{[-\log(1-u) - u]^k}{k!}, \quad k = 0, 1, \ldots,$$

are called *associated signless or absolute Stirling numbers of the first kind*.

More generally, for $r = 1, 2, \ldots$ and $k = 0, 1, \ldots$, let

$$g_{k,r}(u) = \sum_{n=rk}^{\infty} s_r(n, k) \frac{u^n}{n!} = \frac{1}{k!} \left(\log(1+u) - \sum_{j=1}^{r-1} (-1)^{j-1} \frac{u^j}{j} \right)^k, \quad (3.7)$$

and

$$f_{k,r}(u) = \sum_{n=rk}^{\infty} S_r(n, k) \frac{u^n}{n!} = \frac{1}{k!} \left(e^u - \sum_{j=0}^{r-1} \frac{u^j}{j!} \right)^k, \quad (3.8)$$

where for $r = 1$ the sum on the right-hand side of (3.7) is, by convention, equal to zero. The coefficients $s_r(n, k)$ and $S_r(n, k)$ are called *r-associated Stirling numbers of the first and second kind*, respectively. The numbers $|s_r(n,k)| = (-1)^{n-k}s_r(n,k)$, with generating function

$$h_{k,r}(u) = \sum_{n=rk}^{\infty} |s_r(n,k)|\frac{u^n}{n!} = \frac{1}{k!}\left(-\log(1-u) - \sum_{j=1}^{r-1}\frac{u^j}{j}\right)^k, \qquad (3.9)$$

are called *r-associated signless or absolute Stirling numbers of the first kind*. Notice that these numbers for $r = 1$ reduce to the corresponding Stirling numbers $s_1(n,k) = s(n,k)$, $|s_1(n,k)| = |s(n,k)|$ and $S_1(n,k) = S(n,k)$.

Combinatorial interpretations of the r-associated Stirling numbers are given in Section 3.4. Recurrence relations for these numbers, useful for tabulation purposes, are derived in the following theorems.

Theorem 3.1 *The r-associated signless Stirling numbers of the first kind $|s_r(n,k)|$, $n = rk, rk+1, \ldots, k = 0, 1, \ldots$, for fixed r, satisfy the recurrence relation*

$$|s_r(n+1,k)| = n|s_r(n,k)| + (n)_{r-1}|s_r(n-r+1,k-1)|, \qquad (3.10)$$

for $n = rk - 1, rk, \ldots, k = 1, 2, \ldots$, with initial conditions

$$|s_r(0,0)| = 1, \ |s_r(n,0)| = 0, \ n > 0, \ |s_r(n,k)| = 0, \ n < rk.$$

Also,

$$|s_{r+1}(n,k)| = \sum_{j=0}^{k}(-1)^j \frac{(n)_{rj}}{j!r^j}|s_r(n-rj,k-j)|, \qquad (3.11)$$

for $r = 1, 2, \ldots$, with $|s_1(n,k)| = |s(n,k)|$, the signless Stirling number of the first kind.

Proof Differentiating both members of (3.9), we get the expression

$$(1-u)\frac{d}{du}h_{k,r}(u) = u^{r-1}h_{k-1,r}(t),$$

which implies

$$\sum_{n=rk}^{\infty}|s_r(n,k)|\frac{u^{n-1}}{(n-1)!} = \sum_{n=rk}^{\infty}|s_r(n,k)|\frac{u^n}{(n-1)!}$$
$$+ \sum_{n=r(k-1)}^{\infty}|s_r(n,k-1)|\frac{u^{n+r-1}}{n!},$$

or equivalently,

$$\sum_{n=rk-1}^{\infty}|s_r(n+1,k)|\frac{u^n}{n!} = \sum_{n=rk}^{\infty}n|s_r(n,k)|\frac{u^n}{n!}$$
$$+ \sum_{n=rk-1}^{\infty}(n)_{r-1}|s_r(n-r+1,k-1)|\frac{u^n}{n!}.$$

Equating the coefficients of $u^n/n!$ of both members of the last expression, (3.10) is deduced.

Writing (3.9) in the form

$$h_{k,r+1}(u) = \frac{1}{k!}\left(\left[-\log(1-u) - \sum_{i=1}^{r-1}\frac{u^i}{i}\right] - \frac{u^r}{r}\right)^k$$

and using Newton's binomial formula, we get

$$h_{k,r+1}(u) = \frac{1}{k!}\sum_{j=0}^{k}(-1)^j\binom{k}{j}\frac{u^{rj}}{r^j}(k-j)!h_{k-j,r}(u),$$

and so

$$\sum_{n=(r+1)k}^{\infty}|s_{r+1}(n,k)|\frac{u^n}{n!} = \sum_{j=0}^{k}\frac{(-1)^j u^{rj}}{j!r^j}\sum_{m=r(k-j)}^{\infty}|s_r(m,k-j)|\frac{u^m}{m!}$$

$$= \sum_{n=rk}^{\infty}\left\{\sum_{j=0}^{k}(-1)^j\frac{(n)_{rj}}{j!r^j}|s_r(n-rj,k-j)|\right\}\frac{u^n}{n!}.$$

Equating the coefficients of $u^n/n!$, expression (3.11) is deduced.

Theorem 3.2 *The r-associated Stirling numbers of the second kind $S_r(n,k)$, $n = rk, rk+1, \ldots, k = 0, 1, \ldots$, for fixed r, satisfy the recurrence relation*

$$S_r(n+1,k) = kS_r(n,k) + \binom{n}{r-1}S_r(n-r+1,k-1), \qquad (3.12)$$

for $n = rk-1, rk, \ldots, k = 1, 2, \ldots$, with initial conditions

$$S_r(0,0) = 1, \; S_r(n,0) = 0, \; n > 0, \; S_r(n,k) = 0, \; n < rk.$$

Also,

$$S_{r+1}(n,k) = \sum_{j=0}^{k}(-1)^j\frac{(n)_{rj}}{j!(r!)^j}S_r(n-rj,k-j), \qquad (3.13)$$

for $r = 1, 2, \ldots$, with $S_1(n,k) = S(n,k)$, the Stirling number of the second kind.

Proof Differentiation of both members of (3.8) yields the expression

$$\frac{d}{dt}f_{k,r}(u) = kf_{k,r}(u) + \frac{u^{r-1}}{(r-1)!}f_{k-1,r}(u),$$

which implies

$$\sum_{n=rk}^{\infty}S_r(n,k)\frac{u^{n-1}}{(n-1)!} = \sum_{n=rk}^{\infty}S_r(n,k)\frac{u^n}{(n-1)!}$$

$$+ \sum_{n=r(k-1)}^{\infty}S_r(n,k-1)\frac{u^{n+r-1}}{(r-1)!n!}.$$

Hence,

$$\sum_{n=rk-1}^{\infty} S_r(n+1,k)\frac{u^n}{n!} = \sum_{n=rk}^{\infty} nS_r(n,k)\frac{u^n}{n!}$$
$$+ \sum_{n=rk-1}^{\infty} \binom{n}{r-1} S_r(n-r+1,k-1)\frac{u^n}{n!},$$

and equating the coefficients of $u^n/n!$ of both members, (3.12) is established.

Writing (3.8) in the form

$$f_{k,r+1}(u) = \frac{1}{k!}\left(\left[e^u - \sum_{i=0}^{r-1} \frac{u^i}{i!}\right] - \frac{u^r}{r!}\right)^k$$

and using Newton's binomial formula, we get

$$f_{k,r+1}(u) = \frac{1}{k!}\sum_{j=0}^{k}(-1)^j \binom{k}{j} \frac{u^{rj}}{(r!)^j}(k-j)! f_{k-j,r}(u).$$

Thus,

$$\sum_{n=(r+1)k}^{\infty} S_{r+1}(n,k)\frac{u^n}{n!} = \sum_{j=0}^{k} \frac{(-1)^j u^{rj}}{j!(r!)^j} \sum_{m=r(k-j)}^{\infty} S_r(m, k-j)\frac{u^m}{m!}$$
$$= \sum_{n=rk}^{\infty} \left\{ \sum_{j=0}^{k}(-1)^j \frac{(n)_{rj}}{j!(r!)^j} S_r(n-rj, k-j) \right\} \frac{u^n}{n!},$$

and equating the coefficients of $u^n/n!$, (3.13) is deduced.

Tables of the r-associated Stirling numbers can be constructed by using recurrence relations (3.10) and (3.12). The associated signless Stirling numbers of the first kind $|s_2(n,k)|$, $n = 2k, 2k+1, \ldots$, $k = 0, 1, \ldots$, satisfy the recurrence relation

$$|s_2(n+1,k)| = n|s_2(n,k)| + n|s_2(n-1,k-1)|,$$

for $k = 1, 2, \ldots, [(n+1)/2]$, $n = 1, 2, \ldots$, with initial conditions

$$|s_2(0,0)| = 1, \; |s_2(n,0)| = 0, \; n > 0, \; |s_2(n,k)| = 0, \; k > n/2.$$

Using it, Table 3.1, for $k = 1, 2, \ldots, [n/2]$, $n = 2, 3, \ldots, 12$, is constructed. Also, the associated Stirling numbers of the second kind $S_2(n,k)$, $n = 2k, 2k+1, \ldots$, $k = 0, 1, \ldots$, satisfy the recurrence relation

$$S_2(n+1,k) = kS_2(n,k) + nS_2(n-1,k-1),$$

for $k = 1, 2, \ldots, [(n+1)/2]$, $n = 1, 2, \ldots$, with initial conditions

$$S_2(0,0) = 1, \; S_2(n,0) = 0, \; n > 0, \; S_2(n,k) = 0, \; k > n/2.$$

Using it, Table 3.2, for $k = 1, 2, \ldots, [n/2]$, $n = 2, 3, \ldots, 14$, is constructed.

Table 3.1 Associated signless Stirling numbers of the first kind $|s_2(n,k)|$

k \ n	1	2	3	4	5	6
2	1					
3	2					
4	6	3				
5	24	20				
6	120	130	15			
7	720	924	210			
8	5040	7308	2380	105		
9	40320	64224	26432	2520		
10	362880	623376	303660	44100	945	
11	3628800	6636960	3678840	705320	34650	
12	39916800	76998240	47324376	11098780	866250	10395

Table 3.2 Associated Stirling numbers of the second kind $S_2(n,k)$

k \ n	1	2	3	4	5	6	7
2	1						
3	1						
4	1	3					
5	1	10					
6	1	25	15				
7	1	56	105				
8	1	119	490	105			
9	1	246	1918	1260			
10	1	501	6825	9450	945		
11	1	1012	22935	56980	17325		
12	1	2035	74316	302995	190575	10395	
13	1	4082	235092	1487200	1636635	270270	
14	1	8177	731731	6914908	12122110	4099095	135135

Example 3.1 Suppose that n distinguishable balls are distributed into k distinguishable urns. Find the number of different distributions in which each urn is occupied by at least r balls.

The enumerator for occupancy of the jth urn is $\sum_{i=r}^{\infty} x_j^i u^i/i!$, $j = 1, 2, \ldots, k$, and so the enumerator for occupancy of the k urns is given by

$$\prod_{j=1}^{k} \left(\sum_{i=r}^{\infty} \frac{x_j^i u^i}{i!} \right) = \prod_{j=1}^{k} \left(e^{x_j u} - \sum_{i=0}^{r-1} \frac{x_j^i u^i}{i!} \right).$$

Setting $x_j = 1$, $j = 1, 2, \ldots, k$, we deduce the exponential generating function for occupancy of the k urns as

$$\sum_{n=rk}^{\infty} A_r(n,k) \frac{u^n}{n!} = \left(e^u - \sum_{i=0}^{r-1} \frac{u^i}{i!}\right)^k.$$

Comparing it with the generating function (3.8), we deduce that the number of different distributions of n distinguishable balls into k distinguishable urns in which each urn is occupied by at least r balls equals $A_r(n,k) = k! S_r(n,k)$.

3.3 ASSOCIATED GENERALIZED FACTORIAL COEFFICIENTS

Consider the triangular recurrence relation of the generalized factorial coefficients:

$$C(n+1, k; s) = (sk - n)C(n, k; s) + sC(n, k-1; s), \qquad (3.14)$$

for $k = 1, 2, \ldots, n+1$, $n = 0, 1, \ldots$, with initial conditions

$$C(0, 0; s) = 1, \ C(n, 0; s) = 0, \ n > 0, \ C(n, k; s) = 0, \ k > n.$$

Setting $k = n + 1$, and since $C(n, n+1; s) = 0$, it reduces to

$$C(n+1, n+1; s) = s\, C(n, n; s), \ n = 0, 1, \ldots,$$

with $C(0, 0; s) = 1$. Hence, $C(n, n; s) = s^n$. Further, for $k = n$, (3.14) reduces to

$$C(n+1, n; s) = sC(n, n-1; s) + (s)_2 s^{n-1} n, \ n = 1, 2, \ldots.$$

Thus,

$$C(n, n-1; s) = \sum_{r=1}^{n-1} \{s^{n-r-1} C(r+1, r; s) - s^{n-r} C(r, r-1; s)\}$$

$$= (s)_2 s^{n-2} \sum_{r=1}^{n-1} r$$

and so

$$C(n, n-1; s) = (s)_2 s^{n-2} \binom{n}{2}, \ n = 2, 3, \ldots.$$

Also, for $k = n - 1$, (3.14) reduces to

$$C(n+1, n-1; s) = sC(n, n-2; s) + (s)_3 s^{n-2} \binom{n}{2}$$

$$+ 3(s)_2^2 s^{n-3} \binom{n}{3}, \ n = 2, 3, \ldots.$$

Therefore,

$$C(n, n-2; s) = \sum_{r=2}^{n-1} \{s^{n-r-1} C(r+1, r-1; s) - s^{n-r} C(r, r-2; s)\}$$

$$= (s)_3 s^{n-3} \sum_{r=2}^{n-1} \binom{r}{2} + 3(s)_2^2 s^{n-4} \sum_{r=3}^{n-1} \binom{r}{3}$$

and so

$$C(n, n-2; s) = (s)_3 s^{n-3} \binom{n}{3} + 3(s)_2^2 s^{n-4} \binom{n}{4}, \quad n = 3, 4, \ldots.$$

In general, $C(n, n-k; s)$ may be expressed as a sum of binomials of n as

$$C(n, n-k; s) = \sum_{j=0}^{k} C_2(k+j, j; s) s^{n-k-j} \binom{n}{k+j}, \qquad (3.15)$$

for $n = k+1, k+2, \ldots$. A generating function of the coefficients $C_2(n, k; s)$, $n = 2k, 2k+1, \ldots$, for fixed k, is obtained as follows. Expression (3.15) may be written, equivalently, as

$$C(n, k; s) = \sum_{j=0}^{k} C_2(n-j, k-j; s) s^j \binom{n}{j}, \quad C_2(n, k; s) = 0, \ k > n/2.$$

Multiplying it by t^k and summing for $k = 0, 1, \ldots, n$, we get

$$\sum_{k=0}^{n} C(n, k; s) t^k = \sum_{k=0}^{n} \sum_{j=0}^{k} C_2(n-j, k-j; s) s^j \binom{n}{j} t^k$$

$$= \sum_{j=0}^{n} \binom{n}{j} (st)^j \sum_{k=j}^{n} C_2(n-j, k-j; s) t^{k-j},$$

and setting $a_n(t; s) = \sum_{k=0}^{[n/2]} C_2(n, k; s) t^k$, we deduce the expression

$$\sum_{k=0}^{n} C(n, k; s) t^k = \sum_{j=0}^{n} \binom{n}{j} (st)^j a_{n-j}(t; s).$$

Multiplying the last expression by $u^n/n!$ and summing for $n = 0, 1, \ldots$, we find

$$\sum_{n=0}^{\infty} \sum_{k=0}^{n} C(n, k; s) t^k \frac{u^n}{n!} = \sum_{n=0}^{\infty} \sum_{j=0}^{n} \binom{n}{j} (st)^j a_{n-j}(t; s) \frac{u^n}{n!}$$

$$= \sum_{j=0}^{\infty} \frac{(sut)^j}{j!} \sum_{n=j}^{\infty} a_{n-j}(t; s) \frac{u^{n-j}}{(n-j)!}.$$

Hence,

$$f_2(t,u;s) = \sum_{k=0}^{\infty} \sum_{n=2k}^{\infty} C_2(n,k;s)\frac{u^n}{n!}t^k = \sum_{n=0}^{\infty} \sum_{k=0}^{[n/2]} C_2(n,k;s)t^k \frac{u^n}{n!}$$

$$= e^{-sut} \sum_{n=0}^{\infty} \sum_{k=0}^{n} C(n,k;s)t^k \frac{u^n}{n!} = e^{t[(1+u)^s - 1 - su]}.$$

Further, expanding the function $e^{t[(1+u)^s - 1 - u]}$ into powers of t and equating the coefficients of t^k in the resulting expression, it follows that

$$f_{k,2}(u;s) = \sum_{n=2k}^{\infty} C_2(n,k;s)\frac{u^n}{n!} = \frac{[(1+u)^s - 1 - su]^k}{k!}, \qquad (3.16)$$

for $k = 0, 1, \ldots$. The coefficients $C_2(n,k;s)$ are called *associated generalized factorial coefficients*.

More generally, for $r = 1, 2, \ldots$, and $k = 0, 1, \ldots$, let

$$f_{k,r}(u;s) = \sum_{n=rk}^{\infty} C_r(n,k;s)\frac{u^n}{n!} = \frac{1}{k!}\left((1+u)^s - \sum_{j=0}^{r-1}\binom{s}{j}u^j\right)^k. \qquad (3.17)$$

The coefficients $C_r(n,k;s)$ are called *r-associated generalized factorial coefficients*. For $r = 1$ these numbers reduce to $C_1(n,k;s) = C(n,k;s)$. Recurrence relations for $C_r(n,k;s)$ are given in the following theorem.

Theorem 3.3 *The r-associated generalized factorial coefficients $C_r(n,k;s)$, $n = rk, rk+1, \ldots, k = 0, 1, \ldots,$ for fixed r, satisfy the recurrence relation*

$$C_r(n+1,k;s) = (sk-n)C_r(n,k;s) + \binom{n}{r-1}(s)_r C_r(n-r+1,k-1;s), \quad (3.18)$$

for $n = rk - 1, rk, \ldots, k = 1, 2, \ldots,$ with initial conditions

$$C_r(0,0;s) = 1, \ C_r(n,0;s) = 0, \ n > 0, \ C_r(n,k;s) = 0, \ n < rk.$$

Also,

$$C_{r+1}(n,k;s) = \sum_{j=0}^{k}(-1)^j \frac{(n)_{rj}}{j!}\binom{s}{r}^j C_r(n-rj, k-j; s), \qquad (3.19)$$

for $r = 1, 2, \ldots,$ with $C_1(n,k;s) = C(n,k;s)$, the generalized factorial coefficient.

Proof Differentiating both members of (3.17) with respect to u and using Pascal's triangle, we get for the generating function $f_{k,r}(u;s)$ the expression

$$(1+u)\frac{d}{du}f_{k,r}(u;s) = skf_{k,r}(u;s) + \frac{(s)_r}{(r-1)!}u^{r-1}f_{k-1,r}(u;s),$$

which expanded into powers of u yields

$$\sum_{n=rk}^{\infty} C_r(n,k;s)\frac{u^{n-1}}{(n-1)!} + \sum_{n=rk}^{\infty} C_r(n,k;s)\frac{u^n}{(n-1)!}$$

$$= \sum_{n=rk}^{\infty} skC_r(n,k;s)\frac{u^n}{n!} + \frac{(s)_r}{(r-1)!}\sum_{n=r(k-1)}^{\infty} C_r(n,k-1;s)\frac{u^{n+r-1}}{n!}.$$

Equating the coefficients of $u^n/n!$ in both members, (3.18) is readily deduced.

Writing (3.17) in the form

$$f_{k,r+1}(u;s) = \frac{1}{k!}\left(\left[(1+u)^s - \sum_{j=0}^{r-1}\binom{s}{i}u^i\right] - \binom{s}{r}u^r\right)^k$$

and using Newton's binomial formula, we get the expression

$$f_{k,r+1}(u;s) = \frac{1}{k!}\sum_{j=0}^{k}(-1)^j\binom{k}{j}\binom{s}{r}^j u^{rj}(k-j)!f_{k-j,r}(u;s),$$

which expanded into powers of u yields

$$\sum_{n=(r+1)k}^{\infty} C_{r+1}(n,k;s)\frac{u^m}{n!} = \sum_{j=0}^{k}\frac{(-1)^j}{j!}\binom{s}{r}^j \sum_{m=r(k-j)}^{\infty} C_r(m,k-j;s)\frac{u^{m+rj}}{m!}$$

$$= \sum_{n=rk}^{\infty}\left\{\sum_{j=0}^{k}(-1)^j\frac{(n)_{rj}}{j!}\binom{s}{r}^j C_r(n-rj,k-j;s)\right\}\frac{u^n}{n!}.$$

Equating the coefficients of $u^n/n!$, (3.19) is established.

Tables of the r-associated generalized factorial coefficients can be constructed by using recurrence relation (3.18). The associated generalized factorial coefficients $C_2(n,k;s)$, $n = 2k, 2k+1, \ldots$, $k = 0, 1, \ldots$, satisfy the recurrence relation

$$C_2(n+1,k;s) = (sk-n)C_2(n,k;s) + n(s)_2 C_2(n-1,k-1;s),$$

for $k = 1, 2, \ldots, [(n+1)/2]$, $n = 1, 2, \ldots$, with initial conditions

$$C_2(0,0;s) = 1, \quad C_2(n,0;s) = 0, \quad n > 0, \quad C_2(n,k;s) = 0, \quad k > n/2.$$

Using it, Table 3.3, for $k = 1, 2, \ldots, [n/2]$, $n = 2, 3, \ldots, 8$, is constructed.

Table 3.3 Associated generalized factorial coefficients $C_2(n,k;s)$

k n	1	2	3	4
2	$(s)_2$			
3	$(s)_3$			
4	$(s)_4$	$3(s)_2^2$		
5	$(s)_5$	$10(s)_2(s)_3$		
6	$(s)_6$	$15(s)_2(s)_4 + 10(s)_3^2$	$15(s)_2^3$	
7	$(s)_7$	$21(s)_2(s)_5 + 35(s)_3(s)_4$	$210(s)_2^2(s)_3$	
8	$(s)_8$	$28(s)_2(s)_6 + 42(s)_3(s)_5$ $+35(s)_4^2$	$210(s)_2^2(s)_4$ $+280(s)_2(s)_3^2$	$105(s)_2^4$

Example 3.2 Suppose that n like balls are distributed into k distinguishable urns, each with s distinguishable cells. Find the number of different distributions in which each urn is occupied by at least r balls.

Consider first the case of cells with capacity limited to one ball. Then the enumerator for occupancy of the jth urn is

$$\sum_{i=r}^{s} \binom{s}{i} x_j^i u^i, \quad j=1,2,\ldots,k,$$

and so the enumerator for occupancy of the k urns is given by

$$\prod_{j=1}^{k}\left[\sum_{i=r}^{s}\binom{s}{i}x_j^i u^i\right] = \prod_{j=1}^{k}\left[(1+x_j u)^s - \sum_{i=0}^{r-1}\binom{s}{i}x_j^i u^i\right].$$

Setting $x_j = 1$, $j = 1, 2, \ldots, k$, we deduce the generating function for occupancy of the k urns as

$$\sum_{n=rk}^{sk} A_r(n,k;s) u^n = \left((1+u)^s - \sum_{i=0}^{r-1}\binom{s}{i}u^i\right)^k.$$

A comparison of it with the generating function (3.17) implies that the number of different distributions of n like balls into k distinguishable urns, each with s distinguishable cells of capacity limited to one ball, so that each urn is occupied by r balls equals $A_r(n,k;s) = k! C_r(n,k;s)/n!$.

Further, assume that the cells are of unlimited capacity. Then the enumerator for occupancy of the k urns is given by

$$\prod_{j=1}^{k}\left[\sum_{i=r}^{\infty}\binom{s+i-1}{i}x_j^i u^i\right] = \prod_{j=1}^{k}\left[(1-x_j u)^{-s} - \sum_{i=0}^{r-1}\binom{s+i-1}{i}x_j^i u^i\right].$$

The generating function for occupancy of the k urns is obtained by putting $x_j = 1$, $j = 1, 2, \ldots, k$, as

$$\sum_{n=rk}^{\infty} B_r(n,k;s)u^n = \left((1-u)^{-s} - \sum_{i=0}^{r-1}\binom{-s}{i}(-u)^i\right)^k.$$

Comparing it with the generating function (3.17), we conclude that the number of different distributions of n like balls into k distinguishable urns, each with s distinguishable cells of unlimited capacity, so that each urn is occupied by r balls equals $B_r(n,k;s) = k!|C_r(n,k;-s)|/n!$, where $|C_r(n,k;-s)| = (-1)^n C_r(n,k;-s)$.

3.4 UNIVERSAL GENERATING FUNCTIONS

The discussion of the enumeration of partitions and permutations of a finite set into a given number of subsets and cycles, respectively, started in Section 2.5, is continued in this section with the derivation of suitable universal generating functions. A generating function for the number of partitions of a finite set into subsets of specified or unspecified number, such that the number of subsets of any specific cardinality belongs to a set of nonnegative integers may be constructed. Specifically, let $A = (a_{i,j})$, $i = 1, 2, \ldots, j = 0, 1, \ldots$, be an infinite matrix with elements $a_{i,j} = 0$ or 1. The matrix A determines a sequence K_i, $i = 1, 2, \ldots$, of subsets of nonnegative integers and vice versa as follows: For a specific $i = 1, 2, \ldots$, $K_i = \{j : a_{i,j} = 1\}$ is the set of indices j of the elements of the ith row of the matrix A that are equal to 1.

Let us denote by $S(n,k;A)$ the number of partitions of a finite set of n elements into k subsets, such that the number of subsets containing i elements belongs to the set $K_i = \{j : a_{i,j} = 1\}$, for $k = 1, 2, \ldots, n$ and $n = 1, 2, \ldots$. In addition, let $S(0,0;A) = 1$ and $S(n,0;A) = 0$, for $n = 0, 1, \ldots$. Also, let us denote by $B(n;A) = \sum_{k=0}^{n} S(n,k;A)$ the total number of partitions of a finite set of n elements, such that the number of subsets containing j elements belongs to the set $K_i = \{j : a_{i,j} = 1\}$, for $n = 1, 2, \ldots$ and set $B(0;A) = 1$. Note that in particular for $a_{i,j} = 1$, $i = 1, 2, \ldots, j = 0, 1, \ldots$, whence $K_i = \{0, 1, \ldots\}$, $i = 1, 2, \ldots$, $S(n,k;A) = S(n,k)$, the Stirling number of the second kind and $B(n;A) = B_n$, the Bell number.

A bivariate generating function for the sequence $S(n,k;A)$, $k = 0, 1, \ldots, n$, $n = 0, 1, \ldots$, is derived in the next theorem. In particular, a generating function of the sequence $B(n;A)$, $n = 0, 1, \ldots$, is deduced.

Theorem 3.4 *Let $A = (a_{i,j})$, $i = 1, 2, \ldots, j = 0, 1, \ldots$, be an infinite matrix with elements $a_{i,j} = 0$ or 1 and $S(n,k;A)$ be the number of partitions of a finite set of*

n elements into k subsets, such that the number k_i of subsets containing i elements belongs to the set $K_i = \{j : a_{i,j} = 1\}$, $i = 1, 2, \ldots$. Then

$$F(t, u; \boldsymbol{A}) = \sum_{n=0}^{\infty} \sum_{k=0}^{n} S(n, k; \boldsymbol{A}) u^k \frac{t^n}{n!} = \prod_{i=1}^{\infty} \left\{ \sum_{j=0}^{\infty} a_{i,j} \frac{u^j}{j!} \left(\frac{t^i}{i!} \right)^j \right\}. \quad (3.20)$$

Also, if $B(n; \boldsymbol{A})$ is the total number of partitions of a finite set of n elements, such that the number k_i of subsets containing i elements belongs to the set $K_i = \{j : a_{i,j} = 1\}$, $i = 1, 2, \ldots$, then

$$F(t; \boldsymbol{A}) = \sum_{n=0}^{\infty} B(n; \boldsymbol{A}) \frac{t^n}{n!} = \prod_{i=1}^{\infty} \left\{ \sum_{j=0}^{\infty} \frac{a_{i,j}}{j!} \left(\frac{t^i}{i!} \right)^j \right\}. \quad (3.21)$$

Proof The number $S(n, k; \boldsymbol{A})$, on using Theorem 2.22, is given by the sum

$$S(n, k; \boldsymbol{A}) = \sum \frac{n!}{k_1!(1!)^{k_1} k_2!(2!)^{k_2} \cdots k_n!(n!)^{k_n}},$$

where the summation is extended over all nonnegative integer solutions of the equations

$$k_1 + 2k_2 + \cdots + nk_n = n, \quad k_1 + k_2 + \cdots + k_n = k, \quad (3.22)$$

with

$$k_i \in K_i = \{j : a_{i,j} = 1\}, \quad i = 1, 2, \ldots. \quad (3.23)$$

Consequently,

$$F(t, u; \boldsymbol{A}) = \sum_{n=0}^{\infty} \sum_{k=0}^{n} S(n, k; \boldsymbol{A}) u^k \frac{t^n}{n!} = \sum_{n=0}^{\infty} \sum_{k=0}^{n} \left\{ \sum \prod_{i=1}^{n} \frac{u^{k_i}}{k_i!} \left(\frac{t^i}{i!} \right)^{k_i} \right\},$$

where in the inner sum the summation is extended over all nonnegative integer solutions of the equations (3.22) that satisfy (3.23). Introducing the elements $a_{i,j}$ of the matrix $\boldsymbol{A}$ in this sum, we get the expression

$$F(t, u; \boldsymbol{A}) = \sum_{n=0}^{\infty} \sum_{k=0}^{n} \left\{ \sum \prod_{i=1}^{n} a_{i,k_i} \frac{u^{k_i}}{k_i!} \left(\frac{t^i}{i!} \right)^{k_i} \right\},$$

where in the inner sum the summation is extended over all nonnegative integer solutions of the equations (3.22). Notice that this transformation of the inner sum incorporates the condition (3.23), adding zero terms and thus eliminating the concern of selecting the solutions of (3.22) that satisfy (3.23). Since this inner sum is summed over all $k = 0, 1, \ldots, n$, $n = 0, 1, \ldots$, it follows that

$$F(t, u; \boldsymbol{A}) = \prod_{i=1}^{\infty} \left\{ \sum_{k_i=0}^{\infty} a_{i,k_i} \frac{u^{k_i}}{k_i!} \left(\frac{t^i}{i!} \right)^{k_i} \right\} = \prod_{i=1}^{\infty} \left\{ \sum_{j=0}^{\infty} a_{i,j} \frac{u^j}{j!} \left(\frac{t^i}{i!} \right)^j \right\}.$$

Putting $u = 1$ in (3.20) and since $B(n; \boldsymbol{A}) = \sum_{k=0}^{n} S(n, k; \boldsymbol{A})$, we deduce (3.21).

Example 3.3 *Partitions into subsets of sizes no smaller than r.* Let $S_r(n,k)$ be the number of partitions of a finite set of n elements into k subsets of sizes no smaller than r. The generating function

$$F_r(t,u) = \sum_{n=0}^{\infty} \sum_{k=0}^{n} S_r(n,k) u^k \frac{t^n}{n!},$$

may be deduced from (3.20) as follows. The number k_i, of subsets containing i elements, belongs to the set $K_i = \{0\}$, for $i = 1, 2, \ldots, r-1$ and to the set $K_i = \{0, 1, \ldots\}$, for $i = r, r+1, \ldots$. Consequently, $a_{i,0} = 1$, $i = 1, 2, \ldots$, $a_{i,j} = 0$, $i = 1, 2, \ldots, r-1$, $a_{i,j} = 1$, $i = r, r+1, \ldots$, $j = 1, 2, \ldots$, and so, by (3.20),

$$F_r(t,u) = \sum_{n=0}^{\infty} \sum_{k=0}^{n} S_r(n,k) u^k \frac{t^n}{n!} = \prod_{i=r}^{\infty} \exp\left(\frac{ut^i}{i!}\right)$$

$$= \exp\left\{u\left(e^t - \sum_{i=0}^{r-1} \frac{t^i}{i!}\right)\right\}.$$

The generating function

$$f_{k,r}(t) = \sum_{n=rk}^{\infty} S_r(n,k) \frac{t^n}{n!},$$

since $F_r(t,u) = \sum_{k=0}^{\infty} f_{k,r}(t) u^k$, is given by

$$f_{k,r}(t) = \sum_{n=rk}^{\infty} S_r(n,k) \frac{t^n}{n!} = \frac{1}{k!}\left(e^t - \sum_{i=0}^{r-1} \frac{t^i}{i!}\right)^k.$$

Therefore, by virtue of (3.8), the number $S_r(n,k)$ of partitions of a finite set of n elements into k subsets of sizes no smaller than r equals the r-associated Stirling number of the second kind.

Example 3.4 *Partitions into subsets of even sizes and into subsets of odd sizes.* (a) Let A_n be the number of partitions of a finite set of n elements, into subsets of even sizes. The generating function

$$F_0(t) = \sum_{n=0}^{\infty} A_n \frac{t^n}{n!}$$

may be deduced from (3.21) as follows. The number k_{2i}, of subsets each containing $2i$ elements, belongs to the set $K_{2i} = \{0, 1, \ldots\}$, while the number k_{2i-1}, of subsets each containing $2i-1$ elements, belongs to the set $K_{2i-1} = \{0\}$, $i = 1, 2, \ldots$. Consequently, $a_{i,0} = 1$, $a_{2i,j} = 1$, $a_{2i-1,j} = 0$, $i = 1, 2, \ldots$, $j = 1, 2, \ldots$, and so, according to (3.21),

$$F_0(t) = \sum_{n=0}^{\infty} A_n \frac{t^n}{n!} = \exp\{(\cosh t - 1)\},$$

where $\cosh t = (e^t + e^{-t})/2$ is the hyperbolic cosine. Differentiating this generating function, we get the differential equation

$$\frac{dF_0(t)}{dt} = F_0(t) \sinh t,$$

where $\sinh t = (e^t - e^{-t})/2$ is the hyperbolic sine. Expanding it into powers of t, it follows that

$$\sum_{n=1}^{\infty} A_n \frac{t^{n-1}}{(n-1)!} = \left(\sum_{r=0}^{\infty} A_r \frac{t^r}{r!}\right) \cdot \left(\sum_{k=1}^{\infty} \frac{t^{2k-1}}{(2k-1)!}\right)$$

$$= \sum_{n=1}^{\infty} \left\{\sum_{k=1}^{[n/2]} \binom{n-1}{2k-1} A_{n-2k}\right\} \frac{t^{n-1}}{(n-1)!}.$$

Hence, the numbers A_n, $n = 0, 1, \ldots$, satisfy the recurrence relation

$$A_n = \sum_{k=1}^{[n/2]} \binom{n-1}{2k-1} A_{n-2k},$$

with initial conditions $A_0 = 1$, $A_1 = 0$.

(b) Let C_n be the number of partitions of a finite set of n elements, into subsets of odd sizes. The generating function

$$F_1(t) = \sum_{n=0}^{\infty} C_n \frac{t^n}{n!}$$

may also be deduced from (3.21). Specifically, the number k_{2i-1}, of subsets each containing $2i - 1$ elements, belongs to the set $K_{2i-1} = \{0, 1, \ldots\}$, while the number k_{2i}, of subsets each containing $2i$ elements, belongs to the set $K_{2i} = \{0\}$, $i = 1, 2, \ldots$. Consequently, $a_{i,0} = 1$, $a_{2i-1,j} = 1$, $a_{2i,j} = 0$, $i = 1, 2, \ldots$, $j = 1, 2, \ldots$, and so, according to (3.21),

$$F_1(t) = \sum_{n=0}^{\infty} C_n \frac{t^n}{n!} = \exp\{\sinh t\}.$$

Differentiating this generating function, we get the differential equation

$$\frac{dF_1(t)}{dt} = F_1(t) \cosh t,$$

which expanded into powers of t yields

$$\sum_{n=1}^{\infty} C_n \frac{t^{n-1}}{(n-1)!} = \left(\sum_{r=0}^{\infty} C_r \frac{t^r}{r!}\right) \cdot \left(\sum_{k=0}^{\infty} \frac{t^{2k}}{(2k)!}\right)$$

$$= \sum_{n=1}^{\infty} \left\{\sum_{k=0}^{[(n-1)/2]} \binom{n-1}{2k} C_{n-2k-1}\right\} \frac{t^{n-1}}{(n-1)!}.$$

Hence, the numbers C_n, $n = 0, 1, \ldots$, satisfy the recurrence relation

$$C_n = \sum_{k=0}^{[(n-1)/2]} \binom{n-1}{2k-1} C_{n-2k-1},$$

with initial condition $C_0 = 1$.

A generating function for the number of permutations of a finite set that are decomposed into cycles of specified or unspecified number, such that the number of cycles of any specific length belongs to a set of nonnegative integers, analogous to that for the number of partitions, may be similarly deduced.

Theorem 3.5 *Let $\boldsymbol{A} = (a_{i,j})$, $i = 1, 2, \ldots$, $j = 0, 1, \ldots$, be an infinite matrix with elements $a_{i,j} = 0$ or 1 and $c(n, k; \boldsymbol{A})$ be the number of permutations of a finite set of n elements that are decomposed into k cycles, such that the number k_i of cycles of length i belongs to the set $K_i = \{j : a_{i,j} = 1\}$, $i = 1, 2, \ldots$. Then*

$$G(t, u; \boldsymbol{A}) = \sum_{n=0}^{\infty} \sum_{k=0}^{n} c(n, k; \boldsymbol{A}) u^k \frac{t^n}{n!} = \prod_{i=1}^{\infty} \left\{ \sum_{j=0}^{\infty} a_{i,j} \frac{u^j}{j!} \left(\frac{t^i}{i}\right)^j \right\}. \quad (3.24)$$

Also, if $b(n; \boldsymbol{A})$ is the total number of permutations of a finite set of n elements that are decomposed into cycles, such that the number k_i of cycles of length i belongs to the set $K_i = \{j : a_{i,j} = 1\}$, $i = 1, 2, \ldots$, then

$$G(t; \boldsymbol{A}) = \sum_{n=0}^{\infty} b(n; \boldsymbol{A}) \frac{t^n}{n!} = \prod_{i=1}^{\infty} \left\{ \sum_{j=0}^{\infty} \frac{a_{i,j}}{j!} \left(\frac{t^i}{i}\right)^j \right\}. \quad (3.25)$$

Example 3.5 *Permutations into cycles of length no smaller than r.* Let $c_r(n, k)$ be the number of permutations of a finite set of n elements that are decomposed into k cycles of length no smaller than r. The generating function

$$H_r(t, u) = \sum_{n=0}^{\infty} \sum_{k=0}^{n} c_r(n, k) u^k \frac{t^n}{n!}$$

may be deduced from (3.24) as follows. The number k_i, of cycles of length i, belongs to the set $K_i = \{0\}$, for $i = 1, 2, \ldots, r-1$ and to the set $K_i = \{0, 1, \ldots\}$, for $i = r, r+1, \ldots$. Consequently, $a_{i,0} = 1, i = 1, 2, \ldots, a_{i,j} = 0, i = 1, 2, \ldots, r-1$, $a_{i,j} = 1, i = r, r+1, \ldots, j = 1, 2, \ldots$, and so, by (3.24),

$$H_r(t, u) = \sum_{n=0}^{\infty} \sum_{k=0}^{n} c_r(n, k) u^k \frac{t^n}{n!} = \prod_{i=r}^{\infty} \exp\left(\frac{ut^i}{i}\right)$$

$$= \exp\left\{ u\left(-\log(1-t) - \sum_{i=1}^{r-1} \frac{t^i}{i}\right) \right\}.$$

The generating function

$$h_{k,r}(t) = \sum_{n=rk}^{\infty} c_r(n,k)\frac{t^n}{n!},$$

since $H_r(t,u) = \sum_{k=0}^{\infty} h_{k,r}(t)u^k$, is given by

$$h_{k,r}(t) = \sum_{n=rk}^{\infty} c_r(n,k)\frac{t^n}{n!} = \frac{1}{k!}\left(-\log(1-t) - \sum_{i=1}^{r-1}\frac{t^i}{i}\right)^k.$$

Therefore, by virtue of (3.9), the number $c_r(n,k)$ of permutations of a finite set of n elements that are decomposed into k cycles of length no smaller than r equals $|s_r(n,k)|$, the r-associated signless Stirling number of the first kind.

3.5 GENERALIZED STIRLING NUMBERS

Consider the generalized factorial of t of order n and increments $\boldsymbol{a_0} = (a_0, a_1, a_2, \ldots)$:

$$(t)_{n,\boldsymbol{a_0}} = (t-a_0)(t-a_1)\cdots(t-a_{n-1}),\ n=1,2,\ldots,\ (t)_{0,\boldsymbol{a_0}}=1.$$

Note that this is a polynomial of t of degree n. Executing the multiplications and arranging the terms in ascending order of powers of t, we get

$$(t)_{n,\boldsymbol{a_0}} = \sum_{k=0}^{n} s(n,k;\boldsymbol{a_0})t^k,\ n=0,1,\ldots. \tag{3.26}$$

Inversely, the nth power of t may be expressed in the form of a polynomial of generalized factorials of t of order n and increments $\boldsymbol{a_0} = (a_0, a_1, a_2, \ldots)$. Specifically,

$$t^n = \sum_{k=0}^{n} S(n,k;\boldsymbol{a_0})(t)_{k,\boldsymbol{a_0}},\ n=0,1,\ldots. \tag{3.27}$$

The coefficients $s(n,k;\boldsymbol{a_0})$ and $S(n,k;\boldsymbol{a_0})$ in expansions (3.26) and (3.27) are called *generalized Stirling numbers of the first and second kind*, respectively.

Clearly, $s(0,0;\boldsymbol{a_0}) = S(0,0;\boldsymbol{a_0}) = 1$ and $s(n,k;\boldsymbol{a_0}) = S(n,k;\boldsymbol{a_0}) = 0$, for $k > n$. Further, replacing t by $-t$ in (3.26) and since $(t)_{n,-\boldsymbol{a_0}} = (-1)^n(-t)_{n,\boldsymbol{a_0}}$, we deduce the expression

$$(t)_{n,-\boldsymbol{a_0}} = \sum_{k=0}^{n} |s(n,k;\boldsymbol{a_0})|t^k, \tag{3.28}$$

where the coefficient

$$|s(n,k;\boldsymbol{a_0})| \equiv s(n,k;-\boldsymbol{a_0}) = (-1)^{n-k}s(n,k;\boldsymbol{a_0}), \tag{3.29}$$

for $a_i \geq 0$, $i = 0, 1, \ldots$, as a sum of nonnegative numbers is nonnegative. Specifically, from (3.28) it follows that

$$|s(n, k; \boldsymbol{a}_0)| = \sum a_{i_1} a_{i_2} \cdots a_{i_{n-k}}, \qquad (3.30)$$

where the summation is extended over all $(n-k)$-combinations $\{i_1, i_2, \ldots, i_{n-k}\}$ of the n indices $\{0, 1, \ldots, n-1\}$. The coefficient $|s(n, k; \boldsymbol{a}_0)|$ for $a_i \geq 0$, $i = 0, 1, \ldots$, may be called *generalized signless Stirling number of the first kind*.

Remark 3.1 The generalized Stirling numbers for particular sequences $\boldsymbol{a}_0 = (a_0, a_1, a_2, \ldots)$ reduce to some well-known numbers. Specifically, for $a_i = a$, $i = 0, 1, \ldots$,

$$s(n, k; \boldsymbol{a}_0) = (-1)^{n-k} \binom{n}{k} a^{n-k}, \ S(n, k; \boldsymbol{a}_0) = \binom{n}{k} a^{n-k}.$$

Also, for $a_i = i + r$, $i = 0, 1, \ldots$,

$$s(n, k; \boldsymbol{a}_0) = s(n, k; r), \ S(n, k; \boldsymbol{a}_0) = S(n, k; r),$$

where $s(n, k; r)$ and $S(n, k; r)$ are the noncentral Stirling numbers of the first and second kind, respectively. In particular, for $r = 0$,

$$s(n, k; \boldsymbol{a}_0) = s(n, k), \ S(n, k; \boldsymbol{a}_0) = S(n, k),$$

where $s(n, k)$ and $S(n, k)$ are the Stirling numbers of the first and second kind, respectively.

Remark 3.2 *q-Stirling numbers.* Let x and q be a real numbers, with $q \neq 1$, and k a positive integer. The number $[x]_q = (1 - q^x)/(1 - q)$ is called a *q-number*. Also, the product

$$[x]_{k,q} = [x]_q [x-1]_q \cdots [x-k+1]_q$$

is called a *q-factorial of x of order k*. In particular, $[k]_q! = [1]_q [2]_q \cdots [k]_q$. The *q-binomial coefficient* (or *Gaussian coefficient*) is defined by

$$\begin{bmatrix} x \\ k \end{bmatrix}_q = \frac{[x]_{k,q}}{[k]_q!}.$$

Note that the noncentral q-factorial of t of order n and noncentrality parameter r, on using the relation $[t - r - j]_q = q^{-r-j}([t]_q - [r+j]_q)$, $j = 0, 1, \ldots$, may be expressed as

$$[t - r]_{n,q} = q^{-\binom{n}{2} - rn}([t]_q - [r]_q)([t]_q - [r+1]_q) \cdots ([t]_q - [r+n-1]_q).$$

This is a polynomial of the q-number $[t]_q$ of degree n:

$$[t - r]_{n,q} = q^{-\binom{n}{2} - rn} \sum_{k=0}^{n} s_q(n, k; r)[t]_q^k, \ n = 0, 1, \ldots.$$

Inversely,

$$[t]_q^n = \sum_{k=0}^{n} q^{\binom{k}{2}+rk} S_q(n,k;r)[t-r]_{k,q}, \ n = 0, 1, \ldots.$$

The coefficients $s_q(n,k;r)$ and $S_q(n,k;r)$ are called *noncentral q-Stirling numbers of the first and second kind*, respectively; $s_q(n,k) \equiv s_q(n,k;0)$ and $S_q(n,k) \equiv S_q(n,k;0)$ are the *q-Stirling numbers of the first and second kind*, respectively.

The generalized Stirling numbers, for $a_i = [i+r]_q = (1 - q^{i+r})/(1-q)$, $i = 0, 1, \ldots$, reduce to

$$s(n,k;\boldsymbol{a}_0) = s_q(n,k;r), \ S(n,k;\boldsymbol{a}_0) = S_q(n,k;r).$$

In particular, for $r = 0$,

$$s(n,k;\boldsymbol{a}_0) = s_q(n,k), \ S(n,k;\boldsymbol{a}_0) = S_q(n,k).$$

Several properties of the central and noncentral q-Stirling numbers are provided in the Exercises and Complements.

The expansion of the function $f(t) = (t)_{n,\boldsymbol{a}_0}$ into a Maclaurin series, since $D_t^k(t)_{n,\boldsymbol{a}_0} = 0$ for $k > n$, is given by

$$(t)_{n,\boldsymbol{a}_0} = \sum_{k=0}^{n} \frac{1}{k!} \left[D_t^k(t)_{n,\boldsymbol{a}_0} \right]_{t=0} \cdot t^k, \ n = 0, 1, \ldots,$$

which, by virtue of (3.26), implies

$$s(n,k;\boldsymbol{a}_0) = \frac{1}{k!} \left[D_t^k(t)_{n,\boldsymbol{a}_0} \right]_{t=0}, \ k = 0, 1, \ldots, n, \ n = 0, 1, \ldots. \quad (3.31)$$

The divided differences of a function $f(t)$, which is given at the points $\{t_0, t_1, \ldots, t_n, \ldots\}$, are successively defined by

$$\left[\mathcal{D}_t f(t) \right]_{t=t_i} = \frac{f(t_{i+1}) - f(t_i)}{t_{i+1} - t_i}, \ i = 0, 1, \ldots$$

and

$$\left[\mathcal{D}_t^k f(t) \right]_{t=t_i} = \frac{\left[\mathcal{D}_t^{k-1} f(t) \right]_{t=t_{i+1}} - \left[\mathcal{D}_t^{k-1} f(t) \right]_{t=t_i}}{t_{i+k} - t_i},$$

for $k = 2, 3, \ldots, \ i = 0, 1, \ldots$. Further, a function $f(t)$, for which the divided differences at the points $\boldsymbol{a}_0 = (a_0, a_1, a_2, \ldots)$, $[\mathcal{D}_t^k f(t)]_{t=\boldsymbol{a}_0}$, exist, may be expanded into generalized factorials of t with increments $\boldsymbol{a}_0 = (a_0, a_1, a_2, \ldots)$ as

$$f(t) = \sum_{k=0}^{\infty} \left[\mathcal{D}_t^k f(t) \right]_{t=\boldsymbol{a}_0} \cdot (t)_{k,\boldsymbol{a}_0}, \ n = 0, 1, \ldots,$$

provided the series is absolutely convergent. In the case of the nth power of t, $[\mathcal{D}_t^k t^n]_{t=a_0} = 0$ for $k > 0$ and so

$$t^n = \sum_{k=0}^{n} [\mathcal{D}_t^k t^n]_{t=a_0} \cdot (t)_{k,\boldsymbol{a}_0}, \; n = 0, 1, \ldots,$$

implying

$$S(n, k; \boldsymbol{a}_0) = [\mathcal{D}_t^k t^n]_{t=a_0}, \; k = 0, 1, \ldots, n, \; n = 0, 1, \ldots. \tag{3.32}$$

The generalized Stirling numbers of the first and second kind constitute a pair of orthogonal bivariate sequences. This is shown in the next theorem.

Theorem 3.6 *The generalized Stirling numbers of the first and second kind satisfy the following orthogonality relations:*

$$\sum_{j=k}^{n} s(n, j; \boldsymbol{a}_0) S(j, k; \boldsymbol{a}_0) = \delta_{n,k}, \tag{3.33}$$

and

$$\sum_{j=k}^{n} S(n, j; \boldsymbol{a}_0) s(j, k; \boldsymbol{a}_0) = \delta_{n,k}, \tag{3.34}$$

where $\delta_{n,k}$ is the Kronecker delta.

Proof Expanding the nth order generalized factorial of t, with increments $\boldsymbol{a}_0 = (a_0, a_1, a_2, \ldots)$, into powers of t, using (3.26), and in the resulting expression expanding the powers of t into generalized factorials of t, with increments $\boldsymbol{a}_0 = (a_0, a_1, a_2, \ldots)$, using (3.27), we get the relation

$$(t)_{n,\boldsymbol{a}_0} = \sum_{j=0}^{n} s(n, j; \boldsymbol{a}_0) t^j = \sum_{j=0}^{n} s(n, j; \boldsymbol{a}_0) \sum_{k=0}^{j} S(j, k; \boldsymbol{a}_0)(t)_{k,\boldsymbol{a}_0}$$

$$= \sum_{k=0}^{n} \left\{ \sum_{j=k}^{n} s(n, j; \boldsymbol{a}_0) S(j, k; \boldsymbol{a}_0) \right\} (t)_{k,\boldsymbol{a}_0},$$

which implies (3.33). Similarly, expanding the nth power of t into generalized factorials of t, with increments $\boldsymbol{a}_0 = (a_0, a_1, a_2, \ldots)$, using (3.27), and in the resulting expression expanding the generalized factorials of t into powers of t, using (3.26), we derive (3.34).

An explicit expression of the generalized Stirling numbers of the second kind is deduced in the following theorem.

Theorem 3.7 *The generalized Stirling number of the second kind $S(n, k; \boldsymbol{a}_0)$, $k = 0, 1, \ldots, n$, $n = 0, 1, \ldots$, is given by*

$$S(n, k; \boldsymbol{a}_0) = \sum_{j=0}^{k} \frac{a_j^n}{(a_j)_{j,\boldsymbol{a}_0} (a_j)_{k-j,\boldsymbol{a}_{j+1}}}. \tag{3.35}$$

Proof The kth divided difference of a function $f(t)$ at the points $\boldsymbol{a}_0 = (a_0, a_1, a_2, \ldots)$ is given by

$$\left[\mathcal{D}_t^k f(t)\right]_{t=a_0} = \sum_{j=0}^{k} \frac{f(a_j)}{(a_j)_{j,a_0}(a_j)_{k-j,\boldsymbol{a}_{j+1}}},$$

where $\boldsymbol{a}_{j+1} = (a_{j+1}, a_{j+2}, \ldots)$. Applying this operator on the function $f(t) = t^n$, we get (3.35).

Recurrence relations for the generalized Stirling numbers of the first and second kind are derived in the following theorems.

Theorem 3.8 *The generalized signless Stirling numbers of the first kind $|s(n, k; \boldsymbol{a}_0)|$, $k = 0, 1, \ldots, n = 0, 1, \ldots$, satisfy the triangular recurrence relation*

$$|s(n, k; \boldsymbol{a}_0)| = |s(n-1, k-1; \boldsymbol{a}_0)| + a_{n-1}|s(n-1, k; \boldsymbol{a}_0)|, \qquad (3.36)$$

for $k = 1, 2, \ldots, n$, $n = 1, 2, \ldots$, with initial conditions

$$|s(0, 0; \boldsymbol{a}_0)| = 1, \ |s(n, 0; \boldsymbol{a}_0)| = a_{n-1} a_{n-2} \cdots a_0, \ n > 0, \ |s(n, k; \boldsymbol{a}_0)| = 0, \ k > n.$$

Also, they satisfy the vertical recurrence relation

$$|s(n, k; \boldsymbol{a}_0)| = |s(n-1, k-1; \boldsymbol{a}_0)|$$
$$+ \sum_{j=k}^{n-1} a_{n-1} a_{n-2} \cdots a_j |s(j-1, k-1; \boldsymbol{a}_0)|, \qquad (3.37)$$

for $k = 1, 2, \ldots, n$, $n = 1, 2, \ldots$, with $s(0, 0; \boldsymbol{a}_0) = 1$.

Proof Expanding into powers of t both members of the recurrence relation $(t)_{n,-\boldsymbol{a}_0} = (t + a_{n-1})(t)_{n-1,-\boldsymbol{a}_0}$, using (3.28), and equating the coefficients of t^k in both members of the resulting expression, we deduce recurrence relation (3.36). Its initial conditions follow directly from (3.28).

In (3.36), replacing n by j and then multiplying it by $a_{n-1} a_{n-2} \cdots a_j$, we get the recurrence relation

$$a_{n-1} a_{n-2} \cdots a_j |s(j, k; \boldsymbol{a}_0)| - a_{n-1} a_{n-2} \cdots a_{j-1} |s(j-1, k; \boldsymbol{a}_0)|$$
$$= a_{n-1} a_{n-2} \cdots a_j |s(j-1, k-1; \boldsymbol{a}_0)|,$$

for $j = k, k+1, \ldots$. Summing it for $j = k, k+1, \ldots, n$, and since $|s(k-1, k; \boldsymbol{a}_0)| = 0$, we deduce the vertical recurrence relation (3.37).

Theorem 3.9 *The generalized Stirling numbers of the second kind $S(n, k; \boldsymbol{a}_0)$, $k = 0, 1, \ldots, n, n = 0, 1, \ldots$, satisfy the triangular recurrence relation*

$$S(n, k; \boldsymbol{a}_0) = S(n-1, k-1; \boldsymbol{a}_0) + a_k S(n-1, k; \boldsymbol{a}_0), \qquad (3.38)$$

for $k = 1, 2, \ldots, n$, $n = 1, 2, \ldots$, with initial conditions

$$S(0, 0; \boldsymbol{a}_0) = 1, \ S(n, 0; \boldsymbol{a}_0) = a_0^n, \ n > 0, \ S(n, k; \boldsymbol{a}_0) = 0, \ k > n.$$

Also, they satisfy the vertical recurrence relation

$$S(n,k;\boldsymbol{a}_0) = \sum_{j=k}^{n} a_k^{n-j} S(j-1, k-1; \boldsymbol{a}_0), \qquad (3.39)$$

for $k = 1, , 2, \ldots, n$, $n = 1, 2, \ldots$, *with* $S(0,0;\boldsymbol{a}_0) = 1$.

Proof Expanding both members of $t^n = t \cdot t^{n-1}$ into generalized factorials of t with increments $\boldsymbol{a}_0 = (a_0, a_1, a_2, \ldots)$, using (3.27), we get the relation

$$\sum_{k=0}^{n} S(n,k;\boldsymbol{a}_0)(t)_{k,\boldsymbol{a}_0} = t \sum_{j=0}^{n-1} S(n-1,j;\boldsymbol{a}_0)(t)_{j,\boldsymbol{a}_0},$$

which, on using the recurrence relation $t(t)_{j,\boldsymbol{a}_0} = (t)_{j+1,\boldsymbol{a}_0} + a_j(t)_{j,\boldsymbol{a}_0}$, is written as

$$\sum_{k=0}^{n} S(n,k;\boldsymbol{a}_0)(t)_{k,\boldsymbol{a}_0} = \sum_{k=1}^{n} S(n-1,k-1;\boldsymbol{a}_0)(t)_{k,\boldsymbol{a}_0}$$

$$+ \sum_{k=0}^{n-1} S(n-1,k;\boldsymbol{a}_0)(t)_{k,\boldsymbol{a}_0},$$

implying (3.38). The initial conditions follow directly from (3.27).

In (3.38), replacing n by j and then multiplying it by a_k^{n-j}, we get the recurrence relation

$$a_k^{n-j} S(j,k;\boldsymbol{a}_0) - a_k^{n-j+1} S(j-1,k;\boldsymbol{a}_0) = a_k^{n-j} S(j-1,k-1;\boldsymbol{a}_0),$$

for $j = k, k+1, \ldots$. Summing it for $j = k, k+1, \ldots, n$, and since $S(k-1,k;\boldsymbol{a}_0) = 0$, we deduce the vertical recurrence relation (3.39).

A useful generating function for the generalized Stirling numbers of the second kind,

$$\phi_k(u;\boldsymbol{a}_0) = \sum_{n=k}^{\infty} S(n,k;\boldsymbol{a}_0) u^n,$$

is readily deduced from the triangular recurrence relation (3.38). Indeed, multiplying (3.38) by u^n and summing for $n = k, k+1, \ldots$, we get the recurrence relation

$$\phi_k(u;\boldsymbol{a}_0) = u(1 - a_k u)^{-1} \phi_{k-1}(u;\boldsymbol{a}_0), \quad k = 1, 2, \ldots,$$

with $\phi_0(u;\boldsymbol{a}_0) = (1 - a_0 u)^{-1}$. Iterating it, we deduce the expression

$$\phi_k(u;\boldsymbol{a}_0) = \sum_{n=k}^{\infty} S(n,k;\boldsymbol{a}_0) u^n = u^k \prod_{j=0}^{k}(1 - a_j u)^{-1}. \qquad (3.40)$$

Further, expanding each term of the product in (3.40) into powers of u and equating the coefficients of u^n, we find the expressions

$$S(n,k;\boldsymbol{a}_0) = \sum a_0^{r_0} a_1^{r_1} \cdots a_k^{r_k} = \sum a_{i_1} a_{i_2} \cdots a_{i_{n-k}}, \qquad (3.41)$$

where the summation in the first sum is extended over all nonnegative integer solutions $(r_0, r_1, \ldots, r_k)$ of the equation $r_0 + r_1 + \cdots + r_k = n - k$, while in the second sum is extended over all $(n - k)$-combinations with repetition $\{i_1, i_2, \ldots, i_{n-k}\}$ of the $k + 1$ indices $\{0, 1, \ldots, k\}$.

Setting $u = 1/t$ in (3.40), we get

$$\frac{1}{(t)_{k+1,a_0}} = \sum_{n=k}^{\infty} S(n, k; \boldsymbol{a}_0) \frac{1}{t^{n+1}}. \tag{3.42}$$

Using the orthogonality relation of the generalized Stirling numbers this expansion can be inverted as

$$\frac{1}{t^{k+1}} = \sum_{n=k}^{\infty} |s(n, k; \boldsymbol{a}_0)| \frac{1}{(t)_{n+1,-\boldsymbol{a}_0}}. \tag{3.43}$$

Example 3.6 *Bernoulli trials with varying success probability.* Consider a sequence of independent Bernoulli trials and assume that the probability of success at the nth trial is $p_n = 1/(1 + a_{n-1})$, with $a_{n-1} \geq 0$, $n = 1, 2, \ldots$. Let X_n be the number of successes up to the nth trial. The probability function $p(k; n) = P(X_n = k)$, $k = 0, 1, \ldots, n$, and the binomial moments $b_{(j)}(n) = E\left[\binom{X_n}{j}\right]$, $j = 1, 2, \ldots$, of the random variable X_n may be determined as follows.

Expressing the random variable X_n as a sum of independent zero-one Bernoulli random variables and using (1.73), its probability generating function is readily obtained as

$$G(t) = \sum_{k=0}^{n} p(k; n) t^k = \prod_{i=0}^{n-1}(t + a_i) \Big/ \prod_{i=0}^{n-1}(1 + a_i).$$

A comparison of it with (3.28) entails

$$p(k; n) = \frac{|s(n, k; \boldsymbol{a}_0)|}{(1 + a_0)(1 + a_1) \cdots (1 + a_{n-1})}, \quad k = 0, 1, \ldots, n.$$

The binomial moment generating function of X_n is deduced from the probability generating function $G(t)$ by replacing t by $t + 1$:

$$G(t + 1) = \sum_{j=0}^{n} b_{(j)}(n) t^j = \prod_{i=0}^{n-1}(t + a_i + 1) \Big/ \prod_{i=0}^{n-1}(1 + a_i).$$

Hence, by (3.28),

$$b_{(j)}(n) = \frac{|s(n, j; \boldsymbol{a}_0 + 1)|}{(1 + a_0)(1 + a_1) \cdots (1 + a_{n-1})}, \quad j = 1, 2, \ldots, n.$$

Further, let W_k be the number of trials until the occurrence of the kth success. The probability function $q(n; k) = P(W_k = n)$, $n = k, k+1, \ldots$, is connected with $p(k; n) = P(X_n = k)$ by $q(n; k) = p(k-1; n-1) p_n$ and so

$$q(n; k) = \frac{|s(n-1, k-1; \boldsymbol{a}_0)|}{(1 + a_0)(1 + a_1) \cdots (1 + a_{n-1})}, \quad n = k, k+1, \ldots.$$

Example 3.7 (*Continuation*). Consider a sequence of Bernoulli trials and assume that the probability of success at any trial given that k successes occur in the previous trials is p_k, $k = 0, 1, \ldots$. Let W_k be the number of trials until the occurrence of the kth success. The probability function $q(n; k) = P(W_k = n)$, $n = k, k+1, \ldots$, and ascending binomial moments $b_{[j]}(k) = E[\binom{W_k+j-1}{j}]$, $j = 1, 2, \ldots$, may be derived as follows.

Let Y_i be the number of trials right after the $(i-1)$st success and up to the ith success, $i = 1, 2, \ldots, k$. Then, Y_i has a geometric distribution with probability generating function $p_{i-1}t(1 - q_{i-1}t)^{-1}$, $q_{i-1} = 1 - p_{i-1}$, $i = 1, 2, \ldots, k$. Therefore, the probability generating function of $W_k = Y_1 + Y_2 + \cdots + Y_k$ is given by

$$H(t) = \sum_{n=k}^{\infty} q(n; k) t^n = p_0 p_1 \cdots p_{k-1} t^k \prod_{i=0}^{k-1} (1 - q_i t)^{-1}.$$

Comparing it with (3.42), we conclude that

$$q(n; k) = p_0 p_1 \cdots p_{k-1} S(n-1, k-1; \boldsymbol{q}_0), \; n = k, k+1, \ldots,$$

where $\boldsymbol{q}_0 = (q_0, q_1, q_2, \ldots)$. The ascending binomial moment generating function of W_k is deduced from the probability generating function $H(t)$ by replacing t by $(1-t)^{-1}$:

$$\sum_{j=0}^{\infty} b_{[j]}(k) t^j = \prod_{i=0}^{k-1} (1 - t/p_i)^{-1}.$$

Therefore, by (3.42),

$$b_{[j]}(k) = S(k+j, k; \boldsymbol{c}_0), \; j = 1, 2, \ldots,$$

where $\boldsymbol{c}_0 = (1/p_0, 1/p_1, 1/p_2, \ldots)$.

Further, let X_n be the number of successes up to the nth trial. The probability function $p(k; n) = P(X_n = k)$, $k = 0, 1, \ldots, n$, since $p(k; n) = q(k+1; n+1)/p_k$, is deduced as

$$p(k; n) = p_0 p_1 \cdots p_{k-1} S(n, k; \boldsymbol{q}_0), \; k = 0, 1, \ldots, n.$$

3.6 GENERALIZED LAH NUMBERS

Consider the expansion of the generalized factorial of t of order n and increments $\boldsymbol{a}_0 = (a_0, a_1, a_2, \ldots)$ into generalized factorials of t with increments $\boldsymbol{b}_0 = (b_0, b_1, b_2, \ldots)$:

$$(t)_{n,\boldsymbol{a}_0} = \sum_{k=0}^{n} C(n, k; \boldsymbol{a}_0, \boldsymbol{b}_0)(t)_{k,\boldsymbol{b}_0}, \; n = 0, 1, \ldots. \tag{3.44}$$

Clearly, $C(0, 0; \boldsymbol{a}_0, \boldsymbol{b}_0) = 0$ and $C(n, k; \boldsymbol{a}_0, \boldsymbol{b}_0) = 0$, for $k > n$. Also, $C(n, k; \boldsymbol{a}_0, \boldsymbol{0}) = s(n, k; \boldsymbol{a}_0)$ and $C(n, k; \boldsymbol{0}, \boldsymbol{b}_0) = S(n, k; \boldsymbol{b}_0)$. The coefficient $C(n, k; \boldsymbol{a}_0, \boldsymbol{b}_0)$ is called the *generalized Lah number*.

Remark 3.3 The generalized Lah numbers for particular sequences $\boldsymbol{a}_0 = (a_0, a_1, a_2, \ldots)$ and $\boldsymbol{b}_0 = (b_0, b_1, b_2, \ldots)$ reduce to some known numbers. Specifically, for $a_i = a$, $b_i = b$, $i = 0, 1, \ldots$,

$$C(n, k; \boldsymbol{a}_0, \boldsymbol{b}_0) = \binom{n}{k}(b-a)^{n-k}.$$

Further, for $a_i = ia - c$, $b_i = ib$, $i = 0, 1, \ldots$,

$$C(n, k; \boldsymbol{a}_0, \boldsymbol{b}_0) = a^n b^{-k} C(n, k; s, r), \quad s = b/a, \ r = c/a,$$

where $C(n, k; s, r)$ is the noncentral generalized factorial coefficient. In particular, for $b = -a = 1$,

$$C(n, k; \boldsymbol{a}_0, \boldsymbol{b}_0) = (-1)^n L(n, k; r) = |L(n, k; r)|, \quad r = -c,$$

where $L(n, k; r)$ is the noncentral Lah number. Also, for $c = 0$,

$$C(n, k; \boldsymbol{a}_0, \boldsymbol{b}_0) = a^n b^{-k} C(n, k; s), \quad s = b/a,$$

where $C(n, k; s)$ is the generalized factorial coefficient, and for $b = -a = 1$, $c = 0$,

$$C(n, k; \boldsymbol{a}_0, \boldsymbol{b}_0) = (-1)^n L(n, k) = |L(n, k)|,$$

where $L(n, k)$ is the Lah number.

Remark 3.4 *Generalized q-factorial coefficients.* The noncentral generalized q-factorial of t of order n, scale parameter s and noncentrality parameter r, $[st + r]_{n,q}$, may be expressed in terms of q^s-factorials of t as

$$[st + r]_{n,q} = q^{-\binom{n}{2}+rn} \sum_{k=0}^{n} q^{s\binom{k}{2}} C_q(n, k; s, r)[t]_{k, q^s}.$$

The coefficient $C_q(n, k; s, r)$ is called *noncentral generalized q-factorial coefficient*; $C_q(n, k; s) \equiv C_q(n, k; s, 0)$ is the (central) *generalized q-factorial coefficient*. Also, $L_q(n, k; r) \equiv C_{q^{-1}}(n, k; -1, r)$ and $L_q(n, k) \equiv C_{q^{-1}}(n, k; -1)$ are the *noncentral* and *central q-Lah numbers*, respectively.

Notice that the generalized Lah numbers for $a_i = [ia - c]_q = (1 - q^{ia-c})/(1-q)$, $b_i = [ib]_q = (1 - q^{ib})/(1-q)$, $i = 0, 1, \ldots$, reduce to

$$C(n, k; \boldsymbol{a}_0, \boldsymbol{b}_0) = [a]_q^n [b]_q^{-k} C_{q^a}(n, k; s, r), \quad s = b/a, \ r = c/a,$$

where $C_q(n, k; s, r)$ is the noncentral generalized q-factorial coefficient. In particular, for $b = -a = 1$,

$$C(n, k; \boldsymbol{a}_0, \boldsymbol{b}_0) = [-1]_q^n L_q(n, k; r) = |L_q(n, k; r)|,$$

where $L_q(n, k; r)$ is the noncentral q-Lah number. Also, for $c = 0$,

$$C(n, k; \boldsymbol{a}_0, \boldsymbol{b}_0) = [a]_q^n [b]_q^{-k} C_{q^a}(n, k; s), \quad s = b/a,$$

where $C_q(n,k;s)$ is the generalized q-factorial coefficient, and for $b = -a = 1$, $c = 0$,
$$C(n,k;\boldsymbol{a}_0,\boldsymbol{b}_0) = [-1]_q^n L_q(n,k) = |L_q(n,k)|,$$
where $L_q(n,k)$ is the q-Lah number.

Note that the expansion of the function $f(t) = (t)_{n,\boldsymbol{a}_0}$ into a series of generalized factorials of t, with increments $\boldsymbol{b}_0 = (b_0, b_1, b_2, \ldots)$, and since $\mathcal{D}_t^k(t)_{n,\boldsymbol{a}_0} = 0$ for $k > n$, is given by
$$(t)_{n,\boldsymbol{a}_0} = \sum_{k=0}^{n} \left[\mathcal{D}_t^k(t)_{n,\boldsymbol{a}_0}\right]_{t=b_0} \cdot (t)_{k,\boldsymbol{b}_0},$$
which, by virtue of (3.44), implies
$$C(n,k;\boldsymbol{a}_0,\boldsymbol{b}_0) = \left[\mathcal{D}_t^k(t)_{n,\boldsymbol{a}_0}\right]_{t=b_0}.$$
Further, the kth divided difference of the function $f(t) = (t)_{n,\boldsymbol{a}_0}$ at the points $\boldsymbol{b}_0 = (b_0, b_1, b_2, \ldots)$ is given by
$$C(n,k;\boldsymbol{a}_0,\boldsymbol{b}_0) = \sum_{j=0}^{k} \frac{(b_j)_{n,\boldsymbol{a}_0}}{(b_j)_{j,\boldsymbol{b}_0}(b_j)_{k-j,\boldsymbol{b}_{j+1}}}, \tag{3.45}$$
where $\boldsymbol{b}_{j+1} = (b_{j+1}, b_{j+2}, \ldots)$.

Expanding the nth order generalized factorial of t, with increments $\boldsymbol{a}_0 = (a_0, a_1, a_2, \ldots)$, into powers of t, using (3.26), and in the resulting expression expanding the powers of t into generalized factorials of t, with increments $\boldsymbol{b}_0 = (b_0, b_1, b_2, \ldots)$, using (3.27), we get the relation
$$(t)_{n,\boldsymbol{a}_0} = \sum_{r=0}^{n} s(n,r;\boldsymbol{a}_0) t^r = \sum_{r=0}^{n} s(n,r;\boldsymbol{a}_0) \sum_{k=0}^{r} S(r,k;\boldsymbol{b}_0)(t)_{k,\boldsymbol{b}_0}$$
$$= \sum_{k=0}^{n} \left\{\sum_{r=k}^{n} s(n,r;\boldsymbol{a}_0) S(r,k;\boldsymbol{b}_0)\right\}(t)_{k,\boldsymbol{b}_0},$$
which, by virtue of (3.44), implies
$$C(n,k;\boldsymbol{a}_0,\boldsymbol{b}_0) = \sum_{r=k}^{n} s(n,r;\boldsymbol{a}_0) S(r,k;\boldsymbol{b}_0). \tag{3.46}$$

Expanding into generalized factorials of t, with increments $\boldsymbol{b}_0 = (b_0, b_1, b_2, \ldots)$, both members of the recurrence relation
$$(t)_{n,\boldsymbol{a}_0} = (t - a_{n-1})(t)_{n-1,\boldsymbol{a}_0},$$
using (3.44), and equating the coefficients of $(t)_{k,\boldsymbol{b}_0}$ in both members of the resulting expression, we deduce for the generalized Lah numbers the triangular recurrence relation
$$C(n,k;\boldsymbol{a}_0,\boldsymbol{b}_0) = (b_k - a_{n-1})C(n-1,k;\boldsymbol{a}_0,\boldsymbol{b}_0) + C(n-1,k-1;\boldsymbol{a}_0,\boldsymbol{b}_0), \tag{3.47}$$

for $k = 1, 2, \ldots, n$, $n = 1, 2, \ldots$, with initial conditions

$$C(0, 0; \boldsymbol{a}_0, \boldsymbol{b}_0) = 1, \ C(n, 0; \boldsymbol{a}_0, \boldsymbol{b}_0) = (b_0)_{n, a_0}, \ n > 0,$$
$$C(n, k; \boldsymbol{a}_0, \boldsymbol{b}_0) = 0, \ k > n.$$

Replacing in (3.47) n by j and then multiplying it by $(b_k)_{n-j, a_j}$, we get the recurrence relation

$$(b_k)_{n-j, a_j} C(j-1, k-1; \boldsymbol{a}_0, \boldsymbol{b}_0) = (b_k)_{n-j, a_j} C(j, k; \boldsymbol{a}_0, \boldsymbol{b}_0)$$
$$- (b_k)_{n-j+1, a_{j-1}} C(j-1, k; \boldsymbol{a}_0, \boldsymbol{b}_0).$$

Summing it for $j = k, k+1, \ldots, n$, we deduce the vertical recurrence relation

$$C(n, k; \boldsymbol{a}_0, \boldsymbol{b}_0) = \sum_{j=k}^{n} (b_k)_{n-j, a_j} C(j-1, k-1; \boldsymbol{a}_0, \boldsymbol{b}_0). \tag{3.48}$$

Expanding the nth order generalized factorial of t, with increments $\boldsymbol{a}_0 = (a_0, a_1, a_2, \ldots)$, into generalized factorials of t, with increments $\boldsymbol{b}_0 = (b_0, b_1, b_2, \ldots)$, using (3.44), and in the resulting expression expanding the generalized factorials of t, with increments $\boldsymbol{b}_0 = (b_0, b_1, b_2, \ldots)$, into generalized factorials of t, with increments $\boldsymbol{a}_0 = (a_0, a_1, a_2, \ldots)$, we get the relation

$$(t)_{n, a_0} = \sum_{j=0}^{n} C(n, j; \boldsymbol{a}_0, \boldsymbol{b}_0)(t)_{j, b_0} = \sum_{j=0}^{n} C(n, j; \boldsymbol{a}_0, \boldsymbol{b}_0) \sum_{k=0}^{j} C(j, k; \boldsymbol{b}_0, \boldsymbol{a}_0)(t)_{k, a_0}$$
$$= \sum_{k=0}^{n} \left\{ \sum_{j=k}^{n} C(n, j; \boldsymbol{a}_0, \boldsymbol{b}_0) C(j, k; \boldsymbol{b}_0, \boldsymbol{a}_0) \right\} (t)_{k, a_0},$$

which implies

$$\sum_{j=k}^{n} C(n, j; \boldsymbol{a}_0, \boldsymbol{b}_0) C(j, k; \boldsymbol{b}_0, \boldsymbol{a}_0) = \delta_{n, k}. \tag{3.49}$$

Multiplying the triangular recurrence relation for the generalized Lah numbers by $1/(t)_{n, a_0}$ and summing for $n = k, k+1, \ldots$, we get

$$\frac{1}{(t)_{k+1, b_0}} = \sum_{n=k}^{\infty} C(n, k; \boldsymbol{a}_0, \boldsymbol{b}_0) \frac{1}{(t)_{n+1, a_0}}. \tag{3.50}$$

Example 3.8 *Bernoulli trials with varying success probability.* Consider a sequence of Bernoulli trials and assume that the probability of success at the $(n+1)$st trial, given that k successes occur up to the nth trial, is given by $p_{n,k} = (1 - b_k)/(1 + a_n)$, with $0 \leq b_k \leq 1$, $0 \leq a_n < \infty$, $k = 0, 1, \ldots, n$, $n = 0, 1, \ldots$. Let X_n be the number of successes up to the nth trial. The probability function $p(k; n) = P(X_n = k)$, $k = 0, 1, \ldots, n$, may be derived as follows.

The probabilities $p(k;n)$, $k = 0,1,\ldots,n$, $n = 0,1,\ldots$, satisfy the triangular recurrence relation

$$p(k;n) = \frac{a_{n-1} + b_k}{1 + a_{n-1}} p(k;n-1) + \frac{1 - b_{k-1}}{1 + a_{n-1}} p(k-1;n-1),$$

for $k = 1,2,\ldots,n$, $n = 1,2,\ldots$, with initial conditions

$$p(0;0) = 1, \quad p(0;n) = \frac{(b_0)_{n,-a_0}}{(1)_{n,-a_0}}, \quad n > 0, \quad p(k;0) = 0, \quad k > 0.$$

Setting

$$p(k;n) = \frac{(1)_{k,b_0}}{(1)_{n,-a_0}} c_{n,k}, \quad k = 0,1,\ldots,n, \quad n = 0,1,\ldots,$$

the coefficients $c_{n,k}$, $k = 0,1,\ldots,n$, $n = 0,1,\ldots$, satisfy the triangular recurrence relation

$$c_{n,k} = (a_{n-1} + b_k)c_{n-1,k} + c_{n-1,k-1},$$

for $k = 1,2,\ldots,n$, $n = 1,2,\ldots$, with initial conditions

$$c_{0,0} = 1, \quad c_{n,0} = (b_0)_{n,-a_0}, \quad n > 0, \quad c_{0,k} = 0, \quad k > 0.$$

Comparing the last recurrence with the recurrence relation of the generalized Lah numbers, (3.47), we get $c_{n,k} = C(n,k;-a_0,b_0)$. Therefore

$$p(k;n) = \frac{(1)_{k,b_0}}{(1)_{n,-a_0}} C(n,k;-a_0,b_0), \quad k = 0,1,\ldots,n.$$

3.7 REFERENCE NOTES

The associated Stirling numbers, introduced by J. Riordan (1958), are closely related to the numbers of Ch. Jordan (1933, 1939a) and M. Ward (1934), which are the coefficients of the representation of $s(n, n-k)$ and $S(n, n-k)$ as sums of binomials of n. These numbers were further discussed by L. Carlitz (1971). Recurrence relations and other properties of the r-associated Stirling numbers and the generalized factorial coefficients were discussed by J. Riordan (1958), L. Comtet (1974) and Ch. A. Charalambides (1974a).

The generalized Stirling and Lah numbers were introduced by S. Tauber (1962, 1965) and independently by M. L. Platonov (1964) and subsequently further studied by L. Comtet (1972) and M. L. Platonov (1976). The q-Stirling numbers were examined by L. Carlitz (1933, 1948) and H. W. Gould (1961). The q-Lah numbers were introduced by W. Hahn (1949) and further studied by A. M. Garsia and J. B. Remmel (1986). The more general q-differences of the generalized q-factorial were examined by Ch. A. Charalambides (1996, 2004). Further, D. Crippa and K. Simon (1997) and D. Crippa, K. Simon and P. Trunz (1997) used the q-Stirling numbers in expressing the probability function of certain Markov processes.

The distribution of the number of successes in a sequence of n independent Bernoulli trials with probability of success at any trial depending on the number of previous trials, which is derived in the first part of Example 3.6, was first discussed by S. D. Poisson (1837). The expression of its probability function in terms of the generalized signless Stirling numbers of the first kind was derived by M. L. Platonov (1976). N. Balakrishnan and V. B. Nevzorov (1997) obtained the same distribution as the distribution of the number of records at time n in a sequence of independent observations from a respective sequence of distribution functions, which is positive powers of a continuous distribution function. The probability function of the number of successes in a sequence of n Bernoulli trials with probability of success at any trial depending on the number of previous successes, which is deduced in the second part of Example 3.7, was obtained by M. A. Woodbury (1949) essentially in terms of the generalized Stirling numbers of the second kind. The derivation of the probability function of the number of trials until the occurrence of the kth success, as a convolution of independent and non-identical geometrics, which is presented in the first part of Example 3.7, is due to A. Sen and N. Balakrishnan (1999); their expression, by virtue of (3.35), is also essentially in terms of the Stirling numbers of the second kind.

A rich bibliography of the associated Stirling numbers and the generalized Stirling and Lah numbers can be found in the review paper by Ch. A. Charalambides and J. Singh (1988).

3.8 EXERCISES AND COMPLEMENTS

3.1 Show that the r-associated signless Stirling number of the first kind is given by

$$|s_r(n,k)| = \frac{n!}{k!} \sum \frac{1}{r_1 \cdot r_2 \cdots r_k},$$

where the summation is extended over all compositions of n into k parts of sizes no less than r, that is, over all integer solutions $r_j \geq r$, $j = 1, 2, \ldots, k$, of the equation

$$r_1 + r_2 + \cdots + r_k = n.$$

Alternatively,

$$|s_r(n,k)| = \sum \frac{n!}{k_r! k_{r+1}! \cdots k_n!} \left(\frac{1}{r}\right)^{k_r} \left(\frac{1}{r+1}\right)^{k_{r+1}} \cdots \left(\frac{1}{n}\right)^{k_n},$$

where the summation is extended over all partitions of n into k parts of sizes no less than r, that is, over all nonnegative integer solutions of the equations

$$rk_r + (r+1)k_{r+1} + \cdots + nk_n = n, \quad k_r + k_{r+1} + \cdots + k_n = k.$$

3.2 (*Continuation*). Show that the r-associated Stirling number of the second kind is given by

$$S_r(n,k) = \frac{n!}{k!} \sum \frac{1}{r_1! r_2! \cdots r_k!},$$

where the summation is extended over all compositions of n into k parts of sizes no less than r. Alternatively,

$$S_r(n,k) = \sum \frac{n!}{k_r! k_{r+1}! \cdots k_n!} \left(\frac{1}{r!}\right)^{k_r} \left(\frac{1}{(r+1)!}\right)^{k_{r+1}} \cdots \left(\frac{1}{n!}\right)^{k_n},$$

where the summation is extended over all partitions of n into k parts of sizes no less than r.

3.3 (*Continuation*). Show that the r-associated generalized factorial coefficient is given by

$$C_r(n,k;s) = \frac{n!}{k!} \sum \binom{s}{r_1}\binom{s}{r_2}\cdots\binom{s}{r_k},$$

where the summation is extended over all compositions of n into k parts of sizes no less than r. Alternatively,

$$C_r(n,k;s) = \sum \frac{n!}{k_r! k_{r+1}! \cdots k_n!} \binom{s}{r}^{k_r}\binom{s}{r+1}^{k_{r+1}}\cdots\binom{s}{n}^{k_n},$$

where the summation is extended over all partitions of n into k parts of sizes no less than r.

3.4 Suppose that n distinguishable balls are distributed into k distinguishable urns. Show that the number of different distributions in which each urn contains at most r balls equals $B_r(n,k) = k! T_r(n,k)$, where

$$F_{k,r}(u) = \sum_{n=0}^{rk} T_r(n,k) \frac{u^n}{n!} = \frac{1}{k!}\left(\sum_{i=0}^{r} \frac{u^i}{i!}\right)^k.$$

Further, show that

$$T_r(n+1,k) = k T_r(n,k) - \binom{n}{r} T_r(n-r, k-1),$$

for $n = r, r+1, \ldots, rk$, $k = 1, 2, \ldots$, with $T_r(n,k) = k^n/k!$, $n = 0,1,\ldots,r$, $k = 0, 1, \ldots,$ and

$$k T_r(n,k) = \sum_{j=0}^{m} \binom{n}{j} T_r(n-j, k-1), \quad m = \min\{n, r\}.$$

3.5 Suppose that n like balls are distributed into k distinguishable urns, each with s distinguishable cells of capacity limited to one ball. Show that the number of different distributions in which each urn contains at most r balls equals $B_r(n,k;s) = k! D_r(n,k;s)/n!$, where

$$F_{k,r}(u;s) = \sum_{n=0}^{rk} D_r(n,k;s) \frac{u^n}{n!} = \frac{1}{k!}\left[\sum_{i=0}^{r}\binom{s}{i} u^i\right]^k.$$

Further, show that

$$D_r(n+1,k;s) = (sk-n)D_r(n,k;s) - \binom{n}{r}(s)_{r+1}D_r(n-r,k-1;s),$$

for $n = r, r+1, \ldots, rk$, $k = 1, 2, \ldots$, with $D_r(n,k;s) = (sk)_n/k!$, $n = 0, 1, \ldots, r$, $k = 0, 1, \ldots$, and

$$kD_r(n,k;s) = \sum_{j=0}^{m} \binom{n}{j}(s)_j D_r(n-j,k-1;s), \quad m = \min\{n,s,r\}.$$

3.6 *(Continuation)*. If the cells are of unlimited capacity show that the number of different distributions in which each urn contains at most r balls equals $|B_r(n,k;-s)| = k!|D_r(n,k;-s)|/n!$, where

$$G_{k,r}(u;s) = \sum_{n=0}^{rk} |D_r(n,k;-s)|\frac{u^n}{n!} = \frac{1}{k!}\left[\sum_{i=0}^{r}\binom{s+i-1}{i}u^i\right]^k.$$

Thus, conclude that $|D_r(n,k;-s)| = (-1)^n D_r(n,k;-s)$ and

$$|D_r(n+1,k;-s)| = (sk+n)|D_r(n,k;-s)| - \binom{n}{r}(s+r)_{r+1}|D_r(n-r,k-1;-s)|,$$

for $n = r, r+1, \ldots, rk$, $k = 1, 2, \ldots$, with $|D_r(n,k;-s)| = (sk+n-1)_n/k!$, $n = 0, 1, \ldots, r$, $k = 0, 1, \ldots$, and

$$k|D_r(n,k;-s)| = \sum_{j=0}^{m}\binom{n}{j}(s+j-1)_j|D_r(n-j,k-1;-s)|, \quad m = \min\{n,r\}.$$

3.7 Show that the sequence of generalized signless Stirling numbers of the first kind $|s(n,k;a_0)|$, where $a_k \geq 0$, $k = 0, 1, \ldots, n$, for fixed n, is strong logarithmically concave:

$$|s(n,k;a_0)|^2 > |s(n,k-1;a_0)||s(n,k+1;a_0)|, \quad k = 1, 2, \ldots, n-1.$$

3.8 Let $a_0 = (a_0, a_1, a_2, \ldots)$, where $a_k \geq 0$, $k = 0, 1, \ldots, n$, is a logarithmically concave sequence: $a_k^2 \geq a_{k-1}a_{k+1}$, $k = 1, 2, \ldots, n-1$. Show that the sequence of generalized Stirling numbers of the second kind $S(n,k;a_0)$, $k = 0, 1, \ldots, n$, for fixed n, is strong logarithmically concave:

$$[S(n,k;a_0)]^2 > S(n,k-1;a_0)S(n,k+1;a_0), \quad k = 1, 2, \ldots, n-1.$$

3.9 *q-binomial and negative binomial formulae*. For n a positive integer and x a real number, prove that

$$\prod_{i=1}^{n}(1+xq^{i-1}) = \sum_{k=0}^{n} q^{\binom{k}{2}}\begin{bmatrix}n\\k\end{bmatrix}_q x^k, \quad 0 < q < 1$$

and
$$\prod_{i=1}^{n}(1-xq^{i-1})^{-1} = \sum_{k=0}^{\infty} \begin{bmatrix} n+k-1 \\ k \end{bmatrix}_q x^k, \quad 0 < q < 1.$$

Also, prove that
$$(q^x)^n = \sum_{k=0}^{n}(-1)^k q^{\binom{k}{2}}(1-q)^k \begin{bmatrix} n \\ k \end{bmatrix}_q [x]_{k,q}.$$

3.10 (*Continuation*). For r, s and n positive integers, show that
$$\begin{bmatrix} r+s \\ n \end{bmatrix}_q = \sum_{k=0}^{n} q^{(n-k)(r-k)} \begin{bmatrix} r \\ k \end{bmatrix}_q \begin{bmatrix} s \\ n-k \end{bmatrix}_q$$

and
$$\begin{bmatrix} r+s+n-1 \\ n \end{bmatrix}_q = \sum_{k=0}^{n} q^{r(n-k)} \begin{bmatrix} r+k-1 \\ k \end{bmatrix}_q \begin{bmatrix} s+n-k-1 \\ n-k \end{bmatrix}_q.$$

More generally, for x and y real numbers and n a positive integer, show the *q-Cauchy formula*
$$\begin{bmatrix} x+y \\ n \end{bmatrix}_q = \sum_{k=0}^{n} q^{(n-k)(x-k)} \begin{bmatrix} x \\ k \end{bmatrix}_q \begin{bmatrix} y \\ n-k \end{bmatrix}_q$$

or, equivalently, the *q-Vandermonde formula*
$$[x+y]_{n,q} = \sum_{k=0}^{n} q^{(n-k)(x-k)} \begin{bmatrix} n \\ k \end{bmatrix}_q [x]_{k,q}[y]_{n-k,q}.$$

3.11 *Noncentral q-Stirling numbers of the first kind.* Consider the expansion
$$[t-r]_{n,q} = q^{-\binom{n}{2}-rn} \sum_{k=0}^{n} s_q(n,k;r)[t]_q^k, \quad n = 0,1,\ldots.$$

The coefficient $s_q(n,k;r)$ is the *noncentral q-Stirling number of the first kind*; $s_q(n,k) \equiv s_q(n,k;0)$ is the *q-Stirling number of the first kind*.

(a) Show that
$$s_q(n,k;r) = \sum_{j=k}^{n}(-1)^{j-k} q^{r(n-j)} \binom{j}{k} [r]_q^{j-k} s_q(n,j)$$

and
$$s_q(n,k;r) = \sum_{j=k}^{n} q^{\binom{n-j}{2}+r(n-j)} \begin{bmatrix} n \\ j \end{bmatrix}_q [-r]_{n-j,q} s_q(j,k).$$

(b) Also, show that

$$s_q(n,k;r) = (-1)^{n-k} \sum [i_1+r]_q[i_2+r]_q \cdots [i_{n-k}+r]_q,$$

where the summation is extended over all $(n-k)$-combinations $\{i_1, i_2, \ldots, i_{n-k}\}$ of the n nonnegative integers $\{0, 1, \ldots, n-1\}$. Further, show that

$$s_q(n,k) = (-1)^{n-k} \sum [i_1]_q[i_2]_q \cdots [i_{n-k}]_q,$$

where the summation is extended over all $(n-k)$-combinations $\{i_1, i_2, \ldots, i_{n-k}\}$ of the $n-1$ positive integers $\{1, 2, \ldots, n-1\}$.

(c) Derive the triangular recurrence relation

$$s_q(n+1,k;r) = s_q(n,k-1;r) - [n+r]_q s_q(n,k;r),$$

for $k = 1, 2, \ldots, n+1$, $n = 0, 1, \ldots$, with

$$s_q(0,0;r) = 1, \quad s_q(n,0;r) = q^{\binom{n}{2}+rn}[-r]_{n,q}, \quad n > 0, \quad s_q(n,k;r) = 0, \quad k > n.$$

3.12 *Noncentral q-Stirling numbers of the second kind.* Consider the expansion

$$[t]_q^n = \sum_{k=0}^n q^{\binom{k}{2}+rk} S_q(n,k;r)[t-r]_{k,q}, \quad n = 0, 1, \ldots.$$

The coefficient $S_q(n,k;r)$ is the *noncentral q-Stirling number of the second kind*; $S_q(n,k) \equiv S_q(n,k;0)$ is the *q-Stirling number of the second kind*.

(a) Show that

$$S_q(n,k;r) = \sum_{j=k}^n q^{\binom{j-k}{2}} \begin{bmatrix} j \\ k \end{bmatrix}_q [r]_{j-k,q} S_q(n,j)$$

and

$$S_q(n,k;r) = \sum_{j=k}^n q^{r(j-k)} \binom{n}{j} [r]_q^{n-j} S_q(j,k).$$

(b) Derive the triangular recurrence relation

$$S_q(n+1,k;r) = S_q(n,k-1;r) + [k+r]_q S_q(n,k;r),$$

for $k = 1, 2, \ldots, n+1$, $n = 0, 1, \ldots$, with initial conditions

$$S_q(0,0;r) = 1, \quad S_q(n,0;r) = [r]_q^n, \quad n > 0, \quad S_q(n,k;r) = 0, \quad k > n.$$

(c) Also, derive the orthogonal relations

$$\sum_{j=k}^n s_q(n,j;r) S_q(j,k;r) = \delta_{n,k}, \quad \sum_{j=k}^n S_q(n,j;r) s_q(j,k;r) = \delta_{n,k}.$$

3.13 (*Continuation*). (a) Show that

$$\phi_k(u; r, q) = \sum_{n=k}^{\infty} S_q(n, k; r) u^n = u^k \prod_{j=0}^{k} (1 - [r+j]_q u)^{-1}, \quad k = 1, 2, \ldots$$

and conclude the expansion

$$\frac{1}{[t-r]_{k+1,q}} = q^{\binom{k+1}{2}+r(k+1)} \sum_{n=k}^{\infty} S_q(n, k; r) \frac{1}{[t]_q^{n+1}}$$

and its inverse

$$\frac{1}{[t]_q^{k+1}} = \sum_{n=k}^{\infty} q^{-\binom{n+1}{2}-r(n+1)} s_q(n, k; r) \frac{1}{[t-r]_{n+1,q}}.$$

(b) Further, deduce the vertical recurrence relation

$$S_q(n, k; r) = \sum_{j=k}^{n} [k+r]_q^{n-j} S_q(j-1, k-1; r),$$

and the expression

$$S_q(n, k; r) = \sum [i_1 + r]_q [i_2 + r]_q \cdots [i_{n-k} + r]_q,$$

where the summation is extended over all $(n-k)$-combinations $\{i_1, i_2, \ldots, i_{n-k}\}$, with repetition, of the $k+1$ nonnegative integers $\{0, 1, \ldots, k\}$.

3.14 (*Continuation*). Show that the noncentral q-Stirling number of the second kind $S_q(n, k; r)$, $k = 0, 1, \ldots, n$, $n = 0, 1, \ldots$, is given by the sum

$$S_q(n, k; r) = \frac{1}{[k]_q!} \sum_{j=0}^{k} (-1)^{k-j} q^{\binom{j+1}{2}-(r+j)k} \begin{bmatrix} k \\ j \end{bmatrix}_q [j+r]_q^n.$$

Alternatively,

$$S_q(n, k; r) = \frac{1}{(1-q)^{n-k}} \sum_{j=k}^{n} (-1)^{j-k} q^{r(j-k)} \binom{n}{j} \begin{bmatrix} j \\ k \end{bmatrix}_q.$$

Inverting the last expression, deduce for the noncentral q-Stirling number of the first kind $s_q(n, k; r)$, $k = 0, 1, \ldots, n$, $n = 0, 1, \ldots$, the explicit expression

$$s_q(n, k; r) = \frac{1}{(1-q)^{n-k}} \sum_{j=k}^{n} (-1)^{j-k} q^{\binom{n-j}{2}+r(n-j)} \begin{bmatrix} n \\ j \end{bmatrix}_q \binom{j}{k}.$$

3.15 (*Continuation*). Show that

$$\sum_{n=0}^{\infty} \sum_{k=0}^{n} s_q(n, k; r) t^k \frac{u^n}{[n]_q!} = \prod_{i=1}^{\infty} \frac{1 + u q^{r+i-1}}{1 + (1-(1-q)t) u q^{i-1}}.$$

and

$$\sum_{n=0}^{\infty}\sum_{k=0}^{n} q^{\binom{k}{2}+rk} S_q(n,k;r) t^k \frac{u^n}{n!} = E_q(-t) \sum_{j=0}^{\infty} e^{[j+r]_q u} \frac{t^j}{[j]_q!},$$

where $E_q(t) = \sum_{k=0}^{\infty} q^{\binom{k}{2}} t^k / [k]_q!$ is a q-exponential function.

3.16 *Noncentral generalized q-factorial coefficients.* Consider the expansion

$$[st+r]_{n,q} = q^{-\binom{n}{2}+rn} \sum_{k=0}^{n} q^{s\binom{k}{2}} C_q(n,k;s,r) [t]_{k,q^s}.$$

The coefficient $C_q(n,k;s,r)$ is the *noncentral generalized q-factorial coefficient*; $C_q(n,k;s) \equiv C_q(n,k;s,0)$ is the *generalized q-factorial coefficient*.

(a) Show that

$$C_q(n,k;s,r) = q^{-r(n-k)} \sum_{j=k}^{n} q^{s\binom{j-k}{2}} \begin{bmatrix} j \\ k \end{bmatrix}_{q^s} [r/s]_{j-k,q^s} C_q(n,j;s)$$

and

$$C_q(n,k;s,r) = \sum_{j=k}^{n} q^{\binom{n-j}{2}-r(n-j)} \begin{bmatrix} n \\ j \end{bmatrix}_q [r]_{n-j,q} C_q(j,k;s).$$

(b) Derive the triangular recurrence relation

$$C_q(n+1,k;s,r) = ([sk]_q - [n-r]_q) C_q(n,k;s,r) + [s]_q C_q(n,k-1;s,r),$$

for $k = 1, 2, \ldots, n+1$, $n = 0, 1, \ldots$, with

$$C_q(0,0;s,r) = 1, \ C_q(n,0;s,r) = q^{\binom{n}{2}-rn} [r]_{n,q}, \ n > 0,$$
$$C_q(n,k;s,r) = 0, \ k > n.$$

3.17 *(Continuation).* Show that

$$C_q(n,k;s,r) = \frac{q^{\binom{n}{2}-s\binom{k}{2}-rn}}{[k]_{q^s}!} \sum_{j=0}^{k} (-1)^{k-j} q^{s\binom{k-j}{2}} \begin{bmatrix} k \\ j \end{bmatrix}_{q^s} [sj+r]_{n,q}$$

and

$$C_q(n,k;s,r) = \frac{[s]_q^k}{(1-q)^{n-k}} \sum_{j=k}^{n} (-1)^{j-k} q^{\binom{n-j}{2}-r(n-j)} \begin{bmatrix} n \\ j \end{bmatrix}_q \begin{bmatrix} j \\ k \end{bmatrix}_{q^s}.$$

3.18 *(Continuation).* Show that

$$C_q(n,k;s,s\rho-r) = q^{-s\rho(n-k)} \sum_{j=k}^{n} s_q(n,j;r) S_{q^s}(j,k;\rho) [s]_q^j$$

and
$$\lim_{s \to 0} [s]_q^{-k} C_q(n, k; s, r) = s_q(n, k; -r),$$
$$\lim_{s \to \infty} q^{r(n-k)} [s]_{q^{1/s}}^{-n} C_{q^{1/s}}(n, k; s, s\rho) = S_q(n, k; \rho).$$

3.19 *(Continuation).* Show that
$$\frac{1}{[t]_{k+1,q^s}} = q^{s\binom{k+1}{2}} \sum_{n=k}^{\infty} q^{-\binom{n+1}{2}+r(n+1)} [s]_q C_q(n, k; s, r) \frac{1}{[st+r]_{n+1,q}}.$$

3.20 *Noncentral signless q-Lah numbers.* Consider the expansion
$$[-(t-r)]_{n,q^{-1}} = q^{\binom{n}{2}-rn} \sum_{k=0}^{n} q^{\binom{k}{2}} L_q(n, k; r) [t]_{k,q}.$$

Since $[-(t-r)]_{n,q^{-1}} = [t - r + n - 1]_{n,q} / [-1]_q^n$ and setting
$$|L_q(n, k; r)| = [-1]_q^n L_q(n, k; r),$$
it can be written as
$$[t - r + n - 1]_{n,q} = q^{\binom{n}{2}-rn} \sum_{k=0}^{n} q^{\binom{k}{2}} |L_q(n, k; r)| [t]_{k,q}.$$

The coefficient $|L_q(n, k; r)|$ is the *noncentral signless q-Lah number*. Show that
$$|L_q(n, k; r)| = q^{-\binom{n}{2}+\binom{k}{2}+r(n-k)} \frac{[n]_q!}{[k]_q!} \begin{bmatrix} n - r - 1 \\ n - k \end{bmatrix}_q.$$

In particular, the *signless q-Lah number* is given by
$$|L_q(n, k)| = q^{-\binom{n}{2}+\binom{k}{2}} \frac{[n]_q!}{[k]_q!} \begin{bmatrix} n - 1 \\ k - 1 \end{bmatrix}_q.$$

4

OCCUPANCY DISTRIBUTIONS

4.1 INTRODUCTION

In occupancy theory, balls (objects) of different kinds are distributed into urns (cells) of different kinds. A collection of m balls of a general specification may be denoted, in partition notation, by $(1^{r_1} 2^{r_2} \cdots m^{r_m})$, where $r_j \geq 0$ is the number of different kinds (of balls), each including j like balls, $j = 1, 2, \ldots, m$, with $r_1 + 2r_2 + \cdots + mr_m = m$. Similarly, a collection of n urns of a general specification may be denoted by $(1^{u_1} 2^{u_2} \cdots n^{u_n})$, where $u_i \geq 0$ is the number of different kinds (of urns), each including i like urns, $i = 1, 2, \ldots, n$, with $u_1 + 2u_2 + \cdots + nu_n = n$. Further, the urns may be of limited or unlimited capacity and the balls in the urns may be ordered or unordered. A host of enumerative problems arising from this model were treated through generating functions by P. A. MacMahon (1915, 1916) and combinatorially by J. Riordan (1958). In the sequel, an urn containing at least one ball is referred to as an *occupied urn*. When the balls are randomly distributed into the urns several interesting random variables may be defined. The number K_0 of empty urns as well as the number $K = n - K_0$ of occupied urns and more generally the number K_i of urns occupied by i balls each have been extensively studied.

Consider the particular case of n distinguishable urns ($u_1 = n$ and $u_i = 0$, $i = 2, 3, \ldots, n$) and let Z_j be the number of balls distributed into the jth urn, $j = 1, 2, \ldots, n$. These random variables are called *occupancy numbers*. The reduction of the joint distribution of the occupancy numbers $Z_1, Z_2, \ldots, Z_n$ to a joint conditional distribution of independent random variables $X_1, X_2, \ldots, X_n$, given that

Combinatorial Methods in Discrete Distributions. By Charalambos A. Charalambides
ISBN 0-471-68027-3 Copyright © 2005 John Wiley & Sons, Inc.

$S_n = \sum_{j=1}^{n} X_j = m$, is a powerful technique in the derivation and study of the distribution of the number K_i of urns occupied by i balls each. In particular, it facilitates the study of the distribution of the number $K = n - K_0$ of occupied urns. Aiming to this reduction, a random occupancy model is introduced, by considering a supply of balls of an unspecified number. Under this model, the occupancy distributions are thoroughly studied.

4.2 A RANDOM OCCUPANCY MODEL

Consider a supply of m balls randomly distributed into n distinguishable urns and let Z_j be the number of balls distributed into the jth urn, $j = 1, 2, \ldots, n$. Further, let

$$P(Z_1 = z_1, Z_2 = z_2, \ldots, Z_n = z_n) = p(z_1, z_2, \ldots, z_n; m),$$

with $z_j = 0, 1, \ldots, m$, $j = 1, 2, \ldots, n$ and $z_1 + z_2 + \cdots + z_n = m$, be the joint probability function of the occupancy numbers $Z_1, Z_2, \ldots, Z_n$. The derivation of the probability function of the number K_i, of urns occupied by i balls each, is based on this joint probability function $p(z_1, z_2, \ldots, z_n; m)$.

A *random occupancy model*, with a more probabilistic flavor in it, which under a general condition is equivalent to the preceding model, may be described as follows. Consider a supply of balls randomly distributed into n distinguishable urns. Let X_j be the number of balls distributed into the jth urn, $j = 1, 2, \ldots, n$. These random variables may be called *random occupancy numbers* in distinction to the usual occupancy numbers. Assume that the random occupancy numbers X_j, $j = 1, 2, \ldots, n$, are independently distributed with probability function

$$P(X_j = x) = q_{j,x}, \; x = 0, 1, \ldots, \; j = 1, 2, \ldots, n.$$

Let $S_n = \sum_{j=1}^{n} X_j$ be the total number of balls distributed into the n urns, with

$$P(S_n = m) = q_m(n), \; m = 0, 1, \ldots.$$

Then

$$q_m(n) = \sum q_{1,x_1} q_{2,x_2} \cdots q_{n,x_n},$$

where the summation is extended over all integers $x_j \geq 0$, $j = 1, 2, \ldots, n$, such that $x_1 + x_2 + \cdots + x_n = m$. Further, the conditional joint probability function

$$P(X_1 = x_1, X_2 = x_2, \ldots, X_n = x_n | S_n = m) = q(x_1, x_2, \ldots, x_n | m),$$

with $x_j = 0, 1, \ldots, m$, $j = 1, 2, \ldots, n$ and $x_1 + x_2 + \cdots + x_n = m$, is given by

$$q(x_1, x_2, \ldots, x_n | m) = \frac{q_{1,x_1} q_{2,x_2} \cdots q_{n,x_n}}{q_m(n)}.$$

If

$$P(Z_1 = x_1, Z_2 = x_2, \ldots, Z_n = x_n)$$
$$= P(X_1 = x_1, X_2 = x_2, \ldots, X_n = x_n | S_n = m),$$

that is, if
$$p(x_1, x_2, \ldots, x_n; m) = q(x_1, x_2, \ldots, x_n | m) = \frac{q_{1,x_1} q_{2,x_2} \cdots q_{n,x_n}}{q_m(n)},$$

the two models are equivalent. For the sake of illustration of the random occupancy model, let us consider the following examples.

Example 4.1 *The classical occupancy model.* Consider a supply of m distinguishable balls randomly and independently distributed into n distinguishable urns and assume that the probability for any ball to fall into the jth urn is p_j, $j = 1, 2, \ldots, n$. Then, the joint distribution of the occupancy numbers $Z_1, Z_2, \ldots, Z_n$ is the multinomial with probability function

$$p(z_1, z_2, \ldots, z_n; m) = \frac{m!}{z_1! z_2! \cdots z_n!} p_1^{z_1} p_2^{z_2} \cdots p_n^{z_n},$$

where $z_j = 0, 1, \ldots, m$, $j = 1, 2, \ldots, n$ and $z_1 + z_2 + \cdots + z_n = m$.

A random occupancy model equivalent to this classical occupancy model may be described as follows. Consider a supply of balls randomly distributed into n distinguishable urns and assume that the random variables X_j, $j = 1, 2, \ldots, n$, are independently distributed with Poisson probability function

$$q_{j,x} = e^{-\lambda_j} \frac{\lambda_j^x}{x!}, \ x = 0, 1, \ldots, \ 0 < \lambda_j < \infty, \ j = 1, 2, \ldots, n.$$

Then, the distribution of the sum $S_n = \sum_{j=1}^n X_j$ is also Poisson with probability function

$$q_m(n) = e^{-\lambda} \frac{\lambda^m}{m!}, \ m = 0, 1, \ldots, \ \lambda = \sum_{j=1}^n \lambda_j.$$

Further, the conditional joint probability function of the random variables $X_1, X_2, \ldots, X_n$, given that $S_n = m$, is given by

$$q(x_1, x_2, \ldots, x_n | m) = \frac{m!}{x_1! x_2! \cdots x_n!} p_1^{x_1} p_2^{x_2} \cdots p_n^{x_n},$$

with $x_j = 0, 1, \ldots, m$, $p_j = \lambda_j / \lambda$, $j = 1, 2, \ldots, n$ and $x_1 + x_2 + \cdots + x_n = m$. Thus,

$$p(x_1, x_2, \ldots, x_n; m) = q(x_1, x_2, \ldots, x_n | m)$$

and so the two models are indeed equivalent. The probability function

$$p_k(m, n) = P(K = k | S_n = m), \ k = 0, 1, \ldots, n,$$

and the binomial moments

$$b_{(r)}(m, n) = E\left[\binom{K}{r} \Big| S_n = m\right], \ r = 0, 1, \ldots,$$

of the number K of occupied urns have been obtained in Example 1.13 as

$$p_k(m,n) = \sum_{r=0}^{k}(-1)^{k-r}\binom{n-r}{k-r}S_{n,n-r}$$

and

$$b_{(r)}(m,n) = \sum_{k=0}^{r}(-1)^k\binom{n-k}{r-k}S_{n,k},$$

with

$$S_{n,n-r} = \sum(p_{j_1} + p_{j_2} + \cdots + p_{j_r})^m,$$

where the summation is extended over all r-combinations $\{j_1, j_2, \ldots, j_r\}$ of the n indices $\{1, 2, \ldots, n\}$.

Example 4.2 *A restricted occupancy model.* Consider a supply of m like balls randomly distributed into n distinguishable urns. Assume that the jth urn is divided into s_j distinguishable cells (compartments), $j = 1, 2, \ldots, n$, each of capacity limited to one ball and set $s = s_1 + s_2 + \cdots + s_n$. Then, the joint distribution of the occupancy numbers $Z_1, Z_2, \ldots, Z_n$ is the multivariate hypergeometric with probability function

$$p(z_1, z_2, \ldots, z_n; m) = \binom{s_1}{z_1}\binom{s_2}{z_2}\cdots\binom{s_n}{z_n}\bigg/\binom{s}{m},$$

where $z_j = 0, 1, \ldots, m$, $j = 1, 2, \ldots, n$ and $z_1 + z_2 + \cdots + z_n = m$.

A random occupancy model equivalent to this restricted occupancy model may be described as follows. Consider a supply of balls randomly distributed into n distinguishable urns and assume that the random variables X_j, $j = 1, 2, \ldots, n$, are independently distributed with binomial probability function

$$q_{j,x} = \binom{s_j}{x}p^x q^{s_j - x}, \ x = 0, 1, \ldots, s_j, \ j = 1, 2, \ldots, n,$$

where $0 < p < 1$ and $q = 1 - p$. Then, the distribution of the sum $S_n = \sum_{j=1}^{n} X_j$ is also binomial with probability function

$$q_m(n) = \binom{s}{m}p^m q^{s-m}, \ m = 0, 1, \ldots, s, \ s = \sum_{j=1}^{n}s_j.$$

Further, the conditional joint probability function of the random variables $X_1, X_2, \ldots, X_n$, given that $S_n = m$, is given by

$$q(x_1, x_2, \ldots, x_n | m) = \binom{s_1}{x_1}\binom{s_2}{x_2}\cdots\binom{s_n}{x_n}\bigg/\binom{s}{m},$$

with $x_j = 0, 1, \ldots, m$, $j = 1, 2, \ldots, n$ and $x_1 + x_2 + \cdots + x_n = m$. Thus,

$$p(x_1, x_2, \ldots, x_n; m) = q(x_1, x_2, \ldots, x_n | m)$$

and so the two models are indeed equivalent. The probability function

$$p_k(m,n) = P(K = k | S_n = m), \quad k = 0, 1, \ldots, n,$$

and the binomial moments

$$b_{(r)}(m,n) = E\left[\binom{K}{r} \Big| S_n = m\right], \quad r = 0, 1, \ldots,$$

of the number K of occupied urns have been obtained in Example 1.14 as

$$p_k(m,n) = \sum_{r=0}^{k} (-1)^{k-r} \binom{n-r}{k-r} S_{n,n-r}$$

and

$$b_{(r)}(m,n) = \sum_{k=0}^{r} (-1)^k \binom{n-k}{r-k} S_{n,k},$$

with

$$S_{n,n-r} = \sum \frac{(s_{j_1} + s_{j_2} + \cdots + s_{j_r})_m}{(s)_m},$$

where the summation is extended over all r-combinations $\{j_1, j_2, \ldots, j_r\}$ of the n indices $\{1, 2, \ldots, n\}$.

Example 4.3 *A pseudo-contagious occupancy model.* Consider a supply of m like balls randomly distributed into n distinguishable urns. Assume that the jth urn is divided into s_j distinguishable cells of unlimited capacity, $j = 1, 2, \ldots, n$, and set $s = s_1 + s_2 + \cdots + s_n$. Then, the joint distribution of the occupancy numbers $Z_1, Z_2, \ldots, Z_n$ is the multivariate negative hypergeometric with probability function

$$p(z_1, z_2, \ldots, z_n; m) = \frac{\binom{s_1 + z_1 - 1}{z_1}\binom{s_2 + z_2 - 1}{z_2} \cdots \binom{s_n + z_n - 1}{z_n}}{\binom{s + m - 1}{m}},$$

where $z_j = 0, 1, \ldots, m$, $j = 1, 2, \ldots, n$ and $z_1 + z_2 + \cdots + z_n = m$.

A random occupancy model equivalent to this restricted occupancy model is deduced by assuming that the random variables X_j, $j = 1, 2, \ldots, n$, are independently distributed with negative binomial probability function

$$q_{j,x} = \binom{s_j + x - 1}{x} p^{s_j} q^x, \quad x = 0, 1, \ldots, \; j = 1, 2, \ldots, n,$$

where $0 < p < 1$ and $q = 1 - p$. Then, the distribution of the sum $S_n = \sum_{j=1}^{n} X_j$ is also negative binomial with probability function

$$q_m(n) = \binom{s + m - 1}{m} p^s q^m, \quad m = 0, 1, \ldots, \; s = \sum_{j=1}^{n} s_j.$$

Further, the conditional joint probability function $q(x_1, x_2, \ldots, x_n|m)$ of the random variables $X_1, X_2, \ldots, X_n$, given that $S_n = m$, equals $p(x_1, x_2, \ldots, x_n; m)$ and so the two models are equivalent. The probability function

$$p_k(m, n) = P(K = k|S_n = m), \ k = 0, 1, \ldots, n,$$

and the binomial moments

$$b_{(r)}(m, n) = E\left[\binom{K}{r}\bigg|S_n = m\right], \ r = 0, 1, \ldots,$$

of the number K of occupied urns may be obtained as (cf. Example 1.14)

$$p_k(m, n) = \sum_{r=0}^{k}(-1)^{k-r}\binom{n-r}{k-r}S_{n,n-r}$$

and

$$b_{(r)}(m, n) = \sum_{k=0}^{r}(-1)^{k}\binom{n-k}{r-k}S_{n,k},$$

with

$$S_{n,n-r} = \sum \frac{(s_{j_1} + s_{j_2} + \cdots + s_{j_r} + m - 1)_m}{(s + m - 1)_m},$$

where the summation is extended over all r-combinations $\{j_1, j_2, \ldots, j_r\}$ of the n indices $\{1, 2, \ldots, n\}$.

In the sequel, the study of occupancy distributions focuses on the particular case of the random occupancy model in which the random variables $X_j, j = 1, 2, \ldots, n$, are independent and identically distributed.

4.3 OCCUPANCY DISTRIBUTIONS

Consider a supply of balls randomly distributed into n distinguishable urns. Let X_j be the number of balls distributed into the jth urn, $j = 1, 2, \ldots, n$. Assume that the random variables $X_j, j = 1, 2, \ldots, n$, are independently and identically distributed, with common probability function

$$P(X = x) = q_x, \ x = 0, 1, \ldots. \tag{4.1}$$

Then

$$P(S_n = m) = q_m(n) = \sum q_{x_1} q_{x_2} \cdots q_{x_n}, \ m = 0, 1, \ldots, \tag{4.2}$$

where the summation is extended over all integers $x_j \geq 0, j = 1, 2, \ldots, n$, such that $x_1 + x_2 + \cdots + x_n = m$, is the n-fold convolution of $q_x, x = 0, 1, \ldots$. Further, the conditional joint probability function of the random variables $X_1, X_2, \ldots, X_n$, given that $S_n = m$, is given by

$$q(x_1, x_2, \ldots, x_n | m) = \frac{q_{x_1} q_{x_2} \cdots q_{x_n}}{q_m(n)}, \quad (4.3)$$

with $x_j = 0, 1, \ldots, m$, $j = 1, 2, \ldots, n$ and $x_1 + x_2 + \cdots + x_n = m$. The additional assumption that the random variables X_j, $j = 1, 2, \ldots, n$, are identically distributed leads to considerably simplified and more manageable expressions for the probability function and factorial moments of the occupancy distributions. Under this random occupancy model, let K_i be the number of urns occupied by i balls each. Also, let

$$p_k(m, n, i) = P(K_i = k | S_n = m), \quad k = 0, 1, \ldots, n, \quad (4.4)$$

and

$$b_{(r)}(m, n, i) = E\left[\binom{K_i}{r} \bigg| S_n = m\right], \quad r = 0, 1, \ldots, \quad (4.5)$$

with $b_{(0)}(m, n, i) = 1$. The probability function (4.4) and its binomial moments (4.5) may be expressed in terms of finite differences of the convolution (4.2).

Theorem 4.1 (a) *Under the random occupancy model with $q_i > 0$, the conditional probability $p_k(m, n, i)$, that k urns are occupied by i balls each, given that m balls are distributed into the n distinguishable urns, $k = 0, 1, \ldots, n$, is given by*

$$p_k(m, n, i) = \binom{n}{k} \frac{\left[\Delta_u^{n-k} q_i^{n-u} q_{m-i(n-u)}(u)\right]_{u=0}}{q_m(n)}. \quad (4.6)$$

(b) *The rth binomial moment $b_{(r)}(m, n, i)$, $r = 0, 1, \ldots$, of the probability function $p_k(m, n, i)$, $k = 0, 1, \ldots, n$, is given by*

$$b_{(r)}(m, n, i) = \binom{n}{r} \frac{q_i^r q_{m-ir}(n-r)}{q_m(n)}. \quad (4.7)$$

(c) *Under the random occupancy model with $q_i > 0$, the conditional probability $R_k(m, n, i)$, that at least k urns are occupied by i balls each, given that m balls are distributed into the n distinguishable urns, $k = 0, 1, \ldots, n$, is given by*

$$R_k(m, n, i) = n \binom{n-1}{k-1} \frac{\left[\Delta_u^{n-k} q_i^{n-u} q_{m-i(n-u)}(u)/(n-u)\right]_{u=0}}{q_m(n)}. \quad (4.8)$$

Proof (a) Under the random occupancy model, let A_j be the event that the jth urn contains i balls, $j = 1, 2, \ldots, n$, given that m balls are distributed into the n urns. Then $p_k(m, n, i)$ is the probability that k among the n events $A_1, A_2, \ldots, A_n$ occur. Further, for any r indices $\{j_1, j_2, \ldots, j_r\}$ from the set $\{1, 2, \ldots, n\}$ and $\{j_{r+1}, j_{r+2}, \ldots, j_n\} = \{1, 2, \ldots, n\} - \{j_1, j_2, \ldots, j_r\}$, it follows from (4.3) that

$$P(A_{j_1} A_{j_2} \cdots A_{j_r}) = \frac{q_i^r}{q_m(n)} \sum q_{x_{j_{r+1}}} q_{x_{j_{r+2}}} \cdots q_{x_{j_n}},$$

where the summation is extended over all $x_{j_s} \geq 0$, $s = r+1, r+2, \ldots, n$, such that $x_{j_{r+1}} + x_{j_{r+2}} + \cdots + x_{j_n} = m - ir$. Then, on using (4.2), we find

$$p_r = P(A_{j_1} A_{j_2} \cdots A_{j_r}) = \frac{q_i^r q_{m-ir}(n-r)}{q_m(n)}, \quad r = 1, 2, \ldots, n.$$

Therefore, the events $A_1, A_2, \ldots, A_n$ are exchangeable, and so applying Corollary 1.5, we get for the probability $p_k(m, n, i)$ expression (4.6).

(b) The rth binomial moment $b_{(r)}(m, n, i)$, of the number K_i of urns occupied by i balls each, by applying Corollary 1.6, is obtained as

$$b_{(r)}(m, n, i) = \binom{n}{r} p_r = \binom{n}{r} \frac{q_i^r q_{m-ir}(n-r)}{q_m(n)}, \quad r = 0, 1, \ldots,$$

and (4.7) is established.

(c) The conditional probability $R_k(m, n, i)$ equals the probability that at least k among the n events $A_1, A_2, \ldots, A_n$ occur. Since these events are exchangeable, applying Corollary 1.7, we deduce (4.8) and the proof of the theorem is completed.

Consider the number K of occupied urns. Since $K = n - K_0$, the conditional probability function

$$p_k(m, n) = P(K = k | S_n = m), \quad k = 0, 1, \ldots, n,$$

its binomial moments

$$b_{(r)}(m, n) = E\left[\binom{K}{r} \bigg| S_n = m\right], \quad r = 0, 1, \ldots,$$

and the conditional distribution function

$$P_k(m, n) = P(K \leq k | S_n = m), \quad k = 0, 1, \ldots, n,$$

may be deduced from (4.6), (4.7) and (4.8), respectively.

Corollary 4.1 (a) *Under the random occupancy model with $q_0 > 0$, the conditional probability $p_k(m, n)$, that k urns are occupied, given that m balls are distributed into the n distinguishable urns, $k = 0, 1, \ldots, n$, is given by*

$$p_k(m, n) = \binom{n}{k} \frac{[\Delta_u^k q_0^{n-u} q_m(u)]_{u=0}}{q_m(n)}. \tag{4.9}$$

(b) *The rth binomial moment $b_{(r)}(m, n)$, $r = 0, 1, \ldots,$ of the probability function $p_k(m, n)$, $k = 0, 1, \ldots, n$, is given by*

$$b_{(r)}(m, n) = \binom{n}{r} \frac{[\Delta_u^r q_0^{n-u} q_m(u)]_{u=n-r}}{q_m(n)}. \tag{4.10}$$

(c) *Under the random occupancy model with $q_0 > 0$, the conditional probability $P_k(m, n)$, that at most k urns are occupied, given that m balls are distributed into the n distinguishable urns, $k = 0, 1, \ldots, n$, is given by*

$$P_k(m, n) = n \binom{n-1}{k} \frac{[\Delta_u^k q_0^{n-u} q_m(u)/(n-u)]_{u=0}}{q_m(n)}. \tag{4.11}$$

Proof (a) The probability function (4.9) follows directly from (4.6) by putting $i = 0$ and replacing $n - k$ by k.

(b) The jth binomial moment of the number K_0 of empty urns, by virtue of (4.7), is given by

$$E\left[\binom{K_0}{j} \Big| S_n = m\right] = \binom{n}{j} \frac{q_0^j q_m(n-j)}{q_m(n)}.$$

Further, the rth binomial moment of $K = n - K_0$ may be written as

$$b_{(r)}(m, n) = E\left[\binom{K}{r} \Big| S_n = m\right] = E\left[\binom{n - K_0}{r} \Big| S_n = m\right]$$

$$= (-1)^r E\left[\binom{K_0 - n + r - 1}{r} \Big| S_n = m\right]$$

and since, by Theorem 1.11,

$$E\left[\binom{K_0 - n + r - 1}{r} \Big| S_n = m\right] = \sum_{j=0}^{r}(-1)^{r-j}\binom{n-j}{r-j} E\left[\binom{K_0}{j} \Big| S_n = m\right],$$

it follows that

$$b_{(r)}(m, n) = \sum_{j=0}^{r}(-1)^j \binom{n-j}{r-j}\binom{n}{j} \frac{q_0^j q_m(n-j)}{q_m(n)}$$

$$= \binom{n}{r}\sum_{j=0}^{r}(-1)^{r-j}\binom{r}{j} \frac{q_0^{n-(n-r+j)} q_m(n-r+j)}{q_m(n)}.$$

The last expression, on introducing the finite difference operator, yields (4.10).

(c) The conditional probability $P_k(m, n)$, that at most k urns are occupied, given that m balls are distributed into the n distinguishable urns, equals the conditional probability $R_{n-k}(m, n, 0)$, that at least $n - k$ urns are empty, given that m balls are distributed into the n distinguishable urns. Thus, $P_k(m, n)$ may be deduced from (4.8) by putting $i = 0$ and replacing $n - k$ by k.

Example 4.4 *Stochastic models in statistical physics.* Consider a physical system composed of m particles such as electrons, neutrons, protons or photons. In statistical mechanics it is assumed that there are n distinct energy levels in which each particle can be. The overall energy level of a system of particles may be described by an ordered n-tuple $(z_1, z_2, \ldots, z_n)$, where z_j is the number of particles in the jth energy level, $j = 1, 2, \ldots, n$. According to whether the particles are distinguishable or indistinguishable and obey or do not obey the Pauli exclusion principle, three models have been studied. Recall that the Pauli exclusion principle states that there cannot be more than one particle in any particular energy level.

The *Maxwell-Boltzmann* model (or statistics in physicists' terminology) assumes that the particles are distinguishable and do not obey the Pauli exclusion principle. The classical theory of gases (at low densities and not very low temperatures) was based on this model. In modern theory it was shown beyond doubt that the Maxwell-Boltzmann model does not describe the behavior of any known particles.

Further, two far more plausible models were proposed. In both the *Bose-Einstein* and the *Fermi-Dirac* models the particles are assumed to be indistinguishable. In addition, in the first model the particles are not required to obey the Pauli exclusion principle, while in the second model they are assumed to satisfy the Pauli exclusion principle. Particles having integer spin, such as photons and pions, obey the assumptions of the Bose-Einstein model, while particles with half-integer spin, such as electrons and protons, obey the assumptions of the Fermi-Dirac model.

Clearly, the general stochastic model is an occupancy model, with the particles considered as balls and the distinct energy levels as distinguishable urns. Thus, if the number X of particles that are in any specific energy level obeys a general probability function $q_x = P(X = x)$, $x = 0, 1, \ldots$, then the conditional probability function of the number K_i of energy levels with i particles in each, given that the physical system is composed of m particles, is given by (4.6). In particular, the conditional probability function of the number K of energy levels with at least one particle, given that the system is composed of m particles, is given by (4.9). In the next section it will be seen that the assumptions of the Maxwell-Boltzmann, Bose-Einstein and Fermi-Dirac models are equivalent to the assumption that X obeys a Poisson, geometric and (zero-one) Bernoulli distributions, respectively.

In statistical inference, the most widely used family of discrete distributions, as a population (parent) distribution, is the power series distribution introduced by A. Noack (1950). Its probability function is defined by

$$q_x = P(X = x) = [g(\theta)]^{-1} c_x \theta^x, \; x = 0, 1, \ldots, \tag{4.12}$$

where the function $g(\theta)$ admits a power series expansion,

$$g(\theta) = \sum_{x=0}^{\infty} c_x \theta^x, \; 0 < \theta < \rho,$$

and the coefficient $c_x = [D_\theta^x g(\theta)]_{\theta=0}/x! \geq 0$ does not involve the parameter θ, constitutes a wide class of discrete distributions. The Poisson, binomial (with a known number of trials), negative binomial (with a known number of successes) and logarithmic distributions belong in this class. For $q_0 > 0$, without any loss of generality it is assumed that $c_0 = 1$. Its n-fold convolution obeys also a power series distribution with probability function

$$q_m(n) = P(S_n = m) = [g(\theta)]^{-n} c_m(n) \theta^m, \; m = 0, 1, \ldots, \tag{4.13}$$

with
$$c_m(n) = \sum c_{x_1} c_{x_2} \cdots c_{x_n}, \; m = 0, 1, \ldots,$$
where the summation is extended over all integers $x_j \geq 0$, $j = 1, 2, \ldots, n$, such that $x_1 + x_2 + \cdots + x_n = m$. Alternatively,
$$c_m(n) = \frac{[D_\theta^m (g(\theta))^n]_{\theta=0}}{m!}.$$

Another wide class of discrete distributions, emerging in occupancy theory, is the factorial series distribution introduced by S. Berg (1974). Its probability function is defined by
$$p_k = P(K = k) = [f(n)\}]^{-1} c_k \cdot (n)_k, \; k = 0, 1, \ldots, \quad (4.14)$$
where the function $f(n)$ admits a factorial series expansion
$$f(n) = \sum_{k=0}^{\infty} c_k \cdot (n)_k, \text{ for integer values of } n,$$
and the coefficient $c_k = [\Delta_u^k f(u)]_{u=0}/k! \geq 0$ does not involve the parameter n. The binomial and negative binomial distributions (with known probability of success), the hypergeometric and negative (inverse) hypergeometric and more generally the Pólya and inverse Pólya distributions belong in this class. Further, the conditional probability that k urns are occupied, given that m balls are distributed into n distinguishable urns, when the number of balls distributed into any specific urn obeys a power series distribution, is a factorial series distribution. More precisely, by introducing expressions (4.12) and (4.13) into Corollary 4.1 the following corollary is deduced.

Corollary 4.2 (a) *Under the random occupancy model with power series probabilities,*
$$q_x = P(X = x) = [g(\theta)]^{-1} c_x \theta^x, \; x = 0, 1, \ldots, \; 0 < \theta < \rho,$$
the conditional probability $p_k(m, n)$, that k urns are occupied, given that m balls are distributed into the n distinguishable urns, $k = 0, 1, \ldots, n$, is given by
$$p_k(m, n) = \binom{n}{k} \frac{[\Delta_u^k c_m(u)]_{u=0}}{c_m(n)}. \quad (4.15)$$

(b) *The rth binomial moment $b_{(r)}(m, n)$, $r = 0, 1, \ldots, n$, of the probability function $p_k(m, n)$, $k = 0, 1, \ldots, n$, is given by*
$$b_{(r)}(m, n) = \binom{n}{r} \frac{[\Delta_u^r c_m(u)]_{u=n-r}}{c_m(n)}, \quad (4.16)$$
where $c_m(n) = [D_\theta^m (g(\theta))^n]_{\theta=0}/m!$.

4.4 PARTICULAR OCCUPANCY DISTRIBUTIONS

4.4.1 Classical Occupancy Distribution

Assume that the number X_j, of balls distributed into the jth urn, $j = 1, 2, \ldots, n$, obeys a *Poisson distribution* with

$$q_x = P(X = x) = e^{-\lambda}\frac{\lambda^x}{x!}, \; x = 0, 1, \ldots, \; 0 < \lambda < \infty. \tag{4.17}$$

Then, its u-fold convolution obeys a Poisson distribution with

$$q_m(u) = P(S_u = m) = e^{-u\lambda}\frac{(u\lambda)^m}{m!}, \; m = 0, 1, \ldots. \tag{4.18}$$

Note that the conditional joint probability function of the random variables $X_1, X_2, \ldots, X_n$, given that $S_n = m$, is given by

$$q(x_1, x_2, \ldots, x_n|m) = \frac{m!}{x_1!x_2!\cdots x_n!} \cdot \frac{1}{n^m},$$

with $x_j = 0, 1, \ldots, m$, $j = 1, 2, \ldots, n$ and $x_1 + x_2 + \cdots + x_n = m$. Therefore, this particular random occupancy model, with the assumption that the number of balls distributed into any specific urn obeys a Poisson distribution, is equivalent to the stochastic model in which a fixed number m of distinguishable balls are randomly distributed into n distinguishable urns (Maxwell-Boltzmann model, classical occupancy model.) Upon introducing expressions (4.17) and (4.18) into Theorem 4.1 the following corollary is deduced.

Corollary 4.3 (a) *Under the random occupancy model with Poisson probabilities, the conditional probability $p_k(m, n, i)$, that k urns are occupied by i balls each, given that m balls are distributed into the n distinguishable urns, $k = 0, 1, \ldots, n$, is given by*

$$p_k(m, n, i) = \binom{n}{k}\left[\Delta_u^{n-k}\frac{(m)_{i(n-u)}u^{m-i(n-u)}}{(i!)^{n-u}n^m}\right]_{u=0}$$

$$= \binom{n}{k}\sum_{j=0}^{n-k}(-1)^{n-k-j}\binom{n-k}{j}\frac{(m)_{i(n-j)}j^{m-i(n-j)}}{(i!)^{n-j}n^m}. \tag{4.19}$$

(b) *The rth binomial moment $b_{(r)}(m, n, i)$, $r = 0, 1, \ldots$, of the probability function $p_k(m, n, i)$, $k = 0, 1, \ldots, n$, is given by*

$$b_{(r)}(m, n, i) = \binom{n}{r}\frac{(m)_{ir}(n-r)^{m-ir}}{(i!)^r n^m}. \tag{4.20}$$

Similarly, from Corollary 4.1, the *classical occupancy distribution* is deduced in the following corollary.

Corollary 4.4 (a) *Under the random occupancy model with Poisson probabilities, the conditional probability $p_k(m,n)$, that k urns are occupied, given that m balls are distributed into the n distinguishable urns, $k = 0, 1, \ldots, n$, is given by*

$$p_k(m,n) = \frac{(n)_k S(m,k)}{n^m}, \qquad (4.21)$$

where $S(m,k) = [\Delta_u^k u^m]_{u=0}/k!$ is the Stirling number of the second kind.

(b) *The rth binomial moment $b_{(r)}(m,n)$, $r = 0, 1, \ldots$, of the probability function $p_k(m,n)$, $k = 0, 1, \ldots, n$, is given by*

$$b_{(r)}(m,n) = \frac{(n)_r S(m,r;n-r)}{n^m}, \qquad (4.22)$$

where $S(m,r;n-r) = [\Delta_u^r u^m]_{u=n-r}/r!$ is the noncentral Stirling number of the second kind.

The expected value of the classical occupancy distribution, on using (4.22) and (2.15), is obtained as

$$E(K) = n \sum_{j=1}^{\infty} \binom{m}{j} \left(\frac{1}{n}\right)^j \left(1 - \frac{1}{n}\right)^{m-j} S(j,1)$$

and since $S(j,1) = 1$, $j = 1, 2, \ldots$,

$$E(K) = n\left[1 - \left(1 - \frac{1}{n}\right)^m\right].$$

Similarly,

$$E\left[\binom{K}{2}\right] = (n)_2 \sum_{j=2}^{m} \binom{m}{j} \left(\frac{1}{n}\right)^j \left(1 - \frac{2}{n}\right)^{m-j} S(j,2)$$

and since, by (2.38),

$$S(j,2) = \sum_{i=1}^{j-1} \binom{j-1}{i} S(i,1) = \sum_{i=1}^{j-1} \binom{j-1}{i} = 2^{j-1} - 1, \quad j = 2, 3, \ldots,$$

it follows that

$$E\left[\binom{K}{2}\right] = \binom{n}{2} \sum_{j=1}^{m} \binom{m}{j} \left(\frac{2}{n}\right)^j \left(1 - \frac{2}{n}\right)^{m-j}$$
$$- (n)_2 \sum_{j=1}^{m} \binom{m}{j} \left(\frac{1}{n}\right)^j \left(1 - \frac{2}{n}\right)^{m-j}.$$

172   OCCUPANCY DISTRIBUTIONS

Thus, carrying out the summations,

$$E\left[\binom{K}{2}\right] = \binom{n}{2}\left[1 - \left(\frac{2}{n}\right)^m\right] - (n)_2\left[\left(1 - \frac{1}{n}\right)^m - \left(1 - \frac{2}{n}\right)^m\right].$$

Clearly, the variance of the classical occupancy distribution equals

$$V(K) = (n)_2\left[1 - \left(1 - \frac{2}{n}\right)^m\right] - 2(n)_2\left[\left(1 - \frac{1}{n}\right)^m - \left(1 - \frac{2}{n}\right)^m\right]$$
$$+ n\left[1 - \left(1 - \frac{1}{n}\right)^m\right] - n^2\left[1 - 2\left(1 - \frac{1}{n}\right)^m + \left(1 - \frac{1}{n}\right)^{2m}\right],$$

which after some algebraic manipulations reduces to

$$V(K) = n\left(1 - \frac{1}{n}\right)^m\left[1 - \left(1 - \frac{1}{n}\right)^m\right]$$
$$- (n)_2\left(1 - \frac{1}{n}\right)^m\left[\left(1 - \frac{1}{n}\right)^m - \left(1 - \frac{1}{n-1}\right)^m\right].$$

Example 4.5 *Usable numbers in a sample of random numbers.* Consider a file of n serially numbered cards, $\{1, 2, \ldots, n\}$, that refer to customers' accounts, one account per card. Suppose that a random sample of size m is selected from the file by employing a suitable table of random numbers. This selection is equivalent to a random selection of the m numbers from the set $\{1, 2, \ldots, n\}$ with replacement. Further, the n^m possible samples are equally probable. Since some of the selected numbers may be duplicates of others, not all m numbers are usable. The number K of usable numbers in the sample is a random variable. Its probability function and moments are of interest.

Considering the n cards as n distinguishable urns and interpreting the selection of the jth card as a placement of a ball into the jth urn, then K is the number of occupied urns. Clearly, the joint probability function of the occupancy numbers $Z_1, Z_2, \ldots, Z_n$ is a multinomial with parameters m and $p_j = 1/n, j = 1, 2, \ldots, n$. This is equivalent to the assumption that for a random sample of an unspecified size the random variables $X_1, X_2, \ldots, X_n$ are independently and identically distributed with a Poisson distribution. Therefore, the probability function and factorial moments of K are given by (4.21) and (4.22), respectively.

In practice, only a subset $\{i_1, i_2, \ldots, i_r\}$ of not necessarily serially numbered cards refer to customers' accounts. This subset is referred to as the *target set*. In this case, among the selected numbers, there may be numbers not referring to customers' accounts. The number Y of usable numbers, among m selected numbers, is a random variable with a binomial distribution

$$P(Y = y) = \binom{m}{y}\left(\frac{r}{n}\right)^y\left(1 - \frac{r}{n}\right)^{m-y}, \quad y = 0, 1, \ldots, m.$$

The probability function of the number K of usable numbers given that y numbers belong in the target set, according to (4.21), is

$$P(K = k | Y = y) = \frac{(r)_k S(y, k)}{r^y}, \quad k = 0, 1, \ldots, r,$$

where $S(y, k) = [\Delta_u^k u^y]_{u=0}/k!$ is the Stirling number of the second kind. Thus

$$P(K = k) = \sum_{y=0}^{m} P(Y = y) P(K = k | Y = y)$$

$$= \frac{(r)_k}{n^m k!} \left[\Delta_u^k \sum_{y=0}^{m} \binom{m}{y} u^y (n - r)^{m-y} \right]_{u=0}$$

and so

$$P(K = k) = \frac{(r)_k}{n^m k!} \left[\Delta_u^k (u + n - r)^m \right]_{u=0}, \quad k = 0, 1, \ldots, r,$$

which, upon introducing the noncentral Stirling number of the second kind $S(m, k; r)$ $= [\Delta_u^k u^m]_{u=r}$, may be expressed as

$$P(K = k) = \frac{(r)_k S(m, k; n - r)}{n^m}, \quad k = 0, 1, \ldots, r.$$

Also, according to (4.22),

$$E\left[\binom{K}{j} \middle| Y = y\right] = \frac{(r)_j S(y, j; r - j)}{r^y}, \quad j = 0, 1, \ldots, r$$

and so

$$E\left[\binom{K}{j}\right] = \sum_{y=0}^{m} P(Y = y) E\left[\binom{K}{j} \middle| Y = y\right]$$

$$= \frac{(r)_j}{n^m j!} \left[\Delta_u^j \sum_{y=0}^{m} \binom{m}{y} u^y (n - r)^{m-y} \right]_{u=r-j}$$

$$= \frac{(r)_j}{n^m j!} \left[\Delta_u^j (u + n - r)^m \right]_{u=r-j}.$$

Therefore,

$$E\left[\binom{K}{j}\right] = \frac{(r)_j S(m, j; n - j)}{n^m}, \quad j = 0, 1, \ldots, r.$$

4.4.2 Restricted Occupancy Distribution

Assume that the number X_j, of balls distributed into the jth urn, $j = 1, 2, \ldots, n$, obeys a *Bernoulli distribution* with

$$q_x = P(X = x) = p^x q^{1-x}, \quad x = 0, 1,$$

where $0 < p < 1$ and $q = 1 - p$. Then, its u-fold convolution obeys a binomial distribution with

$$q_m(u) = P(S_u = m) = \binom{u}{m} p^m q^{u-m}, \quad m = 0, 1, \ldots, u.$$

Note that the conditional joint probability function of the random variables $X_1, X_2, \ldots, X_n$, given that $S_n = m$, is given by

$$q(x_1, x_2, \ldots, x_n | m) = 1 \bigg/ \binom{n}{m},$$

with $x_j = 0, 1$, $j = 1, 2, \ldots, n$ and $x_1 + x_2 + \cdots + x_n = m$. Therefore, this particular random occupancy model, with the assumption that the number of balls distributed into any specific urn obeys a Bernoulli distribution, is equivalent to the stochastic model in which a fixed number m of like (indistinguishable) balls are randomly distributed into n distinguishable urns, each with capacity limited to one ball (Fermi-Dirac model.)

A more general random occupancy model is deduced by assuming that the number X_j, of balls distributed into the jth urn, $j = 1, 2, \ldots, n$, obeys a *binomial distribution* with

$$q_x = P(X = x) = \binom{s}{x} p^x q^{s-x}, \quad x = 0, 1, \ldots, s, \tag{4.23}$$

where $0 < p < 1$ and $q = 1 - p$. Then, its u-fold convolution obeys a binomial distribution with

$$q_m(u) = P(S_u = m) = \binom{su}{m} p^m q^{su-m}, \quad m = 0, 1, \ldots, su. \tag{4.24}$$

Note that the conditional joint probability function of the random variables $X_1, X_2, \ldots, X_n$, given that $S_n = m$, is given by

$$q(x_1, x_2, \ldots, x_n | m) = \binom{s}{x_1}\binom{s}{x_2}\cdots\binom{s}{x_n} \bigg/ \binom{sn}{m},$$

with $x_j = 0, 1, \ldots, m$, $j = 1, 2, \ldots, n$ and $x_1 + x_2 + \cdots + x_n = m$. Therefore, this particular random occupancy model, with the assumption that the number of balls distributed into any specific urn obeys a binomial distribution, is equivalent to the stochastic model in which a fixed number m of like (indistinguishable) balls are randomly distributed into n distinguishable urns, each with s distinguishable compartments of capacity limited to one ball (restricted occupancy model.) Upon introducing expressions (4.23) and (4.24) into Theorem 4.1 the following corollary is deduced.

Corollary 4.5 (a) *Under the random occupancy model with binomial probabilities, the conditional probability $p_k(m, n, i)$, that k urns are occupied by i balls each, given* that m balls are distributed into the n distinguishable urns, $k = 0, 1, \ldots, n$, is given by

$$p_k(m,n,i) = \binom{n}{k}\left[\Delta_u^{n-k}\frac{(m)_{i(n-u)}(s)_i^{n-u}(su)_{m-i(n-u)}}{(i!)^{n-u}(sn)_m}\right]_{u=0}$$

$$= \binom{n}{k}\sum_{j=0}^{n-k}(-1)^{n-k-j}\binom{n-k}{j}\frac{(m)_{i(n-j)}(s)_i^{n-j}(sj)_{m-i(n-j)}}{(i!)^{n-j}(sn)_m}. \quad (4.25)$$

(b) *The rth binomial moment* $b_{(r)}(m, n, i)$, $r = 0, 1, \ldots,$ *of the probability function* $p_k(m, n, i)$, $k = 0, 1, \ldots, n$, *is given by*

$$b_{(r)}(m,n,i) = \binom{n}{r}\frac{(m)_{ir}(s)_i^r(sn-sr)_{m-ir}}{(i!)^r(sn)_m}. \quad (4.26)$$

Similarly, from Corollary 4.1, the *restricted occupancy distribution* is deduced in the following corollary.

Corollary 4.6 (a) *Under the random occupancy model with binomial probabilities, the conditional probability* $p_k(m, n)$, *that k urns are occupied, given that m balls are distributed into the n distinguishable urns*, $k = 0, 1, \ldots, n$, *is given by*

$$p_k(m,n) = \frac{(n)_k C(m,k;s)}{(sn)_m}, \quad (4.27)$$

where $C(m, k; s) = [\Delta_u^k(su)_m]_{u=0}/k!$ is the generalized factorial coefficient.

(b) *The rth binomial moment* $b_{(r)}(m, n)$, $r = 0, 1, \ldots,$ *of the probability function* $p_k(m, n)$, $k = 0, 1, \ldots, n$, *is given by*

$$b_{(r)}(m,n) = \frac{(n)_r C(m,r;s,s(n-r))}{(sn)_m}, \quad (4.28)$$

where $C(m, r; s, s(n-r)) = [\Delta_u^r(su)_m]_{u=n-r}/r!$ is the noncentral generalized factorial coefficient.

The expected value of the restricted occupancy distribution, on using (4.28) and (2.56), is deduced as

$$E(K) = n\sum_{j=1}^{m}\binom{m}{j}\frac{(s(n-1))_{m-j}}{(sn)_m}C(j,1;s)$$

and since $C(j, 1; s) = (s)_j$, $j = 1, 2, \ldots,$ it follows, by Vandermonde's formula, that

$$E(K) = n\left[1 - \frac{(s(n-1))_m}{(sn)_m}\right].$$

Similarly,

$$E\left[\binom{K}{2}\right] = (n)_2\sum_{j=2}^{m}\binom{m}{j}\frac{(s(n-2))_{m-j}}{(sn)_m}C(j,2;s)$$

and since, by (2.71),

$$C(j,2;s) = \sum_{i=1}^{j-1} \binom{j-1}{i}(s)_{j-i}C(i,1;s) = s\sum_{i=1}^{j-1}\binom{j-1}{i}(s)_i(s-1)_{j-1-i}$$

$$= s[(2s-1)_{j-1} - (s-1)_{j-1}] = \frac{1}{2}(2s)_j - (s)_j,$$

it follows that

$$E\left[\binom{K}{2}\right] = \binom{n}{2}\sum_{j=1}^{m}\binom{m}{j}\frac{(2s)_j(s(n-2))_{m-j}}{(sn)_m}$$

$$- (n)_2\sum_{j=1}^{m}\binom{m}{j}\frac{(s)_j(s(n-2))_{m-j}}{(sn)_m}.$$

Hence, by carrying out the summations,

$$E\left[\binom{K}{2}\right] = \binom{n}{2}\left[1 - \frac{(s(n-2))_m}{(sn)_m}\right] - (n)_2\left[\frac{(s(n-1))_m}{(sn)_m} - \frac{(s(n-2))_m}{(sn)_m}\right].$$

Clearly, the variance of the restricted occupancy distribution equals

$$V(K) = (n)_2\left[1 - \frac{(s(n-2))_m}{(sn)_m}\right] - 2(n)_2\left[\frac{(s(n-1))_m}{(sn)_m} - \frac{(s(n-2))_m}{(sn)_m}\right]$$

$$+ n\left[1 - \frac{(s(n-1))_m}{(sn)_m}\right] - n^2\left[1 - \frac{2(s(n-1))_m}{(sn)_m} + \left(\frac{(s(n-1))_m}{(sn)_m}\right)^2\right],$$

which after some algebraic manipulations reduces to

$$V(K) = n\frac{(s(n-1))_m}{(sn)_m}\left[1 - \frac{(s(n-1))_m}{(sn)_m}\right]$$

$$- (n)_2\frac{(s(n-1))_m}{(sn)_m}\left[\frac{(s(n-1))_m}{(sn)_m} - \frac{(s(n-2))_m}{(s(n-1))_m}\right].$$

Example 4.6 *Hemacytometer counts.* Assume that a total of m red blood cells are randomly distributed into the n squares of the hemacytometer grid. Each of the n squares is divided into s compartments (smaller squares) with capacity limited to one blood cell. Let Z_j be the number of red blood cells accommodated into the jth square, $j = 1, 2, \ldots, n$. Then the joint probability function of the random variables $Z_1, Z_2, \ldots, Z_n$ is given by

$$p(z_1, z_2, \ldots, z_n; m) = \binom{s}{z_1}\binom{s}{z_2}\cdots\binom{s}{z_n}\bigg/\binom{sn}{m},$$

where $z_j = 0, 1, \ldots, m$, $j = 1, 2, \ldots, n$ and $z_1 + z_2 + \cdots + z_n = m$. A test of the adequacy of the suggested model can be based on the number K_0 of empty squares.

The probability function of K_0, according to Corollary 4.6, is given by

$$P(K_0 = k) = \frac{(n)_{n-k} C(m, n-k; s)}{(sn)_m}, \quad k = 0, 1, \ldots, n,$$

where $C(m, k; s) = [\Delta_u^k (su)_m]_{u=0}/k!$ is the generalized factorial coefficient. Then, the null hypothesis that the joint distribution of red blood cells in the hemacytometer is the multivariate hypergeometric, with $s_j = s$, $j = 1, 2, \ldots, n$, is rejected if the value k_0 of K_0 exceeds a critical value c_α. The critical value (positive integer) c_α, for a given level of significance α, is determined by

$$P(K_0 \geq c_\alpha) = \sum_{k=c_\alpha}^{n} \frac{(n)_{n-k} C(m, n-k; s)}{(sn)_m} \leq \alpha.$$

Example 4.7 *Formation of a central committee.* A federation of n scientific societies is to form a central committee of m members. Assume that each society submits a list of s unordered nominations. In a random selection of the m members, let K be the number of societies represented in the central committee. Find the probability function and the binomial moments of the random variable K.

The m positions of the central committee may be considered as m like (indistinguishable) balls and the n different lists, each with s unordered nominations as n distinguishable urns, each with s distinguishable compartments. Since no candidate can be selected to more than one position of the central committee, the compartments are considered to be of capacity limited to one ball. Consequently, the probability function and the binomial moments of the random variable K are given by (4.27) and (4.28), respectively.

4.4.3 Pseudo-contagious Occupancy Distribution

Assume that the number X_j, of balls distributed into the jth urn, $j = 1, 2, \ldots, n$, obeys a *geometric distribution* with

$$q_x = P(X = x) = pq^x, \quad x = 0, 1, \ldots,$$

where $0 < p < 1$ and $q = 1 - p$. Then, its u-fold convolution obeys a negative binomial distribution with

$$q_m(u) = P(S_u = m) = \binom{u + m - 1}{m} p^u q^m, \quad m = 0, 1, \ldots.$$

Note that the conditional joint probability function of the random variables $X_1, X_2, \ldots, X_n$, given that $S_n = m$, is given by

$$q(x_1, x_2, \ldots, x_n | m) = 1 \bigg/ \binom{n + m - 1}{m},$$

with $x_j = 0, 1, \ldots, m$, $j = 1, 2, \ldots, n$ and $x_1 + x_2 + \cdots + x_n = m$. Therefore, this particular random occupancy model, with the assumption that the number of

178 OCCUPANCY DISTRIBUTIONS

balls distributed into any specific urn obeys a geometric distribution, is equivalent to the stochastic model in which a fixed number m of like (indistinguishable) balls are randomly distributed into n distinguishable urns of unlimited capacity (Bose-Einstein model.)

A more general random occupancy model is deduced by assuming that the number X_j, of balls distributed into the jth urn, $j = 1, 2, \ldots, n$, obeys a *negative binomial distribution* with

$$q_x = P(X = x) = \binom{s+x-1}{x} p^s q^x, \; x = 0, 1, \ldots, \quad (4.29)$$

where $0 < p < 1$ and $q = 1 - p$. Then, its u-fold convolution obeys a negative binomial distribution with

$$q_m(u) = P(S_u = m) = \binom{su+m-1}{m} p^{su} q^m, \; m = 0, 1, \ldots . \quad (4.30)$$

Note that the conditional joint probability function of the random variables $X_1, X_2, \ldots, X_n$, given that $S_n = m$, is given by

$$q(x_1, x_2, \ldots, x_n | m) = \frac{\binom{s+x_1-1}{x_1}\binom{s+x_2-1}{x_2} \cdots \binom{s+x_n-1}{x_n}}{\binom{sn+m-1}{m}},$$

with $x_j = 0, 1, \ldots, m$, $j = 1, 2, \ldots, n$ and $x_1 + x_2 + \cdots + x_n = m$. Therefore, this particular random occupancy model, with the assumption that the number of balls distributed into any specific urn obeys a negative binomial distribution, is equivalent to the stochastic model in which a fixed number m of like (indistinguishable) balls are randomly distributed into n distinguishable urns, each with s distinguishable compartments of unlimited capacity (pseudo-contagious occupancy model.) Upon introducing expressions (4.29) and (4.30) into Theorem 4.1 the following corollary is deduced.

Corollary 4.7 (a) *Under the random occupancy model with negative binomial probabilities, the conditional probability $p_k(m, n, i)$, that k urns are occupied by i balls each, given that m balls are distributed into the n distinguishable urns, $k = 0, 1, \ldots, n$, is given by*

$$p_k(m, n, i) = \binom{n}{k} \left[\Delta_u^{n-k} \frac{(m)_{i(n-u)}(-s)_i^{n-u}(-su)_{m-i(n-u)}}{(i!)^{n-u}(-sn)_m} \right]_{u=0}$$

$$= \binom{n}{k} \sum_{j=0}^{n-k} (-1)^{n-k-j} \binom{n-k}{j} \frac{(m)_{i(n-j)}(-s)_i^{n-j}(-sj)_{m-i(n-j)}}{(i!)^{n-j}(-sn)_m}.$$

$$(4.31)$$

(b) *The rth binomial moment $b_{(r)}(m, n, i)$, $r = 0, 1, \ldots$, of the probability function $p_k(m, n, i)$, $k = 0, 1, \ldots, n$, is given by*

$$b_{(r)}(m, n, i) = \binom{n}{r} \frac{(m)_{ir}(-s)_i^r(-sn+sr)_{m-ir}}{(i!)^r(-sn)_m}. \tag{4.32}$$

Similarly, from Corollary 4.1, the *pseudo-contagious occupancy distribution* is deduced in the following corollary.

Corollary 4.8 (a) *Under the random occupancy model with negative binomial probabilities, the conditional probability $p_k(m, n)$, that k urns are occupied, given that m balls are distributed into the n distinguishable urns, $k = 0, 1, \ldots, n$, is given by*

$$p_k(m, n) = \frac{(n)_k |C(m, k; -s)|}{(sn + m - 1)_m}, \tag{4.33}$$

where $|C(m, k; -s)| = [\Delta_u^k (su + m - 1)_m]_{u=0}/k!$ is the absolute generalized factorial coefficient.

(b) *The rth binomial moment $b_{(r)}(m, n)$, $r = 0, 1, \ldots$, of the probability function $p_k(m, n)$, $k = 0, 1, \ldots, n$, is given by*

$$b_{(r)}(m, n) = \frac{(n)_r |C(m, r; -s, -s(n-r))|}{(sn + m - 1)_m}, \tag{4.34}$$

where $|C(m, r; -s, -s(n-r))| = [\Delta_u^r (su + m - 1)_m]_{u=n-r}/r!$ is the noncentral absolute generalized factorial coefficient.

The mean and variance of the pseudo-contagious occupancy distribution, on using (4.34) and following the steps of derivation of the corresponding moments of the restricted occupancy distribution, are obtained as

$$E(K) = n\left[1 - \frac{(s(n-1)+m-1)_m}{(sn+m-1)_m}\right]$$

and

$$V(K) = n\frac{(s(n-1)+m-1)_m}{(sn+m-1)_m}\left[1 - \frac{(s(n-1)+m-1)_m}{(sn+m-1)_m}\right]$$
$$- (n)_2 \frac{(s(n-1)+m-1)_m}{(sn+m-1)_m}\left[\frac{(s(n-1)+m-1)_m}{(sn+m-1)_m} - \frac{(s(n-2)+m-1)_m}{(s(n-1)+m-1)_m}\right],$$

respectively.

4.4.4 Restricted Bose-Einstein Occupancy Distribution

Notice first that the replacement of the assumption that the number X_j, of balls distributed into the jth urn, $j = 1, 2, \ldots, n$, obeys a zero-one Bernoulli or a geometric distribution by the assumption that it obeys a positive or a negative binomial distribution divides each urn into s distinguishable compartments with limited or unlimited capacity. Further, a restriction on the capacity of each urn can be expressed by choosing a suitable truncated version of the distribution of the random variable X_j, $j = 1, 2, \ldots, n$. Thus, the assumption of geometric probabilities truncated to the right at the point s, that is, with

$$q_x = P(X = x) = \frac{pq^x}{1 - q^{s+1}}, \quad x = 0, 1, \ldots, s, \tag{4.35}$$

where $0 < p < 1$ and $q = 1 - p$, imposes the restriction that the capacity of each urn is limited to s balls. Clearly, the probability generating function of this distribution is given by

$$g(t) = \sum_{x=0}^{s} q_x t^x = \frac{p(1 - q^{s+1} t^{s+1})}{(1 - q^{s+1})(1 - qt)},$$

and thus the probability generating function of its n-fold convolution is

$$g_n(t) = \sum_{m=0}^{sn} q_m(n) t^m = \frac{p^n (1 - q^{s+1} t^{s+1})^n}{(1 - q^{s+1})^n (1 - qt)^n}.$$

Therefore, the probability function of the n-fold convolution, on using the number

$$L(m, n; s) = \sum_{j=0}^{n} (-1)^j \binom{n}{j} \binom{n + m - j(s+1) - 1}{n - 1}, \tag{4.36}$$

of m-combinations of n, with repetition and the restriction that each element appears at most s times, which has generating function

$$L_n(t; s) = \sum_{m=0}^{sn} L(m, n; s) t^m = \frac{(1 - t^{s+1})^n}{(1 - t)^n}, \tag{4.37}$$

is obtained as

$$q_m(n) = \frac{L(m, n; s) p^n q^m}{(1 - q^{s+1})^n}, \quad m = 0, 1, \ldots, sn. \tag{4.38}$$

Further, the conditional joint probability function of the random variables $X_1, X_2, \ldots, X_n$, given that $S_n = m$, is given by

$$q(x_1, x_2, \ldots, x_n | m) = \frac{1}{L(m, n; s)}, \tag{4.39}$$

with $x_j = 0, 1, \ldots, s$, $j = 1, 2, \ldots, n$ and $x_1 + x_2 + \cdots + x_n = m$. It is worth noting that the same random occupancy model is deduced by assuming that the number X_j, of balls distributed into the jth urn, $j = 1, 2, \ldots, n$, obeys a discrete uniform distribution on the set $\{0, 1, \ldots, s\}$,

$$q_x = P(X = x) = \frac{1}{s + 1}, \quad x = 0, 1, \ldots, s. \tag{4.40}$$

Indeed, since
$$h(t) = \sum_{x=0}^{s} q_x t^x = \frac{(1-t^{s+1})}{(s+1)(1-t)},$$
the n-fold convolution of this distribution has generating function
$$h_n(t) = \sum_{m=0}^{sn} q_m(n) t^m = \frac{(1-t^{s+1})^n}{(s+1)^n (1-t)^n}.$$

Hence,
$$q_m(n) = \frac{L(m,n;s)}{(s+1)^n}, \quad m = 0, 1, \ldots, sn, \tag{4.41}$$

and the conditional joint probability function of the random variables $X_1, X_2, \ldots, X_n$, given that $S_n = m$, is given by (4.39). Since, by (4.37),
$$\sum_{m=0}^{sk} \left[\Delta_u^k L(m,u;s) \right]_{u=0} t^m = \left[\Delta_u^k \frac{(1-t^{s+1})^u}{(1-t)^u} \right]_{u=0} = t^k \frac{(1-t^s)^k}{(1-t)^k},$$
it follows that
$$\left[\Delta_u^k L(m,u;s) \right]_{u=0} = L(m-k, k; s-1). \tag{4.42}$$

Therefore, upon introducing (4.35) and (4.38) or (4.40) and (4.41) into Corollary 4.1 and using (4.42), we deduce the *restricted Bose-Einstein occupancy distribution* in the following corollary. Clearly, lifting the restriction, that is letting $s \to \infty$, this distribution reduces to the Bose-Einstein occupancy distribution (cf. Exercise 4.13).

Corollary 4.9 (a) *Under the random occupancy model with geometric probabilities truncated from the right at the point s, (or with discrete uniform probabilities on the set $\{0, 1, \ldots, s\}$), the conditional probability $p_k(m, n; s)$, that k urns are occupied, given that m balls are distributed into the n distinguishable urns, $k = 0, 1, \ldots, n$, is given by*
$$p_k(m, n; s) = \binom{n}{k} \frac{L(m-k, k; s-1)}{L(m, n; s)}, \tag{4.43}$$
where $L(m, n; s)$ is number of the m-combinations of n with repetition and the restriction that each element appears at most s times, which is given by (4.36).
(b) *The rth binomial moment $b_{(r)}(m, n; s)$, $r = 0, 1, \ldots$, of the probability function $p_k(m, n; s)$, $k = 0, 1, \ldots, n$, is given by*
$$b_{(r)}(m, n; s) = \binom{n}{r} \sum_{k=0}^{r} (-1)^{r-k} \binom{r}{k} \frac{L(m, n-r+k; s)}{L(m, n; s)}. \tag{4.44}$$

The expected value of the restricted Bose-Einstein occupancy distribution is deduced from (4.44), by setting $r = 1$, as

$$E(K) = n\left[1 - \frac{L(m, n-1; s)}{L(m, n; s)}\right].$$

Further, putting $r = 2$ into (4.44), it follows that

$$E\left[\binom{K}{2}\right] = \binom{n}{2}\left[1 - 2\frac{L(m, n-1; s)}{L(m, n; s)} + \frac{L(m, n-2; s)}{L(m, n; s)}\right].$$

Therefore, the variance of this distribution is given by

$$V(K) = (n)_2\left[1 - 2\frac{L(m, n-1; s)}{L(m, n; s)} + \frac{L(m, n-2; s)}{L(m, n; s)}\right]$$
$$+ n\left[1 - \frac{L(m, n-1; s)}{L(m, n; s)}\right] - n^2\left[1 - \frac{L(m, n-1; s)}{L(m, n; s)}\right]^2,$$

which after a simple algebra reduces to

$$V(K) = n\frac{L(m, n-1; s)}{L(m, n; s)}\left[1 - \frac{L(m, n-1; s)}{L(m, n; s)}\right]$$
$$- (n)_2 \frac{L(m, n-1; s)}{L(m, n; s)}\left[\frac{L(m, n-1; s)}{L(m, n; s)} - \frac{L(m, n-2; s)}{L(m, n-1; s)}\right].$$

Example 4.8 *Consistency of product ratings.* Suppose that n judges rate a product on the scale $\{0, 1, \ldots, s\}$ and assign to it a total of m points. Let k_i be the number of judges who give the product i points, $i = 0, 1, \ldots, s$. Whereas the average rating is m/n, it seems reasonable to use the variance of the ratings $\sum_{i=0}^{s}(i - m/n)^2(k_i/n)$ as a measure of consistency. In order to test the consistency of such ratings, an appropriate stochastic model is required.

A model that seems to be suitable for the case is the restricted occupancy. Precisely, a test on whether the points given by the judges are randomly distributed among them may be based on the number K_i of judges who give the product i points, $i = 0, 1, \ldots, s$. Considering the judges as distinguishable urns of capacity limited to s balls and interpreting the assignment of i points by the jth judge as a placement of i like balls into the jth urn, then K_i is the number of urns containing i balls each, $i = 1, 2, \ldots, s$. The probability function and binomial moments of $K = n - K_0$, according to Corollary 4.9, are given by (4.43) and (4.44), respectively. Further, the probability function and binomial moments of K_i are given in Exercise 4.16.

4.5 STATISTICAL APPLICATIONS

There is a wide range of applications of occupancy distributions in many fields such as cluster analysis, biology, computer science and statistical physics. In this section several statistical applications are discussed in the form of examples. In the first two examples the *zero frequency test* for continuous and discrete populations is presented.

Example 4.9 *The zero frequency test for a continuous population.* The χ^2 test is proposed for testing the null hypothesis H_0 that a sample comes from a population with a continuous distribution function $F(x) = F_0(x)$. This test, due to the approximations involved in the derivation of the sampling distribution, can only be applied when a sample of large size is available. When the sample size is small an admissible test is the one based on the zero frequency, which can be described as follows. Consider the n intervals $(a_{j-1}, a_j]$, $j = 1, 2, \ldots, n$, with $-\infty = a_0 < a_1 < \cdots < a_{n-1} < a_n = \infty$, such that $F_0(a_j) - F_0(a_{j-1}) = 1/n$, $j = 1, 2, \ldots, n$. Suppose that a sample of a small size m is available and let z_j be the number of observations that belong in the interval $(a_{j-1}, a_j]$, $j = 1, 2, \ldots, n$, so that $z_1 + z_2 + \cdots + z_n = m$. The test of the null hypothesis $H_0 : F(x) = F_0(x)$, is based on the number K_0 of intervals containing no observation. The rejection region is defined by $K_0 > c_a$, where the critical value (positive integer) c_a for a given level of significance a, on using Corollary 4.4 and since $K_0 = n - K$, is determined by

$$P(K_0 \geq c_a) = P(K \leq n - c_a) = \sum_{k=0}^{n-c_a} \frac{(n)_k S(m, k)}{n^m} \leq a$$

and

$$P(K_0 \geq c_a - 1) = P(K \leq n - c_a + 1) = \sum_{k=0}^{n-c_a+1} \frac{(n)_k S(m, k)}{n^m} > a.$$

Tables of the upper $a = 0.01$ and $a = 0.05$ points c_a have been constructed by M. Csörgo and I. Guttman (1962). Further, as an illustration they considered the following sample of $m = 10$ observations:

$$\{481, 506, 527, 661, 501, 572, 561, 501, 487, 524\},$$

which are burning times in seconds of floating smoke pots. In order to test the null hypothesis H_0 that the sample comes from a normal population with mean $\mu = 500$ and standard deviation $\sigma = 50$, consider the following $n = 4$ intervals $I_1 = (-\infty, 466.28]$, $I_2 = (466.28, 500]$, $I_3 = (500, 533.72]$ and $I_4 = (533.72, \infty)$. Note that the interval I_1 is empty, I_2 contains 2 observations, I_3 contains 5 observations and I_4 contains 3 observations and so $K_0 = 1$. From the table of the upper $a = 0.01$ points the critical value is $c_{0.01} = 2$ and since $K_0 < c_{0.01}$ the null hypothesis is not rejected.

Example 4.10 *The zero frequency test for a discrete population.* The χ^2 test cannot be applied either in a continuous or in a discrete population if the sample size is not large enough. A small sample size test for testing the null hypothesis H_0 that a sample comes from a population with a Poisson distribution, especially when its mean is small, may be carried out by considering the conditional probability of the observations given their total. This conditional distribution is a multinomial with the number of trials equal to the sum m of the observations and probabilities $p_j = 1/n$, $j = 1, 2, \ldots, n$, where n is the number of observations. In terms of occupancy theory, n is the number of urns and m is the total number of balls randomly distributed into the urns. A zero observation corresponds to an empty urn and the zero frequency is the number K_0 of empty urns. The rejection region is then defined by $K_0 > c_a$, where the critical value (positive integer) c_a for a given level of significance a is computed as in Example 4.9.

It may be noted that a similar small size test for testing the null hypothesis H_0 that a sample comes from a population with a specific discrete distribution can be carried out, where certain nuisance parameters can be avoided by considering conditional probabilities. Thus, if the distribution specified by the null hypothesis is a general binomial with parameters s and p (where s is a positive integer and $0 < p < 1$ (binomial distribution) or $s, p < 0$ (negative binomial distribution)), then the conditional probability of the observations given their total is a general multivariate hypergeometric with parameter s, where s is a positive integer (hypergeometric) or $s < 0$ (negative hypergeometric), and probability function

$$\binom{s}{x_1}\binom{s}{x_2}\cdots\binom{s}{x_n}\bigg/\binom{sn}{m}.$$

Then, according to Corollary 4.6, the probability function of the number $K = n - K_0$ of occupied urns is given by (4.27). The rejection region is defined by $K_0 > c_a$, where the critical value (positive integer) c_a for a given level of significance a, since $K_0 = n - K$, is determined by

$$P(K_0 \geq c_a) = P(K \leq n - c_a) = \sum_{k=0}^{n-c_a} \frac{(n)_k C(m, k; s)}{(sn)_m} \leq a$$

and

$$P(K_0 \geq c_a - 1) = P(K \leq n - c_a + 1) = \sum_{k=0}^{n-c_a+1} \frac{(n)_k C(m, k; s)}{(sn)_m} > a.$$

The *estimation of the number of classes* in a finite population is an interesting field of applications of occupancy distributions. Assume that a random sample is drawn from a population containing an unknown but finite number n of classes (types) of elements (individuals). Let X_j be the number of elements of the jth class in the sample, $j = 1, 2, \ldots, n$, and $S_n = \sum_{j=1}^n X_j$. Further, for a fixed sample size m, let Z_j be the number of elements of the jth class in the sample, $j = 1, 2, \ldots, n$. A uniformly minimum variance unbiased (UMVU) estimator of the number n of classes can be based on the number K of classes observed in the sample. The probability function of K (occupancy distribution) depends on the joint probability function of the occupancy numbers $Z_1, Z_2, \ldots, Z_n$, which equals the conditional joint probability function of the random variables $X_1, X_2, \ldots, X_n$, given that $S_n = m$.

The most widely used family of discrete distributions as population-class distribution is the power series distribution with probability function

$$q_x = P(X = x) = [g(\theta)]^{-1} c_x \theta^x, \quad x = 0, 1, \ldots, \quad 0 < \theta < \rho,$$

where $g(\theta) = \sum_{x=0}^{\infty} c_x \theta^x$, $0 < \theta < \rho$. Then

$$q_m(n) = P(S_n = m) = [g(\theta)]^{-n} c_m(n) \theta^m, \quad m = 0, 1, \ldots, \quad 0 < \theta < \rho,$$

where $c_m(n) = [D_\theta^m (g(\theta))^n]_{\theta=0}/m!$, $m = 0, 1, \ldots$. In this case

$$P(Z_1 = x_1, Z_2 = x_2, \ldots, Z_n = x_n) = \frac{c_{x_1} c_{x_2} \cdots c_{x_n}}{c_m(n)},$$

with $x_j = 0, 1, \ldots, m$, $j = 1, 2, \ldots, n$ and $x_1 + x_2 + \cdots + x_n = m$. Let K_i be the number of classes appearing i times in the sample, $i = 1, 2, \ldots, m$. Since the number of divisions of the set of the n classes into $m+1$ subsets, with the ith subset containing $k_i \geq 0$ classes (appearing i times in the sample), $i = 0, 1, \ldots, m$, is given by

$$\frac{n!}{k_0! k_1! \cdots k_m!},$$

with $k_0 = n - (k_1 + k_2 + \cdots + k_m)$, the joint probability function of $K_1, K_2, \ldots, K_m$ is

$$P(K_1 = k_1, K_2 = k_2, \ldots, K_m = k_m) = \frac{(n)_k}{c_m(n)} \cdot \frac{c_1^{k_1} c_2^{k_2} \cdots c_m^{k_m}}{k_1! k_2! \cdots k_m!},$$

where $k_i = 0, 1, \ldots, n$, $i = 1, 2, \ldots, m$, with $\sum_{i=1}^{m} i k_i = m$ and $\sum_{i=1}^{m} k_i = n - k_0 = k$. In the indicated factorization the function $g(k; n) = (n)_k / c_m(n)$ depends on $k_1, k_2, \ldots, k_m$ only through $k = \sum_{i=1}^{m} k_i$ and the function $h(k_1, k_2, \ldots, k_m) = (c_1^{k_1} c_2^{k_2} \cdots c_m^{k_m})/(k_1! k_2! \cdots k_m!)$ does not depend on n. Thus, the number of classes observed in the sample, $K = \sum_{i=1}^{m} K_i$, is a sufficient statistic for the unknown number n of classes of the population. Further, the probability function of K, according to Corollary 4.2, is given by (4.15). It is not difficult to verify that the only function $g(K)$ with zero expectation, $E[g(K)] = 0$, for all $n = 1, 2, \ldots$, is $g(K) \equiv 0$, which implies that the sufficient statistic K is also complete for n. Then, by the classical theorems of Rao-Blackwell and Lehmann-Scheffé, a UMVU estimator of n based on K exists when $m \geq n$. More generally, the parametric function $(n)_j, j = 1, 2, \ldots, n$, is estimable if, in addition to $m \geq n$, the observed value of K is $k \geq j$. The derivation of the UMVU estimator of $(n)_j$, $j = 1, 2, \ldots, k$, may be carried out noting that the expected value $E[g_j(K)]$, where

$$g_j(k) = (k)_j \frac{[\Delta_u^{k-j} c_m(u)]_{u=j}}{[\Delta_u^k c_m(u)]_{u=0}}, \quad j = 1, 2, \ldots, k,$$

is equal to $E[g_j(K)] = (n)_j$, $j = 1, 2, \ldots, k$. Indeed, from (4.15) and since

$$(k)_j \binom{n}{k} = (n)_j \binom{n-j}{k-j},$$

it follows that

$$E[g_j(K)] = \frac{(n)_j}{c_m(n)} \sum_{k=j}^{n} \binom{n-j}{k-j} [\Delta_u^{k-j} c_m(u)]_{u=j}$$

$$= \frac{(n)_j}{c_m(n)} [E_u^{n-j} c_m(u)]_{n=j} = (n)_j.$$

Therefore, the UMVU estimator of $(n)_j$ is given by $\hat{n}_j(K)$, $j = 1, 2, \ldots, k$, where

$$\hat{n}_j(k) = (k)_j \frac{[\Delta_u^{k-j} c_m(u)]_{u=j}}{[\Delta_u^k c_m(u)]_{u=0}}, \quad j = 1, 2, \ldots, k.$$

Further, the variance of the UMVU estimator $\hat{n}_1(K)$ of n is

$$v \equiv V[\hat{n}_1(K)] = E\{\hat{n}_1(K)[\hat{n}_1(K) - 1]\} - E[\hat{n}_1(K)]\{E[\hat{n}_1(K)] - 1\}$$
$$= E\{\hat{n}_1(K)[\hat{n}_1(K) - 1]\} - (n)_2.$$

Thus, the UMVU estimator $\hat{v}(K)$ of the variance v of $\hat{n}_1(K)$ is given by

$$\hat{v}(K) = \hat{n}_1(K)[\hat{n}_1(K) - 1] - \hat{n}_2(K).$$

Summarizing, the following theorem is shown.

Theorem 4.2 *Consider a population with an unknown number n of classes of elements and assume that a random sample of size m is drawn from it. Let X_j be the number of elements of the jth class in the sample, $j = 1, 2, \ldots, n$, and let K be the number of observed classes in the sample, given that $S_n = \sum_{j=1}^{n} X_j = m$. Suppose that $X_1, X_2, \ldots, X_n$ are independently and identically distributed random variables with common probability function*

$$q_x = P(X = n) = [g(\theta)]^{-1} c_x \theta^x, \quad x = 0, 1, \ldots, \quad 0 < \theta < \rho.$$

Then, if $m \geq n$ and the observed value of K is $k \geq j$, a UMVU estimator of $(n)_j$ exists and is given by $\hat{n}_j(K)$, where

$$\hat{n}_j(k) = (k)_j \frac{[\Delta_u^{k-j} c_m(u)]_{u=j}}{[\Delta_u^k c_m(u)]_{u=0}}, \quad j = 1, 2, \ldots, \quad (4.45)$$

with $c_m(n) = [D_\theta^m (g(\theta))^n]_{\theta=0}/m!$. In particular, the UMVU estimator of n is given by $\hat{n}(K) = \hat{n}_1(K)$ and the UMVU estimator of its variance is given by $\hat{v}(K) = \hat{n}_1(K)[\hat{n}_1(K) - 1] - \hat{n}_2(K)$.

Applications to problems with a particular population-class distribution are discussed in the following examples.

Example 4.11 *Capture-recapture sampling scheme.* A general sampling method from a population containing a finite number n of individuals is the capture-recapture scheme. It is one of the most popular methods for obtaining statistical information about the size of mobile populations. Such populations arise in a variety of biological situations relating to population growth, ecological adaptation, natural selection and evolution. In a capture-recapture sampling scheme, an experimenter captures individuals from the population, marks, releases them and recaptures individuals from the same population. Records are kept of the number of times Z_j the jth individual of the population is caught, $j = 1, 2, \ldots, n$, after m individuals are caught. Let K be the number of different individuals caught. The joint distribution of $Z_1, Z_2, \ldots, Z_n$ is a multinomial with parameters m and $p_j = 1/n, j = 1, 2, \ldots, n$. Clearly, this is equivalent to the assumption that for a random sample of an unspecified sample size drawn from the same population, the random variables $X_1, X_2, \ldots, X_n$ are independently and identically distributed with a Poisson probability function,

$$q_x = P(X = x) = e^{-\lambda} \frac{\lambda^x}{x!}, \quad x = 0, 1, \ldots, \quad 0 < \lambda < \infty.$$

This is a power series distribution with series function $g(\lambda) = e^\lambda$ and its u-fold convolution is also a Poisson distribution with

$$q_m(u) = P(S_u = m) = e^{-u\lambda} \frac{(u\lambda)^m}{m!}, \quad m = 0, 1, \ldots.$$

Its mth coefficient function is $c_m(u) = u^m/m!, m = 0, 1, \ldots$. Thus, from Theorem 4.2, it follows that the UMVU estimator of $(n)_j$ is given by $\hat{n}_j(K) \equiv \hat{n}_j(K; m)$, where

$$\hat{n}_j(k; m) = \frac{S(m, k-j; j)}{S(m, k)}, \quad j = 1, 2, \ldots,$$

with $S(m, k; r) = [\Delta_u^k u^m]_{u=r}/k!$ and $S(m, k) = [\Delta_u^k u^m]_{u=0}/k!$ the noncentral and the central Stirling numbers of the second kind, respectively. According to (2.15), the noncentral Stirling numbers of the second kind are expressed in terms of the central Stirling numbers of the second kind by

$$S(m, k-j; j) = \sum_{i=k}^{k+j} \binom{i-j}{i-k} (j)_{i-k} S(m, i-j).$$

Using this expression for $j = 1$ and $j = 2$, we get the UMVU estimator of n as

$$\hat{n}(k; m) = k + \frac{S(m, k-1)}{S(m, k)}$$

and the UMVU estimator of $(n)_2$ as

$$\hat{n}_2(k; m) = k(k-1) + \frac{(k-1)S(m, k-1)}{S(m, k)} + \frac{S(m, k-2)}{S(m, k)}$$
$$= k(k-1) + \frac{S(m+1, k-2)}{S(m, k)}.$$

The UMVU estimator of the variance of $\hat{n}(k;m)$, by Theorem 4.2, is given by

$$\hat{v}(k;m) = \left[k + \frac{S(m,k-1)}{S(m,k)}\right] \cdot \left[(k-1) + \frac{S(m,k-1)}{S(m,k)}\right]$$
$$- k(k-1) - \frac{(k-1)S(m,k-1)}{S(m,k)} - \frac{S(m,k-2)}{S(m,k)},$$

which, after some algebraic manipulations, is simplified to

$$\hat{v}(k;m) = \frac{kS(m,k-1)}{S(m,k)} + \left[\frac{S(m,k-1)}{S(m,k)}\right]^2 - \frac{S(m,k-2)}{S(m,k)}.$$

Note that, since the Stirling numbers of the second kind, for fixed m, constitute a strong logarithmic concave sequence of k, $[S(m,k)]^2 > S(m,k+1)S(m,k-1)$, $k = 1, 2, \ldots$, the last two terms of this estimator sum to a positive number.

Example 4.12 *Missing dies in an old series of coins.* Consider a collection of m coins $\{c_{i_1}, c_{i_2}, \ldots, c_{i_m}\}$ belonging in an old series of n different obverse dies of coins $\{c_1, c_2, \ldots, c_n\}$ and let k be the number different obverse dies represented in the collection. Since in old series n is unknown, the degree of certainty that the collection is complete and an estimate of the number of missing obverse dies are of interest in numismatics.

An evaluation of the degree of certainty of this event may be given under the following assumptions. The m coins in the collection $\{c_{i_1}, c_{i_2}, \ldots, c_{i_m}\}$ are assumed to be selected from the set $\{c_1, c_2, \ldots, c_n\}$ at random and with replacement. Further, the n^m possible samples are equally probable. Although there are factors that may have militated against random selection of coins and uniform distribution (equally probable) of dies in the ground, these assumptions may be considered as approximately valid. Also, it seems that the general results, derived under these assumptions, agree fairly well with experience.

Considering the n different obverse dies of coins as n distinguishable urns and interpreting the selection of coin c_j as a placement of a ball into the jth urn, then the number K of occupied urns counts the different obverse dies in the collection. Clearly, according to the assumption made above, the joint probability function of the occupancy numbers $Z_1, Z_2, \ldots, Z_n$ is a multinomial with parameters m and $p_j = 1/n$, $j = 1, 2, \ldots, n$. Therefore, by Corollary 4.4, the probability function of K is given by

$$p_k(m,n) = \frac{(n)_k S(m,k)}{n^m}, \quad k = 0, 1, \ldots, n,$$

where $S(m,k) = [\Delta_u^k u^m]_{u=0}/k!$ is the Stirling number of the second kind. Also, according to Example 4.11, the UMVU estimator of the number $k_0 = n - k$ of missing obverse dies, which is $\hat{k}_0(k;m) = \hat{n}(k;m) - k$, is given by

$$\hat{k}_0(k;m) = \frac{S(m,k-1)}{S(m,k)}.$$

Example 4.13 *Sampling without replacement.* Consider a population containing n classes each including s elements. Suppose that a random sample of a fixed size m is drawn without replacement from this population and let Z_j be the number of elements of the jth class in the sample, $j = 1, 2, \ldots, n$. Then, the joint distribution of $Z_1, Z_2, \ldots, Z_n$ is a multivariate hypergeometric with parameters m and $s_j = s$, $j = 1, 2, \ldots, n$. Clearly, this is equivalent to the assumption that, for a random sample of an unspecified sample size drawn from the same population, the random variables $X_1, X_2, \ldots, X_n$ are independently and identically distributed with a binomial probability function

$$q_x = P(X = x) = \binom{s}{x} p^x q^{s-x}, \quad x = 0, 1, \ldots, s,$$

where $q = 1 - p$ and $0 < p < 1$. This is a power series distribution with parameter $\theta = p/q$ and series function $g(\theta) = (1 + \theta)^s$, and its u-fold convolution is also a binomial distribution with

$$q_m(u) = P(S_n = m) = \binom{su}{m} p^m q^{su-m}, \quad m = 0, 1, \ldots, su.$$

Its mth coefficient function is $c_m(u) = (su)_m/m!$, $m = 0, 1, \ldots, su$. Thus, from Theorem 4.2, it follows that the UMVU estimator of $(n)_j$ is given by $\hat{n}_j(K) \equiv \hat{n}_j(K; m, s)$, where

$$\hat{n}_j(k; m, s) = \frac{C(m, k - j; s, sj)}{C(m, k; s)},$$

with $C(m, k-j; s, sj) = [\Delta_u^k (su)_m]_{u=j}/k!$ and $C(m, k; s) = [\Delta_u^k (su)_m]_{u=0}/k!$, the noncentral and the central generalized factorial coefficients, respectively. According to (2.55), the noncentral generalized factorial coefficients are expressed in terms of the central generalized factorial coefficients by

$$C(m, k - j; s, sj) = \sum_{i=k}^{k+j} \binom{i - j}{i - k} (j)_{i-k} C(m, i - j; s).$$

Using this expression for $j = 1$ and $j = 2$, we get the UMVU estimator of n as

$$\hat{n}(k; m, s) = k + \frac{C(m, k - 1; s)}{C(m, k; s)}$$

and the UMVU estimator of $(n)_2$ as

$$\hat{n}_2(k; m, s) = k(k - 1) + \frac{(k - 1)C(m, k - 1; s)}{C(m, k; s)} + \frac{C(m, k - 2; s)}{C(m, k; s)}.$$

The UMVU estimator of the variance of $\hat{n}(k; m, s)$, by Theorem 4.2, is deduced as

$$\hat{v}(k; m, s) = \frac{kC(m, k - 1; s)}{C(m, k; s)} + \left[\frac{C(m, k - 1; s)}{C(m, k; s)}\right]^2 - \frac{C(m, k - 2; s)}{C(m, k; s)}.$$

It is worth noting that, since $C(m, k; s)$, $k = 0, 1, \ldots$, for fixed m and $s > 0$, is a strong logarithmic concave sequence of k,

$$[C(m, k; s)]^2 > C(m, k+1; s)C(m, k-1, s), \quad k = 0, 1, \ldots,$$

the last two terms of this estimator sum to a positive number.

Example 4.14 *Sampling with replacement and addition of elements.* Consider a population containing n classes each including s elements. Suppose that a random sample of a fixed size m is drawn from it by returning to the population each drawn element together with one element of the same class and let Z_j be the number of elements of the jth class in the sample, $j = 1, 2, \ldots, n$. Then, the joint distribution of $Z_1, Z_2, \ldots, Z_n$ is a multivariate negative (waiting-time) hypergeometric with parameters m and $s_j = s$, $j = 1, 2, \ldots, n$. This sampling model is equivalent to the assumption that, for a random sample of an unspecified sample size drawn from the same population, the random variables $X_1, X_2, \ldots, X_n$ are independently and identically distributed with a negative binomial probability function

$$q_x = P(X = x) = \binom{s + x - 1}{x} p^s q^x, \quad x = 0, 1, \ldots,$$

where $q = 1 - p$ and $0 < p < 1$. This is a power series distribution with parameter $\theta = q$ and series function $g(q) = (1 - q)^{-s}$, and its u-fold convolution is also a negative binomial distribution with

$$q_m(u) = \binom{su + m - 1}{m} p^{su} q^m, \quad m = 0, 1, \ldots.$$

Its mth coefficient function is $c_m(u) = (-1)^m (-su)_m / m!$, $m = 0, 1, \ldots$. Consequently, from Theorem 4.2, it follows that the UMVU estimator of the parametric function $(n)_j$ is given by $\hat{n}_j(K) \equiv \hat{n}_j(K; m, -s)$, with

$$\hat{n}_j(k; m, -s) = \frac{|C(m, k - j; -s, -sj)|}{|C(m, k; -s)|},$$

where $|C(m, k; -s, -sj)|$ and $|C(m, k; -s)|$ are the absolute noncentral and the central generalized factorial coefficients, respectively. Then, proceeding as in Example 4.11, the UMVU estimator of n is obtained as

$$\hat{n}(k; m, -s) = k + \frac{|C(m, k - 1; -s)|}{|C(m, k; -s)|}$$

and the UMVU estimator of the variance of $\hat{n}(k; m, -s)$ is deduced as

$$\hat{v}(k; m, -s) = \frac{k|C(m, k - 1; -s)|}{|C(m, k; -s)|} + \left[\frac{|C(m, k - 1; -s)|}{|C(m, k; -s)|}\right]^2 - \frac{|C(m, k - 2; -s)|}{|C(m, k; -s)|}.$$

Since $|C(m, k; -s)|$, $k = 0, 1, \ldots$, for fixed m and $s > 0$, is a strong logarithmic concave sequence of k,

$$|C(m, k; -s)|^2 > |C(m, k+1; -s)||C(m, k-1; -s)|, \quad k = 0, 1, \ldots,$$

the last two terms sum to a positive number.

4.6 A GENERAL RANDOM OCCUPANCY MODEL

The random occupancy model studied in the preceding sections allows for the balls distributed into the urns to be either distinguishable or indistinguishable (like). Specifically, the assumption that the number X of balls distributed into any specific urn obeys a Poisson distribution is equivalent to the assumption that the balls distributed into the urns are distinguishable. Also, the assumption that X obeys a Bernoulli or geometric and more generally a binomial or negative binomial distribution is equivalent to the assumption that the balls are indistinguishable. A *general random occupancy model*, which allows for the balls distributed into the urns to be of the general specification $(1^{r_1} 2^{r_2} \cdots m^{r_m})$, may be described as follows.

Consider a supply of balls randomly distributed into n distinguishable urns. Assume that the number R of different kinds of balls distributed into the urns is a Poisson random variable with probability function

$$P(R = r) = e^{-\lambda} \frac{\lambda^r}{r!}, \ r = 0, 1, \ldots, \ 0 < \lambda < \infty. \tag{4.46}$$

Further, suppose that the number $X_{i,j}$ of balls of the ith kind distributed into the jth urn is a random variable with known probability function

$$P(X_{i,j} = x) = q_x, \ x = 0, 1, \ldots, \tag{4.47}$$

for all $i = 1, 2, \ldots$ and $j = 1, 2, \ldots, n$. Assuming that the occupancy of each urn is independent of the others, the probability function of the number $Y_i = \sum_{j=1}^n X_{i,j}$ of balls of the ith kind distributed into the n urns, $P(Y_i = y_i) = q_{y_i}(n)$, $y_i = 0, 1, \ldots$, $i = 1, 2, \ldots$, is given by the sum

$$q_{y_i}(n) = \sum q_{x_{i,1}} q_{x_{i,2}} \cdots q_{x_{i,n}}, \ i = 1, 2, \ldots, \tag{4.48}$$

where the summation is extended over all integers $x_{i,j} \geq 0$, $j = 1, 2, \ldots, n$ such that $\sum_{j=1}^n x_{i,j} = y_i$. Also, the total number $S_R = \sum_{i=1}^R Y_i$ of balls distributed into the n urns has a compound Poisson distribution with probability generating function

$$P_{S_R}(t) = e^{-\lambda[1 - P_Y(t)]}, \ P_Y(t) = \sum_{y=0}^{\infty} q_y(n) t^y.$$

Consider, in addition, the number R_y of those variables among $Y_1, Y_2, \ldots, Y_R$ that are equal to y, for $y = 0, 1, \ldots$. Clearly, R_y is the number of different kinds of

192 OCCUPANCY DISTRIBUTIONS

balls of multiplicity y distributed into the n urns and $\sum_{y=0}^{\infty} R_y = R$. Further, R_y, $y = 0, 1, \ldots$, are stochastically independent random variables with Poisson probability function

$$P(R_y = r) = e^{-\lambda q_y(n)} \frac{[\lambda q_y(n)]^r}{r!}, \quad r = 0, 1, \ldots \quad (y = 0, 1, \ldots) \quad (4.49)$$

and $\sum_{y=1}^{\infty} y R_y = S_R$ [cf. Feller (1968, pp. 291-292)]. Note that the assumption that $R = r$ different kinds of balls are distributed into the n urns implies $R_0 = 0$. Also, if the total number of balls distributed into the n urns is assumed to be $S_R = m$, then $R_{m+1} = 0, R_{m+2} = 0, \ldots$. The inverse is not necessarily true.

Under this stochastic model, let K be the number of occupied urns and

$$\begin{aligned} p_k &\equiv p_k(r_1, r_2, \ldots, r_m; m, r, n) \\ &= P(K = k | S_R = m, R = r, R_1 = r_1, R_2 = r_2, \ldots, R_m = r_m), \end{aligned} \quad (4.50)$$

for $k = 1, 2, \ldots, n$, with $r_1 + 2r_2 + \cdots + mr_m = m$ and $r_1 + r_2 + \cdots + r_m = r$. Also, let

$$\begin{aligned} b_{(j)} &\equiv b_{(j)}(r_1, r_2, \ldots, r_m; m, r, n) \\ &= E\left[\binom{K}{j} \Big| S_R = m, R = r, R_1 = r_1, R_2 = r_2, \ldots, R_m = r_m\right], \end{aligned} \quad (4.51)$$

for $j = 1, 2, \ldots$. The probability function (4.50) and its binomial moments (4.51) may be expressed in terms of finite differences of the convolutions (4.48). The derivation of these expressions is facilitated by the following lemma.

Lemma 4.1 *Consider the general random occupancy model and assume that $S_R = m$ balls of $R = r$ different kinds, among which $R_y = r_y$ balls are of multiplicity y for $y = 1, 2, \ldots, m$, with $r_1 + 2r_2 + \cdots + mr_m = m$ and $r_1 + r_2 + \cdots + r_m = r$, are randomly distributed into the n distinguishable urns. Let A_j be the event that the jth urn remains empty, $j = 1, 2, \ldots, n$. Then the events $A_1, A_2, \ldots, A_n$ are exchangeable, with*

$$P(A_{j_1} A_{j_2} \cdots A_{j_k}) = \frac{q_0^{rk} [q_1(n-k)]^{r_1} [q_2(n-k)]^{r_2} \cdots [q_m(n-k)]^{r_m}}{[q_1(n)]^{r_1} [q_2(n)]^{r_2} \cdots [q_m(n)]^{r_m}}, \quad (4.52)$$

for every k-combination $\{j_1, j_2, \ldots, j_k\}$ of the n indices $\{1, 2, \ldots, n\}$ and $k = 1, 2, \ldots, n$.

Proof Note that the event $\{S_R = m, R = r, R_1 = r_1, R_2 = r_2, \ldots, R_m = r_m\}$, with $r_1 + 2r_2 + \cdots + mr_m = m$ and $r_1 + r_2 + \cdots + r_m = r$, is equivalent to the event $\{R_0 = 0, R_1 = r_1, R_2 = r_2, \ldots, R_m = r_m, R_{m+1} = 0, R_{m+2} = 0, \ldots\}$. Consequently, by the independence of R_y, $y = 0, 1, \ldots$, it follows that

$$\begin{aligned} &P(S_R = m, R_1 = r_1, R_2 = r_2, \ldots, R_m = r_m | R = r) \\ &= \frac{P(R_0 = 0) \prod_{y=1}^{m} P(R_y = r_y) \prod_{y=m+1}^{\infty} P(R_y = 0)}{P(R = r)}, \end{aligned}$$

which, upon using (4.46) and (4.49), entails the expression

$$P(S_R = m, R_1 = r_1, R_2 = r_2, \ldots, R_m = r_m | R = r)$$
$$= \frac{r!}{r_1! r_2! \cdots r_m!} [q_1(n)]^{r_1} [q_2(n)]^{r_2} \cdots [q_m(n)]^{r_m}.$$

Further, consider the number $Z_i = \sum_{s=k+1}^{n} X_{i,j_s}$ of balls of the ith kind distributed into the $n - k$ urns $\{j_{k+1}, j_{k+2}, \ldots, j_n\}$, $i = 1, 2, \ldots$, where $\{j_{k+1}, j_{k+2}, \ldots, j_n\}$ $= \{1, 2, \ldots, n\} - \{j_1, j_2, \ldots, j_k\}$. Then, $P(Z_i = z) = q_z(n - k)$, $z = 0, 1, \ldots$, $i = 1, 2, \ldots$. Assume that $X_{i,j_s} = 0$, $s = 1, 2, \ldots, k$, $i = 1, 2, \ldots$, and let U_z be the number of those variables, among $Z_1, Z_2, \ldots, Z_R$, that are equal to z, for $z = 0, 1, \ldots$. The random variables U_z, $z = 0, 1, \ldots$ are stochastically independent with Poisson probability function

$$P(U_z = u) = e^{-\lambda q_z(n-k)} \frac{[\lambda q_z(n-k)]^u}{u!}, \quad u = 0, 1, \ldots \ (z = 0, 1, \ldots)$$

and $\sum_{z=0}^{\infty} U_z = R$, $\sum_{z=1}^{\infty} z U_z = S_R$. Then, the probability $P(A_{j_1} A_{j_2} \cdots A_{j_k})$ is equal to the conditional probability of the event

$$\{X_{i,j_s} = 0, \ s = 1, 2, \ldots, k, \ i = 1, 2, \ldots, R, \ U_1 = r_1, U_2 = r_2, \ldots, U_m = r_m\},$$

given the event $\{S_R = m, R = r, R_1 = r_1, R_2 = r_2, \ldots, R_m = r_m\}$, with $r_1 + 2r_2 + \cdots + mr_m = m$ and $r_1 + r_2 + \cdots + r_m = r$, which, in turn, is equal to the probability

$$P(X_{i,j_s} = 0, \ s = 1, 2, \ldots, k, \ i = 1, 2, \ldots, r,$$
$$U_1 = r_1, U_2 = r_2, \ldots, U_m = r_m | R = r)$$
$$= \frac{r! q_0^{rk}}{r_1! r_2! \cdots r_m!} [q_1(n-k)]^{r_1} [q_2(n-k)]^{r_2} \cdots [q_m(n-k)]^{r_m}$$

divided by the probability

$$P(S_R = m, R_1 = r_1, R_2 = r_2, \ldots, R_m = r_m | R = r)$$
$$= \frac{r!}{r_1! r_2! \cdots r_m!} [q_1(n)]^{r_1} [q_2(n)]^{r_2} \cdots [q_m(n)]^{r_m}.$$

Hence (4.52) is established.

Theorem 4.3 (a) *Under the general random occupancy model, the conditional probability $p_k \equiv p_k(r_1, r_2, \ldots, r_m; m, r, n)$ that k urns are occupied, given that $S_R = m$ balls of $R = r$ different kinds, among which $R_y = r_y$ balls are of multiplicity y for $y = 1, 2, \ldots, m$, with $r_1 + 2r_2 + \cdots + mr_m = m$ and $r_1 + r_2 + \cdots + r_m = r$, are randomly distributed into the n distinguishable urns, $k = 1, 2, \ldots, n$, is given by*

$$p_k = \binom{n}{k} \frac{[\Delta_u^k q_0^{r(n-u)} [q_1(u)]^{r_1} [q_2(u)]^{r_2} \cdots [q_m(u)]^{r_m}]_{u=0}}{[q_1(n)]^{r_1} [q_2(n)]^{r_2} \cdots [q_m(n)]^{r_m}}. \quad (4.53)$$

(b) *The jth binomial moment* $b_{(j)} \equiv b_{(j)}(r_1, r_2, \ldots, r_m; r, n)$, $j = 1, 2, \ldots,$ *of the probability function* p_k, $k = 1, 2, \ldots, n$, *is given by*

$$b_{(j)} = \binom{n}{j} \frac{\left[\Delta_u^j q_0^{r(n-u)} [q_1(u)]^{r_1} [q_2(u)]^{r_2} \cdots [q_m(u)]^{r_m}\right]_{u=n-j}}{[q_1(n)]^{r_1} [q_2(n)]^{r_2} \cdots [q_m(n)]^{r_m}}. \tag{4.54}$$

Proof (a) Consider the general random occupancy model and assume that $S_R = m$ balls of $R = r$ different kinds, among which $R_y = r_y$ balls are of multiplicity y for $y = 1, 2, \ldots, m$, with $r_1 + 2r_2 + \cdots + mr_m = m$ and $r_1 + r_2 + \cdots + r_m = r$, are randomly distributed into the n distinguishable urns. Let A_j be the event that the jth urn remains empty, $j = 1, 2, \ldots, n$. The conditional probability p_k, that k urns are occupied, is equal to the probability that exactly $n - k$ among the n events $A_1, A_2, \ldots, A_n$ occur. Since, by Lemma 4.1, the events $A_1, A_2, \ldots, A_n$ are exchangeable, on using Corollary 1.5, it follows that

$$p_k = \binom{n}{n-k} \sum_{s=n-k}^{n} (-1)^{s-n+k} \binom{k}{n-s} P(A_{j_1} A_{j_2} \cdots A_{j_s})$$

$$= \binom{n}{k} \sum_{j=0}^{k} (-1)^{k-j} \binom{k}{j} \frac{q_0^{r(n-j)} [q_1(j)]^{r_1} [q_2(j)]^{r_2} \cdots [q_m(j)]^{r_m}}{[q_1(n)]^{r_1} [q_2(n)]^{r_2} \cdots [q_m(n)]^{r_m}}.$$

The last expression, upon introducing the finite difference operator, yields (4.53).

(b) The ith binomial moment

$$\beta_{(i)} = E\left[\binom{K_0}{i} \bigg| S_R = m, R = r, R_1 = r_1, R_2 = r_2, \ldots, R_m = r_m\right],$$

of the number K_0 of empty urns is equal to the conditional probability that k urns remain empty, given that $S_R = m$ balls of $R = r$ different kinds, among which $R_y = r_y$ balls are of multiplicity y, for $y = 1, 2, \ldots, m$ with $r_1 + 2r_2 + \cdots + mr_m = m$ and $r_1 + r_2 + \cdots + r_m = r$, are randomly distributed into n distinguishable urns. Thus, applying Corollary 1.6 and using Lemma 4.1, we get

$$\beta_{(i)} = \binom{n}{i} \frac{q_0^{ri} [q_1(n-i)]^{r_1} [q_2(n-i)]^{r_2} \cdots [q_m(n-i)]^{r_m}}{[q_1(n)]^{r_1} [q_2(n)]^{r_2} \cdots [q_m(n)]^{r_m}}.$$

Further, the jth binomial moment $b_{(j)}$ of the number $K = n - K_0$, of occupied urns, may be written as

$$b_{(j)} = E\left[\binom{K}{j} \bigg| S_R = m, R = r, R_1 = r_1, R_2 = r_2, \ldots, R_m = r_m\right]$$

$$= (-1)^j E\left[\binom{K_0 - n + j - 1}{j} \bigg| S_R = m, R = r, R_1 = r_1, \ldots, R_m = r_m\right]$$

$$\equiv (-1)^j \beta_{[j]}(n).$$

and since, by Theorem 1.11,

$$\beta_{[j]}(n) = \sum_{i=0}^{j}(-1)^{j-i}\binom{n-i}{j-i}\beta_{(i)},$$

it follows that

$$b_{(j)} = \binom{n}{j}\sum_{i=0}^{j}(-1)^i\binom{j}{i}\frac{q_0^{ri}[q_1(n-i)]^{r_1}[q_2(n-i)]^{r_2}\cdots[q_m(n-i)]^{r_m}}{[q_1(n)]^{r_1}[q_2(n)]^{r_2}\cdots[q_m(n)]^{r_m}}.$$

The last expression implies (4.54).

An interesting corollary of Theorem 4.3 is deduced when it is assumed that $S_R = m$ and $R = r$, $R_s = r$, $R_y = 0$, $y \neq s$, so that $m = rs$.

Corollary 4.10 (a) *Under the general random occupancy model, the conditional probability $p_k(s; r, n)$ that k urns are occupied, given that s balls from each of r kinds are distributed into the n distinguishable urns, $k = 1, 2, \ldots, n$, is given by*

$$p_k(s; r, n) = \binom{n}{k}\frac{\left[\Delta_u^k q_0^{r(n-u)}[q_s(u)]^r\right]_{u=0}}{[q_s(n)]^r}. \qquad (4.55)$$

(b) *The jth binomial moment $b_{(j)}(s; r, n)$, $j = 1, 2, \ldots$, of the probability function $p_k(s; r, n)$, $k = 1, 2, \ldots, n$, is given by*

$$b_{(j)}(s; r, n) = \binom{n}{j}\frac{\left[\Delta_u^j q_0^{r(n-u)}[q_s(u)]^r\right]_{u=n-j}}{[q_s(n)]^r}. \qquad (4.56)$$

Some interesting special occupancy distributions are deduced by specifying the probability function q_x of the number X of balls distributed into any specific urn. Consider, first, the case of a geometric probability function:

$$q_x = P(X = x) = pq^x, \quad x = 0, 1, \ldots,$$

where $0 < p < 1$ and $q = 1 - p$. Its u-fold convolution obeys a negative binomial distribution with

$$q_y(u) = P(Y = y) = \binom{u+y-1}{y}p^u q^y, \quad y = 0, 1, \ldots.$$

Then, according to Lemma 4.1, the conditional probability that k specified urns remain empty, given that $S_R = m$ balls of $R = r$ different kinds, among which $R_y = r_y$ balls are of multiplicity y for $y = 1, 2, \ldots, m$, with $r_1 + 2r_2 + \cdots + mr_m = m$ and $r_1 + r_2 + \cdots + r_m = r$, are randomly distributed into the n distinguishable urns, $k = 1, 2, \ldots, n$, is given by

$$P(A_{j_1}A_{j_2}\cdots A_{j_k}) = \prod_{y=1}^{m}\binom{n-k+y-1}{y}^{r_y} \Big/ \prod_{y=1}^{m}\binom{n+y-1}{y}^{r_y}.$$

196 OCCUPANCY DISTRIBUTIONS

Therefore, the general random occupancy model, with the assumption of geometric probabilities, is equivalent to the occupancy model, where a fixed number m of balls of the general specification $(1^{r_1} 2^{r_2} \cdots m^{r_m})$, with $r_1 + 2r_2 + \cdots + mr_m = m$ and $r_1 + r_2 + \cdots + r_m = r$, are randomly distributed into n distinguishable urns.

From Theorem 4.3 and Corollary 4.10, with the assumption of geometric probabilities, the following corollary is deduced.

Corollary 4.11 (a) *Under the general random occupancy model with geometric probabilities, the conditional probability $p_k \equiv p_k(r_1, r_2, \ldots, r_m; m, r, n)$ that k urns are occupied, given that $S_R = m$ balls of $R = r$ different kinds, among which $R_y = r_y$ balls are of multiplicity y for $y = 1, 2, \ldots, m$, with $r_1 + 2r_2 + \cdots + mr_m = m$ and $r_1 + r_2 + \cdots + r_m = r$, are randomly distributed into the n distinguishable urns, $k = 1, 2, \ldots, n$, is given*

$$p_k = \binom{n}{k} \left[\Delta_u^k \prod_{y=1}^{m} \binom{u+y-1}{y}^{r_y} \right]_{u=0} \bigg/ \prod_{y=1}^{m} \binom{n+y-1}{y}^{r_y}. \qquad (4.57)$$

Also, the jth binomial moment $b_{(j)} \equiv b_{(j)}(r_1, r_2, \ldots, r_m; r, n)$, $j = 1, 2, \ldots$, of the probability function p_k, $k = 1, 2, \ldots, n$, is given by

$$b_{(j)} = \binom{n}{j} \left[\Delta_u^j \prod_{y=1}^{m} \binom{u+y-1}{y}^{r_y} \right]_{u=n-j} \bigg/ \prod_{y=1}^{m} \binom{n+y-1}{y}^{r_y}. \qquad (4.58)$$

(b) *In particular, under the general random occupancy model with geometric probabilities, the conditional probability $p_k(s; r, n)$ that k urns are occupied, given that s balls from each of r kinds are distributed into the n distinguishable urns, $k = 1, 2, \ldots, n$, is given by*

$$p_k(s; r, n) = \binom{n}{k} \left[\Delta_u^k \binom{u+s-1}{s}^{r} \right]_{u=0} \bigg/ \binom{n+s-1}{s}^{r}. \qquad (4.59)$$

Also, the jth binomial moment $b_{(j)}(s; r, n)$, $j = 1, 2, \ldots$, of the probability function $p_k(s; r, n)$, $k = 1, 2, \ldots, n$, is given by

$$b_{(j)}(s; r, n) = \binom{n}{j} \left[\Delta_u^j \binom{u+s-1}{s}^{r} \right]_{u=n-j} \bigg/ \binom{n+s-1}{s}^{r}. \qquad (4.60)$$

Consider, now, the case of a Bernoulli probability function:

$$q_x = P(X = x) = p^x q^{1-x}, \quad x = 0, 1,$$

where $0 < p < 1$ and $q = 1 - p$. Its u-fold convolution obeys a binomial distribution with

$$q_y(u) = P(Y = y) = \binom{u}{y} p^y q^{u-y}, \ y = 0, 1, \ldots, u.$$

Then, according to Lemma 4.1, the conditional probability that k specified urns remain empty, given that $S_R = m$ balls of $R = r$ different kinds, among which $R_y = r_y$

balls are of multiplicity y for $y = 1, 2, \ldots, m$, with $r_1 + 2r_2 + \cdots + mr_m = m$ and $r_1 + r_2 + \cdots + r_m = r$, are randomly distributed into the n distinguishable urns, $k = 1, 2, \ldots, n$, is given by

$$P(A_{j_1} A_{j_2} \cdots A_{j_k}) = \prod_{y=1}^{m} \binom{n-k}{y}^{r_y} \Big/ \prod_{y=1}^{m} \binom{n}{y}^{r_y}.$$

Therefore, the general random occupancy model, with the assumption of Bernoulli probabilities, is equivalent to the occupancy model, where a fixed number m of balls of the general specification $(1^{r_1} 2^{r_2} \cdots m^{r_m})$, with $r_1 + 2r_2 + \cdots + mr_m = m$ and $r_1 + r_2 + \cdots + r_m = r$, are randomly distributed into n distinguishable urns, each of capacity limited to one ball from each kind.

From Theorem 4.3 and Corollary 4.10, with the assumption of Bernoulli probabilities, the following corollary is deduced.

Corollary 4.12 (a) *Under the general random occupancy model with Bernoulli probabilities, the conditional probability $p_k \equiv p_k(r_1, r_2, \ldots, r_m; m, r, n)$ that k urns are occupied, given that $S_R = m$ balls of $R = r$ different kinds, among which $R_y = r_y$ balls are of multiplicity y for $y = 1, 2, \ldots, m$, with $r_1 + 2r_2 + \cdots + mr_m = m$ and $r_1 + r_2 + \cdots + r_m = r$, are randomly distributed into the n distinguishable urns, $k = 1, 2, \ldots, n$, is given by*

$$p_k = \binom{n}{k} \left[\Delta_u^k \prod_{y=1}^{m} \binom{u}{y}^{r_y} \right]_{u=0} \Big/ \prod_{y=1}^{m} \binom{n}{y}^{r_y}. \tag{4.61}$$

Also, the jth binomial moment $b_{(j)} \equiv b_{(j)}(r_1, r_2, \ldots, r_m; r, n)$, $j = 1, 2, \ldots$, of the probability function p_k, $k = 1, 2, \ldots, n$, is given by

$$b_{(j)} = \binom{n}{j} \left[\Delta_u^j \prod_{y=1}^{m} \binom{u}{y}^{r_y} \right]_{u=n-j} \Big/ \prod_{y=1}^{m} \binom{n}{y}^{r_y}. \tag{4.62}$$

(b) *In particular, under the general random occupancy model with Bernoulli probabilities, the conditional probability $p_k(s; r, n)$ that k urns are occupied, given that s balls from each of r kinds are distributed into the n distinguishable urns, $k = 1, 2, \ldots, n$, is given by*

$$p_k(s; r, n) = \binom{n}{k} \left[\Delta_u^k \binom{u}{s}^r \right]_{u=0} \Big/ \binom{n}{s}^r. \tag{4.63}$$

Also, the jth binomial moment $b_{(j)}(s; r, n)$, $j = 1, 2, \ldots$, of the probability function $p_k(s; r, n)$, $k = 1, 2, \ldots, n$, is given by

$$b_{(j)}(s; r, n) = \binom{n}{j} \left[\Delta_u^j \binom{u}{s}^r \right]_{u=n-j} \Big/ \binom{n}{s}^r. \tag{4.64}$$

198   OCCUPANCY DISTRIBUTIONS

Extensions and/or imposition of restrictions on this general random occupancy model can be achieved and/or incorporated by replacing the assumption that the number X obeys a geometric or Bernoulli distribution by the assumption that it obeys a negative binomial distribution,

$$q_x = P(X = x) = \binom{s+x-1}{x} p^s q^x, \quad x = 0, 1, \ldots,$$

or binomial distribution,

$$q_x = P(X = x) = \binom{s}{x} p^x q^{s-x}, \quad x = 0, 1, \ldots, s.$$

These assumptions correspond to the assumption that each urn is divided into s distinguishable compartments with unlimited capacity (negative binomial distribution) or with capacity limited to one ball from each kind (binomial distribution). Under the general random occupancy model with negative binomial probabilities,

$$p_k = \binom{n}{k} \left[\Delta_u^k \prod_{y=1}^m \binom{su+y-1}{y}^{r_y} \right]_{u=0} \bigg/ \prod_{y=1}^m \binom{sn+y-1}{y}^{r_y}$$

and

$$b_{(j)} = \binom{n}{j} \left[\Delta_u^j \prod_{y=1}^m \binom{su+y-1}{y}^{r_y} \right]_{u=n-j} \bigg/ \prod_{y=1}^m \binom{sn+y-1}{y}^{r_y},$$

while, with binomial probabilities,

$$p_k = \binom{n}{k} \left[\Delta_u^k \prod_{y=1}^m \binom{su}{y}^{r_y} \right]_{u=0} \bigg/ \prod_{y=1}^m \binom{sn}{y}^{r_y} \quad (4.65)$$

and

$$b_{(j)} = \binom{n}{j} \left[\Delta_u^j \prod_{y=1}^m \binom{su}{y}^{r_y} \right]_{u=n-j} \bigg/ \prod_{y=1}^m \binom{sn}{y}^{r_y}. \quad (4.66)$$

Further, a restriction on the capacity of each urn or each compartment can be expressed by choosing a suitable truncated version of the distribution of X. Thus, the assumption of geometric probabilities truncated to the right at the point s, that is, with

$$q_x = P(X = x) = \frac{pq^x}{1 - q^{s+1}}, \quad x = 0, 1, \ldots, s,$$

whence

$$q_y(u) = P(Y = y) = \frac{L(y, u; s) p^u q^y}{(1 - q^{s+1})^u}, \quad y = 0, 1, \ldots, su,$$

where

$$L(y, u; s) = \sum_{j=0}^{u} (-1)^j \binom{u}{j} \binom{u+y-j(s+1)-1}{j},$$

imposes the restriction that the capacity of each urn is limited to s balls from each kind. The corresponding occupancy distribution can be similarly deduced.

Example 4.15 *Multiple capture-recapture sampling scheme.* Consider a population containing an unknown but finite number n of individuals. A multiple capture-recapture sampling scheme, which constitutes an extension of the simple capture-recapture sampling scheme of Example 4.11, is carried out in a sequence of r samplings. An experimenter, at the ith sampling, captures m_i individuals from the population, without replacement, marks and releases them, $i = 1, 2, \ldots, r$; the total number of individuals caught is $m = m_1 + m_2 + \cdots + m_r$. The probability function and moments of the number K of individuals caught in the r samplings are of interest.

Consider the random variable $Z_{i,j}$ assuming the values 1 or 0 according to whether at the ith sampling the jth individual is caught or not, respectively, $i = 1, 2, \ldots, r$, $j = 1, 2, \ldots, n$. Clearly, the joint probability function of $Z_{i,1}, Z_{i,2}, \ldots, Z_{i,n}$ is given by

$$p_i(z_1, z_2, \ldots, z_n; m_i) = 1 \Big/ \binom{n}{m_i},$$

where $z_j = 0, 1, j = 0, 1, \ldots, n$, with $z_1 + z_2 + \cdots + z_n = m_i, i = 1, 2, \ldots, r$. Note that if $X_{i,1}, X_{i,2}, \ldots, X_{i,n}$ are independent zero-one Bernoulli random variables, the sum $Y_i = \sum_{j=1}^{n} X_{i,j}$ obeys a binomial distribution and the conditional joint probability function of $X_{i,1}, X_{i,2}, \ldots, X_{i,n}$, given $Y_i = m_i$, equals $p_i(z_1, z_2, \ldots, z_n; m_i)$. Thus, considering the n individuals as n distinguishable urns and interpreting a capture of the jth individual as a placement of a ball into the jth urn, $j = 1, 2, \ldots, n$, the multiple capture-recapture sampling scheme is equivalent to a random occupancy model with Bernoulli probabilities. Let r_y be the number of those sample sizes, among $m_1, m_2, \ldots, m_r$, that are equal to y, for $y = 1, 2, \ldots, m$. Then, according to Corollary 4.12, the probability function and binomial moments of the number K of different individuals caught are given by (4.61) and (4.62), respectively.

Example 4.16 *(Continuation)* The probability function of the number K of individuals caught in a multiple capture-recapture sampling can be used to derive a uniformly minimum variance unbiased (UMVU) estimator of the population size n.

Note first that K is a sufficient statistic for the unknown number n of individuals. Indeed, let $Z_j = \sum_{i=1}^{r} Z_{i,j}$, be the number of balls distributed into the jth urn, $j = 1, 2, \ldots, n$. Clearly, the joint probability function of the occupancy numbers $Z_1, Z_2, \ldots, Z_n$ is given by

$$P(Z_1 = z_1, Z_2 = z_2, \ldots, Z_n = z_n) = \frac{\binom{r}{z_1}\binom{r}{z_2}\cdots\binom{r}{z_n}}{\binom{n}{m_1}\binom{n}{m_2}\cdots\binom{n}{m_r}},$$

where $z_j = 0, 1, \ldots, m$, $j = 1, 2, \ldots, n$, with $z_1 + z_2 + \cdots + z_n = m$. Further, if K_i is the number of individuals appearing i times in the sample, $i = 1, 2, \ldots, m$, then

$$P(K_1 = k_1, K_2 = k_2, \ldots, K_m = k_m) = \frac{(n)_k}{\prod_{i=1}^{m} \binom{n}{i}^{r_i}} \cdot \prod_{i=1}^{m} \binom{r}{i}^{k_i} / k_i!,$$

where $k_i = 0, 1, \ldots, n$, $i = 1, 2, \ldots, n$, with $\sum_{i=1}^{m} i k_i = m$ and $\sum_{i=1}^{m} k_i = k$. The indicated factorization implies that $K = \sum_{i=1}^{m} K_i$ is a sufficient statistic for the unknown number n of individuals. Also, the probability function of K, (4.61), as a factorial series distribution, is a complete family. Thus, by the classical theorems of Rao-Blackwell and Lehmann-Scheffé, a UMVU estimator of n based on K exists only when $m \geq n$, a condition imposed by the probability function of K. More generally, the parametric function $(n)_j$, $j = 1, 2, \ldots, n$, is estimable if, in addition to $m \geq n$, the observed value of K is $k \geq j$.

The derivation of the UMVU estimator of $(n)_j$, say $\hat{n}_j(K)$, may be carried out by noting that the unbiasedness condition $E[\hat{n}_j(K)] = (n)_j$, on using the probability function (4.61) of K, and since

$$\prod_{y=1}^{n} \binom{n}{y}^{r_y} = \left[E_u^{n-j} \prod_{y=1}^{n} \binom{u}{y}^{r_y} \right]_{u=j} = \sum_{k=j}^{n} \binom{n-j}{k-j} \left[\Delta_u^{k-j} \prod_{y=1}^{m} \binom{u}{y}^{r_y} \right]_{u=j}$$

and

$$(n)_j \binom{n-j}{k-j} = (k)_j \binom{n}{k},$$

may be expressed as

$$\sum_{k=1}^{n} \hat{n}_j(k) \binom{n}{k} \left[\Delta_u^k \prod_{y=1}^{m} \binom{u}{y}^{r_y} \right]_{u=0} = (n)_j \sum_{k=j}^{n} \binom{n-j}{k-j} \left[\Delta_u^{k-j} \prod_{y=1}^{m} \binom{u}{y}^{r_y} \right]_{u=j}$$

$$= \sum_{k=j}^{n} (k)_j \binom{n}{k} \left[\Delta_u^{k-j} \prod_{y=1}^{m} \binom{u}{y}^{r_y} \right]_{u=j}.$$

Therefore,

$$\hat{n}_j(k) = (k)_j \left[\Delta_u^{k-j} \prod_{y=1}^{m} \binom{u}{y}^{r_y} \right]_{u=j} \bigg/ \left[\Delta_u^k \prod_{y=1}^{m} \binom{u}{y}^{r_y} \right]_{u=0}, \quad j = 1, 2, \ldots, k$$

and $\hat{n}_j(k) = 0$, $k < j$. In particular, the UMVU estimator of n is $\hat{n}_1(K)$. Further, the variance of $\hat{n}_1(K)$ is

$$v = V[\hat{n}_1(K)] = E\{\hat{n}_1(K)[\hat{n}_1(K) - 1]\} - E[\hat{n}_1(K)]\{E[\hat{n}_1(K)] - 1\}$$
$$= E\{\hat{n}_1(K)[\hat{n}_1(K) - 1]\} - (n)_2$$

and so the UMVU estimator $\hat{v}(K)$ of the variance v of $\hat{n}_1(K)$ is given by

$$\hat{v}(K) = \hat{n}_1(K)[\hat{n}_1(K) - 1] - \hat{n}_2(K).$$

Example 4.17 *Distribution of different lottery numbers drawn in a given number of drawings.* In lotto every week s numbers are drawn without replacement from a set of n numbers, $\{1, 2, \ldots, n\}$. A gambler bets a fixed amount of money on a subset of s numbers, $\{i_1, i_2, \ldots, i_s\}$, of his own choice. If $\{i_1, i_2, \ldots, i_s\}$ is the result of the drawing, the gambler wins an amount of money depending on the total number of bets and on the number of winners. The distribution of the number K of different numbers drawn in r weeks is of interest to the gambler.

In order to derive the probability function and moments of K, the rs numbers drawn in r weeks may be considered as balls belonging to r different kinds with s balls in each kind. Also, the n numbers of the set $\{1, 2, \ldots, n\}$ may be considered as n distinguishable urns and the drawing of the rs numbers as a distribution of the rs balls into the n urns. Thus, the probability function and binomial moments of K are given by (4.63) and (4.64), respectively.

Example 4.18 *Formation of committees from grouped candidates.* Assume that there are m vacant positions in r committees. More specifically, $r_y \geq 0$ committees have y vacant positions, $y = 1, 2, \ldots, m$, so that $r_1 + 2r_2 + \cdots + mr_m = m$ and $r_1 + r_2 + \cdots + r_m = r$. Further, assume that for these m positions there are m priority-ordered nominations by each of n different groups. In a random selection of the m members of the r committees, find the probability function and moments of the number K of groups represented in at least one committee.

The m vacant positions of the r committees may be considered as m balls of the specification $(1^{r_1} 2^{r_2} \cdots m^{r_m})$, with $r_1 + 2r_2 + \cdots + mr_m = m$, $r_1 + r_2 + \cdots + r_m = r$, and the n different groups as n distinguishable urns. Thus, the placement of a ball of the ith kind into the jth urn corresponds to the selection of the jth group to occupy a position in the ith committee. Since it is assumed that the nominations are priority-ordered, the selection of the jth group uniquely determines the candidate who occupies a position in the ith committee. Further, there is no restriction on the number of balls of the same kind that each urn may accommodate. Therefore, according to Corollary 4.11, the probability function and binomial moments of the numbers K of groups represented in at least one committee are given by (4.57) and (4.58), respectively.

In the particular case of the formation of r committees each of size s, the probability function and binomial moments of the number K of groups represented in at least one committee are given by (4.59) and (4.60), respectively.

Example 4.19 *Formation of committees from individual candidates.* As in Example 4.18, assume that there are m vacant positions in r committees, among which $r_y \geq 0$ committees have y vacant positions, $y = 1, 2, \ldots, m$, so that $r_1 + 2r_2 + \cdots + mr_m = m$ and $r_1 + r_2 + \cdots + r_m = r$. Further, assume that for these m positions there are n individual nominations. Clearly, no candidate can be selected to more than one position of the same committee, while it can participate in more than one committee. In a random selection of the m members of the r committees, find the probability function and moments of the number K of candidates participating in at least one committee.

As in Example 4.18, the m vacant positions in the r committees may be considered as m balls of the specification $(1^{r_1} 2^{r_2} \cdots m^{r_m})$, with $r_1 + 2r_2 + \cdots + mr_m = m$, $r_1 + r_2 + \cdots + r_m = r$, and the n individual candidates as n distinguishable urns. Thus, the placement of a ball of the ith kind in the jth urn corresponds to the selection of the jth candidate for a position in the ith committee. The clarification that no candidate can be selected to more than one position of the same committee, while it can participate in more than one committee, implies that each urn may accommodate at most one ball from each kind. Then, according to Corollary 4.12, the probability function and binomial moments of the number K of candidates participating in at least one committee are given by (4.61) and (4.62), respectively.

In the particular case of the formation of r committees each of size s, the probability function and binomial moments of the number K of candidates participating in at least one committee are given by (4.63) and (4.64), respectively.

Example 4.20 *Formation of committees from grouped candidates with unordered nominations.* As in Example 4.18, assume that there are m vacant positions in r committees, among which $r_y \geq 0$ committees have y vacant positions, $y = 1, 2, \ldots, m$, so that $r_1 + 2r_2 + \cdots + mr_m = m$ and $r_1 + r_2 + \cdots + r_m = r$. Further, suppose that for these m positions there are s unordered nominations by each of n different groups. In a random selection of the m members of the r committees, find the probability function and moments of the number K of groups represented in at least one committee.

In this case the n different groups each with s nominations may be considered as n distinguishable urns each with s distinguishable compartments of capacity limited to one ball. Then the probability function and binomial moments of the number K of groups represented in at least one committee are given by (4.65) and (4.66), respectively.

4.7 REFERENCE NOTES

The random occupancy model, on which the presentation of occupancy distributions is based, was inspired by the papers of D. E. Barton and F. N. David (1959a,b); these papers were included in the chapter on occupancy in the book of F. N. David and D. E. Barton (1962). Considering a supply of balls randomly distributed into distinguishable urns and assuming that the number of balls distributed into any specific urn is a random variable obeying a Poisson, binomial or negative binomial law, derived the classical, restricted and pseudo-contagious occupancy distributions; these distributions are deduced in Corollaries 4.4, 4.6 and 4.8. As K.-T. Fang (1985) commented, the reduction of the joint distribution of the random occupancy numbers to a joint conditional distribution of independent random variables, given their sum, is one of the most powerful techniques in the study of occupancy distributions. The occupancy distribution, when the number of balls distributed into any specific urn obeys a general discrete probability law, was studied in Ch. A. Charalambides (1986a, 1997).

Particular cases were discussed in Ch. A. Charalambides (1979, 1981, 1983). The distribution of red blood cells in the hemacytometer, discussed in Example 4.6, was examined by M. E. Turner (1957). The general random occupancy model, presented in Section 4.6, was studied in Ch. A. Charalambides (2001). It is worth noting at this point that an interesting derivation of occupancy distributions was provided by M. Cerasoli (1983) by assuming that the total number of balls distributed into the urns is a Poisson stochastic process.

As has been noted in the introduction, in the classical approach of the occupancy theory, a host of enumerative problems were treated mostly through generating functions by P. A. MacMahon (1915, 1916) and combinatorially by J. Riordan (1958). Also, N. L. Johnson and S. Kotz (1977) devoted a chapter to occupancy theory containing a wealth of information and an extensive list of references. As a compact sequel of this book, the paper of S. Kotz and N. Balakrishnan (1997) surveys some new developments in urn models, and in particular in occupancy models. Moreover, a section on occupancy distributions is included in N. L. Johnson, S. Kotz and A. W. Kemp (1992). Additional information on occupancy distributions is contained in the review paper of N. L. Johnson and S. Kotz (1988). The classical occupancy distribution (4.21) was first obtained by A. De Moivre (1718). But in the literature it is known as Arfwedson distribution or Stevens-Craig distribution. W. L. Stevens (1937) and G. Arfwedson (1951) also derived this distribution, while C. C. Craig (1953) used it in estimating the size of a population by means of capture-recapture data. S. Kullback (1937) and G. B. Price (1946) with the multinomial distribution as background obtained the occupancy distribution of Example 4.1, and as a particular case the classical occupancy distribution. The distribution of the number of usable numbers in a sample of random numbers, presented in Example 4.5, was discussed by H. L. Jones (1959). The generating function of the maximum of occupancy numbers, given in Exercise 4.9, was derived by P. I. Richards (1968). Exercises 4.14 and 4.15 were inspired by the work of B. M. Hill (1974).

The restricted occupancy distribution (4.27), apart from the introduction of the generalized factorial coefficients, was first derived by V. Romanovsky (1934). A different restricted occupancy distribution, which is included as Exercise 4.16, was studied via generating functions by J. E. Freund and A. N. Pozner (1956) and combinatorially by K.-T. Fang (1982). Example 4.8, in which a particular case of this distribution is applied, is discussed in the first of these papers. The randomized occupancy distribution of Exercise 4.18 was studied by W. L. Harkness (1970).

The zero frequency test for a continuous population, discussed in Example 4.9, was first proposed by F. N. David (1950) and obtained independently by M. Okamoto (1952). The zero frequency tests for Poisson and binomial populations, presented in Example 4.10, were studied by C. R. Rao and I. M. Chakravarti (1956). Examples 4.11, 4.13 and 4.14 on the estimation methods of the number of classes in a finite population are based on the papers of L. A. Goodman (1949), B. Harris (1968), S. Berg (1974, 1975) and Ch. A. Charalambides (1981). Example 4.12 on missing dies in an old series of coins is inspired by the paper of W. E. McGovern (1980). Des Raj and S. H. Khamis (1958) considered the estimator of the mean of a finite population, which was based on the distinct units of a sample of fixed size. In the assessment of these estimators, the classical occupancy distribution (4.21) emerged.

The derivation of the exact distribution in several committee problems and discussion of their applications can be found in N. Mantel and B. S. Pasternack (1968), A. M. Gittelsohn (1969), D. A. Sprott (1969), C. White (1971), N. Mantel (1974) and P. Davies (1983). The sampling distribution (4.61) in the multiple capture-recapture sampling, discussed in Example 4.15, was derived by S. Berg (1976). J. G. Leite and C. A. de B. Pereira (1987) deduced the same distribution by using an equivalent occupancy model. A different expression of the probability function of this distribution was previously obtained by P. K. Pathak (1964).

The stochastic model presented in Exercise 4.17 was proposed by D. B. Mertz and R. B. Davies (1968) to describe the phenomenon of adult flour beetles cannibalizing their own pupae. A closely related modified occupancy distribution was derived by V. R. R. Uppuluri and J. A. Carpenter (1971) by embedding an absorbing Markov chain. Also, the modified occupancy distribution of Exercise 4.21 was derived as an extended Stirling distribution by K. Nishimura and M. Sibuya (1997) under a random walk.

Approximations of occupancy distributions are extensively discussed in a separate chapter in the book of N. L. Johnson and S. Kotz (1977). Also, the book of V. F. Kolchin, B. A. Sevastyanov and V. P. Chistyakov (1978) is devoted on this subject. The contribution of L. Holst (1979, 1980, 1986, 1988) on this topic is worth separate mention. Using the reduction of the joint distribution of the occupancy numbers to a joint conditional distribution of independent random variables, given their sum (random occupancy model), furnished a simple derivation of the characteristic function and deduced limit laws for occupancy distributions.

4.8 EXERCISES AND COMPLEMENTS

4.1 Consider the random occupancy model and assume that the number X_j of balls distributed into the jth urn obeys a Poisson distribution with $q_{j,x} = P(X_j = x) = e^{-\lambda_j} \lambda_j^x / x!$, $x = 0, 1, \ldots$, $0 < \lambda_j < \infty$, $j = 1, 2, \ldots, n$. Let $S_n = \sum_{j=1}^{n} X_j$ be the total number of balls distributed into the n urns and set $p_j = \lambda_j/\lambda$, $j = 1, 2, \ldots, n$, with $\lambda = \sum_{j=0}^{n} \lambda_j$.

Show that the rth binomial moment of the number K_i of urns occupied by i balls each, given that $S_n = m$, is given by

$$b_{(r)} = \sum \frac{m!}{(i!)^r (m-ir)!} p_{j_1}^i p_{j_2}^i \cdots p_{j_r}^i (1 - p_{j_1} - p_{j_2} - \cdots - p_{j_r})^{m-ir},$$

where the summation is extended over all r-combinations $\{j_1, j_2, \ldots, j_r\}$ of the n indices $\{1, 2, \ldots, n\}$.

In particular, conclude that

$$E(K_i | S_n = m) = \sum_{j=1}^{n} \binom{m}{i} p_j^i (1 - p_j)^{m-i}.$$

4.2 (*Continuation*). Assume that the number X_j of balls distributed into the jth urn obeys a binomial distribution with $q_{j,x} = P(X_j = x) = \binom{s_j}{x} p^x q^{s_j-x}$, $x = 0, 1, \ldots, s_j$, $j = 1, 2, \ldots, n$, where $0 < p < 1$ and $q = 1 - p$. Setting $s = \sum_{j=0}^{n} s_j$, show that the rth binomial moment of the number K_i of urns occupied by i balls each, given that $S_n = m$, is given by

$$b_{(r)} = \sum \frac{m!}{(i!)^r (m-ir)!} (s_{j_1})_i (s_{j_2})_i \cdots (s_{j_r})_i$$
$$\times (s - s_{j_1} - s_{j_2} - \cdots - s_{j_r})_{m-ir} / (s)_m,$$

where the summation is extended over all r-combinations $\{j_1, j_2, \ldots, j_r\}$ of the n indices $\{1, , 2, \ldots, n\}$. In particular, conclude that

$$E(K_i | S_n = m) = \sum_{j=1}^{n} \binom{m}{i} \frac{(s_j)_i (s - s_j)_{m-i}}{(s)_m}.$$

4.3 (*Continuation*). Assume that the number X_j of balls distributed into the jth urn obeys a negative binomial distribution with $q_{j,x} = P(X_j = x) = \binom{s_j + x - 1}{x} p^{s_j} q^x$, $x = 0, 1, \ldots, j = 1, 2, \ldots, n$, where $0 < p < 1$ and $q = 1-p$. Setting $s = \sum_{j=0}^{n} s_j$, show that the rth binomial moment of the number K_i of urns occupied by i balls each, given that $S_n = m$, is given by

$$b_{(r)} = \sum \frac{m!}{(i!)^r (m-ir)!} (s_{j_1} + i - 1)_i (s_{j_2} + i - 1)_i \cdots (s_{j_r} + i - 1)_i$$
$$\times (s - s_{j_1} - s_{j_2} - \cdots - s_{j_r} + m - ir - 1)_{m-ir} / (s + m - 1)_m,$$

where the summation is extended over all r-combinations $\{j_1, j_2, \ldots, j_r\}$ of the n indices $\{1, 2, \ldots, n\}$. In particular, conclude that

$$E(K_i | S_n = m) = \sum_{j=1}^{n} \binom{m}{i} \frac{(s_j + i - 1)_i (s - s_j + m - i - 1)_{m-i}}{(s + m - 1)_m}.$$

4.4 Consider the random occupancy model and assume that the number X_j of balls distributed into the jth urn obeys a discrete distribution with $q_x = P(X = x)$, $x = 0, 1, \ldots, j = 1, 2, \ldots, n$. If $q_x = e^{-\lambda} \lambda^x / x!$, $x = 0, 1, \ldots$, show that

$$E(K_i | S_n = m) = n p_{m,i}(1/n)$$

and

$$V(K_i | S_n = m) = n p_{m,i}(1/n)[1 - p_{m-i,i}(1/(n-1))]$$
$$- n^2 p_{m,i}(1/n)[p_{m,i}(1/n) - p_{m-i,i}(1/(n-1))],$$

where

$$p_{m,i}(1/n) = \binom{m}{i} \left(\frac{1}{n}\right)^i \left(1 - \frac{1}{n}\right)^{m-i}.$$

is a binomial probability function. In particular, conclude that

$$E(K_0|S_n = m) = n\left(1 - \frac{1}{n}\right)^m$$

and

$$V(K_0|S_n = m) = n\left(1 - \frac{1}{n}\right)^m\left[1 - \left(1 - \frac{1}{n-1}\right)^m\right]$$
$$- n^2\left(1 - \frac{1}{n}\right)^m\left[\left(1 - \frac{1}{n}\right)^m - \left(1 - \frac{1}{n-1}\right)^m\right].$$

4.5 (*Continuation*). If $q_x = P(X = x) = \binom{s}{x}p^x q^{s-x}$, $x = 0, 1, \ldots, s$, where $0 < p < 1$ and $q = 1 - p$, show that

$$E(K_i|S_n = m) = np_{m,i}(sn, s)$$

and

$$V(K_i|S_n = m) = np_{m,i}(sn, s)[1 - p_{m-i,i}(s(n-1), s)]$$
$$- n^2 p_{m,i}(sn, s)[p_{m,i}(sn, s) - p_{m-i,i}(s(n-1), s)],$$

where

$$p_{m,i}(sn, s) = \binom{m}{i}\frac{(s)_i(s(n-1))_{m-i}}{(sn)_m} = \binom{s}{i}\binom{s(n-1)}{m-i}\bigg/\binom{sn}{m}$$

is a hypergeometric probability function. In particular, conclude that

$$E(K_0|S_n = m) = n\frac{(s(n-1))_m}{(sn)_m}$$

and

$$V(K_0|S_n = m) = n\frac{(s(n-1))_m}{(sn)_m}\left[1 - \frac{(s(n-2))_m}{(s(n-1))_m}\right]$$
$$- n^2\frac{(s(n-1))_m}{(sn)_m}\left[\frac{(s(n-1))_m}{(sn)_m} - \frac{(s(n-2))_m}{(s(n-1))_m}\right].$$

4.6 Under the random occupancy model, with $q_x = P(X_j = x)$, $x = 0, 1, \ldots$, $j = 1, 2, \ldots, n$, consider the bivariate probability mass function $p_{k,m}(n, i) = P(K_i = k, S_n = m)$, $k = 0, 1, \ldots, n$, $m = 0, 1, \ldots$. Show that

$$g_{n,i}(t, u) = \sum_{m=0}^{\infty}\sum_{k=0}^{n} p_{k,m}(n, i) t^k u^m = [g(u) + q_i u^i(t - 1)]^n,$$

where $g(u) = \sum_{j=0}^{\infty} q_j u^j$. In particular, if $q_x = e^{-\lambda}\lambda^x/x!$, $x = 0, 1, \ldots$, conclude that

$$g_{n,i}(t, u) = \sum_{m=0}^{\infty} \sum_{k=0}^{n} p_{k,m}(n, i) t^k u^m = \left[e^{\lambda(u-1)} + e^{-\lambda} \frac{(\lambda u)^i}{i!} (t - 1) \right]^n,$$

while, if $q_x = P(X = x) = \binom{s}{x} p^x q^{s-x}$, $x = 0, 1, \ldots, s$, where $0 < p < 1$ and $q = 1 - p$, deduce that

$$g_{n,i}(t, u) = \sum_{m=0}^{\infty} \sum_{k=0}^{n} p_{k,m}(n, i) t^k u^m$$

$$= \left[(1 + p(u - 1))^s + \binom{s}{i} (pu)^i q^{s-i} (t - 1) \right]^n.$$

Also, if $q_x = P(X = x) = (1 - q^{s+1})^{-1} pq^x$, $x = 0, 1, \ldots, s$, where $0 < p < 1$ and $q = 1 - p$, conclude that

$$g_{n,i}(t, u) = \sum_{m=0}^{\infty} \sum_{k=0}^{n} p_{k,m}(n, i) t^k u^m$$

$$= \frac{p^n}{(1 - q^{s+1})^n} \left[\frac{1 - (qu)^{s+1}}{1 - qu} + (qu)^i (t - 1) \right]^n.$$

4.7 Consider the random occupancy model and assume that the number X_j of balls distributed into the jth urn obeys a discrete distribution, with $q_x = P(X = x)$, $x = 0, 1, \ldots$, $j = 1, 2, \ldots, n$. Let L_i be the number of urns occupied by at least i balls each and set

$$b_{(r)} = E\left[\binom{L_i}{r} \Big| S_n = m \right], \quad r = 0, 1, \ldots.$$

If $q_x = e^{-\lambda}\lambda^x/x!$, $x = 0, 1, \ldots$, show that

$$b_{(r)} = \frac{(n)_r}{n^m} \sum_{j=ir}^{m} \binom{m}{j} (n - r)^{m-j} S_i(j, r),$$

where $S_i(j, r)$ is the i-associated Stirling number of the second kind.

4.8 (*Continuation*). If $q_x = \binom{s}{x} p^x q^{s-x}$, $x = 0, 1, \ldots$, where s is a positive integer and $0 < p < 1$, $q = 1 - p$ (binomial distribution) or s is negative integer and $p < 0$, $q = 1 - p$ (negative binomial distribution), show that

$$b_{(r)} = \frac{(n)_r}{(sn)_m} \sum_{j=ir}^{m} \binom{m}{j} (s(n - r))_{m-j} C_i(j, r; s),$$

where $C_i(j, r; s)$ is the i-associated generalized factorial coefficient.

4.9 *Minimum and maximum of occupancy numbers.* Consider a supply of m distinguishable balls randomly and independently distributed into n distinguishable urns and assume that the probability for any ball to fall into the jth urn is $p_j, j = 1, 2, \ldots, n$. Let Z_j be the number of balls distributed into the jth urn, $j = 1, 2, \ldots, n$ and let $Z_{(1)} = \min\{Z_1, Z_2, \ldots, Z_n\}$, $Z_{(n)} = \max\{Z_1, Z_2, \ldots, Z_n\}$. Further, consider the distribution functions $P(z; m, n) = P(Z_{(1)} \leq z)$, $Q(z; m, n) = P(Z_{(n)} \leq z)$, for $z = 0, 1, \ldots$, and the exponential generating functions

$$G(u; z, n) = \sum_{m=0}^{\infty} P(z; m, n) \frac{u^m}{m!}, \quad H(u; z, n) = \sum_{m=0}^{\infty} Q(z.m, n) \frac{u^m}{m!}.$$

Show that

$$G(u; z, n) = e^u - \prod_{j=1}^{n} \sum_{x=z+1}^{\infty} \frac{(up_j)^x}{x!}, \quad H(u; z, n) = \prod_{j=1}^{n} \sum_{x=0}^{z} \frac{(up_j)^x}{x!}.$$

In particular, for $p_j = 1/n, j = 1, 2, \ldots, n$, conclude that

$$P(Z_{(1)} = r) = \frac{n!\{S_r(m, n) - S_{r+1}(m, n)\}}{n^m}$$

and

$$P(Z_{(n)} = r) = \frac{n!\{T_{r+1}(m, n) - T_r(m, n)\}}{n^m},$$

where $S_r(m, n)$ is the r-associated Stirling number of the second kind and $n! T_r(m, n)$ is the number of distributions of m distinguishable balls into n distinguishable urns in which each urn contains at most r balls (cf. Exercise 3.4).

4.10 (*Continuation*). Consider a supply of m like balls randomly distributed into n distinguishable urns. Assume that the jth urn is divided into s_j distinguishable cells, $j = 1, 2, \ldots, n$, each of capacity limited to one ball, and set $c = s_1 + s_2 + \cdots + s_n$. Show that

$$G(u; z, n) = e^u - \frac{1}{(c)_m} \prod_{j=1}^{n} \sum_{x=z+1}^{\infty} \binom{s_j}{x} u^x$$

and

$$H(u; z, n) = \frac{1}{(c)_m} \prod_{j=1}^{n} \sum_{x=0}^{z} \binom{s_j}{x} u^x.$$

In particular, for $s_j = s, j = 1, 2, \ldots, n$, conclude that

$$P(Z_{(1)} = r) = \frac{n!\{C_r(m, n; s) - C_{r+1}(m, n; s)\}}{(sn)_m}$$

and

$$P(Z_{(n)} = r) = \frac{n!\{D_{r+1}(m, n; s) - D_r(m, n; s)\}}{(sn)_m},$$

where $C_r(m, n; s)$ is the r-associated generalized factorial coefficient and $n!D_r(m, n; s)/m!$ is the number of distributions of m like balls into n distinguishable urns, with s distinguishable cells of capacity limited to one ball, so that each urn contains at most r balls (cf. Exercise 3.5).

4.11 *(Continuation).* If the cells are of unlimited capacity, show that

$$G(u; z, n) = e^u - \frac{1}{(c+m-1)_m} \prod_{j=1}^{n} \sum_{x=z+1}^{\infty} \binom{s_j + x - 1}{x} u^x$$

and

$$H(u; z, n) = \frac{1}{(c+m-1)_m} \prod_{j=1}^{n} \sum_{x=0}^{z} \binom{s_j + x - 1}{x} u^x.$$

In particular, for $s_j = s$, $j = 1, 2, \ldots, n$, conclude that

$$P(Z_{(1)} = r) = \frac{n!\{|C_r(m, n; -s)| - |C_{r+1}(m, n; -s)|\}}{(sn + m - 1)_m}$$

and

$$P(Z_{(n)} = r) = \frac{n!\{|D_{r+1}(m, n; -s)| - |D_r(m, n; -s)|\}}{(sn + m - 1)_m},$$

where $|C_r(m, n; -s)| = (-1)^n C_r(m, n; -s)$ and $n!|D_r(m, n; -s)|/m!$ is the number of distributions of m like balls into n distinguishable urns, with s distinguishable cells of unlimited capacity, so that each urns contains at most r balls (cf. Exercise 3.6).

4.12 *Bose-Einstein occupancy model.* Suppose that m indistinguishable balls are randomly distributed into n distinguishable urns of unlimited capacity. Show that the probability $p_k(m, n, i)$ that k urns are occupied by i balls each is given by

$$p_k(m, n, i) = \frac{\binom{n}{k}}{\binom{n+m-1}{m}} \sum_{j=1}^{n-k} (-1)^{n-k-j} \binom{n-k}{j} \binom{m - i(n-j) + j - 1}{j - 1}$$

and the rth binomial moment $b_{(r)}(m, n, i)$ of the number of urns occupied by i balls each is given by

$$b_{(r)}(m, n, i) = \binom{n}{r} \binom{n - r + m - ir - 1}{n - r - 1} \bigg/ \binom{n + m - 1}{m}.$$

4.13 *(Continuation).* Show that the probability $p_k(m, n)$ that k urns are occupied is given by

$$p_k(m, n) = \binom{n}{k} \binom{m-1}{k-1} \bigg/ \binom{n+m-1}{m}, \quad k = 1, 2, \ldots, n$$

and the rth binomial moment $b_{(r)}(m, n)$ of the number of occupied urns is given by

$$b_{(r)}(m, n) = \binom{n}{r}\binom{n+m-r-1}{m-r} \Big/ \binom{n+m-1}{m}, \quad r = 0, 1, \ldots.$$

4.14 *(Continuation)*. Let Z_j be the number of balls distributed into the jth urn, $j = 1, 2, \ldots, n$, and consider the corresponding order statistics $Z_{(k)}$, $k = 1, 2, \ldots, n$. Show that

$$P(Z_{(k)} \geq i) = \frac{\binom{n-1}{k-1}}{\binom{n+m-1}{m}} \sum_{j=0}^{k-1} (-1)^{k-j-1} \frac{n}{n-j} \binom{k-1}{j} \binom{n+m-i(n-j)-1}{n-1}.$$

In particular, conclude that

$$P(Z_{(1)} \geq i) = \binom{n+m-in-1}{n-1} \Big/ \binom{n+m-1}{m}$$

and

$$P(Z_{(n)} \leq i) = \sum_{j=0}^{n} (-1)^j \binom{n}{j} \binom{n+m-(i+1)j-1}{n-1} \Big/ \binom{n+m-1}{m}.$$

4.15 *(Continuation)*. Suppose that m indistinguishable balls are randomly distributed into n distinguishable urns subject to the condition that each urn must contain at least one ball. Show that

$$P(Z_{(k)} > i) = \frac{\binom{n-1}{k-1}}{\binom{m-1}{n-1}} \sum_{j=0}^{k-1} (-1)^{k-j-1} \frac{n}{n-j} \binom{k-1}{j} \binom{m-i(n-j)-1}{n-1}.$$

In particular, conclude that

$$P(Z_{(1)} > i) = \binom{m-in-1}{n-1} \Big/ \binom{m-1}{n-1}$$

and

$$P(Z_{(n)} \leq i) = \sum_{j=0}^{n} (-1)^j \binom{n}{j} \binom{m-ij-1}{n-1} \Big/ \binom{m-1}{n-1}.$$

4.16 Suppose that m indistinguishable balls are randomly distributed into n distinguishable urns, each of capacity limited to s balls. Show that the probability $p_k(m, n, i)$ that k urns are occupied by i balls each is given by

$$p_k(m, n, i) = \binom{n}{k} \sum_{j=0}^{n-k} (-1)^{n-k-j} \binom{n-k}{j} \frac{L(m - i(n-j), j; s)}{L(m, n; s)}.$$

and the rth binomial moment $b_{(r)}(m, n, i)$ of the number of urns occupied by i balls each is given by

$$b_{(r)}(m, n, i) = \binom{n}{r} \frac{L(m - ir, n - r; s)}{L(m, n; s)},$$

where

$$L(m, n; s) = \sum_{j=0}^{n} (-1)^j \binom{n}{j} \binom{n + m - j(s+1) - 1}{n - 1}$$

is the number of m-combinations of n, with repetition and the restriction that each element appears at most s times.

4.17 *Cannibalism of adult flour beetles.* A stochastic model proposed to describe the phenomenon of adult flour beetles of a certain genus cannibalizing their own pupae is based on the following assumptions. All adults are equally predacious and act independently of one another in their predatory behavior. A prey once attacked is regarded as a corpse; that is, no provision is made for mortality not associated with cannibalism or for the survival of wounded pupae. All pupae prey and prey corpses are equally vulnerable and corpses are capable of sustaining an unlimited number of attacks during the period of vulnerability. Under these assumptions, consider a population of r adult predators and a population of n pupae prey. Let $S(t)$ be the number of attacks generated up to time t by the population of adult predators and $K_0(t)$ be the number of pupae prey surviving at time t, with both numbers being random variables. Show that

$$P[K_0(t) = k | S(t) = m] = \frac{(n)_{n-k} S(m, n-k)}{n^m}, \quad k = 0, 1, \ldots, n,$$

where $S(m, k) = [\Delta_u^k u^m]_{u=0}/k!$ is the Stirling number of the second kind. Further, if $S_i(t)$ is the number of attacks up to time t of the ith adult predator, $i = 1, 2, \ldots, r$, and

$$f(v; t) = \sum_{m=0}^{\infty} P[K_0(t) = k] v^k,$$

$$g(u; t) = \sum_{s=0}^{\infty} P[S_i(t) = s] u^s, \quad i = 1, 2, \ldots, r,$$

derive the expressions

$$P[K_0(t) = k] = \binom{n}{k} [\Delta_u^{n-k} (g(u/n; t))^r]_{u=0}, \quad k = 0, 1, \ldots$$

and

$$f(v; t) = [(v + \Delta_u)^n (g(u/n; t))^r]_{u=0}.$$

4.18 *Randomized occupancy distribution.* Consider the random occupancy model and assume that the number X_j of balls distributed into the jth urn obeys a Poisson distribution with $q_x = P(X = x) = e^{-\lambda}\lambda^x/x!$, $x = 0, 1, \ldots$, $0 < \lambda < \infty$, $j = 1, 2, \ldots, n$. Further, assume that each ball has probability p of staying in an urn and probability $q = 1 - p$ of falling through. Show that the conditional probability $p_k(m, n; p)$ that k urns are occupied, given that m balls are distributed into the n urns, $k = 0, 1, \ldots, n$, is given by

$$p_k(m, n; p) = \frac{(n)_k S(m, k; nq/p)}{(n/p)^m},$$

where $S(m, k; r) = [\Delta_u^k u^m]_{u=r}/k!$ is the noncentral Stirling number of the second kind.

4.19 *Extended random occupancy model.* Consider a supply of balls randomly distributed into $n + r$ distinguishable urns, among which n are labeled as *target* urns and the other r as *control* urns. Let X_j be the number of balls distributed into the jth urn, $j = 1, 2, \ldots, n+r$. Assume that the random variables $X_j, j = 1, 2, \ldots, n+r$ are independently and identically distributed with common probability function $P(X = x) = q_x$, $x = 0, 1, \ldots$. Also, let S_{n+r} be the total number of balls distributed into the $n + r$ urns with $P(S_{n+r} = m) = q_m(n+r)$, $m = 0, 1, \ldots$. Under this extended random occupancy model, let K_i be the number of target urns occupied by i balls each and

$$p_k(m, n, r, i) = P(K_i = k | S_{n+r} = m),$$

$$b_{(j)}(m, n, i) = E\left[\binom{K_i}{j}\bigg| S_{n+r} = m\right].$$

(a) Show that the conditional probability $p_k(m, n, r, i)$ that k target urns are occupied by i balls each, given that m balls are distributed into the $n + r$ urns, $k = 0, 1, \ldots, n$, is given by

$$p_k(m, n, r, i) = \binom{n}{k} \frac{[\Delta_u^{n-k} q_i^{n+r-u} q_{m-i(n+r-u)}(u)]_{u=r}}{q_m(n+r)}.$$

(b) Further, show that the jth binomial moment $b_{(j)}(m, n, i)$, $j = 0, 1, \ldots$, of the probability function $p_k(m, n, r, i)$, $k = 0, 1, \ldots, n$, is given by

$$b_{(j)}(m, n, i) = \binom{n}{j} \frac{q_i^j q_{m-ij}(n+r-j)}{q_m(n+r)}.$$

(c) Also, show that the generating function of the bivariate probability function $p_{k,m}(n, r, i) = P(K_i = k, S_{n+r} = m)$, $k = 0, 1, \ldots, n$, $m = 0, 1, \ldots$, is given by

$$g_{n,r,i}(t, u) = \sum_{m=0}^{\infty} \sum_{k=0}^{n} p_{k,m}(n, r, i) t^k u^m = [g(u)]^r [g(u) + q_i u^i(t-1)]^n,$$

where $g(u) = \sum_{j=0}^{\infty} q_j u^j$.

4.20 (*Continuation*). Under the extended random occupancy model, let K be the number of target urns occupied by at least one ball each and

$$p_k(m, n, r) = P(K = k | S_{n+r} = m), \quad b_{(j)}(m, n, r) = E\left[\binom{K}{j} \bigg| S_{n+r} = m\right].$$

(a) Show that the conditional probability $p_k(m, n, r)$ that k target urns are occupied, given that m balls are distributed into the $n + r$ urns, $k = 0, 1, \ldots, n$, is given by

$$p_k(m, n, r) = \binom{n}{k} \frac{[\Delta_u^k q_0^{n+r-u} q_m(u)]_{u=r}}{q_m(n+r)}.$$

(b) Further, show that the jth binomial moment $b_{(j)}(m, n, r)$, $j = 0, 1, \ldots$, of the probability function $p_k(m, n, r)$, $k = 0, 1, \ldots, n$, is given by

$$b_{(j)}(m, n, r) = \binom{n}{j} \frac{[\Delta_u^j q_0^{n+r-u} q_m(u)]_{u=n+r-j}}{q_m(n+r)}.$$

4.21 (*Continuation*). Consider the extended random occupancy model and assume that the number X_j of balls distributed into the jth urn, $j = 1, 2, \ldots, n+r$, obeys a Poisson distribution with $q_x = P(X = x) = e^{-\lambda} \lambda^x / x!$, $x = 0, 1, \ldots$, $0 < \lambda < \infty$.

(a) Show that the conditional probability $p_k(m, n, r)$ that k target urns are occupied, given that m balls are distributed into the $n + r$ urns, $k = 0, 1, \ldots, n$, is given by

$$p_k(m, n, r) = \frac{(n)_k S(m, k; r)}{(n+r)^m},$$

where $S(m, k; r) = [\Delta_u^k u^m]_{u=r}/k!$ is the noncentral Stirling number of the second kind.

(b) Further, show that the jth binomial moment $b_{(j)}(m, n, r)$, $j = 0, 1, \ldots$, of the probability function $p_k(m, n, r)$, $k = 0, 1, \ldots, n$, is given by

$$b_{(j)}(m, n, r) = \frac{(n)_j S(m, j; n+r-j)}{(n+r)^m}.$$

4.22 (*Continuation*). Consider the extended random occupancy model and assume that the number X_j of balls distributed into the jth urn, $j = 1, 2, \ldots, n+r$, obeys a binomial distribution with $q_x = P(X = x) = \binom{s}{x} p^x q^{s-x}$, $x = 0, 1, \ldots, s$, where $0 < p < 1$ and $q = 1 - p$.

(a) Show that the conditional probability $p_k(m, n, r)$ that k target urns are occupied, given that m balls are distributed into the $n + r$ urns, $k = 0, 1, \ldots, n$, is given by

$$p_k(m, n, r) = \frac{(n)_k C(m, k; s, rs)}{(sn + sr)_m},$$

where $C(m, k; s, rs) = [\Delta_u^k (su)_m]_{u=r}/k!$ is the noncentral generalized factorial coefficient.

(b) Further, show that the jth binomial moment $b_{(j)}(m,n,r)$, $j = 0, 1, \ldots$, of the probability function $p_k(m,n,r)$, $k = 0, 1, \ldots, n$, is given by

$$b_{(j)}(m,n,r) = \frac{(n)_j C(m,j;s,(n+r-j)s)}{(sn+sr)_m}.$$

4.23 Under the general random occupancy model, let

$$P_k \equiv P_k(r_1, r_2, \ldots, r_m; r, m, n)$$
$$= P(K \leq k | S_R = m, R = r, R_1 = r_1, R_2 = r_2, \ldots, R_m = r_m).$$

Show that

$$P_k = n\binom{n-1}{k} \frac{[\Delta_u^k q_0^{r(n-k)} [q_1(u)]^{r_1} [q_2(u)]^{r_2} \cdots [q_m(u)]^{r_m}/(n-u)]_{u=0}}{[q_1(n)]^{r_1} [q_2(u)]^{r_2} \cdots [q_m(n)]^{r_m}}.$$

4.24 Under the random occupancy model with $q_i > 0$ and $q_j > 0$, $i \neq j$, show that the joint probability function $p_{k,r}(m,n;i,j) = P(K_i = k, K_j = r | S_n = m)$, $k = 0, 1, \ldots, n-r$, $r = 0, 1, \ldots, n$ and the joint binomial moments $b_{(k,r)}(m,n;i,j)$ $= E[\binom{K_i}{k}\binom{K_j}{r}|S_n = m]$, $k = 0, 1, \ldots, n-r$, $r = 0, 1, \ldots, n$, are given by

$$p_{k,r}(m,n;i,j) = \binom{n}{k,r}$$
$$\times \frac{[\Delta_v^{n-k-r} \Delta_u^v q_i^{k+v-u} q_j^{n-k-v} q_{m-i(k+v-u)-j(n-k-v)}(u)]_{u=0,v=0}}{q_m(n)}$$

and

$$b_{(k,r)}(m,n;i,j) = \binom{n}{k,r} \frac{q_i^k q_j^r q_{m-ik-jr}(n-k-r)}{q_m(n)},$$

where

$$\binom{n}{k,r} = \frac{n!}{k! r! (n-k-r)!}.$$

4.25 (*Continuation*). Assume the number of balls distributed into any specific urn obeys a Poisson distribution with $q_x = P(X = x) = e^{-\lambda}\lambda^x/x!$, $x = 0, 1, \ldots$, $0 < \lambda < \infty$. Show that

$$p_{k,r}(m,n;i,j) = \binom{n}{k,r}$$
$$\times \left[\Delta_v^{n-k-r} \Delta_u^v \frac{(m)_{i(k+v-u)+j(n-k-v)} u^{m-i(k+v-u)-j(n-k-v)}}{(i!)^{k+v-u}(j!)^{n-k-v} n^m} \right]_{u=0,v=0}$$

and

$$b_{(k,r)}(m,n;i,j) = \binom{n}{k,r} \frac{(m)_{ik+jr}(n-k-r)^{m-ik-jr}}{(i!)^k (j!)^r n^m}.$$

5
SEQUENTIAL OCCUPANCY DISTRIBUTIONS

5.1 INTRODUCTION

The systematic presentation of the occupancy theory, started in Chapter 4 with the study of occupancy distributions, is continued with the discussion of sequential occupancy distributions. Specifically, when m balls of a general specification are randomly distributed into n distinguishable urns, the *occupancy distribution* of the number $K = n - K_0$ of occupied urns and more generally the number K_i of urns occupied by i balls each, $i = 0, 1, \ldots, m$, was extensively studied and a variety of applications was provided. In this chapter, the study moves on to the case of sequentially distributed balls into the n distinguishable urns. When the distribution of balls is made at random, several random variables of probabilistic and statistical interest may be defined. The *waiting-time* (or *sequential*) *occupancy distribution* of the number W_k of balls required to be distributed until k urns are occupied (by at least one ball each) is thoroughly examined. Also, the study of the *complementary waiting-time occupancy distribution* of the number V_r of balls required to be distributed until r balls are placed into previously occupied urns is considered. Instead, the distribution of the number $L_r = V_r - r$ of occupied urns until r balls are placed into previously occupied urns, which is more convenient in statistical inference, is discussed.

The reduction of the joint distribution of the occupancy numbers to a joint conditional distribution of independent random variables, given their sum, is a powerful technique in the derivation and study, not only of occupancy distributions, but also of sequential occupancy distributions. For this reason, the waiting-time and the

---

Combinatorial Methods in Discrete Distributions. By Charalambos A. Charalambides
ISBN 0-471-68027-3 Copyright © 2005 John Wiley & Sons, Inc.

complementary waiting-time occupancy distributions are presented within the framework of a sequential random occupancy model.

5.2 A SEQUENTIAL RANDOM OCCUPANCY MODEL

Consider a supply of balls randomly and sequentially distributed into n distinguishable urns and let Z_j be the number of balls distributed into the jth urn, $j = 1, 2, \ldots, n$. For a fixed number m of balls, let us denote the joint probability function of the occupancy numbers $Z_1, Z_2, \ldots, Z_n$ by $p(z_1, z_2, \ldots, z_n; m)$, with $z_j = 0, 1, \ldots, m$, $j = 1, 2, \ldots, n$ and $z_1 + z_2 + \cdots + z_n = m$. Further, consider the *random occupancy model*, introduced in Section 4.2, and suppose now that balls are randomly and sequentially distributed into the n distinguishable urns. Let X_j be the number of balls distributed into the jth urn, $j = 1, 2, \ldots, n$. The random variables X_j, $j = 1, 2, \ldots, n$, are assumed to be independently distributed with probability function

$$P(X_j = x) = q_{j,x}, \; x = 0, 1, \ldots, \; j = 1, 2, \ldots, n.$$

Let $S_n = \sum_{j=1}^n X_j$ be the total number of balls distributed into the n urns, with

$$P(S_n = m) = q_m(n), \; m = 0, 1, \ldots.$$

Then

$$q_m(n) = \sum q_{1,x_1} q_{2,x_2} \cdots q_{n,x_n},$$

where the summation is extended over all integers $x_j \geq 0, j = 1, 2, \ldots, n$, such that $x_1 + x_2 + \cdots + x_n = m$. Also, the conditional joint probability function of the random variables $X_1, X_2, \ldots, X_n$, given that $S_n = m$, is

$$q(x_1, x_2, \ldots, x_n | m) = \frac{q_{1,x_1} q_{2,x_2} \cdots q_{n,x_n}}{q_m(n)},$$

with $x_j = 0, 1, \ldots, m, \; j = 1, 2, \ldots, n$ and $x_1 + x_2 + \cdots + x_n = m$. As was noted in Section 4.2, if

$$p(x_1, x_2, \ldots, x_n; m) = \frac{q_{1,x_1} q_{2,x_2} \cdots q_{n,x_n}}{q_m(n)},$$

the two models are equivalent. Further, the derivation of the probability function of the number W_k of balls required to be distributed until k urns are occupied, as well as that of the number L_r of occupied urns until r balls are placed into previously occupied urns, is based on this joint probability function. Clearly, if

$$q_m(k, n) = P(W_k = m), \; m = k, k+1, \ldots,$$

$$c_k(r, n) = P(L_r = k), \; k = 1, 2, \ldots, n,$$

and

$$P_k(m, n) = P(K \leq k | S_n = m), \; k = 1, 2, \ldots, n,$$

with $K = n - K_0$ the number of occupied urns, then
$$q_m(k,n) = P_{k-1}(m-1,n) - P_{k-1}(m,n), \ m = k, k+1, \ldots,$$
and
$$c_k(r,n) = P_k(r+k,n) - P_{k-1}(r+k-1,n), \ k = 1, 2, \ldots, n.$$

For the sake of illustration of the sequential occupancy model, let us consider the following examples.

Example 5.1 *The sequential classical occupancy model.* Consider a supply of balls randomly and sequentially distributed into n distinguishable urns and assume that the probability for any ball to fall into the jth urn is $p_j, j = 1, 2, \ldots, n$. Let W_k be the number of balls required to be distributed until k urns are occupied. The probability function and moments of the *waiting-time classical occupancy distribution* of W_k may be derived as follows.

The probability $P_k(m,n) = P(K \leq k | S_n = m), \ k = 1, 2, \ldots, n$, has been obtained in Example 1.13 as
$$P_k(m,n) = \sum_{r=0}^{k}(-1)^{k-r}\binom{n-r-1}{k-r}S_{n,n-r},$$
with
$$S_{n,n-r} = \sum (p_{j_1} + p_{j_2} + \cdots + p_{j_r})^m,$$
where the summation is extended over all r-combinations $\{j_1, j_2, \ldots, j_r\}$ of the n indices $\{1, 2, \ldots, n\}$. Thus, the probability function $q_m(k,n) = P(W_k = n), m = k, k+1, \ldots$, upon using the expression $q_m(k,n) = P_{k-1}(m-1,n) - P_{k-1}(m,n)$, $m = k, k+1, \ldots$, is obtained as
$$q_m(k,n) = \sum_{r=0}^{k-1}(-1)^{k-r-1}\binom{n-r-1}{k-r-1}T_{n,n-r},$$
with
$$T_{n,n-r} = \sum [1-(p_{j_1}+p_{j_2}+\cdots+p_{j_r})](p_{j_1}+p_{j_2}+\cdots+p_{j_r})^{m-1},$$
where the summation is extended over all r-combinations $\{j_1, j_2, \ldots, j_r\}$ of the n indices $\{1, 2, \ldots, n\}$.

The ith ascending binomial moment of W_k,
$$b_{[i]}(k,n) = E\left[\binom{W_k+i-1}{i}\right] = \sum_{m=n}^{\infty}\binom{m+i-1}{i}q_m(k,n),$$
may be derived by noting that the ith ascending binomial moment of the geometric distribution $pq^{m-1}, m = 1, 2, \ldots$, with $p = 1 - (p_{j_1}+p_{j_2}+\cdots+p_{j_r})$ and $q = p_{j_1}+p_{j_2}+\cdots+p_{j_r}$ is given by
$$a_{[i]} = \frac{1}{[1-(p_{j_1}+p_{j_2}+\cdots+p_{j_r})]^i}.$$

Therefore,
$$b_{[i]}(k,n) = \sum_{r=0}^{k-1}(-1)^{k-r-1}\binom{n-r-1}{k-r-1}U_{n,n-r},$$

with
$$U_{n,n-r} = \sum \frac{1}{[1-(p_{j_1}+p_{j_2}+\cdots+p_{j_r})]^i},$$

where the summation is extended over all r-combinations $\{j_1, j_2, \ldots, j_r\}$ of the n indices $\{1, 2, \ldots, n\}$. Also, the ith descending binomial moment $b_{(i)}(k,n) = E\left[\binom{W_k}{i}\right]$ may similarly been obtained as

$$b_{(i)}(k,n) = \sum_{r=0}^{k-1}(-1)^{k-r-1}\binom{n-r-1}{k-r-1}V_{n,n-r},$$

with
$$V_{n,n-r} = \sum \frac{(p_{j_1}+p_{j_2}+\cdots+p_{j_r})^{i-1}}{[1-(p_{j_1}+p_{j_2}+\cdots+p_{j_r})]^i},$$

where the summation is extended over all r-combinations $\{j_1, j_2, \ldots, j_r\}$ of the n indices $\{1, 2, \ldots, n\}$.

Example 5.2 (*Continuation*). Let L_r be the number of occupied urns until r balls are placed into previously occupied urns. The probability function of the *complementary waiting-time classical occupancy distribution* of L_r may be derived as follows.

The probability function $c_k(r,n) = P(L_r = k)$, $k = 1, 2, \ldots, n$, is expressed in terms of probabilities $P_k(m,n) = P(K \leq k|S_n = m)$, $k = 1, 2, \ldots, n$, by $c_k(r,k) = P_k(r+k,n) - P_{k-1}(r+k-1,n)$, $k = 1, 2, \ldots, n$. Denoting by $S_{n,n-i}(m)$ the general term $S_{n,n-i}$ in the expression of the probability $P_k(m,n)$, it follows that (cf. Example 5.1)

$$P_k(r+k,n) = \sum_{i=0}^{k}(-1)^{k-i}\binom{n-i-1}{k-i}S_{n,n-i}(r+k)$$

and

$$P_{k-1}(r+k-1,n) = \sum_{i=0}^{k-1}(-1)^{k-i-1}\binom{n-i-1}{k-i-1}S_{n,n-i}(r+k-1)$$

$$= -\sum_{i=0}^{k}(-1)^{k-i}\binom{n-i-1}{k-i}\frac{k-i}{n-k}S_{n,n-i}(r+k-1).$$

Thus, setting
$$T_{n,n-i} = S_{n,n-i}(r+k) + \frac{k-i}{n-k}S_{n,n-i}(r+k-1),$$

we get
$$c_k(r,n) = \sum_{i=0}^{k}(-1)^{k-i}\binom{n-i-1}{k-i}T_{n,n-i}$$
with
$$T_{n,n-i} = \sum \left(p_{j_1} + p_{j_2} + \cdots + p_{j_i} + \frac{k-i}{n-k}\right)(p_{j_1} + p_{j_2} + \cdots + p_{j_i})^{r+k-1},$$
where the summation is extended over all i-combinations $\{j_1, j_2, \ldots, j_i\}$ of the n indices $\{1, 2, \ldots, n\}$.

Example 5.3 *A sequential restricted occupancy model.* Consider a supply of like balls randomly and sequentially distributed into n distinguishable urns. Assume that the jth urn is divided into s_j distinguishable cells, $j = 1, 2, \ldots, n$, each of capacity limited to one ball and set $s = s_1 + s_2 + \cdots + s_n$. Let W_k be the number of balls required to be distributed until k urns are occupied. The probability function and moments of the *waiting-time restricted occupancy distribution* of W_k may be derived as follows.

The probability $P_k(m,n) = P(K \leq k | S_n = m)$, $k = 1, 2, \ldots, n$, has been obtained in Example 1.14 as
$$P_k(m,n) = \sum_{r=0}^{k}(-1)^{k-r}\binom{n-r-1}{k-r}S_{n,n-r},$$
with
$$S_{n,n-r} = \sum \frac{(s_{j_1} + s_{j_2} + \cdots + s_{j_r})_m}{(s)_m},$$
where the summation is extended over all r-combinations $\{j_1, j_2, \ldots, j_r\}$ of the n indices $\{1, 2, \ldots, n\}$. Thus, the probability function $q_m(k,n) = P(W_k = m)$, $m = k, k+1, \ldots$, on using the expression $q_m(k,n) = P_{k-1}(m-1,n) - P_{k-1}(m,n)$, $m = k, k+1, \ldots$, is deduced as
$$q_m(k,n) = \sum_{r=0}^{k-1}(-1)^{k-r-1}\binom{n-r-1}{k-r-1}T_{n,n-r},$$
with
$$T_{n,n-r} = \sum \frac{[s - (s_{j_1} + s_{j_2} + \cdots + s_{j_r})](s_{j_1} + s_{j_2} + \cdots + s_{j_r})_{m-1}}{(s)_m},$$
where the summation is extended over all r-combinations $\{j_1, j_2, \ldots, j_r\}$ of the n indices $\{1, 2, \ldots, n\}$.

The ith ascending binomial moment of the waiting-time hypergeometric distribution $(s-u)(u)_{m-1}/(s)_m$, $m = 1, 2, \ldots, u+1$, with $u = s_{j_1} + s_{j_2} + \cdots + s_{j_r}$ is given by
$$a_{[i]} = \frac{(s+i)_i}{(s - s_{j_1} - s_{j_2} - \cdots - s_{j_r} + i)_i},$$

and so the ith ascending binomial moment $b_{[i]}(k,n) = E\left[\binom{W_k+i-1}{i}\right]$ is given by

$$b_{[i]}(k,n) = \sum_{r=0}^{k-1}(-1)^{k-r-1}\binom{n-r-1}{k-r-1}U_{n,n-r},$$

with

$$U_{n,n-r} = \sum \frac{(s+i)_i}{(s-s_{j_1}-s_{j_2}-\cdots-s_{j_r}+i)_i},$$

where the summation is extended over all r-combinations $\{j_1, j_2, \ldots, j_r\}$ of the n indices $\{1, 2, \ldots, n\}$.

Example 5.4 (*Continuation*). Let L_r be the number of occupied urns until r balls are placed into previously occupied urns. The probability function of the *complementary waiting-time restricted occupancy distribution* of L_r may be obtained as follows.

The probability function $c_k(r, n) = P(L_r = k)$, $k = 1, 2, \ldots, n$, is expressed in terms of the probabilities $P_k(m, n) = P(K \leq k | S_n = m)$, $k = 1, 2, \ldots, n$, by

$$c_k(r, n) = P_k(r+k, n) - P_{k-1}(r+k-1, n), \quad k = 1, 2, \ldots, n.$$

Thus, proceeding as in Example 5.2, we deduce the expression

$$c_k(r, n) = \sum_{i=0}^{k}(-1)^{k-i}\binom{n-i-1}{k-i}T_{n,n-i}$$

with

$$T_{n,n-i} = \sum \left[s_{j_1} + s_{j_2} + \cdots + s_{j_i} - r - k + 1 + \frac{k-i}{n-k}(s - r - k + 1)\right]$$
$$\times (s_{j_1} + s_{j_2} + \cdots + s_{j_i})_{r+k-1}/(s)_{r+k},$$

where the summation is extended over all i-combinations $\{j_1, j_2, \ldots, j_i\}$ of the n indices $\{1, 2, \ldots, n\}$.

In the sequel, the study of the sequential occupancy distributions focuses on the particular case of the random occupancy model in which the random variables X_j, $j = 1, 2, \ldots, n$, are independently and identically distributed, with common probability function

$$P(X = x) = q_x, \ x = 0, 1, \ldots. \tag{5.1}$$

In this case,

$$q_m(n) = P(S_n = m) = \sum q_{x_1} q_{x_2} \cdots q_{x_n}, \ m = 0, 1, \ldots, \tag{5.2}$$

where the summation is extended over all integers $x_j \geq 0$, $j = 1, 2, \ldots, n$, such that $x_1 + x_2 + \cdots + x_n = m$, is the n-fold convolution of q_x, $x = 0, 1, \ldots$.

5.3 SEQUENTIAL OCCUPANCY DISTRIBUTIONS

Consider the random occupancy model and assume now that balls are randomly and sequentially distributed into the n distinguishable urns until k urns are occupied and let W_k be the number of balls required. Further, consider the probability function

$$q_m(k,n) = P(W_k = m), \ m = k, k+1, \ldots \quad (5.3)$$

and its rth ascending binomial moment

$$b_{[r]}(k,n) = E\left[\binom{W_k + r - 1}{r}\right], \ r = 0, 1, \ldots. \quad (5.4)$$

Note that for a discrete waiting-time distribution the ascending binomial moments, in general, can be computed more easily than any other moments. This computation is facilitated by the consideration of the conditional probability

$$h_m(u,n) = \frac{n q_0^{n-u-1} q_1 q_{m-1}(u)}{m q_m(n)}, \ m = 1, 2, \ldots, \quad (5.5)$$

that one ball is distributed at the mth allocation into an empty urn, among the n urns, and $m-1$ balls are distributed into u specific urns (among the other $n-1$ urns), given that m balls are distributed into the n urns. Further, consider the rth ascending binomial moment of the sequence of probabilities $h_m(u,n)$, $m = 1, 2, \ldots$,

$$a_{[r]}(u,n) = \sum_{m=1}^{\infty} \binom{m+r-1}{r} h_m(u,n), \ r = 0, 1, \ldots. \quad (5.6)$$

Notice that the probabilities $h_m(u,n)$, $m = 1, 2, \ldots$, do not necessarily add to unity.

In the next theorem the probability function, (5.3), and the ascending binomial moments, (5.4), of the *waiting-time occupancy distribution* are expressed in terms of finite differences of the convolution probabilities (5.2) and the ascending binomial moments (5.6), respectively.

Theorem 5.1 (a) *Under the sequential random occupancy model with $q_0 > 0$, the probability $q_m(k,n)$, that m balls are required to be distributed into the n distinguishable urns until k urns are occupied, $m = k, k+1, \ldots$, is given by*

$$q_m(k,n) = \binom{n-1}{k-1} \frac{n q_1 [\Delta_u^{k-1} q_0^{n-u-1} q_{m-1}(u)]_{u=0}}{m q_m(n)}. \quad (5.7)$$

(b) *If the conditional probability (5.5) is a polynomial in u of degree at most $m-1$, then the rth ascending binomial moment $b_{[r]}(k,n)$, $r = 0, 1, \ldots$, of the probability function $q_m(k,n)$, $m = k, k+1, \ldots$, is given by*

$$b_{[r]}(k,n) = \binom{n-1}{k-1}\left[\Delta_u^{k-1} a_{[r]}(u,n)\right]_{u=0}, \quad (5.8)$$

where $a_{[r]}(u,n)$, $r = 0, 1, \ldots$, is given by (5.6).

Proof (a) Clearly, the conditional probability that one ball is distributed at any specific allocation into any specific empty urn and $m-1$ balls are distributed into the other $n-1$ urns, given that m balls are distributed into the n distinguishable urns, is

$$\frac{q_1 q_{m-1}(n-1)}{q_m(n)}.$$

Thus, dividing it by the number m of allocations and then multiplying the resulting expression by the number n of urns, we deduce the conditional probability

$$h_m(n) = \frac{n q_1 q_{m-1}(n-1)}{m q_m(n)},$$

that one ball is distributed at the mth allocation into an empty urn and $m-1$ balls are distributed into the other $n-1$ urns, given that m balls are distributed into the n distinguishable urns. Further, the probability $q_m(k,n)$ that m balls are required to be distributed into the n distinguishable urns until k urns are occupied equals the product

$$q_m(k,n) = h_m(n) p_{k-1}(m-1, n-1),$$

where $p_{k-1}(m-1, n-1)$ is the conditional probability that $k-1$ urns are occupied, given that $m-1$ balls are distributed into the $n-1$ distinguishable urns. Since, by Corollary 4.1,

$$p_{k-1}(m-1, n-1) = \binom{n-1}{k-1} \frac{\left[\Delta_u^{k-1} q_0^{n-u-1} q_{m-1}(u)\right]_{u=0}}{q_{m-1}(n-1)},$$

expression (5.7) is established.

(b) From the definition of $b_{[r]}(k,n)$, on using (5.7) and (5.5), it follows that

$$b_{[r]}(k,n) = \binom{n-1}{k-1} \left[\Delta_u^{k-1} \sum_{m=k}^{\infty} \binom{m+r-1}{r} h_m(u,n)\right]_{u=0}.$$

Further, the assumption that the probability $h_m(u,n)$ is a polynomial in u of degree at most $m-1$ implies $\Delta_u^{k-1} h_m(u,n) = 0$, $m=1,2,\ldots,k-1$, and hence

$$b_{[r]}(k,n) = \binom{n-1}{k-1} \left[\Delta_u^{k-1} \sum_{m=1}^{\infty} \binom{m+r-1}{r} h_m(u,n)\right]_{u=0}.$$

Thus, by (5.6), expression (5.8) is deduced.

Remark 5.1 Under the sequential random occupancy model, the probability $q_m(k,n)$ that m balls are required to be distributed into the n distinguishable urns until k urns are occupied, may be expressed in terms of the conditional probability $P_k(m,n)$, that at most k urns are occupied, given that m balls are distributed into the n distinguishable urns. Specifically, $q_m(k,n) = P_{k-1}(m-1, n) - P_{k-1}(m, n)$ and using Corollary 4.1(c) we deduce the expression

$$q_m(k,n) = n \binom{n-1}{k-1} \left[\Delta_u^{k-1} \frac{1}{n-u} \left(\frac{q_0^{n-u} q_{m-1}(u)}{q_{m-1}(n)} - \frac{q_0^{n-u} q_m(u)}{q_m(n)}\right)\right]_{u=0},$$

which is equivalent to (5.7). Indeed,

$$\frac{q_0^{n-u}q_{m-1}(u)}{q_{m-1}(n)} - \frac{q_0^{n-u}q_m(u)}{q_m(n)}$$

is the conditional probability that $m-1$ balls are distributed into u specified urns and at the mth allocation one ball is distributed into any of the other $n-u$ urns, given that m balls are distributed into the n distinguishable urns. The same probability is equal to

$$\frac{(n-u)q_0^{n-u-1}q_1 q_{m-1}(u)}{m q_m(n)}$$

and so

$$\frac{q_0^{n-u}q_{m-1}(u)}{q_{m-1}(n)} - \frac{q_0^{n-u}q_m(u)}{q_m(n)} = \frac{(n-u)q_0^{n-u-1}q_1 q_{m-1}(u)}{m q_m(n)}.$$

The number W_k of balls required to be distributed into n distinguishable urns until a predetermined number k of urns are occupied may be expressed as a sum

$$W_k = Y_1 + Y_2 + \cdots + Y_k, \ k = 1, 2, \ldots,$$

where Y_j is the number of balls required to be distributed into n distinguishable urns after the $(j-1)$st urn, empty before, receives one ball and until the jth urn, empty before, receives one ball, $j = 1, 2, \ldots$. Clearly, $P(Y_1 = 1) = 1$. Further, some light into the structure of the distribution of W_k can be shed by the conditional distribution of Y_{k+1} given W_k. In the next theorem the conditional probability function

$$q_{r;m}(k, n) = P(Y_{k+1} = r | W_k = m), \ r = 1, 2, \ldots, \tag{5.9}$$

is expressed in terms of the convolution probabilities $q_m(u)$, $m = 0, 1, \ldots$.

Theorem 5.2 *Under the sequential random occupancy model with $q_0 > 0$, the conditional probability $q_{r;m}(k, n)$, that r additional balls are required to be distributed into the n distinguishable urns until the $(k+1)$st urn is occupied, given that m balls are distributed until k urns are occupied, $r = 1, 2, \ldots$, is given by*

$$q_{r;m}(k, n) = \frac{(n-k)q_1 q_{m+r-1}(k)/q_{m+r}(n)}{(m+r)q_0 q_m(k)/q_m(n)}. \tag{5.10}$$

Proof Consider first any of the $\binom{n}{k}$ subsets $\{j_1, j_2, \ldots, j_k\}$ of the set of urns $\{1, 2, \ldots, n\}$ and let $\{j_{k+1}, j_{k+2}, \ldots, j_n\} = \{1, 2, \ldots, n\} - \{j_1, j_2, \ldots, j_k\}$. Under the sequential random occupancy model, let $B_{k-1,m}$ be the event that $k-1$ urns, among the k urns $\{j_1, j_2, \ldots, j_k\}$, are occupied and one ball is distributed at the mth allocation into the other urn, given that m balls are distributed into the k urns $\{j_1, j_2, \ldots, j_k\}$. Also, let $A_{j_s,m}$ be the event that the j_sth urn remains empty, given that m balls are distributed into the n urns, $s = k+1, k+2, \ldots, n$. Further, consider any urn, say, j_{k+1}, from the set $\{j_{k+1}, j_{k+2}, \ldots, j_n\}$ and let $C_{k+1, m+r}$ be the event that one ball is distributed into the urn j_{k+1} at the $(m+r)$th allocation and $m+r-1$ balls are distributed into the k urns $\{j_1, j_2, \ldots, j_k\}$, given that $m+r$ balls are distributed into the $k+1$ urns $\{j_1, j_2, \ldots, j_k, j_{k+1}\}$. Then

$$P(W_k = m) = \binom{n}{k} P(B_{k-1,m}) P(A_{j_{k+1},m} A_{j_{k+2},m} \cdots A_{j_n,m})$$

and

$$P(W_{k+1} = m+r, W_k = m) = \binom{n}{k}(n-k) P(B_{k-1,m})$$
$$\times P(A_{j_{k+2},m+r} A_{j_{k+3},m+r} \cdots A_{j_n,m+r}) P(C_{k+1,m+r}).$$

Therefore,

$$P(W_{k+1} - W_k = r | W_k = m)$$
$$= \frac{(n-k) P(A_{j_{k+2},m+r} A_{j_{k+3},m+r} \cdots A_{j_n,m+r}) P(C_{k+1,m+r})}{P(A_{j_{k+1},m} A_{j_{k+2},m} \cdots A_{j_n,m})}$$

and since

$$P(A_{j_{k+1},m} A_{j_{k+2},m} \cdots A_{j_n,m}) = \frac{q_0^{n-k} q_m(k)}{q_m(n)},$$

$$P(A_{j_{k+2},m+r} A_{j_{k+3},m+r} \cdots A_{j_n,m+r}) = \frac{q_0^{n-k-1} q_{m+r}(k+1)}{q_{m+r}(n)},$$

$$P(C_{k+1,m+r}) = \frac{q_1 q_{m+r-1}(k)}{(m+r) q_{m+r}(k+1)},$$

(5.10) is deduced.

In sequential occupancy theory, in addition to the *waiting-time* occupancy distribution of the number W_k of balls required to be distributed into n distinguishable urns until k urns are occupied, the *complementary waiting-time* distribution of the number V_r of balls required to be distributed into n distinguishable urns, until r balls are placed into previously occupied urns, is of probabilistic and statistical interest. Clearly, with $U_m \equiv K|S_n = m$, the number of occupied urns given that m balls are distributed into n distinguishable urns, it holds true that

$$W_k \leq m \text{ if and only if } U_m \geq k,$$

while

$$V_r \leq m \text{ if and only if } m - U_m \geq r,$$

justifying the adjective complementary waiting-time to the distribution of V_r. In sequential statistical inference, instead of V_r, it is more convenient to consider the number $L_r = V_r - r$ of occupied urns, among n distinguishable urns, until r balls are placed into previously occupied urns. The probability function,

$$c_k(r,n) = P(L_r = k), \quad k = 1, 2, \ldots, n,$$

of the *complementary waiting-time distribution* is expressed in terms of finite differences of the convolution probabilities (5.2) in the following theorem.

Theorem 5.3 *Under the sequential random occupancy model with $q_0 > 0$, the probability $c_k(r, n)$, that k among the n distinguishable urns are occupied until r balls are distributed into previously occupied urns, $k = 1, 2, \ldots, n$, is given by*

$$c_k(r, n) = \binom{n}{k} \frac{q_{r+k}(k) \left[\Delta_u^k q_0^{n-u} q_{r+k-1}(u) \right]_{u=0}}{q_{r+k-1}(k) q_{r+k}(n)}. \qquad (5.11)$$

Proof Under the sequential random occupancy model, let $B_{k,m,n}$ be the event that k urns are occupied, given that m balls are distributed into the n distinguishable urns. Without any loss of generality we may assume that $\{j_1, j_2, \ldots, j_k\}$ is the set of occupied urns so that $\{j_{k+1}, j_{k+2}, \ldots, j_n\} = \{1, 2, \ldots, n\} - \{j_1, j_2, \ldots, j_k\}$ is the set of empty urns. Let $A_{j_s, m}$ be the event that the j_sth urn remains empty, given that m balls are distributed into the n urns, $s = k+1, k+2, \ldots, n$. Then

$$c_k(r, n) = P(B_{k, r+k-1, n})$$
$$\times P(A_{j_{k+1}, r+k} A_{j_{k+2}, r+k} \cdots A_{j_n, r+k} | A_{j_{k+1}, r+k-1} A_{j_{k+2}, r+k-1} \cdots A_{j_n, r+k-1})$$

and since $A_{j_s, r+k} \subseteq A_{j_s, r+k-1}$, $s = k+1, k+2, \ldots, n$,

$$c_k(r, n) = P(B_{k, r+k-1, n}) \frac{P(A_{j_{k+1}, r+k} A_{j_{k+2}, r+k} \cdots A_{j_n, r+k})}{P(A_{j_{k+1}, r+k-1} A_{j_{k+2}, r+k-1} \cdots A_{j_n, r+k-1})}.$$

Introducing into it the expression

$$P(B_{k, r+k-1, n}) = \binom{n}{k} \frac{\left[\Delta_u^k q_0^{n-u} q_{r+k-1}(u) \right]_{u=0}}{q_{r+k-1}(n)},$$

which is deduced from Corollary 4.1, and the expressions

$$P(A_{j_{k+1}, r+k} A_{j_{k+2}, r+k} \cdots A_{j_n, r+k}) = \frac{q_0^{n-k} q_{r+k}(k)}{q_{r+k}(n)},$$

$$P(A_{j_{k+1}, r+k-1} A_{j_{k+2}, r+k-1} \cdots A_{j_n, r+k-1}) = \frac{q_0^{n-k} q_{r+k-1}(k)}{q_{r+k-1}(n)},$$

(5.11) is deduced.

5.4 PARTICULAR SEQUENTIAL OCCUPANCY DISTRIBUTIONS

5.4.1 Sequential Classical Occupancy Distributions

The sequential classical occupancy model is equivalent to the sequential random occupancy model with the assumption that the number X_j, of balls distributed into the jth urn, $j = 1, 2, \ldots, n$, obeys a *Poisson distribution* with (cf. Section 4.4.1)

$$q_x = P(X = x) = e^{-\lambda} \frac{\lambda^x}{x!}, \ x = 0, 1, \ldots, \ 0 < \lambda < \infty.$$

Its u-fold convolution obeys also a Poisson distribution with

$$q_m(u) = P(S_u = m) = e^{-u\lambda}\frac{(u\lambda)^m}{m!}, \ m = 0, 1, \ldots.$$

In the following corollary of Theorem 5.1, the probability function,

$$q_m(k, n) = P(W_k = m), \ m = k,, k+1, \ldots,$$

and ascending binomial moments,

$$b_{[r]}(k, n) = E\left[\binom{W_k + r - 1}{r}\right], \ r = 0, 1, \ldots,$$

of the *waiting-time classical occupancy distribution* are derived.

Corollary 5.1 (a) *Under the sequential random occupancy model with Poisson probabilities, the probability* $q_m(k, n)$, *that m balls are required to be distributed into the n distinguishable urns until k urns are occupied*, $m = k, k+1, \ldots$, *is given by*

$$q_m(k, n) = \frac{(n)_k S(m-1, k-1)}{n^m}, \tag{5.12}$$

where $S(m-1, k-1) = [\Delta_u^{k-1} u^{m-1}]_{u=0}/(k-1)!$ *is the Stirling number of the second kind.*

(b) *The rth ascending binomial moment* $b_{[r]}(k, n)$, $r = 0, 1, \ldots$, *of the probability function* $q_m(k, n)$, $m = k, k+1, \ldots$, *is given by*

$$b_{[r]}(k, n) = \binom{n-1}{k-1}\sum_{j=0}^{k-1}(-1)^{k-j-1}\binom{k-1}{j}\left(1 - \frac{j}{n}\right)^{-r-1}. \tag{5.13}$$

Proof (a) Expression (5.12) follows from (5.7) by noting that

$$[\Delta_u^{k-1} u^{m-1}]_{u=0} = (k-1)!S(m-1, k-1).$$

(b) The conditional probability (5.5) reduces, in this case, to

$$h_m(u, n) = \left(\frac{u}{n}\right)^{m-1}, \ m = 1, 2, \ldots,$$

which is a polynomial in u of degree $m - 1$. Its rth ascending binomial moment,

$$a_{[r]}(u, n) = \sum_{m=1}^{\infty}\binom{m+r-1}{r}\left(\frac{u}{n}\right)^{m-1} = \sum_{i=0}^{\infty}\binom{r+i}{i}\left(\frac{u}{n}\right)^i,$$

on using the negative binomial expansion

$$\sum_{i=0}^{\infty}\binom{n+i-1}{i}t^i = (1-t)^{-n}, \ |t| < 1$$

and since $u < n$, is obtained as

$$a_{[r]}(u, n) = \left(1 - \frac{u}{n}\right)^{-r-1}.$$

Hence, by (5.8),

$$b_{[r]}(k, n) = \binom{n-1}{k-1}\left[\Delta_u^{k-1}\left(1 - \frac{u}{n}\right)^{-r-1}\right]_{u=0}$$

$$= \binom{n-1}{k-1}\sum_{j=0}^{k-1}(-1)^{k-j-1}\binom{k-1}{j}\left(1 - \frac{j}{n}\right)^{-r-1}$$

and the proof of the corollary is completed.

The mean of the waiting-time classical occupancy distribution is deduced from (5.13) as

$$E(W_k) = \binom{n-1}{k-1}\sum_{j=0}^{k-1}(-1)^{k-j-1}\binom{k-1}{j}\left(1 - \frac{j}{n}\right)^{-2}.$$

It is worth noting that this sum of terms of alternating sign may be reduce to a sum of positive terms. Specifically, using Pascal's triangle

$$\binom{k-1}{j} = \binom{k-2}{j} + \binom{k-2}{j-1},$$

and since

$$\sum_{j=0}^{k-1}(-1)^{k-j-1}\binom{k-1}{j}\left(1 - \frac{j}{n}\right)^{-1} = \binom{n-1}{k-1}^{-1},$$

it follows that

$$E(W_k) - E(W_{k-1}) = \frac{n}{n-k+1}, \quad k = 1, 2, \ldots,$$

with $E(W_0) = 0$. Therefore,

$$E(W_k) = n\sum_{j=1}^{k}\frac{1}{n-j+1}.$$

Similarly, for the second ascending binomial moment

$$E\left[\binom{W_k+1}{2}\right] = \binom{n-1}{k-1}\sum_{j=0}^{k-1}(-1)^{k-j-1}\binom{k-1}{j}\left(1 - \frac{j}{n}\right)^{-3},$$

we deduce the recurrence relation

$$E\left[\binom{W_k+1}{2}\right] - E\left[\binom{W_{k-1}+1}{2}\right] = \frac{n}{n-k+1}E(W_k), \quad k = 1, 2, \ldots,$$

with $E\left[\binom{W_0+1}{2}\right] = 0$, Consequently,

$$E\left[\binom{W_k+1}{2}\right] = \sum_{j=1}^{k}\sum_{i=1}^{j} \frac{n^2}{(n-j+1)(n-i+1)}.$$

Thus, the variance of the waiting-time classical occupancy distribution is given by

$$V(W_k) = 2\sum_{j=1}^{k}\sum_{i=1}^{j} \frac{n^2}{(n-j+1)(n-i+1)} - \sum_{j=1}^{k} \frac{n}{n-j+1}$$

$$- \left(\sum_{j=1}^{k} \frac{n}{n-j+1}\right)^2$$

$$= \sum_{j=1}^{k} \frac{n^2}{(n-j+1)^2} - \sum_{j=1}^{k} \frac{n}{n-j+1}$$

and so

$$V(W_k) = n\sum_{j=1}^{k} \frac{j-1}{(n-j+1)^2}.$$

The conditional probability function

$$q_{r;m}(k,n) = P(Y_{k+1} = r | W_k = m), \quad r = 1, 2, \ldots,$$

where $W_k = Y_1 + Y_2 + \cdots + Y_k$, $k = 1, 2, \ldots$, is deduced in the following corollary of Theorem 5.2.

Corollary 5.2 *Under the sequential random occupancy model with Poisson probabilities, the conditional probability $q_{r;m}(k,n)$, that r additional balls are required to be distributed into the n distinguishable urns until the $(k+1)$st urn is occupied, given that m balls are distributed until k urns are occupied, $r = 1, 2, \ldots$, is given by*

$$q_{r;m}(k,n) = \left(\frac{n-k}{n}\right)\left(\frac{k}{n}\right)^{r-1}. \tag{5.14}$$

Note that, according to Corollary 5.2, W_k is expressed as a sum of k independent geometric random variables Y_j with varying success probability $p_j = (n-j+1)/n$, $j = 1, 2, \ldots, k$. Using this expression, the mean and variance of W_k, obtained above, are verified.

The probability function

$$c_k(r,n) = P(L_r = k), \quad k = 1, 2, \ldots, n,$$

of the *complementary waiting-time classical occupancy distribution* is deduced in the following corollary of Theorem 5.3.

Corollary 5.3 *Under the sequential random occupancy model with Poisson probabilities, the probability $c_k(r,n)$, that k among the n distinguishable urns are occupied until r balls are distributed into previously occupied urns, $k = 1, 2, \ldots, n$, is given by*

$$c_k(r,n) = \frac{(n)_k k S(r+k-1,k)}{n^{r+k}}, \qquad (5.15)$$

where $S(m,k) = [\Delta_u^k u^m]_{u=0}/k!$ is the Stirling number of the second kind.

Example 5.5 *Usable numbers in a sequential sample of random numbers.* Consider a file of n serially numbered cards, $\{1, 2, \ldots, n\}$, that refer to customers' accounts, one account per card. In Example 4.5 a random sample of size m is assumed to be chosen from the file, by employing a suitable table of random numbers. Then, eliminating from the sample the duplicates, the number K of usable numbers is expressed as the number of occupied urns in a classical occupancy model and its distribution is deduced. Note that when a sample of fixed size m is selected, the problem arises as to how large m should be in order to obtain a reasonable number k of usable numbers. Selection of too few or too many random numbers may require additional selection or elimination of too many duplicates, which is somewhat expensive and time-consuming. Therefore, it is preferable to select sequentially a sample from the file $\{1, 2, \ldots, n\}$ until a fixed number k of usable numbers is obtained. In this case the number W_k of random numbers required to be selected is a random variable. Its probability function and moments are of interest.

As in Example 4.5, considering the n cards as n distinguishable urns and interpreting the selection of the jth card as a placement of a ball into the jth urn, W_k is the number of balls required to be distributed into n distinguishable urns until k urns are occupied. Thus, its probability function and binomial moments are given by (5.12) and (5.13), respectively.

If the number r, say, of duplicates is required to be fixed, then a sample is sequentially selected from the file $\{1, 2, \ldots, n\}$ until r duplicates are drawn. Then, the number L_r of usable numbers in the sample is a random variable with probability function given by (5.5).

When only a subset $\{i_1, i_2, \ldots, i_\nu\}$ of not necessarily serially numbered cards refer to customers' accounts, the probability functions of W_k and L_r may be deduced as (see Exercises 5.8 and 5.13)

$$P(W_k = m) = \frac{(\nu)_k S(m-1, k-1; n-\nu)}{n^m}, \quad m = k, k+1, \ldots,$$

and

$$P(L_r = k) = \frac{(\nu)_k (k+n-\nu) S(r+k-1, k; n-\nu)}{n^{r+k}}, \quad k = 0, 1, \ldots, \nu,$$

respectively, where $S(m, k; r) = [\Delta_u^k u^m]_{u=r}/k!$ is the noncentral Stirling number of the second kind.

5.4.2 Sequential Restricted Occupancy Distributions

The sequential restricted occupancy model is equivalent to the sequential random occupancy model with the assumption that the number X_j, of balls distributed into the jth urn, $j = 1, 2, \ldots, n$, obeys a *binomial distribution* with (cf. Section 4.4.2)

$$q_x = P(X = x) = \binom{s}{x} p^x q^{s-x}, \quad x = 0, 1, \ldots, s,$$

where $0 < p < 1$ and $q = 1 - p$. Its u-fold convolution obeys a binomial distribution with

$$q_m(u) = P(S_u = m) = \binom{su}{m} p^m q^{su-m}, \quad m = 0, 1, \ldots, su.$$

The probability function and ascending binomial moments of the *waiting-time restricted occupancy distribution* are derived in the following corollary of Theorem 5.1.

Corollary 5.4 (a) *Under the sequential random occupancy model with binomial probabilities, the probability $q_m(k, n; s)$, that m balls are required to be distributed into the n distinguishable urns until k urns are occupied, $m = k, k+1, \ldots$, is given by*

$$q_m(k, n; s) = \frac{(n)_k s C(m-1, k-1; s)}{(sn)_m}, \tag{5.16}$$

where $C(m-1, k-1; s) = [\Delta_u^{k-1}(su)_{m-1}]_{u=0}/(k-1)!$ is the generalized factorial coefficient.

(b) *The rth ascending binomial moment $b_{[r]}(k, n; s)$, $r = 0, 1, \ldots$, of the probability function $q_m(k, n; s)$, $m = k, k+1, \ldots$, is given by*

$$b_{[r]}(k, n; s) = \binom{n-1}{k-1} \sum_{j=0}^{k-1} (-1)^{k-j-1} \binom{k-1}{j} \frac{(sn+r)_{r+1}}{(sn-sj+r)_{r+1}}. \tag{5.17}$$

Proof (a) Expression (5.16) follows from (5.7) by noting that

$$[\Delta_u^{k-1}(su)_{m-1}]_{u=0} = (k-1)! C(m-1, k-1; s).$$

(b) The conditional probability (5.5) reduces, in this case, to

$$h_m(u, n; s) = \frac{(su)_{m-1}}{(sn-1)_{m-1}}, \quad m = 1, 2, \ldots,$$

which is a polynomial in u of degree $m-1$. Its rth ascending binomial moment,

$$a_{[r]}(u, n; s) = \sum_{m=1}^{\infty} \binom{m+r-1}{r} \frac{(su)_{m-1}}{(sn-1)_{m-1}} = \sum_{i=0}^{\infty} \binom{r+i}{i} \frac{(su)_i}{(sn-1)_i},$$

on using the combinatorial identity

$$\sum_{i=0}^{\infty}\binom{n+i-1}{i}\frac{(x)_i}{(z-n)_i}=\frac{(z)_n}{(z-x)_n},$$

with $n=r+1$, $x=su$ and $z=sn+r$, is obtained as

$$a_{[r]}(u,n;s)=\frac{(sn+r)_{r+1}}{(sn-su+r)_{r+1}}.$$

Hence, by (5.8),

$$b_{[r]}(k,n;s)=\binom{n-1}{k-1}\left[\Delta_u^{k-1}\frac{(sn+r)_{r+1}}{(sn-su+r)_{r+1}}\right]_{u=0}$$

$$=\binom{n-1}{k-1}\sum_{j=0}^{k-1}(-1)^{k-j-1}\binom{k-1}{j}\frac{(sn+r)_{r+1}}{(sn-sj+r)_{r+1}}$$

and the proof of the corollary is completed.

The mean of the waiting-time restricted occupancy distribution is deduced from (5.17) as

$$E(W_k)=\binom{n-1}{k-1}\sum_{j=0}^{k-1}(-1)^{k-j-1}\binom{k-1}{j}\frac{(sn+1)_2}{(sn-sj+1)_2}.$$

A reduction of this sum to a sum of positive terms may be accomplished as follows. Using Pascal's triangle

$$\binom{k-1}{j}=\binom{k-2}{j}+\binom{k-2}{j-1},$$

and since

$$\sum_{j=0}^{k-1}(-1)^{k-j-1}\binom{k-1}{j}\frac{n}{n-j}=\binom{n-1}{k-1}^{-1},$$

we get, after some algebraic manipulations, the recurrence relation

$$E(W_k)-\frac{n-k+1}{n+1/s-k+1}E(W_{k-1})=\frac{n+1/s}{n+1/s-k+1},\quad k=1,2,\ldots,$$

with $E(W_0)=0$. Multiplying both members by $(n+1/s)_k$, we deduce the recurrence relation

$$\frac{(n+1/s)_k}{(n)_k}E(W_k)-\frac{(n+1/s)_{k-1}}{(n)_{k-1}}E(W_{k-1})=(n+1/s)\frac{(n+1/s)_{k-1}}{(n)_k},$$

for $k = 1, 2, \ldots$, which, iterated, entails

$$E(W_k) = \frac{(n+1/s)(n)_k}{(n+1/s)_k} \sum_{j=1}^{k} \frac{(n+1/s)_{j-1}}{(n)_j}.$$

Therefore,

$$E(W_k) = (n+1/s) \sum_{j=1}^{k} \frac{(n-j)_{k-j}}{(n+1/s-j+1)_{k-j+1}}.$$

Similarly, for the second ascending binomial moment

$$E\left[\binom{W_k+1}{2}\right] = \binom{n-1}{k-1} \sum_{i=0}^{k-1} (-1)^{k-j-1} \binom{k-1}{j} \frac{(sn+2)_3}{(sn-sj+1)_3},$$

we get the recurrence relation

$$E\left[\binom{W_k+1}{2}\right] - \frac{n-k+1}{n+2/s-k+1} E\left[\binom{W_{k-1}+1}{2}\right]$$
$$= \frac{n+2/s}{n+2/s-k+1} E(W_k),$$

for $k = 1, 2, \ldots$, with $E\left[\binom{W_0+1}{2}\right] = 0$. Thus, proceeding as above, we find

$$E\left[\binom{W_k+1}{2}\right] = \sum_{j=1}^{k} \sum_{i=1}^{j} \frac{(n+2/s)(n-j)_{k-j}}{(n+2/s-j+1)_{k-j+1}} \cdot \frac{(n+1/s)(n-i)_{j-1}}{(n+1/s-i+1)_{j-i+1}}.$$

The conditional probability function

$$q_{r;m}(k, n; s) = P(Y_{k+1} = r | W_k = m), \ r = 1, 2, \ldots,$$

where $W_k = Y_1 + Y_2 + \cdots Y_k$, $k = 1, 2, \ldots$, is deduced in the following corollary of Theorem 5.2.

Corollary 5.5 *Under the sequential random occupancy model with binomial probabilities, the conditional probability $q_{r;m}(k, n; s)$, that r additional balls are required to be distributed into the n distinguishable urns until the $(k+1)$st urn is occupied, given that m balls are distributed until k urns are occupied, $r = 1, 2, \ldots, sk - m + 1$, is given by*

$$q_{r;m}(k, n; s) = \frac{(sn-sk)(sk-m)_{r-1}}{(sn-m)_r}. \tag{5.18}$$

Note that (5.18) is the probability function of a waiting-time hypergeometric distribution. Specifically, it is the probability that r drawings are required until the first white ball is drawn, when balls are sequentially drawn without replacement from an urn containing $sk - m$ black and $sn - sk$ white balls.

The probability function of the *complementary waiting-time restricted occupancy distribution* is deduced in the following corollary of Theorem 5.3.

Corollary 5.6 *Under the sequential random occupancy model with binomial probabilities, the probability $c_k(r, n; s)$, that k among the n distinguishable urns are occupied until r balls are distributed into previously occupied urns, $k = 1, 2, \ldots, n$, is given by*

$$c_k(r, n; s) = \frac{(n)_k (sk - r - k + 1) C(r + k - 1, k; s)}{(sn)_{r+k}}, \quad (5.19)$$

where $C(m, k; s) = [\Delta_u^k (su)_m]_{u=0}/k!$ is the generalized factorial coefficient.

Example 5.6 *Coupon collector's problem.* For the promotion of sales of an industrial product, the manufacturer has printed n different coupons in s identical series. A coupon is placed in each package of the product. A customer who collects a fixed number k of different coupons gets a given number of packages of the product free, while in the exceptional case of collecting a complete series of coupons, the prize is a special gift from the company. The distribution of the number W_k of packages of the product a customer requires to purchase until he/she collects k different coupons is of interest to the customer.

Considering the n different coupons as n distinguishable urns with s distinguishable cells of capacity limited to one coupon and interpreting a collection of coupons as a distribution of like balls into the urns, W_k is the number of balls required to be distributed into the urns until k urns are occupied. Then, the probability function and factorial moments of W_k are given by (5.16) and (5.17), respectively.

Another distribution that may be of interest to the customer is the distribution of number L_r of different coupons collected until a fixed number r of duplicates are collected. Clearly, its probability function is given by (5.19). This distribution can be used to estimate the unknown (to the customer) number n of coupons the manufacturer has printed. A uniformly minimum variance unbiased estimator of n is derived in Example 5.8.

5.4.3 Sequential Pseudo-contagious Occupancy Distributions

The sequential pseudo-contagious occupancy model is equivalent to the sequential random occupancy model with the assumption that the number X_j, of balls distributed into the jth urn, $j = 1, 2, \ldots, n$, obeys a *negative binomial distribution* with (cf. Section 4.4.3)

$$q_x = P(X = x) = \binom{s + x - 1}{x} p^s q^x, \quad x = 0, 1, \ldots,$$

where $0 < p < 1$ and $q = 1 - p$. Its u-fold convolution obeys a negative binomial distribution with

$$q_m(u) = P(S_u = m) = \binom{su + m - 1}{m} p^{su} q^m, \quad m = 0, 1, \ldots.$$

234 SEQUENTIAL OCCUPANCY DISTRIBUTIONS

The probability function and ascending binomial moments of the *waiting-time pseudo-contagious occupancy distribution* are derived in the following corollary of Theorem 5.1.

Corollary 5.7 (a) *Under the sequential random occupancy model with negative binomial probabilities, the probability $q_m(k, n; -s)$, that m balls are required to be distributed into the n distinguishable urns until k urns are occupied, $m = k, k+1, \ldots$, is given by*

$$q_m(k, n; -s) = \frac{s(n)_k |C(m-1, k-1; -s)|}{(sn+m-1)_m}, \quad (5.20)$$

where $|C(m-1, k-1; -s)| = [\Delta_u^{k-1}(su+m-2)_{m-1}]_{u=0}/(k-1)!$ is the absolute generalized factorial coefficient.

(b) *The rth ascending binomial moment $b_{[r]}(k, n; -s)$, $r = 0, 1, \ldots$, of the probability function $q_m(k, n; -s)$, $m = k, k+1, \ldots$, is given by*

$$b_{[r]}(k, n; -s) = \binom{n-1}{k-1} \sum_{j=0}^{k-1} (-1)^{k-j-1} \binom{k-1}{j} \frac{(sn)_{r+1}}{(sn-sj)_{r+1}}. \quad (5.21)$$

Proof (a) Since

$$(-1)^{m-1} \left[\Delta_u^{k-1}(-su)_{m-1}\right]_{m=0} = (k-1)! |C(m-1, k-1; -s)|,$$

expression (5.20) follows directly from (5.7).

(b) The conditional probability (5.5) reduces, in this case, to

$$h_m(u, n; -s) = \frac{(su+m-2)_{m-1}}{(sn+m-1)_{m-1}} = \frac{(-su)_{m-1}}{(-sn-1)_{m-1}}, \quad m = 1, 2, \ldots,$$

which is a polynomial in u of degree $m-1$. Its rth ascending binomial moment,

$$a_{[r]}(u, n; -s) = \sum_{m=1}^{\infty} \binom{m+r-1}{r} \frac{(-su)_{m-1}}{(-sn-1)_{m-1}}$$

$$= \sum_{i=0}^{\infty} \binom{r+i}{i} \frac{(-su)_i}{(-sn-1)_i},$$

on using the combinatorial identity

$$\sum_{i=0}^{\infty} \binom{n+i-1}{i} \frac{(x)_i}{(z-n)_i} = \frac{(z)_n}{(z-x)_n},$$

with $n = r+1$, $x = -su$ and $z = -sn+r$, is obtained as

$$a_{[r]}(u, n; -s) = \frac{(-sn+r)_{r+1}}{(-sn+su+r)_{r+1}} = \frac{(sn)_{r+1}}{(sn-su)_{r+1}}.$$

Hence, by (5.8),

$$b_{[r]}(k,n;-s) = \binom{n-1}{k-1}\left[\Delta_u^{k-1}\frac{(sn)_{r+1}}{(sn-su)_{r+1}}\right]_{u=0}$$

$$= \binom{n-1}{k-1}\sum_{j=0}^{k-1}(-1)^{k-j-1}\binom{k-1}{j}\frac{(sn)_{r+1}}{(sn-sj)_{r+1}},$$

which completes the proof of the corollary.

The mean of the waiting-time pseudo-contagious occupancy distribution,

$$E(W_k) = \binom{n-1}{k-1}\sum_{j=0}^{k-1}(-1)^{k-j-1}\binom{k-1}{j}\frac{(sn)_2}{(sn-sj)_2},$$

may be expressed by a sum of positive terms. Specifically, proceeding as in the corresponding derivation of the mean of the waiting-time restricted occupancy distribution, we deduce the expression

$$E(W_k) = (n-1/s)\sum_{j=1}^{k}\frac{(n-j)_{k-j}}{(n-1/s-j+1)_{k-j+1}}.$$

Similarly, the second ascending binomial moment

$$E\left[\binom{W_k+1}{2}\right] = \binom{n-1}{k-1}\sum_{j=0}^{k-1}(-1)^{k-j-1}\binom{k-1}{j}\frac{(sn)_3}{(sn-sj)_3},$$

may be expressed as

$$E\left[\binom{W_k+1}{2}\right] = \sum_{j=1}^{k}\sum_{i=1}^{j}\frac{(n-2/s)(n-j)_{k-j}}{(n-2/s-j+1)_{k-j+1}}\cdot\frac{(n-1/s)(n-i)_{j-i}}{(n-1/s-i+1)_{j-i+1}}.$$

The conditional probability function

$$q_{r;m}(k,n;-s) = P(Y_{k+1}=r|W_k=m), \quad r=1,2,\ldots,$$

where $W_k = Y_1 + Y_2 + \cdots + Y_k$, $k=1,2,\ldots$, is deduced in the following corollary of Theorem 5.2.

Corollary 5.8 *Under the sequential random occupancy model with negative binomial probabilities, the conditional probability* $q_{r;m}(k,n;-s)$, *that r additional balls are required to be distributed into the n distinguishable urns until the* $(k+1)$*st urn is occupied, given that m balls are distributed until k urns are occupied,* $r=1,2,\ldots$, *is given by*

$$q_{r;m}(k,n;-s) = \frac{(sn-sk)(sk+m+r-2)_{r-1}}{(sn+m+r-1)_r}. \tag{5.22}$$

236 SEQUENTIAL OCCUPANCY DISTRIBUTIONS

Note that (5.22) is the probability function of a waiting-time hypergeometric distribution, known as Waring distribution. Specifically, it is the probability that r drawings are required until the first white ball is drawn, when balls are drawn one after the other from an urn containing $sk + m$ black and $sn - sk$ white balls, returning after each drawing the drawn ball together with another ball of the same color.

The probability function of the *complementary waiting-time pseudo-contagious occupancy distribution* is deduced in the following corollary of Theorem 5.3.

Corollary 5.9 *Under the sequential random occupancy model with negative binomial probabilities, the probability $c_k(r, n; -s)$, that k among n distinguishable urns are occupied until r balls are distributed into previously occupied urns, $k = 1, 2, \ldots, n$, is given by*

$$c_k(r, n; -s) = \frac{(n)_k (sk + r + k - 1) |C(r + k - 1, k; -s)|}{(sn + r + k - 1)_{r+k}}, \qquad (5.23)$$

where $|C(m, k; -s)| = [\Delta_u^k (su + m - 1)_m]_{u=0} / k!$ is the absolute generalized factorial coefficient.

5.5 STATISTICAL APPLICATIONS

The presentation of applications of occupancy distributions in the estimation of the number of classes in a finite population, started in Section 4.5, is continued in this section. According to Theorem 4.2, a uniformly minimum variance unbiased (UMVU) estimator of the number n of classes in a finite population based on the number K of observed classes in a random sample of a fixed size m exists only when $m \geq n$.

A sequential sample estimation of the number of classes requires no restriction. Specifically, assume that a random sample is sequentially drawn from a population, with an unknown number n of classes, until r elements are drawn from previously observed classes. A UMVU estimator of n can be based on the number L_r of observed classes in the sequentially drawn sample. More precisely, suppose that the population-class distribution is a power series distribution with probability function

$$q_x = P(X = x) = [g(\theta)]^{-1} c_x \theta^x, \quad x = 0, 1, \ldots, \quad 0 < \theta < \rho,$$

where $g(\theta) = \sum_{x=0}^{\infty} c_x \theta^x$, $0 < \theta < \rho$; then

$$q_m(n) = P(S_n = m) = [g(\theta)]^{-n} c_m(n) \theta^m, \quad n = 0, 1, \ldots, \quad 0 < \theta < \rho,$$

with $c_m(n) = [D_\theta^m (g(\theta))^n]_{\theta=0} / m!$, $m = 0, 1, \ldots$. Further, the number L_r of observed classes in the sequentially drawn sample is a complete and sufficient statistic for the unknown number n of classes in the population. Indeed, the probability function of L_r may readily be deduced from (5.11) as

$$P(L_r = k) = \binom{n}{k} \frac{c_{r+k}(k) [\Delta_u^k c_{r+k-1}(u)]_{u=0}}{c_{r+k-1}(k) c_{r+k}(n)}, \quad k = 1, 2, \ldots, n,$$

which is a complete family. Further, assume that $L_r = k$ and let K_i be the number of classes appearing i times in the sample, $i = 1, 2, \ldots, r + k$. Then,

$$P(K_1 = k_1, K_2 = k_2, \ldots, K_{r+k} = k_{r+k} | L_r = k)$$

$$= \frac{k! c_{r+k-1}(k)}{c_{r+k}(k) \left[\Delta_u^k c_{r+k-1}(u)\right]_{u=0}} \cdot \prod_{i=1}^{r+k} c_i^{k_i} / k_i!,$$

where $k_i = 0, 1, \ldots, n, i = 1, 2, \ldots, r+k$, with $\sum_{i=1}^{r+k} i k_i = r + k$ and $\sum_{i=1}^{r+k} k_i = k$. This conditional probability function does not involve the unknown parameter n and so L_r is a sufficient statistics for n. Consequently, a UMVU estimator of n based on L_r exists without any restriction. More generally, the parametric function $(n)_j$, $j = 1, 2, \ldots, n$, is estimable if the observed value of L_r is $k \geq j$. The derivation of the UMVU estimator of $(n)_j$, $j = 1, 2, \ldots, k$, is facilitated by the evaluation of the expected value $E[g_j(L_r)]$, where

$$g_j(k) = (k)_j \frac{\left[\Delta_u^{k-j} c_{r+k-1}(u)\right]_{u=j}}{\left[\Delta_u^k c_{r+k-1}(u)\right]_{u=0}}, \quad j = 1, 2, \ldots, k.$$

Upon using the identity

$$(k)_j \binom{n}{k} = (n)_j \binom{n-j}{k-j},$$

$E[g_j(L_r)]$ is expressed as

$$E[g_j(L_r)] = (n)_j \sum_{k=j}^{n} \binom{n-j}{k-j} \frac{c_{r+k}(k) \left[\Delta_u^{k-j} c_{r+k-1}(u)\right]_{u=j}}{c_{r+k-1}(k) c_{r+k}(n)}.$$

Setting $\nu = n - j$, $\kappa = k - j$ and $\rho = r + j$, the general term of this sum takes the form

$$c_\kappa(\rho, \nu, j) = \binom{\nu}{\kappa} \frac{c_{\rho+\kappa}(\kappa + j) \left[\Delta_u^\kappa c_{\rho+\kappa-1}(u)\right]_{u=j}}{c_{\rho+\kappa-1}(\kappa + j) c_{\rho+\kappa}(\nu + j)},$$

for $\kappa = 0, 1, \ldots, \nu$, which is a legitimate probability function. Specifically, consider an extended occupancy model where balls are sequentially distributed into $\nu + j$ distinguishable urns, among which ν are labeled as *target* urns and the other j as *control* urns. Let L_ρ be the number of occupied target urns, until ρ balls are placed into previously occupied target urns and into the control urns. If the distribution of the number X of balls distributed into any specific urn is a power series distribution, then $P(L_\rho = \kappa) = c_\kappa(\rho, \nu, j)$, $\kappa = 0, 1, \ldots, \nu$ (see Exercise 5.13), and since $\sum_{\kappa=0}^\nu P(L_\rho = \kappa) = 1$,

$$E[g_j(L_r)] = (n)_j.$$

Therefore, the UMVU estimator of $(n)_j$ is $\hat{n}_j(L_r)$, where

$$\hat{n}_j(k) = (k)_j \frac{\left[\Delta_u^{k-j} c_{r+k-1}(u)\right]_{u=j}}{\left[\Delta_u^k c_{r+k-1}(u)\right]_{u=0}}, \quad j = 1, 2, \ldots, k.$$

Further, the variance of the UMVU estimator $\hat{n}_1(L_r)$ of n is

$$v \equiv V[\hat{n}_1(L_r)] = E\{\hat{n}_1(L_r)[\hat{n}_1(L_r) - 1]\} - E[\hat{n}_1(L_r)]\{E[\hat{n}_1(L_r)] - 1\}$$
$$= E\{\hat{n}_1(L_r)[\hat{n}_1(L_r) - 1]\} - (n)_2.$$

Thus, the UMVU estimator $\hat{v}(L_r)$ of the variance v of $\hat{n}_1(L_r)$ is given by

$$\hat{v}(L_r) = \hat{n}_1(L_r)[\hat{n}_1(L_r) - 1] - \hat{n}_2(L_r).$$

Summarizing, the following theorem is shown.

Theorem 5.4 *Consider a population with an unknown number n of classes of elements and assume that a random sample is sequentially drawn from it until r elements are drawn from previously observed classes. Let X_j be the number of elements of the jth class in the sample, $j = 1, 2, \ldots, n$, and let L_r be the number of observed classes in the sample. Suppose that $X_1, X_2, \ldots, X_n$ are independently and identically distributed random variables with common probability function*

$$q_x = P(X = x) = [g(\theta)]^{-1} c_x \theta^x, \quad x = 0, 1, \ldots, \quad 0 < \theta < \rho.$$

Then, if the observed value of L_r is $k \geq j$, a UMVU estimator of $(n)_j$ exists and is given by $\hat{n}_j(L_r)$, where

$$\hat{n}_j(k) = (k)_j \frac{\left[\Delta_u^{k-j} c_{r+k-1}(u)\right]_{u=j}}{\left[\Delta_u^k c_{r+k-1}(u)\right]_{u=0}}, \quad j = 1, 2, \ldots, k, \tag{5.24}$$

with $c_m(n) = [D_\theta^m (g(\theta))^n]_{\theta=0}/m!$. In particular, the UMVU estimator of n is given by $\hat{n}(L_r) = \hat{n}_1(L_r)$ and the UMVU estimator of its variance is given by $\hat{v}(L_r) = \hat{n}_1(L_r)[\hat{n}_1(L_r) - 1] - \hat{n}_2(L_r)$.

Example 5.7 *Sequential capture-recapture sampling scheme.* Consider a population containing a finite number n of individuals. Assume that a random sample is sequentially drawn from it, by the capture-recapture scheme, until r marked individuals are caught. Let L_r be the number of different individuals caught. This sampling model is equivalent to the assumption that, for a random sample of an unspecified sample size drawn from the same population, the random variables $X_1, X_2, \ldots, X_n$ are independently and identically distributed with a Poisson probability function

$$q_x = P(X = x) = e^{-\lambda} \frac{\lambda^x}{x!}, \quad x = 0, 1, \ldots, \quad 0 < \lambda < \infty.$$

Then,

$$q_m(u) = P(S_u = m) = e^{-u\lambda} \frac{(u\lambda)^m}{m!}, \quad m = 0, 1, \ldots,$$

and so the mth coefficient function is $c_m(u) = u^m/m!$, $m = 0, 1, \ldots$. Thus, from Theorem 5.4, it follows that the UMVU estimator of the parametric function $(n)_j$ is given by $\hat{n}_j(L_r) \equiv \hat{n}_j(L_r; r)$, where

$$\hat{n}_j(k; r) = \frac{S(r + k - 1, k - j; j)}{S(r + k - 1, k)},$$

where $S(m, k; r) = [\Delta_u^k u^m]_{u=r}/k!$ and $S(m, k) = [\Delta_u^k u^m]_{u=0}/k!$ are the noncentral and the central Stirling numbers of the second kind, respectively. These numbers are connected by the relation

$$S(r+k-1, k-j; j) = \sum_{i=k}^{k+j} \binom{i-j}{i-k} (j)_{i-k} S(r+k-1, i-j),$$

which may be deduced from (2.15) by replacing n, k, r and j by $r+k-1$, $k-j$, j and $i-j$, respectively, and taking into account that $(j)_{i-k} = 0$, for $k+j < i \leq k+j+r-1$. Using this expression for $j = 1$ and $j = 2$, we get the UMVU estimator of n as

$$\hat{n}(k; r) = k + \frac{S(r+k-1, k-1)}{S(r+k-1, k)}$$

and the UMVU estimator of $(n)_2$ as

$$\hat{n}_2(k; r) = k(k-1) + \frac{(k-1)S(r+k-1, k-1)}{S(r+k-1, k)} + \frac{S(r+k-1, k-2)}{S(r+k-1, k)}$$

$$= k(k-1) + \frac{S(r+k, k-1)}{S(r+k-1, k)}.$$

The UMVU estimator of the variance of $\hat{n}(L_r; r)$, by Theorem 5.4, is deduced as

$$\hat{v}(k; r) = \frac{kS(r+k-1, k-1)}{S(r+k-1, k)} + \left[\frac{S(r+k-1, k-1)}{S(r+k-1, k)}\right]^2 - \frac{S(r+k-1, k-2)}{S(r+k-1, k)}.$$

Example 5.8 *Sequential sampling without replacement.* Consider a population containing n classes each including s elements. Assume that a random sample is sequentially drawn from it without replacement, until r elements are drawn from previously observed classes. This sampling model is equivalent to the assumption that for a random sample of an unspecified sample size drawn from the same population, the random variables $X_1, X_2, \ldots, X_n$ are independently and identically distributed with a binomial probability function

$$q_x = P(X = x) = \binom{s}{x} p^x q^{s-x}, \quad x = 0, 1, \ldots, s,$$

where $q = 1 - p$ and $0 < p < 1$. This is a power series distribution with parameter $\theta = p/q$ and series function $g(\theta) = (1 + \theta)^s$, and its u-fold convolution is also a power series with coefficient function $c_m(u) = (su)_m/m!$, $m = 0, 1, \ldots, su$.

Thus, from Theorem 5.4, it follows that the UMVU estimator of $(n)_j$ is given by $\hat{n}_j(L_r) \equiv \hat{n}_j(L_r; r, s)$, where

$$\hat{n}_j(k; r, s) = \frac{C(r+k-1, k-j; s, sj)}{C(r+k-1, k; s)},$$

with $C(m, k; s, sj) = [\Delta_u^k (su)_m]_{u=j}/k!$ and $C(m, k; s) = [\Delta_u^k (su)_m]_{u=0}/k!$ being the noncentral and the central generalized factorial coefficients, respectively. These numbers are connected by the relation

$$C(r + k - 1, k - j; s, sj) = \sum_{i=k}^{k+j} \binom{i-j}{i-k} (j)_{i-k} C(r + k - 1, i - j; s),$$

which may be deduced from (2.55) by replacing n, k, ρ and j by $r+k-1$, $k-j$, j and $i-j$, respectively, and taking into account that $(j)_{i-k} = 0$, for $k+j < i \leq k+j+r-1$. Then, using this expression for $j = 1$ and $j = 2$, we get the UMVU estimator of n as

$$\hat{n}(k; r, s) = k + \frac{C(r + k - 1, k - 1; s)}{C(r + k - 1, k; s)}$$

and the UMVU estimator of $(n)_2$ as

$$\hat{n}_2(k; r, s) = k(k-1) + \frac{(k-1)C(r + k - 1, k - 1; s)}{C(r + k - 1, k; s)}$$
$$+ \frac{C(r + k - 1, k - 2; s)}{C(r + k - 1, k; s)}.$$

The UMVU estimator of the variance of $\hat{n}(L_r; r, s)$, by Theorem 5.4, is deduced as

$$\hat{v}(k; r, s) = \frac{kC(r + k - 1, k - 1; s)}{C(r + k - 1, k; s)} + \left[\frac{C(r + k - 1, k - 1; s)}{C(r + k - 1, k; s)}\right]^2$$
$$- \frac{C(r + k - 1, k - 2; s)}{C(r + k - 1, k; s)}.$$

Example 5.9 *Sequential sampling with replacement and addition of elements.* Consider a population containing n classes each including s elements. Assume that a random sample is sequentially drawn from it, by returning each drawn element to the population together with one element of the same class, and until r elements are drawn from previously observed classes. This is equivalent to the assumption that for a random sample of an unspecified sample size drawn from the same population, the random variables $X_1, X_2, \ldots, X_n$ are independently and identically distributed with a negative binomial probability function

$$q_x = P(X = x) = \binom{s + x - 1}{x} p^s q^x, \quad x = 0, 1, \ldots,$$

where $q = 1 - p$ and $0 < p < 1$. This is a power series distribution with parameter q and series function $g(q) = (1-q)^{-s}$, and its u-fold convolution is also a power series with coefficient function $c_m(u) = (-1)^m (-su)_m/m!$, $m = 0, 1, \ldots$. Therefore, from Theorem 5.4, it follows that the UMVU estimator of $(n)_j$ is given by $\hat{n}_j(L_r) \equiv \hat{n}_j(L_r; r, -s)$, where

$$\hat{n}_j(k; r, -s) = \frac{|C(r + k - 1, k - j; s, sj)|}{|C(r + k - 1, k; s)|},$$

where $|C(m, k; -s, -sj)|$ and $|C(m, k; -s)|$ are the absolute noncentral and the central generalized factorial coefficients, respectively. Then, proceeding as in Example 5.8, the UMVU estimator of n is derived as

$$\hat{n}(k; r, -s) = k + \frac{|C(r+k-1, k-1; -s)|}{|C(r+k-1, k; -s)|}$$

and the UMVU estimator of the variance of $\hat{n}(k; r, -s)$ is deduced as

$$\hat{v}(k; r, -s) = \frac{k|C(r+k-1, k-1; -s)|}{|C(r+k-1, k; -s)|} + \left[\frac{|C(r+k-1, k-1; -s)|}{|C(r+k-1, k; -s)|}\right]^2 - \frac{|C(r+k-1, k-2; -s)|}{|C(r+k-1, k; -s)|}.$$

5.6 A REDUCED SEQUENTIAL OCCUPANCY MODEL

The development of occupancy theory, under a general random occupancy model, started in Section 4.6, is continued here with the discussion of sequential occupancy distributions. Recall that a general random occupancy model was introduced by considering balls of a general specification randomly distributed into n distinguishable urns. Further, it is assumed that the number R of different kinds of balls distributed into the urns is a Poisson random variable with $E(R) = \lambda$. Also, the number $X_{i,j}$ of balls of the ith kind distributed into the jth urn is supposed to obey a general but known probability law:

$$P(X_{i,j} = x) = q_x, \ x = 0, 1, \ldots, \quad (5.25)$$

for all $i = 1, 2, \ldots$ and $j = 1, 2, \ldots, n$. Then, under the assumption of the independence of $X_{i,1}, X_{i,2}, \ldots, X_{i,n}$ for each $i = 1, 2, \ldots$, the probability function of $Y_i = \sum_{j=1}^n X_{i,j}$ is

$$q_{y_i}(n) = P(Y_i = y_i) = \sum q_{x_{i,1}} q_{x_{i,2}} \cdots q_{x_{i,n}}, \quad (5.26)$$

where the summation is extended over all integers $x_{i,j} \geq 0$, $j = 1, 2, \ldots, n$, such that $\sum_{j=1}^n x_{i,j} = y_i$. Finally, the number R_y of balls of multiplicity y distributed into the n urns is a Poisson random variable with $E(R_y) = \lambda q_y(n)$, $y = 0, 1, \ldots$; the random variables R_y, $y = 0, 1, \ldots$, are stochastically independent.

In this section, the study is moving on to the case of sequentially distributed balls into n distinguishable urns. If the balls are allocated into the urns one after the other, then the distribution of the number W_k of balls required to be distributed until k urns are occupied is of interest. The probability function $P(W_k = m)$, $m = k, k+1, \ldots$, may be deduced by connecting it to the probability $p_{k-1}(r_1, r_2, \ldots, r_{m-1}; m-1, r, n-1)$, derived in Section 4.6, and the probability $h_m(n)$ that at the mth allocation one ball is placed into an empty urn. This ball can be either from a kind that has not already appeared or from a kind that appeared j times in the first $m-1$ allocations. Thus, the specification of the m balls required to be distributed until k urns are occupied will be $(1^{r_1+1}2^{r_2}\cdots m^{r_m})$, or $(1^{r_1-1}2^{r_2+1}\cdots m^{r_m})$, or $(1^{r_1}2^{r_2-1}3^{r_3+1}\cdots m^{r_m})$ and so on. This method of derivation does not lead to a manageable expression for the required probability function. Further, the probability function $P(W_k = m)$, $m = k, k+1, \ldots$, may also be deduced by connecting it with the distribution function $P_k(r_1, r_2, \ldots, r_m; m, r, n)$ (cf. Section 5.2) but again similar problems are encountered.

A *reduced sequential occupancy model*, in which balls of specification (s^R) are sequentially distributed in batches of s like balls into n distinguishable urns, leads to manageable expressions for the probability function of certain sequential occupancy distributions. Under the reduced sequential occupancy model, let $W_{k;s}$ be the number of kinds of balls required to be distributed into the n distinguishable urns until k urns are occupied. The probability function

$$q_r(s; k, n) = P(W_{k;s} = r), \quad r = k, k+1, \ldots,$$

and its jth ascending binomial moment

$$b_{[j]}(s; k, n) = E\left[\binom{W_{k;s} + j - 1}{j}\right], \quad j = 0, 1, \ldots,$$

are expressed in terms of finite differences of the convolution probabilities (5.26) in the following theorem.

Theorem 5.5 (a) *Under the reduced sequential occupancy model with $q_0 > 0$, the probability $q_r(s; k, n)$, that r kinds of balls are required to be distributed into the n distinguishable urns until k urns are occupied, $r = k, k+1, \ldots$, is given by*

$$q_r(s; k, n) = \binom{n-1}{k-1}$$
$$\times \left[\Delta_u^{k-1} \frac{n}{n-u}\left(1 - \frac{q_0^{n-u} q_s(u)}{q_s(n)}\right)\left(\frac{q_0^{n-u} q_s(u)}{q_s(n)}\right)^{r-1}\right]_{u=0}. \quad (5.27)$$

(b) *The jth ascending binomial moment of this distribution is*

$$b_{[j]}(s; k, n) = \binom{n-1}{k-1}\left[\Delta_u^{k-1} \frac{n}{n-u}\left(1 - \frac{q_0^{n-u} q_s(u)}{q_s(n)}\right)^{-j}\right]_{u=0}. \quad (5.28)$$

Proof (a) Under the reduced sequential occupancy model, let A_j be the event that the jth urn remains empty, $j = 1, 2, \ldots, n$, given that r kinds of balls are distributed into the n urns. Also, let $U_{r;s}$ be the number of occupied urns. Then $P(U_{r;s} < k) = P(n - U_{r;s} > n - k)$ is the probability that at least $n - k + 1$ among the n events $A_1, A_2, \ldots, A_n$ occur. Since these events are exchangeable,

$$P(A_{j_1} A_{j_2} \cdots A_{j_i}) = \left(\frac{q_0^i q_s(n-i)}{q_s(n)}\right)^r, \quad i = 1, 2, \ldots, n,$$

applying Corollary 1.7, we get

$$P(U_{r;s} < k) = \binom{n-1}{k-1}\left[\Delta_u^{k-1}\frac{n}{n-u}\left(\frac{q_0^{n-u}q_s(u)}{q_s(n)}\right)^r\right]_{u=0}.$$

Further, $P(W_{k;s} > r) = P(U_{r;s} < k)$ and since

$$q_r(s;k,n) = P(W_{k;s} = r) = P(W_{k;s} > r-1) - P(W_{k;s} > r),$$

expression (5.27) is readily deduced.

(b) The probability function $q_r(s;k,n)$, $r = k, k+1, \ldots$, according to (5.27), is a linear function of k geometric probability functions. The jth ascending binomial moment of the geometric probability function pq^{r-1}, $r = 1, 2, \ldots$, equals p^{-j} and so, with $p = 1 - q_0^{n-u}q_s(u)/q_s(n)$, (5.28) is established.

Some special occupancy distributions are deduced by specifying the probability function q_x, $x = 0, 1, \ldots$, of the number X of balls of any specific kind distributed into any specific urn. Thus, if

$$q_x = P(X = x) = pq^x, \quad x = 0, 1, \ldots,$$

then

$$q_y(u) = P(Y = y) = \binom{u+y-1}{y}p^u q^y, \quad y = 0, 1, \ldots,$$

and, from Theorem 5.5, the following corollary is deduced.

Corollary 5.10 (a) *Under the reduced sequential occupancy model with geometric probabilities, the probability $q_r(s;k,n)$, that r kinds of balls are required to be distributed into n distinguishable urns until k urns are occupied, $r = k, k+1, \ldots$, is given by*

$$q_r(s;k,n) = \binom{n-1}{k-1}$$

$$\times \left[\Delta_u^{k-1}\frac{n}{n-u}\left(1 - \frac{(u+s-1)_s}{(n+s-1)_s}\right)\left(\frac{(u+s-1)_s}{(n+s-1)_s}\right)^{r-1}\right]_{u=0}. \quad (5.29)$$

(b) *The jth ascending binomial moment of this distribution is*

$$b_{[j]}(s;k,n) = \binom{n-1}{k-1}\left[\Delta_u^{k-1}\frac{n}{n-u}\left(1 - \frac{(u+s-1)_s}{(n+s-1)_s}\right)^{-j}\right]_{u=0}. \quad (5.30)$$

In the case where the number X of balls of any specific kind distributed into any specific urn obeys a Bernoulli distribution,

$$q_x = P(X = x) = p^x q^{1-x}, \quad x = 0, 1, \ldots,$$

the number Y of balls of any specific kind distributed into the n urns obeys a binomial distribution with

$$q_y(n) = P(Y = y) = \binom{n}{y} p^y q^{n-y}, \ y = 0, 1, \ldots, n.$$

Then, the following corollary of Theorem 5.5 is readily obtained.

Corollary 5.11 (a) *Under the reduced sequential occupancy model with Bernoulli probabilities, the probability $q_r(s; k, n)$, that r kinds of balls are required to be distributed into n distinguishable urns until k urns are occupied, $r = k, k+1, \ldots$, is given by*

$$q_r(s; k, n) = \binom{n-1}{k-1} \left[\Delta_u^{k-1} \frac{n}{n-u} \left(1 - \frac{(u)_s}{(n)_s}\right) \left(\frac{(u)_s}{(n)_s}\right)^{r-1} \right]_{u=0}. \quad (5.31)$$

(b) *The jth ascending binomial moment of this distribution is*

$$b_{[j]}(s; k, n) = \binom{n-1}{k-1} \left[\Delta_u^{k-1} \frac{n}{n-u} \left(1 - \frac{(u)_s}{(n)_s}\right)^{-j} \right]_{u=0}. \quad (5.32)$$

Example 5.10 *Waiting-time distribution of lotto drawings.* In lotto every week s numbers are drawn without replacement from a set of n numbers, $\{1, 2, \ldots, n\}$ (cf. Example 4.16). The distribution of the number $W_{k;s}$ of drawings required until k different numbers, among $\{1, 2, \ldots, n\}$, are drawn is of interest to gamblers.

Considering the n numbers $\{1, 2, \ldots, n\}$ as n distinguishable urns and interpreting the drawn of number j at the ith drawing as a placement of a ball of the ith kind into the jth urn, this stochastic model is equivalent to the reduced sequential occupancy model with Bernoulli probabilities. Thus, according to Corollary 5.11, the probability function and ascending binomial moments of $W_{k;s}$ are given by (5.31) and (5.32), respectively.

5.7 REFERENCE NOTES

Under a sequential occupancy model, and assuming that the number of balls distributed into any specific urn obeys a Poisson, binomial or negative binomial law, D. E. Barton and F. N. David (1959c) derived the classical, restricted and pseudo-contagious waiting-time occupancy distributions; these distributions are deduced in Corollaries 5.1, 5.4 and 5.7. The waiting-time and the complementary waiting-time occupancy distributions, when the number of balls distributed into any specific urn obeys a general discrete probability law were studied in Ch. A. Charalambides (1997, 2005b). M. Cerasoli (1983), assuming that the total number of balls distributed into the urns is a Poisson stochastic process, derived an interesting expression of the waiting-time classical occupancy distribution.

F. N. David and D. E. Barton (1962), in the chapter of occupancy, extensively treated sequential occupancy distributions. N. L. Johnson and S. Kotz (1977) provided a wealth of information on sequential occupancy distributions and their applications.

H. B. Nath (1974) derived the waiting-time classical occupancy distribution with unequal probabilities for a ball to fall into different urns; this distribution is presented in Example 5.1. The waiting-time distribution of the number of random numbers required to be selected until k usable numbers are obtained, presented in Example 5.5, was discussed by G. J. Glasser (1963). Examples 5.7, 5.8 and 5.9 on sequential sample estimation of the number of classes in a finite population are based on the papers of L. A. Goodman (1953), J. N. Darroch (1958), B. Harris (1968) and S. Berg (1975, 1987). The estimation of the size of a finite population in a sequential sampling scheme, in which the stopping rule is a linear function of the number of observed classes, was studied by S. Berg (1987). Des Raj and S. H. Khamis (1958) used the waiting-time classical occupancy distribution (5.12) in the assessment of an estimator of the mean of a finite population, which was based on sequential sampling.

Example 5.10 and Exercise 5.15 on the waiting-time to exhaust lottery numbers are based on the paper of J. G. Leite, C. A. de B. Pereira and F. W. Rodrigues (1993).

The particular waiting-time and complementary waiting-time occupancy distributions in an extended sequential occupancy model given in Exercises 5.8, 5.9 and 5.13 were derived by K. Nishimura and M. Sibuya (1997) under a random walk. Exercise 5.14 is merely a rephrasing of a problem studied by B. C. Arnold (1972).

Approximations of waiting-time occupancy distributions are briefly discussed by N. L. Johnson and S. Kotz (1977). Among recent papers dealing with asymptotic results for waiting-time occupancy distributions, we mention here the papers of S. Janson (1983), L. Holst and J. Hüsler (1985) and L. Holst (1986).

5.8 EXERCISES AND COMPLEMENTS

5.1 *A sequential pseudo-contagious occupancy model.* Consider a supply of like balls randomly and sequentially distributed into n distinguishable urns. Assume that the jth urn is divided into s_j distinguishable cells, $j = 1, 2, \ldots, n$, of unlimited capacity and set $s = s_1 + s_2 + \cdots + s_n$. Let W_k be the number of balls required to be distributed until k urns are occupied. Find the probability function and ascending binomial moments of W_k.

5.2 (*Continuation*). Find the probability function of the number L_r of occupied urns until r balls are placed into previously occupied urns.

5.3 *Bose-Einstein sequential occupancy model.* Assume that like balls are sequentially distributed at random into n distinguishable urns of unlimited capacity and let W_k be the number of balls required to be distributed until k urns are occupied.

(a) Show that the probability

$$q_m(k, n) = P(W_k = m), \quad m = k, k+1, \ldots,$$

that m balls are required to be distributed into the n urns until k urns are occupied is given by

$$q_m(k, n) = \binom{n-1}{k-1}\binom{m-2}{k-2} \bigg/ \binom{n+m-1}{m-1}.$$

(b) Further, show that the rth ascending binomial moment $b_{[r]}(k, n)$, $r = 0, 1, \ldots$, of the probability function $q_m(k, n)$, $m = k, k+1, \ldots$, is given by

$$b_{[r]}(k, n) = \binom{n-1}{k-1}\binom{r+k-1}{k-1} \bigg/ \binom{n-r-1}{k-1}.$$

5.4 (*Continuation*). (a) Show that the conditional probability $q_{r;m}(k, n) = P(W_{k+1} - W_k = r | W_k = m)$, $r = 1, 2, \ldots$, is given by

$$q_{r;m}(k, n) = \frac{(n-k)(k+m+r-2)_{r-1}}{(n+m+r-1)_r}.$$

(b) Let L_r be the number of occupied urns until r balls are placed into previously occupied urns and $c_k(r, n) = P(L_r = k)$, $k = 1, 2, \ldots, n$. Show that

$$c_k(r, n) = \binom{n}{k} \frac{2k+r-1}{n+k+r-1} \binom{r+k-2}{k-1} \bigg/ \binom{n+r+k-2}{r+k-1}.$$

5.5 *Waiting-time randomized occupancy distribution.* Consider the sequential random occupancy model and assume that the number X_j of balls distributed into the jth urn obeys a Poisson distribution with $q_x = P(X = x) = e^{-\lambda}\lambda^x/x!$, $x = 0, 1, \ldots$, $0 < \lambda < \infty$, $j = 1, 2, \ldots, n$. Further, assume that each ball has probability p of staying in an urn and probability $q = 1 - p$ of falling through.

(a) Show that the probability function $q_m(k, n; p) = P(W_{k;p} = m)$ of the number $W_{k;p}$ of balls that are required to be distributed into the n urns until k urns are occupied, $m = k, k+1, \ldots$, is given by

$$q_m(k, n; p) = \frac{(n)_k S(m-1, k-1; nq/p)}{(n/p)^m},$$

where $S(m-1, k-1; nq/p) = [\Delta_u^{k-1} u^{m-1}]_{u=nq/p}/(k-1)!$ is the noncentral Stirling number of the second kind.

(b) Also, show that the rth ascending binomial moment $b_{[r]}(k, n; p)$, $r = 0, 1, \ldots$, of $W_{k;p}$ is given by

$$b_{[r]}(k, n; p) = p^{-r} b_{[r]}(k, n),$$

where

$$b_{[r]}(k, n) = \binom{n-1}{k-1} \sum_{j=0}^{k-1} (-1)^{k-j-1} \binom{k-1}{j} \left(1 - \frac{j}{n}\right)^{-r-1}$$

is the rth ascending binomial moment of $W_k \equiv W_{k;1}$.

5.6 (*Continuation*). (a) Show that the random variable $W_{k;p}$ may be expressed as a sum $W_{k;p} = \sum_{i=1}^{k} Y_{j;p}$ of independent random variables with

$$P(Y_{j;p} = y) = p \frac{n-j+1}{n} \left(1 - p \frac{n-j+1}{n}\right)^{y-1}, \quad y = 1, 2, \ldots.$$

(b) Further, show that

$$E(W_{k;p}) = \frac{n}{p} \sum_{j=1}^{k} \frac{1}{n-j+1}$$

and verify that it coincides with the expression $b_{[1]}(k,n;p)$ given in Exercise 5.5.

5.7 *Extended sequential random occupancy model.* Consider a supply of balls randomly and sequentially distributed into $n+r$ distinguishable urns, among which n are labeled as *target* urns and the other r as *control* urns. Let X_j be the number of balls distributed into the jth urn, $j = 1, 2, \ldots, n+r$. Assume that the random variables X_j, $j = 1, 2, \ldots, n+r$, are independently and identically distributed with a common probability function $P(X = x) = q_x$, $x = 0, 1, \ldots$. Also, let S_{n+r} be the total number of balls distributed into the $n+r$ urns, with $P(S_{n+r} = m) = q_m(n+r)$, $m = 0, 1, \ldots$. Under this extended sequential random occupancy model, let W_k be the number of balls required to be distributed into the $n+r$ urns until k target urns are occupied and

$$q_m(k,n,r) = P(W_k = m), \quad b_{[j]}(k,n,r) = E\left[\binom{W_k+j-1}{j}\right].$$

(a) Show that the probability $q_m(k,n,r)$ that m balls are sequentially distributed into the $n+r$ urns until k target urns are occupied, $m = k, k+1, \ldots$, is given by

$$q_m(k,n,r) = \binom{n-1}{k-1} \frac{nq_1 \left[\Delta_u^{k-1} q_0^{n+r-u-1} q_{m-1}(u)\right]_{u=r}}{m q_m(n+r)}.$$

(b) Further, if the probability

$$h_m(u,n,r) = \frac{nq_0^{n+r-u-1} q_1 q_{m-1}(u)}{m q_m(n+r)}, \quad m = 1, 2, \ldots,$$

is a polynomial in u of degree at most $m-1$, then show that

$$b_{[j]}(k,n,r) = \binom{n-1}{k-1} \left[\Delta_u^{k-1} a_{[j]}(u,n,r)\right]_{u=r}, \quad j = 1, 2, \ldots,$$

where

$$a_{[j]}(u,n,r) = \sum_{m=1}^{\infty} \binom{m+j-1}{j} h_m(u,n,r), \quad j = 1, 2, \ldots.$$

5.8 *(Continuation).* Consider the extended sequential random occupancy model and assume that the number X_j of balls distributed into the jth urn, $j = 1, 2, \ldots, n+r$, obeys a Poisson distribution with $q_x = P(X = x) = e^{-\lambda} \lambda^x / x!$, $x = 0, 1, \ldots$, $0 < \lambda < \infty$.

(a) Show that the probability $q_m(k, n, r)$ that m balls are required to be distributed into the $n+r$ distinguishable urns until k target urns are occupied, $m = k, k+1, \ldots$, is given by

$$q_m(k, n, r) = \frac{(n)_k S(m-1, k-1; r)}{(n+r)^m},$$

where $S(m, k; r) = [\Delta_u^k u^m]_{u=r}/k!$ is the noncentral Stirling number of the second kind.

(b) Further, show that the jth ascending binomial moment $b_{[j]}(k, n, r)$, $j = 0, 1, \ldots$, of the probability function $q_m(k, n, r)$, $m = k, k+1, \ldots$, is given by

$$b_{[j]}(k, n, r) = \frac{n}{n+r} \binom{n-1}{k-1} \sum_{i=0}^{k-1} (-1)^{k-i-1} \binom{k-1}{i} \left(1 - \frac{i+r}{n+r}\right)^{-j-1}.$$

5.9 (*Continuation*). Consider the extended sequential random occupancy model and assume that the number X_j of balls distributed into the jth urn, $j = 1, 2, \ldots, n+r$, obeys a binomial distribution with $q_x = P(X = x) = \binom{s}{x} p^x q^{s-x}$, $x = 0, 1, \ldots, s$, where $0 < p < 1$ and $q = 1 - p$.

(a) Show that the probability $q_m(k, n, r)$ that m balls are required to be distributed into the $n+r$ distinguishable urns until k target urns are occupied, $m = k, k+1, \ldots$, is given by

$$q_m(k, n, r) = \frac{s(n)_k C(m-1, k-1; s, rs)}{(sn + sr)_m},$$

where $C(m, k; s, rs) = [\Delta_u^k (su)_m]_{u=r}/k!$ is the noncentral generalized factorial coefficient.

(b) Also, show that the jth ascending binomial moment $b_{[j]}(k, n, r)$, $j = 0, 1, \ldots$, of the probability function $q_m(k, n, r)$, $m = k, k+1, \ldots$, is given by

$$b_{[j]}(k, n, r) = \frac{n}{n+r} \binom{n-1}{k-1} \sum_{i=0}^{k-1} (-1)^{k-i-1} \binom{k-1}{i} \frac{(sn+sr+j)_j}{(sn-si+j)_{j+1}}.$$

5.10 Under the extended sequential random occupancy model with $q_0 > 0$, let $q_{j;m}(k, n, r) = P(W_{k+1} - W_k = j | W_k = m)$ be the conditional probability that j additional balls are required to be distributed into the $n+r$ distinguishable urns until the $(k+1)$st target urn is occupied, given that m balls are sequentially distributed into the $n+r$ distinguishable urns until the kth target urn is occupied. Show that

$$q_{j,m}(k, n, r) = \frac{(n-k) q_1 q_{m+j-1}(k+r)/q_m(k+r)}{(m+j) q_0 q_{m+j}(n+r)/q_m(n+r)}.$$

5.11 (*Continuation*). Under the extended sequential random occupancy model with $q_x = e^{-\lambda} \lambda^x / x!$, $x = 0, 1, \ldots$, $0 < \lambda < \infty$, (a) show that the random variable W_k may be expressed as a sum $W_k = \sum_{j=1}^k Y_i$ of independent random variables with

$$P(Y_i = y) = \frac{n-i+1}{n+r} \left(\frac{r+i-1}{n+r}\right)^{y-1}, \quad y = 1, 2, \ldots.$$

(b) Further, show that

$$E(W_k) = (n+r) \sum_{i=1}^{k} \frac{1}{n-i+1}$$

and verify that it coincides with the particular case of expression $b_{[1]}(k,n,r)$ given in Exercise 5.7, when $q_x = e^{-\lambda}\lambda^x/x!$, $x = 0, 1, \ldots$.

5.12 (*Continuation*). Under the extended sequential random occupancy model with $q_x = \binom{s}{x} p^x q^{s-x}$, $x = 0, 1, \ldots, s$, $q = 1-p$, $0 < p < 1$, show that the random variable W_k may be expressed as a sum $W_k = \sum_{j=1}^{k} Y_j$ of dependent random variables with

$$P(Y_{k+1} = y | W_k = m) = \frac{(sn - sk)(sk + sr - m)_{y-1}}{(sn + sr - m)_y}, \quad y = 1, 2, \ldots.$$

5.13 Consider the extended sequential random occupancy model and assume that balls are sequentially distributed at random into the $n+r$ urns until j balls are placed into previously occupied target urns and into the control urns. Let L_j be the number of occupied target urns and

$$c_k(j, n, r) = P(L_j = k), \quad k = 0, 1, \ldots, n.$$

(a) Show that

$$c_k(j, n, r) = \binom{n}{k} \frac{q_{j+k}(k+r)\left[\Delta_u^k q_0^{n+r-u} q_{j+k-1}(u)\right]_{u=r}}{q_{j+k-1}(k+r) q_{j+k}(n+r)}.$$

(b) If the number X_j of balls distributed into the jth urn, $j = 1, 2, \ldots, n+r$, obeys a Poisson distribution with $q_x = P(X = x) = e^{-\lambda}\lambda^x/x!$, $x = 0, 1, \ldots$, $0 < \lambda < \infty$, conclude that

$$c_k(j, n, r) = \frac{(n)_k (k+r) S(j+k-1, k; r)}{(n+r)^{j+k}},$$

where $S(m, k; r) = [\Delta_u^k u^m]_{u=r}/k!$ is the noncentral Stirling number of the second kind.

(c) If the number X_j of balls distributed into the jth urn, $j = 1, 2, \ldots, n+r$, obeys a binomial distribution with $q_x = P(X = x) = \binom{s}{x} p^x q^{s-x}$, $x = 0, 1, \ldots, s$, where $0 < p < 1$ and $q = 1-p$, conclude that

$$c_k(j, n, r) = \frac{(n)_k (sk + sr - j - k + 1) C(j+k-1, k; s, rs)}{(sn + sr)_{j+k}},$$

where $C(m, k; s, rs) = [\Delta_u^k (su)_m]_{u=r}/k!$ is the noncentral generalized factorial coefficient.

5.14 Under the sequential random occupancy model with Poisson probabilities, let M_r be the number of balls required to be distributed into the n urns until one ball is placed into previously occupied urns within at most r successive ball allocations. Show that
$$P(M_r = m) = \frac{(m-1)(n)_{m-1}}{n^m}, \quad m = 2, 3, \ldots, r,$$
and
$$P(M_r = m) = \frac{(n)_r}{n^r} \cdot \frac{r-1}{n} \left(1 - \frac{r-1}{n}\right)^{m-r-1}, \quad m = r+1, r+2, \ldots.$$

5.15 *Waiting-time to exhaust lottery numbers.* In lotto every week s numbers are drawn without replacement from a set of n numbers, $\{1, 2, \ldots, n\}$ (cf. Example 5.10). According to the records of a gambler, after a certain number of drawings, k numbers $\{j_1, j_2, \ldots, j_k\}$ were not drawn. Let $T_{k;s}$ be the number of drawings required until all k numbers are drawn. Find the probability function
$$P(T_{k;s} = r), \quad r = k, k+1, \ldots,$$
and ascending binomial moments
$$E\left[\binom{T_{k;s} + j - 1}{j}\right], \quad j = 0, 1, \ldots.$$

6

CONVOLUTIONS OF TRUNCATED DISTRIBUTIONS

6.1 INTRODUCTION

The classical discrete distributions are defined on the set of nonnegative integers $\{0, 1, \ldots, s, \ldots\}$ or on a finite subset of it, say, $\{0, 1, \ldots, s\}$. Such distributions often arise in practice with no zero value observed. For example, while it is a simple matter to count, over a given period of time, the number of workers in a factory sustaining one, two or more accidents, the number of workers who do not have an accident cannot be enumerated owing to the factory population fluctuating during this period. In this case the distribution of the number of accidents per worker arises with the zero value unobserved.

The power series distribution, with probability function

$$q_x = P(X = x) = [g(\theta)]^{-1} c_x \theta^x, \; x = 0, 1, \ldots,$$

where the function $g(\theta)$ admits a power series expansion $g(\theta) = \sum_{x=0}^{\infty} c_x \theta^x$, $0 < \theta < \rho$, and the coefficient $c_x = [D_\theta^x g(\theta)]_{\theta=0}/x! \geq 0$ does not involve the parameter θ, constitutes a wide class of discrete distributions. The Poisson, binomial (with a known number of trials), negative binomial (with a known number of successes) and logarithmic distributions belong in this class. In the case of a power series distribution in which the zero value cannot be observed, the distribution of the population reduces to the zero truncated power series distribution, with probability function

$$P(Y = x) = P(X = x | X > 0) = [g(\theta) - 1]^{-1} c_x \theta^x, \; x = 1, 2, \ldots.$$

Combinatorial Methods in Discrete Distributions. By Charalambos A. Charalambides
ISBN 0-471-68027-3 Copyright © 2005 John Wiley & Sons, Inc.

Let $Y_1, Y_2, \ldots, Y_n$ be a random sample from a population with a zero truncated power series distribution. Then the sum $T_n = \sum_{j=1}^{n} Y_j$ is a complete sufficient statistic. Its distribution facilitates the derivation of uniformly minimum variance unbiased estimators for estimable functions of the parameter θ.

Consider a random sample $X_1, X_2, \ldots, X_n$ from a population with a general discrete distribution

$$P(X = x) = q_x, \quad x = 0, 1, \ldots,$$

and let $S_n = \sum_{j=1}^{n} X_j$. Further, assuming that $q_0 > 0$, consider a random sample $Y_1, Y_2, \ldots, Y_n$ from the population with the zero truncated discrete distribution

$$P(Y = x) = P(X = x | X > 0) = (1 - q_0)^{-1} q_x, \quad x = 1, 2, \ldots,$$

and let $T_n = \sum_{j=1}^{n} Y_j$. Note that if $q_0 = 0$, the distribution of Y coincides with that of X. Clearly, for $q_0 > 0$, the probability function of T_n may be expressed by the probability function of S_n, suitably renormed. In addition to the probability $P(X > 0) = 1 - q_0$, the conditional probability $P(K = n | S_n = m)$ that, under the random occupancy model (cf. Section 4.2), all n urns are occupied given that m balls are distributed into n distinguishable urns emerges as a renorming factor. Hence, the derivation and the study of the distribution of the n-fold convolution of a truncated discrete distribution within the framework of the random occupancy model can shed some light into its structure.

The present chapter is devoted to the derivation and study of the distribution of the n-fold convolution of truncated discrete distributions within the framework of the random occupancy model.

6.2 ZERO TRUNCATED DISCRETE DISTRIBUTIONS

Consider a supply of balls randomly distributed (simultaneously or sequentially) into n distinguishable urns. Let X_j be the number of balls distributed into the jth urn, $j = 1, 2, \ldots, n$. In the random occupancy model, introduced in Section 4.2, the random variables X_j, $j = 1, 2, \ldots, n$, are assumed to be independently distributed with a known probability function

$$P(X_j = x) = q_{j,x}, \quad x = 0, 1, \ldots, \quad j = 1, 2, \ldots, n.$$

Let $S_n = \sum_{j=1}^{n} X_j$, be the total number of balls distributed into the n urns, with

$$P(S_n = m) = q_m(n), \quad m = 0, 1, \ldots.$$

Then

$$q_m(n) = \sum q_{1,x_1} q_{2,x_2} \cdots q_{n,x_n},$$

where the summation is extended over all integers $x_j \geq 0$, $j = 1, 2, \ldots, n$, such that $x_1 + x_2 + \cdots + x_n = m$. Further, assuming that $q_{j,0} > 0$, $j = 1, , 2, \ldots, n$, consider the zero truncated distribution with probability function

$$P(Y_j = x) = P(X_j = x | X_j > 0) = (1 - q_{j,0})^{-1} q_{j,x}, \quad x = 1, 2, \ldots,$$

for $j = 1, 2, \ldots, n$, and let $T_n = \sum_{j=1}^{n} Y_j$. Then, the probability function $P(T_n = m) = p_m(n)$, $m = n, n+1, \ldots$, of the n-fold convolution of the zero truncated distribution, equals the conditional probability that m balls are distributed into the n urns, given that no urn remains empty; that is,

$$P(T_n = m) = P(S_n = m | X_1 > 0, X_2 > 0, \ldots, X_n > 0),$$

for $m = n, n+1, \ldots$. Since the distribution $P(Y_j = x) = (1 - q_{j,0})^{-1} q_{j,x}$ is expressed by the distribution $P(X_j = x) = q_{j,x}$, suitably renormed, it is expected that the distribution of T_n may be derived by a suitable renorming of the distribution of S_n. In addition to the probabilities $P(X_j > 0) = 1 - q_{j,0}$, $j = 1, 2, \ldots, n$, the conditional probability

$$P(K = n | S_n = m) = P(X_1 > 0, X_2 > 0, \ldots, X_n > 0 | S_n = m),$$

that all n urns are occupied given that m balls are distributed into the n distinguishable urns, emerges as a renorming factor. Specifically,

$$P(T_n = m) = \frac{P(K = n | S_n = m) P(S_n = m)}{P(X_1 > 0) P(X_2 > 0) \cdots P(X_n > 0)}. \tag{6.1}$$

More generally, let $B_j \subseteq \{0, 1, \ldots, s, \ldots\}$, $j = 1, 2, \ldots, n$, and consider the truncated distribution with probability function

$$P(Y_j = x) = P(X_j = x | X_j \in B_j) = \frac{q_{j,x}}{P(X_j \in B_j)}, \quad x \in B_j,$$

for $j = 1, 2, \ldots, n$. Then, the probability function of the sum $T_n = \sum_{j=1}^{n} Y_j$,

$$P(T_n = m) = P(S_n = m | X_1 \in B_1, X_2 \in B_2, \ldots, X_n \in B_n),$$

is given by

$$P(T_n = m) = \frac{P(X_1 \in B_1, X_2 \in B_2, \ldots, X_n \in B_n | S_n = m) P(S_n = m)}{P(X_1 \in B_1) P(X_2 \in B_2) \cdots P(X_n \in B_n)}.$$

The following examples illustrate this connection.

Example 6.1 *Convolution of zero truncated Poisson distributions.* Assume that the random variables X_j, $j = 1, 2, \ldots, n$, are independently distributed with Poisson probability function

$$q_{j,x} = e^{-\lambda_j} \frac{\lambda_j^x}{x!}, \quad x = 0, 1, \ldots, \quad 0 < \lambda_j < \infty, \quad j = 1, 2, \ldots, n.$$

Then, the distribution of the sum $S_n = \sum_{j=1}^{n} X_j$ is also Poisson with probability function

$$q_m(n) = e^{-\lambda} \frac{\lambda^m}{m!}, \quad m = 0, 1, \ldots, \quad \lambda = \sum_{j=1}^{n} \lambda_j.$$

Further, consider the zero truncated Poisson distribution with probability function

$$P(Y_j = x) = P(X_j = x | X_j > 0) = (e^{\lambda_j} - 1)^{-1} \frac{\lambda_j^x}{x!}, \quad x = 1, 2, \ldots,$$

for $j = 1, 2, \ldots, n$. Note that $P(X_j > 0) = 1 - e^{-\lambda_j}$, $j = 1, 2, \ldots, n$, and that the conditional probability $P(K = n | S_n = m)$, according to Example 4.1, is given by

$$P(K = n | S_n = m) = \lambda^{-m} \sum_{r=0}^{n} (-1)^{n-r} S_{n,n-r},$$

with

$$S_{n,n-r} = \sum (\lambda_{j_1} + \lambda_{j_2} + \cdots + \lambda_{j_r})^m,$$

where the summation is extended over all r-combinations $\{j_1, j_2, \ldots, j_r\}$ of the n indices $\{1, 2, \ldots, n\}$. Therefore, the probability function of $T_n = \sum_{j=1}^{n} Y_j$, by virtue of (6.1), is given by

$$P(T_n = m) = \prod_{j=1}^{n} (e^{\lambda_j} - 1)^{-1} \frac{1}{m!} \sum_{r=0}^{n} (-1)^{n-r} S_{n,n-r}, \quad m = n, n+1, \ldots.$$

Example 6.2 *Convolution of zero truncated binomial distributions.* Assume that the random variables X_j, $j = 1, 2, \ldots, n$, are independently distributed with binomial probability function

$$q_{j,x} = \binom{s_j}{x} p^x q^{s_j - x}, \quad x = 0, 1, \ldots, s_j, \; j = 1, 2, \ldots, n,$$

where $0 < p < 1$ and $q = 1 - p$. Then, the distribution of the sum $S_n = \sum_{j=1}^{n} X_j$ is also binomial with probability function

$$q_m(n) = \binom{s}{m} p^m q^{s-m}, \quad m = 0, 1, \ldots, s = \sum_{j=1}^{n} s_j.$$

Further, consider the zero truncated binomial distribution with probability function

$$P(Y_j = x) = P(X_j = x | X_j > 0) = (1 - q^{s_j})^{-1} \binom{s_j}{x} p^x q^{s_j - x},$$

for $x = 1, 2, \ldots, s_j$ and $j = 1, 2, \ldots, n$. Note that $P(X_j > 0) = 1 - q^{s_j}$, for $j = 1, 2, \ldots, n$, and that the conditional probability $P(K = n | S_n = m)$, according to Example 4.2, is given by

$$P(K = n | S_n = m) = \frac{1}{(s)_m} \sum_{r=0}^{n} (-1)^{n-r} S_{n,n-r},$$

with

$$S_{n,n-r} = \sum (s_{j_1} + s_{j_2} + \cdots + s_{j_r})_m,$$

where the summation is extended over all r-combinations $\{j_1, j_2, \ldots, j_r\}$ of the n indices $\{1, 2, \ldots, n\}$. Therefore, the probability function of $T_n = \sum_{j=1}^{n} Y_j$, by virtue of (6.1), is given by

$$P(T_n = m) = \prod_{j=1}^{n}(q^{-s_j} - 1)^{-1} \frac{(p/q)^m}{m!} \sum_{r=0}^{n}(-1)^{n-r} S_{n,n-r},$$

for $m = n, n+1, \ldots$.

In the sequel, the study of the convolutions of zero truncated distributions focuses on the particular case in which the random variables X_j, $j = 1, 2, \ldots, n$, are independently and identically distributed with common probability function

$$P(X = x) = q_x, \ x = 0, 1, \ldots . \tag{6.2}$$

Then,

$$P(S_n = m) = q_m(n) = \sum q_{x_1} q_{x_2} \cdots q_{x_n}, \ m = 0, 1, \ldots, \tag{6.3}$$

where the summation is extended over all integers $x_j \geq 0$, $j = 1, 2, \ldots, n$, such that $x_1 + x_2 + \cdots + x_n = m$, is the n-fold convolution of q_x, $x = 0, 1, \ldots$. Further, consider the rth binomial moment of S_n,

$$a_{(r)}(n) = E\left[\binom{S_n}{r}\right] = \sum_{m=r}^{\infty} \binom{m}{r} q_m(n), \ r = 0, 1, \ldots . \tag{6.4}$$

If $q_0 > 0$, the probability function

$$p_m(n) = P(T_n = m) = P(S_n = m | X_1 > 0, X_2 > 0, \ldots, X_n > 0),$$

for $m = n, n+1, \ldots$, and its binomial moments

$$b_{(r)}(n) = E\left[\binom{T_n}{r}\right] = \sum_{m=r}^{\infty} \binom{m}{r} p_m(n), \ r = 0, 1, \ldots,$$

are expressed as finite differences of (6.3) and (6.4) in the following theorem.

Theorem 6.1 (a) *Under the random occupancy model and assuming that $q_0 > 0$, the conditional probability $p_m(n)$, that m balls are distributed into the n urns, given that no urn remains empty, $m = n, n+1, \ldots$, is given by*

$$p_m(n) = (1 - q_0)^{-n} \left[\Delta_u^n q_0^{n-u} q_m(u)\right]_{u=0}. \tag{6.5}$$

(b) *The rth binomial moment $b_{(r)}(n)$, $r = 0, 1, \ldots$, of the probability function $p_m(n)$, $m = n, n+1, \ldots$, is given by*

$$b_{(r)}(n) = \sum_{k=0}^{n} \binom{n}{k}(1 - q_0)^{-k} \left[\Delta_u^k a_{(r)}(u)\right]_{u=0}, \tag{6.6}$$

where $a_{(r)}(n)$, $r = 0, 1, \ldots$, is the rth binomial moment of the probability function $q_m(n)$, $m = 0, 1, \ldots$.

Proof (a) Under the random occupancy model, the conditional probability that all n urns are occupied, given that m balls are distributed into the n distinguishable urns may be deduced from Corollary 4.1 as

$$P(K = n|S_n = m) = \frac{[\Delta_u^n q_0^{n-u} q_m(u)]_{u=0}}{q_m(n)}.$$

Further, $P(X_j > 0) = 1 - q_0$, $j = 1, 2, \ldots, n$, and using (6.1) along with (6.3), (6.5) is deduced.

(b) Introducing into the definition of $b_{(r)}(n)$ expression (6.5) of $p_m(n)$, we get

$$b_{(r)}(n) = \sum_{m=r}^{\infty} \binom{m}{r} (1 - q_0)^{-n} [\Delta_u^n q_0^{n-u} q_m(u)]_{u=0}$$

$$= \sum_{m=r}^{\infty} \binom{m}{r} (1 - q_0)^{-n} \sum_{j=0}^{n} (-1)^{n-j} \binom{n}{j} q_0^{n-j} q_m(j).$$

Interchanging the order of summation and using (6.4), we deduce the expression

$$b_{(r)}(n) = (1 - q_0)^{-n} \sum_{j=0}^{n} (-1)^{n-j} \binom{n}{j} q_0^{n-j} a_{(r)}(j).$$

The last expression, upon expanding $q_0^{n-j} = [1 - (1 - q_0)]^{n-j}$ into powers of $1 - q_0$, is transformed to

$$b_{(r)}(n) = \sum_{j=0}^{n} (-1)^{n-j} \binom{n}{j} \sum_{k=j}^{n} (-1)^{n-k} \binom{n-j}{n-k} (1 - q_0)^{-k} a_{(r)}(j)$$

$$= \sum_{k=0}^{n} \binom{n}{k} (1 - q_0)^{-k} \sum_{j=0}^{k} (-1)^{k-j} \binom{k}{j} a_{(r)}(j),$$

which implies (6.6).

As is indicated in the introduction, in parametric statistical inference, the most widely used family of discrete distributions, as a population (parent) distribution, is the power series distribution. A separate presentation of the n-fold convolution of the zero truncated power series distribution is thus justified. Consider a random sample $X_1, X_2, \ldots, X_n$, from a population with the power series probability function

$$q_x = P(X = x) = [g(\theta)]^{-1} c_x \theta^x, \quad x = 0, 1, \ldots, \tag{6.7}$$

where the function $g(\theta)$ admits a power series expansion $g(\theta) = \sum_{x=0}^{\infty} c_x \theta^x$, $0 < \theta < \rho$, and the coefficient $c_x = [D_\theta^x g(\theta)]_{\theta=0}/x! \geq 0$ does not involve the parameter θ. For $q_0 > 0$, without any loss of generality it is assumed that $c_0 = 1$. The distribution of the sum $S_n = \sum_{j=1}^{n} X_j$ is also a power series. Specifically, its probability function is given by

$$q_m(n) = P(S_n = m) = [g(\theta)]^{-n} c_m(n) \theta^m, \quad m = 0, 1, \ldots, \tag{6.8}$$

with
$$c_m(n) = \sum c_{x_1} c_{x_2} \cdots c_{x_n}, \quad m = 0, 1, \ldots, \quad (6.9)$$

where the summation is extended over all integers $x_j \geq 0$, $j = 1, 2, \ldots, n$, such that $x_1 + x_2 + \cdots + x_n = m$. Alternatively, the coefficient function may be expressed as

$$c_m(n) = \frac{[D_\theta^m (g(\theta))^n]_{\theta=0}}{m!}.$$

Clearly, the rth binomial moment of S_n is given by

$$a_{(r)}(n) = E\left[\binom{S_n}{r}\right] = \frac{\theta^r D_\theta^r [g(\theta)]^n}{r![g(\theta)]^n}, \quad r = 0, 1, \ldots. \quad (6.10)$$

Further, consider a random sample $Y_1, Y_2, \ldots, Y_n$ from the population with the zero truncated power series probability function

$$P(Y = x) = P(X = x | X > 0) = [g(\theta) - 1]^{-1} c_x \theta^x, \quad x = 1, 2, \ldots.$$

The distribution of the sum $T_n = \sum_{j=1}^n Y_j$ is again a power series distribution. Its probability function,

$$p_m(n) = P(T_n = m), \quad m = n, n+1, \ldots,$$

and binomial moments,

$$b_{(r)}(n) = E\left[\binom{T_n}{r}\right] = \sum_{m=r}^{\infty} \binom{m}{r} p_m(n), \quad r = 0, 1, \ldots,$$

are given in the following corollary of Theorem 6.1.

Corollary 6.1 (a) *Under the random occupancy model with power series probabilities*
$$q_x = P(X = x) = [g(\theta)]^{-1} c_x \theta^x, \quad x = 0, 1, \ldots, \quad c_0 = 1,$$
the conditional probability $p_m(n)$, that m balls are distributed into the n urns, given that no urn remains empty, $m = n, n+1, \ldots$, is given by

$$p_m(n) = [g(\theta) - 1]^{-n} \left[\Delta_u^n c_m(u)\right]_{u=0} \theta^m, \quad (6.11)$$

where $c_m(n) = [D_\theta^m (g(\theta))^n]_{\theta=0}/m!$.

(b) *The rth binomial moment $b_{(r)}(n)$, $r = 0, 1, \ldots$, of the probability function $p_m(n)$, $m = n, n+1, \ldots$, is given by*

$$b_{(r)}(n) = \sum_{k=0}^n \binom{n}{k} [1 - 1/g(\theta)]^{-k} \left[\Delta_u^k a_{(r)}(u)\right]_{u=0}, \quad (6.12)$$

where $a_{(r)}(n) = [g(\theta)]^{-n} \theta^r D_\theta^r [g(\theta)]^n / r!$.

6.3 SOME PARTICULAR CONVOLUTIONS

6.3.1 Zero Truncated Poisson Distribution

Assume that the random variable X_j, $j = 1, 2, \ldots, n$, obeys a *Poisson distribution* with
$$q_x = P(X = x) = e^{-\lambda}\frac{\lambda^x}{x!}, \ x = 0, 1, \ldots, \ 0 < \lambda < \infty.$$

This is a power series distribution, with series function $g(\lambda) = e^\lambda$, and its u-fold convolution is also a Poisson distribution with
$$q_m(u) = P(S_u = m) = e^{-u\lambda}\frac{(u\lambda)^m}{m!}, \ m = 0, 1, \ldots.$$

Clearly, its mth coefficient function and rth binomial moment are given by
$$c_m(u) = \frac{u^m}{m!}, \ a_{(r)}(u) = \frac{u^r}{r!}\lambda^r.$$

Further, consider a random sample $Y_1, Y_2, \ldots, Y_n$ from the population with the zero truncated Poisson distribution
$$P(Y = x) = P(X = x | X > 0) = (e^\lambda - 1)^{-1}\frac{\lambda^x}{x!}, \ x = 1, 2, \ldots,$$

and let $T_n = \sum_{j=1}^n Y_j$. In the following corollary of Corollary 6.1, the probability function $p_m(n) = P(T_n = m)$, $m = n, n+1, \ldots$, and its rth binomial moment $b_{(r)}(n) = E[\binom{T_n}{r}]$, $r = 0, 1, \ldots$, are deduced.

Corollary 6.2 (a) *Under the random occupancy model with Poisson probabilities, the conditional probability $p_m(n)$, that m balls are distributed into the n urns, given that no urn remains empty, $m = n, n+1, \ldots$, is given by*
$$p_m(n) = (e^\lambda - 1)^{-n} n! S(m, n) \frac{\lambda^m}{m!}, \quad (6.13)$$

where $S(m, n) = [\Delta_u^n u^m]_{u=0}/n!$ is the Stirling number of the second kind.

(b) *The rth binomial moment $b_{(r)}(n)$, $r = 0, 1, \ldots$, of the probability function $p_m(n)$, $m = n, n+1, \ldots$, is given by*
$$b_{(r)}(n) = \frac{\lambda^r}{r!}\sum_{k=0}^n (n)_k S(r, k)(1 - e^{-\lambda})^{-k}. \quad (6.14)$$

6.3.2 Logarithmic Distribution

Assume that the random variable X_j, $j = 1, 2, \ldots, n$, obeys a *logarithmic distribution* with
$$q_x = P(X = x) = [-\log(1 - \theta)]^{-1}\frac{\theta^x}{x}, \ x = 1, 2, \ldots, \ 0 < \theta < 1.$$

Note that $P(X > 0) = 1$, and so $q_0 = 0$. Therefore, Theorem 6.1 cannot be applied. Further, since the logarithmic distribution is a power series distribution, with series function $g(\theta) = -\log(1-\theta)$, the distribution of the sum $S_n = \sum_{j=1}^{n} X_j$ is again a power series, with series function $g_n(\theta) = [-\log(1-\theta)]^n$. Hence, by virtue of the generating function of the signless Stirling numbers of the first kind (cf. Corollary 2.1, for $r = 0$),

$$\sum_{m=n}^{\infty} |s(m,n)| \frac{u^m}{m!} = \frac{[-\log(1-u)]^n}{n!}, \qquad (6.15)$$

the coefficient function is expressed as

$$c_m(n) = \frac{n!|s(m,n)|}{m!}.$$

Consequently,

$$p_m(n) = [-\log(1-\theta)]^{-n} n! |s(m,n)| \frac{\theta^m}{m!}, \quad m = n, n+1, \ldots.$$

The binomial moment generating function of S_n,

$$\sum_{r=0}^{\infty} b_{(r)}(n) u^r = \frac{[-\log(1-\theta(u+1))]^n}{[-\log(1-\theta)]^n},$$

on using the expression

$$-\log(1-\theta(u+1)) = -\log(1-\theta) - \log(1 - \theta(1-\theta)^{-1} u)$$

and, successively, Newton's binomial formula and (6.15), may be expanded as

$$\sum_{r=0}^{\infty} b_{(r)}(n) u^r = \sum_{k=0}^{n} \binom{n}{k} \frac{[-\log(1-\theta(1-\theta)^{-1} u)]^k}{[-\log(1-\theta)]^k}$$

$$= \sum_{r=0}^{\infty} \left\{ \frac{[\theta/(1-\theta)]^r}{r!} \sum_{k=0}^{n} \frac{(n)_k |s(r,k)|}{[-\log(1-\theta)]^k} \right\} u^r,$$

and so

$$b_{(r)}(n) = \frac{[\theta/(1-\theta)]^r}{r!} \sum_{k=0}^{n} \frac{(n)_k |s(r,k)|}{[-\log(1-\theta)]^k}.$$

Summarizing, the following theorem is shown.

Theorem 6.2 (a) *Under the random occupancy model with logarithmic probabilities, the probability $p_m(n)$, that m balls are distributed into the n urns, $m = n, n+1, \ldots$, is given by*

$$p_m(n) = [-\log(1-\theta)]^{-n} n! |s(m,n)| \frac{\theta^m}{m!}, \qquad (6.16)$$

where $|s(m,n)|$ is the signless Stirling number of the first kind.

(b) *The rth binomial moment $b_{(r)}(n)$, $r = 0, 1, \ldots$, of the probability function $p_m(n)$, $m = n, n+1, \ldots$, is given by*

$$b_{(r)}(n) = \frac{[\theta/(1-\theta)]^r}{r!} \sum_{k=0}^{n} \frac{(n)_k |s(r,k)|}{[-\log(1-\theta)]^k}. \tag{6.17}$$

6.3.3 Zero Truncated Binomial Distribution

Assume that the random variable X_j, $j = 1, 2, \ldots, n$, obeys a *binomial distribution* with

$$q_x = P(X = x) = \binom{s}{x} p^x q^{s-x}, \quad x = 0, 1, \ldots, s,$$

where $0 < p < 1$ and $q = 1 - p$. Its u-fold convolution obeys also a binomial distribution with

$$q_m(u) = P(S_u = m) = \binom{su}{m} p^m q^{su-m}, \quad m = 0, 1, \ldots, su.$$

Further, consider a random sample $Y_1, Y_2, \ldots, Y_n$ from the population with the zero truncated binomial distribution

$$P(Y = x) = P(X = x | X > 0)$$
$$= (1 - q^s)^{-1} \binom{s}{x} p^x q^{s-x}, \quad x = 1, 2, \ldots, s,$$

and let $T_n = \sum_{j=1}^{n} Y_j$. In the following corollary of Theorem 6.1, the probability function $p_m(n) = P(T_n = m)$, $m = n, n+1, \ldots$, and its binomial moments $b_{(r)}(n) = E[\binom{T_n}{r}]$, $r = 0, 1, \ldots$, are deduced. Note that the binomial distribution (with known number s of trials) is a power series distribution with series function $g(\theta) = (1 + \theta)^s$, $0 < \theta < \infty$, and parameter $\theta = p/q$ and so Corollary 6.1 can also be used.

Corollary 6.3 (a) *Under the random occupancy model with binomial probabilities, the conditional probability $p_m(n)$, that m balls are distributed into the n urns, given that no urn remains empty, $m = n, n+1, \ldots, sn$, is given by*

$$p_m(n) = (q^{-s} - 1)^{-n} n! C(m, n; s) \frac{(p/q)^m}{m!}, \tag{6.18}$$

where $C(m, n; s) = [\Delta_u^n (su)_m]_{u=0}/n!$ is the generalized factorial coefficient.

(b) *The rth binomial moment $b_{(r)}(n)$, $r = 0, 1, \ldots$, of the probability function $p_m(n)$, $m = n, n+1, \ldots$, is given by*

$$b_{(r)}(n) = \frac{p^r}{r!} \sum_{k=0}^{n} (n)_k C(r, k; s)(1 - q^s)^{-k}. \tag{6.19}$$

6.3.4 Zero Truncated Negative Binomial Distribution

Assume that the random variable X_j, $j = 1, 2, \ldots, n$, obeys a *negative binomial distribution* with

$$q_x = P(X = x) = \binom{s + x - 1}{x} p^s q^x, \quad x = 0, 1, \ldots,$$

where $0 < p < 1$ and $q = 1 - p$. This is a power series distribution with series function $g(q) = (1-q)^{-s}$ and its u-fold convolution is also a negative binomial distribution with

$$q_m(u) = P(S_u = m) = \binom{su + m - 1}{m} p^{su} q^m, \quad m = 0, 1, \ldots.$$

Clearly, its mth coefficient function and rth binomial moment are given by

$$c_m(u) = \frac{(-1)^m (-su)_m}{m!}, \quad a_{(r)}(u) = \frac{(-1)^r (-su)_r}{r!} \left(\frac{q}{p}\right)^r.$$

Further, consider a random sample $Y_1, Y_2, \ldots, Y_n$ from the population with the zero truncated negative binomial distribution

$$P(Y = x) = P(X = x | X > 0)$$
$$= (p^{-s} - 1)^{-1} \binom{s + x - 1}{x} q^x, \quad x = 1, 2, \ldots,$$

and let $T_n = \sum_{j=1}^n Y_j$. In the following corollary of Corollary 6.1, the probability function $p_m(n) = P(T_n = m)$, $m = n, n+1, \ldots$, and its binomial moments $b_{(r)}(n) = E\left[\binom{T_n}{r}\right]$, $r = 0, 1, \ldots$, are deduced.

Corollary 6.4 (a) *Under the random occupancy model with negative binomial probabilities, the conditional probability $p_m(n)$, that m balls are distributed into the n urns, given that no urn remains empty, $m = n, n+1, \ldots$, is given by*

$$p_m(n) = (p^{-s} - 1)^{-n} n! |C(m, n; -s)| \frac{q^m}{m!}, \tag{6.20}$$

where $|C(m, n; -s)| = (-1)^m [\Delta_u^n (-su)_m]_{u=0} / n!$ is the absolute generalized factorial coefficient.

(b) *The rth binomial moment $b_{(r)}(n)$, $r = 0, 1, \ldots$, of the probability function $p_m(n)$, $m = n, n+1, \ldots$, is given by*

$$b_{(r)}(n) = \frac{(q/p)^r}{r!} \sum_{k=0}^n (n)_k |C(r, k; -s)| (1 - p^s)^{-k}. \tag{6.21}$$

6.4 GENERAL TRUNCATED DISCRETE DISTRIBUTIONS

This section is devoted to the study of the convolution of general left or right truncated discrete distributions. This study is restricted to the particular case in which the random variables X_j, $j = 1, 2, \ldots, n$, are independently and identically distributed with common probability function

$$P(X = x) = q_x, \quad x = 0, 1, \ldots.$$

Consider, first, the left truncated distribution with probability function

$$P(Y = x) = P(X = x | X \geq r) = \frac{q_x}{R_r}, \quad x = r, r+1, \ldots,$$

where $R_r = \sum_{x=r}^{\infty} q_x$, $r = 0, 1, \ldots$. Further, consider a random sample $Y_1, Y_2, \ldots, Y_n$ from this distribution and let $T_n = \sum_{j=1}^{n} Y_j$. The probability function

$$p_m(n; r) = P(T_n = m) = P(S_n = m | X_1 \geq r, X_2 \geq r, \ldots, X_n \geq r),$$

for $m = rn, rn + 1, \ldots$, is recursively given in the following theorem.

Theorem 6.3 *Under the random occupancy model and assuming that $q_r > 0$, let $p_m(n; r)$ be the conditional probability that m balls are distributed into the n urns, given that no urn contains less than r balls, $m = rn, rn + 1, \ldots$. Then, for $r = 0, 1, \ldots,$*

$$p_m(n; r+1) = (R_r/R_{r+1})^n \left[\Delta_u^n (q_r/R_r)^{n-u} p_{m-rn+ru}(u; r) \right]_{u=0}, \quad (6.22)$$

where $R_r = \sum_{x=r}^{\infty} q_x$, $r = 0, 1, \ldots,$ and $p_m(n; 0) = q_m(n)$.

Recurrence relation (6.22) is readily deduced from the corresponding recurrence relation for the conditional probability $p(m, n; r)$ that, under the random occupancy model, all n urns are occupied, each by at least r balls, given that m balls are distributed into the n distinguishable urns. Specifically, under the random occupancy model, let L_r be the number of urns each containing at least r balls, $r = 1, 2, \ldots,$ and

$$p(m, n; r) = P(L_r = n | S_n = m)$$
$$= P(X_1 \geq r, X_2 \geq r, \ldots, X_n \geq r | S_n = m).$$

Then, from

$$P(T_n = m) = P(S_n = m | X_1 \geq r, X_2 \geq r, \ldots, X_n \geq r)$$
$$= \frac{P(X_1 \geq r, X_2 \geq r, \ldots, X_n \geq r | S_n = m) P(S_n = m)}{P(X_1 \geq r) P(X_2 \geq r) \cdots P(X_n \geq r)},$$

it follows that

$$p_m(n; r) = \frac{p(m, n; r) q_m(n)}{R_r^n}.$$

GENERAL TRUNCATED DISCRETE DISTRIBUTIONS 263

This relation, on using the recurrence relation for the conditional probability $p(m, n; r) = P(L_r = n | S_n = m)$, derived in the following lemma, implies recurrence relation (6.22).

Lemma 6.1 *Under the random occupancy model and assuming that $q_r > 0$, let $p(m, n; r)$ be the conditional probability that all n urns are occupied, each by at least r balls, given that m balls are distributed into the n distinguishable urns, $r = 1, 2, \ldots$. Then, for $r = 0, 1, \ldots$,*

$$p(m, n; r+1) = \frac{\left[\Delta_u^n q_r^{n-u} q_{m-rn+ru}(u) p(m - rn + ru, u; r)\right]_{u=0}}{q_m(n)}, \quad (6.23)$$

with $p(m, n; 0) = 1$.

Proof Under the random occupancy model, let A_j and B_j be the events that the jth urn contains at least r and exactly r balls, respectively, $j = 1, 2, \ldots, n$. Then

$$p(m, n; r+1) = P(B_1' B_2' \cdots B_n' A_1 A_2 \cdots A_n | \{S_n = m\})$$

is the conditional probability that all n urns are occupied, each by at least $r + 1$ balls, given that m balls are distributed into the n distinguishable urns. Further, let $\{j_1, j_2, \ldots, j_i\}$ be any i indices from the set $\{1, 2, \ldots, n\}$ and $\{j_{i+1}, j_{i+2}, \ldots, j_n\} = \{1, 2, \ldots, n\} - \{j_1, j_2, \ldots, j_i\}$. Also, let $U_{n-i} = X_{j_{i+1}} + X_{j_{i+2}} + \cdots + X_{j_n}$. Then

$$P(B_{j_1} B_{j_2} \cdots B_{j_i} A_1 A_2 \cdots A_n | \{S_n = m\})$$
$$= \frac{P(B_{j_1} B_{j_2} \cdots B_{j_i} A_{j_{i+1}} A_{j_{i+2}} \cdots A_{j_n} \{S_n = m\})}{P(S_n = m)}$$
$$= \frac{P(B_{j_1} B_{j_2} \cdots B_{j_i}) P(U_{n-i} = m - ri)}{P(S_n = m)}$$
$$\times P(A_{j_{i+1}} A_{j_{i+2}} \cdots A_{j_n} | \{U_{n-i} = m - ri\})$$
$$= \frac{q_r^i q_{m-ri}(n-i) p(m - ri, n-i; r)}{q_m(n)}.$$

Therefore, the events $B_1, B_2, \ldots, B_n$ are exchangeable and so applying Corollary 1.5, for $k = 0$, we get for the probability $p(m, n; r)$ expression (6.23), and the proof of the lemma is completed.

In the particular case of a random sample $Y_1, Y_2, \ldots, Y_n$ from the population with the left truncated power series probability function

$$P(Y = x) = P(X = x | X \geq r) = [g_r(\theta)]^{-1} c_x \theta^x, \quad x = r, r+1, \ldots,$$

where $g_r(\theta) = \sum_{x=r}^{\infty} c_x \theta^x$, the probability function of the sum $T_n = \sum_{j=1}^{n} Y_j$,

$$p_m(n; r) = P(T_n = m), \quad m = rn, rn+1, \ldots,$$

is given in the following corollary of Theorem 6.3.

Corollary 6.5 *Under the random occupancy model with power series probabilities*
$$q_x = P(X = x) = [g(\theta)]^{-1} c_x \theta^x, \ x = 0, 1, \ldots,$$
the conditional probability $p_m(n;r)$ that m balls are distributed into the n urns, given that no urn contains less than r balls, $m = rn, rn+1, \ldots$, is given by
$$p_m(n;r) = [g_r(\theta)]^{-n} c_m(n;r) \theta^m, \tag{6.24}$$
with $g_r(\theta) = \sum_{x=r}^{\infty} c_x \theta^x$ and
$$c_m(n; r+1) = \left[\Delta_u^n c_r^{n-u} c_{m-rn+ru}(u;r) \right]_{u=0}, \ r = 0, 1, \ldots, \tag{6.25}$$
where $c_m(n;0) = c_m(n) = [D_\theta^m (g(\theta))^n]_{\theta=0}/m!$.

Example 6.3 *Convolution of left truncated logarithmic distributions.* Consider a random sample $Y_1, Y_2, \ldots, Y_n$ from the population with the left truncated logarithmic probability function
$$P(Y = y) = \left(-\log(1-\theta) - \sum_{j=1}^{r-1} \frac{\theta^j}{j} \right)^{-1} \cdot \frac{\theta^y}{y}, \ y = r, r+1, \ldots.$$

This is a (left truncated) power series distribution and so the distribution of the sum $T_n = \sum_{j=1}^n Y_j$ is also a power series with series function
$$g_{n;r}(\theta) = \left(-\log(1-\theta) - \sum_{j=1}^{r-1} \frac{\theta^j}{j} \right)^n.$$

The coefficient function $c_m(n;r)$, in addition to the recurrence relation (6.25) that it satisfies, with $c_r = 1/r$, can be further explored. Specifically, by virtue of the generating function of the r-associated signless Stirling numbers of the first kind (cf. Section 3.2),
$$\sum_{m=rn}^{\infty} |s_r(m,n)| \frac{u^m}{m!} = \frac{1}{n!} \left(-\log(1-u) - \sum_{j=1}^{r-1} \frac{u^j}{j} \right)^n,$$
it follows that $c_m(n;r) = n!|s_r(m,n)|/m!$ and so
$$P(T_n = m) = \left(-\log(1-\theta) - \sum_{j=1}^{r-1} \frac{\theta^j}{j} \right)^{-n} \cdot \frac{n!|s_r(m,n)|\theta^m}{m!},$$
for $m = rn, nr+1, \ldots$.

Example 6.4 *Convolution of left truncated Poisson distributions.* Consider a random sample $Y_1, Y_2, \ldots, Y_n$ from the population with the left truncated Poisson probability function
$$P(Y = y) = \left(e^\lambda - \sum_{j=0}^{r-1} \frac{\lambda^j}{j!} \right)^{-1} \cdot \frac{\lambda^y}{y!}, \ y = r, r+1, \ldots.$$

GENERAL TRUNCATED DISCRETE DISTRIBUTIONS

The distribution of the sum $T_n = \sum_{j=1}^{n} Y_j$ is a power series with series function

$$g_{n;r}(\lambda) = \left(e^{\lambda} - \sum_{j=0}^{r-1} \frac{\lambda^j}{j!}\right)^n.$$

The coefficient function $c_m(n;r)$, in addition to the recurrence relation (6.25) that it satisfies, with $c_r = 1/r!$, by virtue of the generating function of the r-associated Stirling numbers of the second kind (cf. Section 3.2),

$$\sum_{m=rn}^{\infty} S_r(m,n) \frac{u^m}{m!} = \frac{1}{n!}\left(e^u - \sum_{j=0}^{r-1} \frac{u^j}{j!}\right)^n,$$

is expressed as $c_m(n;r) = n! S_r(m,n)/m!$. Consequently,

$$P(T_n = m) = \left(e^{\lambda} - \sum_{j=0}^{r-1} \frac{\lambda^j}{j!}\right)^{-n} \cdot \frac{n! S_r(m,n) \lambda^m}{m!},$$

for $m = rn, rn+1, \ldots$.

Let us now turn to the case of the right truncated distribution with probability function

$$P(Y = x) = P(X = x | X \leq r) = \frac{q_x}{Q_r}, \quad x = 0, 1, \ldots, r,$$

where $Q_r = \sum_{x=0}^{r} q_x$, $r = 0, 1, \ldots$. Further, consider a random sample $Y_1, Y_2, \ldots, Y_n$ from this distribution and let $W_n = \sum_{j=1}^{n} Y_j$. The probability function

$$q_m(n;r) = P(W_n = m) = P(S_n = m | X_1 \leq r, X_2 \leq r, \ldots, X_n \leq r),$$

for $m = 0, 1, \ldots, rn$, is recursively given in the following theorem.

Theorem 6.4 *Under the random occupancy model and assuming that $q_r > 0$, let $q_m(n;r)$ be the conditional probability that m balls are distributed into the n urns, given that no urn contains more than r balls, $m = 0, 1, \ldots, rn$. Then, for $r = 1, 2, \ldots$,*

$$q_m(n; r-1) = (Q_r/Q_{r-1})^n \left[\Delta_u^n (q_r/Q_r)^{n-u} q_{m-rn+ru}(u;r)\right]_{u=0}, \quad (6.26)$$

where $Q_r = \sum_{x=0}^{r} q_x$, $r = 0, 1, \ldots$, and $q_n(n;0) = q_m(n)$.

In analogy to the derivation of recurrence relation (6.22), recurrence relation (6.26) may be deduced from the corresponding recurrence relation for the conditional probability $q(m,n;r)$ that, under the random occupancy model, all n urns are occupied, each by at most r balls, given that m balls are distributed into the n distinguishable urns. Specifically, from the relation

$$q_m(n;r) = \frac{q(m,n;r) q_m(n)}{Q_r^n}$$

and the recurrence relation for the conditional probability $q(m, n; r)$, derived in the following lemma, (6.26) is readily deduced. The proof of the lemma, similar to that of Lemma 6.1, is omitted.

Lemma 6.2 *Under the random occupancy model and assuming that $q_r > 0$, let $q(m, n; r)$ be the conditional probability that all n urns are occupied, each by at most r balls, given that m balls are distributed into the n distinguishable urns, $r = 1, 2, \ldots$. Then, for $r = 1, 2, \ldots$,*

$$q(m, n; r-1) = \frac{\left[\Delta_u^n q_r^{n-u} q_{m-rn+ru}(u) q(m-rn+ru, u; r)\right]_{u=0}}{q_m(n)}, \quad (6.27)$$

with $q(0, n; 0) = 1$ and $q(m, n; 0) = 0$ for $m > 0$.

In the particular case of a random sample $Y_1, Y_2, \ldots, Y_n$ from the population with the right truncated power series probability function

$$P(Y = x) = P(X = x | X \leq r) = [h_r(\theta)]^{-1} c_x \theta^x, \quad x = 0, 1, \ldots, r,$$

where $h_r(\theta) = \sum_{x=0}^{r} c_x \theta^x$, the probability function of the sum $W_n = \sum_{j=1}^{n} Y_j$,

$$q_m(n; r) = P(W_n = m), \quad m = 0, 1, \ldots, rn,$$

is given in the following corollary of Theorem 6.4.

Corollary 6.6 *Under the random occupancy model with power series probabilities*

$$q_x = P(X = x) = [g(\theta)]^{-1} c_x \theta^x, \quad x = 0, 1, \ldots,$$

the conditional probability $q_m(n; r)$, that m balls are distributed into the n urns given that no urn contains more than r balls, $m = 0, 1, \ldots, rn$, is given by

$$q_m(n; r) = [h_r(\theta)]^{-n} c_m(n; r) \theta^m, \quad (6.28)$$

with $h_r(\theta) = \sum_{x=0}^{r} c_x \theta^x$ and

$$c_m(n; r-1) = \left[\Delta_u^n c_r^{n-u} c_{m-rn+ru}(u; r)\right]_{u=0}, \quad r = 1, 2, \ldots, \quad (6.29)$$

where $c_m(n; 0) = c_m(n) = [D_\theta^m (g(\theta))^n]_{\theta=0}/m!$.

6.5 STATISTICAL APPLICATIONS

6.5.1 Zero Truncated Power Series Distribution

As is noted in the introduction, for a random sample from a population with a truncated power series distribution, the derivation of uniformly minimum variance unbiased estimators, for estimable functions, may be based on the distribution of the sum of the observations. More precisely, consider a random sample $Y_1, Y_2, \ldots, Y_n$ from the population with the zero truncated power series probability function

$$P(Y = y) = [g(\theta) - 1]^{-1} c_y \theta^y, \quad y = 1, 2, \ldots,$$

and let $T_n = \sum_{j=1}^n Y_j$. Since

$$\prod_{j=1}^n P(Y_j = y_j) = g(t_n; \theta) h(y_1, y_2, \ldots, y_n),$$

where the function

$$g(t_n; \theta) = [g(\theta) - 1]^{-n} \theta^{t_n}$$

depends on $y_1, y_2, \ldots, y_n$ only through $t_n = \sum_{j=1}^n y_j$ and the function

$$h(y_1, y_2, \ldots, y_n) = c_{y_1} c_{y_2} \cdots c_{y_n}$$

does not depend on θ, T_n is a sufficient statistic for θ. Clearly, upon using the probability function (6.11) of T_n, it follows that if $E[\phi(T_n)] = 0$ for every $\theta \in \Theta$, then $\phi(T_n) \equiv 0$. Consequently, the statistic T_n is also complete for θ. Consider now a function $f(\theta)$ of the parameter θ and assume that an unbiased estimator $U_n = U_n(Y_1, Y_2, \ldots, Y_n)$ of $f(\theta)$, with finite variance, exists. Then, by the classical theorems of Rao-Blackwell and Lehmann-Scheffé, $\hat{f}(T_n) = E(U_n|T_n)$ is a uniformly minimum variance unbiased (UMVU) estimator of $f(\theta)$. The derivation of a UMVU estimator of the parameter θ or more generally of its rth power θ^r, say $\hat{\theta}_r(T_n)$, $r = 1, 2, \ldots$, may be carried out by noting that the unbiasedness condition $E[\hat{\theta}_r(T_n)] = \theta^r$, $r = 1, 2, \ldots$, for every $\theta \in \Theta$, on using the probability function (6.11) of T_n, may be expressed as

$$\sum_{m=n}^\infty \hat{\theta}_r(m) \frac{[\Delta_u^n c_m(u)]_{u=0} \theta^m}{[g(\theta) - 1]^n} = \sum_{m=n}^\infty \frac{[\Delta_u^n c_m(u)]_{u=0} \theta^{m+r}}{[g(\theta) - 1]^n},$$

where

$$c_m(n) = [D_\theta^m (g(\theta))^n]_{\theta=0} / m!.$$

Therefore,

$$\hat{\theta}_r(m) = \frac{[\Delta_u^n c_{m-r}(u)]_{u=0}}{[\Delta_u^n c_m(u)]_{u=0}}, \quad m \geq n + r,$$

and $\hat{\theta}_r(m) = 0$, $m < n + r$. Clearly, when the support of the population distribution is finite there does not exist a UMVU estimator of θ^r. Note that

$$\hat{\theta}_r(m) = \prod_{j=1}^r \hat{\theta}_1(m - j + 1).$$

Further, the variance of the UMVU estimator $\hat{\theta}_1(T_n)$ of the parameter θ is

$$v \equiv V[\hat{\theta}_1(T_n)] = E\{[\hat{\theta}_1(T_n)]^2\} - \theta^2.$$

Thus, the UMVU estimator $\hat{v}(T_n)$ of the variance of $\hat{\theta}_1(T_n)$ is given by

$$\hat{v}(T_n) = [\hat{\theta}_1(T_n)]^2 - \hat{\theta}_2(T_n) = \hat{\theta}_1(T_n)\{\hat{\theta}_1(T_n) - \hat{\theta}_1(T_n - 1)\}.$$

Summarizing, the following theorem is shown.

Theorem 6.5 *Let $Y_1, Y_2, \ldots, Y_n$ be a random sample from the population with the zero truncated power series distribution*

$$P(Y = y) = [g(\theta) - 1]^{-1} c_y \theta^y, \quad y = 1, 2, \ldots, \tag{6.30}$$

and set $T_n = \sum_{j=1}^n Y_j$. The UMVU estimator of the parameter θ is given by $\hat{\theta}_1(T_n) \equiv \hat{\theta}(T_n)$ and the UMVU estimator of the variance of $\hat{\theta}(T_n)$ is given by $\hat{v}(T_n) = \hat{\theta}(T_n)\{\hat{\theta}(T_n) - \hat{\theta}(T_n - 1)\}$, where

$$\hat{\theta}(m) = \frac{[\Delta_u^n c_{m-1}(u)]_{u=0}}{[\Delta_u^n c_m(u)]_{u=0}}, \quad m \geq n+1, \tag{6.31}$$

and $\hat{\theta}(m) = 0$, $m < n+1$, with $c_m(n) = [D_\theta^m(g(\theta))^n]_{\theta=0}/m!$.

Example 6.5 *UMVU estimation for a zero truncated Poisson distribution.* Consider a random sample $Y_1, Y_2, \ldots, Y_n$ from the population with the zero truncated Poisson distribution:

$$P(Y = y) = (e^\lambda - 1)^{-1} \frac{\lambda^y}{y!}, \quad y = 1, 2, \ldots, \quad 0 < \lambda < \infty.$$

The probability function of $T_n = \sum_{j=1}^n Y_j$ is given in Corollary 6.2 as

$$p_m(n) = (e^\lambda - 1)^{-n} n! S(m, n) \frac{\lambda^m}{m!}, \quad m = n, n+1, \ldots.$$

Note that $c_m(u) = u^m/m!$ and $[\Delta_u^n c_m(u)]_{u=0} = n! S(m, n)/m!$. Therefore, the UMVU estimator of the parameter λ is given by $\hat{\lambda}(T_n)$ and the UMVU estimator of the variance of $\hat{\lambda}(T_n)$ is given by $\hat{v}(T_n) = \hat{\lambda}(T_n)\{\hat{\lambda}(T_n) - \hat{\lambda}(T_n - 1)\}$, where

$$\hat{\lambda}(m) = \frac{mS(m-1, n)}{S(m, n)}, \quad m \geq n+1,$$

and $\hat{\lambda}(m) = 0$, $m < n + 1$.

Example 6.6 *UMVU estimation for a logarithmic distribution.* Consider a random sample $X_1, X_2, \ldots, X_n$ from the population with the logarithmic distribution:

$$q_x = P(X = x) = [-\log(1-\theta)]^{-1} \frac{\theta^x}{x}, \quad x = 1, 2, \ldots, \quad 0 < \theta < 1.$$

The probability function of the complete and sufficient statistic $S_n = \sum_{j=1}^n X_j$ is given in Theorem 6.2 as

$$p_m(n) = [-\log(1-\theta)]^{-n} n! |s(m, n)| \frac{\theta^n}{m!}, \quad m = n, n+1, \ldots.$$

The UMVU estimator $\hat{\theta}_r(S_n)$ of θ^r, $r = 1, 2, \ldots$, on using the unbiasedness condition $E[\hat{\theta}_r(S_n)] = \theta^r$, $r = 1, 2, \ldots$, for every $\theta \in (0, 1)$, together with the probability function of S_n, may be deduced as

$$\hat{\theta}_r(m) = \frac{(m)_r |s(m-r, n)|}{|s(m, n)|}, \quad m \geq n + r,$$

and $\hat{\theta}_r(m) = 0$, $m < n + r$. In particular, the UMVU estimator of the parameter θ is given by $\hat{\theta}_1(S_n) \equiv \hat{\theta}(S_n)$ and the UMVU estimator for the variance of $\hat{\theta}(S_n)$ is given by $\hat{v}(S_n) = \hat{\theta}(S_n)\{\hat{\theta}(S_n) - \hat{\theta}(S_n - 1)\}$, where

$$\hat{\theta}(m) = \frac{m|s(m-1, n)|}{|s(m, n)|}, \quad m \geq n + 1,$$

and $\hat{\theta}(m) = 0$, $m < n + 1$.

Example 6.7 *UMVU estimation for a zero truncated negative binomial distribution.* Consider a random sample $Y_1, Y_2, \ldots, Y_n$ from the population with the zero truncated negative binomial distribution:

$$P(Y = y) = (p^{-s} - 1)^{-1} \binom{s + y - 1}{y} q^y, \quad y = 1, 2, \ldots,$$

where $0 < p < 1$, $q = 1 - p$ and s is assumed to be a known real number. The probability function of the complete and sufficient statistic $T_n = \sum_{j=1}^{n} Y_j$ is given in Corollary 6.4 as

$$p_m(n) = (p^{-s} - 1)^{-n} n! |C(m, n; -s)| \frac{q^m}{m!}, \quad m = n, n + 1, \ldots.$$

Note that $c_m(u) = (-1)^m (-su)_m / m!$ and $[\Delta_u^n c_m(u)]_{u=0} = n!|C(m, n; -s)|/m!$. Therefore, the UMVU estimator of the parameter q is given by $\hat{q}(T_n)$ and the UMVU estimator of the variance of $\hat{q}(T_n)$ is given by $\hat{v}(T_n) = \hat{q}(T_n)\{\hat{q}(T_n) - \hat{q}(T_n - 1)\}$, where

$$\hat{q}(m) = \frac{m|C(m-1, n; -s)|}{|C(m, n; -s)|}, \quad m \geq n + 1,$$

and $\hat{q}(m) = 0$, $m < n + 1$.

It is worth noting that in the case of a random sample $Y_1, Y_2, \ldots, Y_n$ from a population with a zero truncated binomial distribution, the probability function of the complete and sufficient statistic $T_n = \sum_{j=1}^{n} Y_j$ is given in Corollary 6.3 as

$$p_m(n) = (q^{-s} - 1)^{-n} n! C(m, n; s) \frac{(p/q)^m}{m!}, \quad m = n, n + 1, \ldots, sn.$$

Since the support of this distribution is finite, there does not exist a UMVU estimator for the ratio $\theta = p/q$. Clearly, the estimator $\hat{\theta}(T_n)$ of $\theta = p/q$, where

$$\hat{\theta}(m) = \frac{mC(m-1, n; s)}{C(m, n; s)}, \quad n + 1 \leq m \leq sn - 1,$$

is not unbiased. Its relative bias is

$$E\left[\frac{\hat{\theta}(T_n)}{\theta}\right] - 1 = -\left[\frac{\theta^s}{(1+\theta)^s - 1}\right]^n.$$

Since $\theta^s/[(1+\theta)^s - 1] < 1$, it follows that the relative bias approaches zero very fast even for moderate values of the sample size n. Therefore, $\hat{\theta}(T_n)$ is an asymptotically UMVU estimator of $\theta = p/q$.

6.5.2 Left Truncated Power Series Distribution

Consider now the more general case of a random sample $Y_1, Y_2, \ldots, Y_n$ from a population with probability function the power series distribution truncated on the left at the point $y = r - 1$,

$$P(Y = y) = [g_r(\theta)]^{-1} c_y \theta^y, \quad y = r, r+1, \ldots,$$

where $g_r(\theta) = \sum_{x=r}^{\infty} c_x \theta^x$, $r = 0, 1, \ldots$. Clearly, if the truncation point is known, $T_n = \sum_{j=1}^{n} Y_j$ is a complete and sufficient statistic for the parameter θ, while if the truncation point is unknown the pair (Z_n, T_n), where $Z_n = \min\{Y_1, Y_2, \ldots, Y_n\}$ and $T_n = \sum_{j=1}^{n} Y_j$, is a complete and sufficient statistic for the pair of parameters (r, θ). The probability function $p_m(n; r) = P(T_n = m)$, $m = rn, rn + 1, \ldots$, is given in Corollary 6.5. Further, since

$$P(Z_n = z, T_n = m) = P(Z_n \geq z, T_n = m) - P(Z_n \geq z+1, T_n = m)$$
$$= P(Y_1 \geq z, Y_2 \geq z, \ldots, Y_n \geq z, T_n = m)$$
$$- P(Y_1 \geq z+1, Y_2 \geq z+1, Y_n \geq z+1, T_n = m)$$

and

$$P(T_n = m) = P(Y_1 \geq r, Y_2 \geq r, \ldots, Y_n \geq r, T_n = m),$$

the probability function $p_{z,m}(n; r) = P(Z_n = z, T_n = m)$, $m = zn, zn + 1, \ldots$, $z = r, r+1, \ldots$, may be expressed in terms of the probability function $p_m(n; r) = P(T_n = m)$, $m = rn, rn + 1, \ldots$, as

$$p_{z,m}(n; r) = p_m(n; z) - p_m(n; z+1). \tag{6.32}$$

The UMVU estimator of the parameter θ, when the truncation point is known, and the UMVU estimators of each of the parameters r and θ, when the truncation point is unknown, may be derived from the corresponding unbiasedness condition, upon using the probability function of the complete and sufficient statistic T_n and (Z_n, T_n), respectively. Thus, when the truncation point is known, the unbiasedness condition $E[\hat{\theta}(T_n; r)] = \theta$, upon using the probability function (6.24) of T_n, may be expressed as

$$\sum_{m=rn}^{\infty} \hat{\theta}(m; r) \frac{c_m(n; r)\theta^m}{[g_r(\theta)]^n} = \sum_{m=rn}^{\infty} \frac{c_m(n; r)\theta^{m+1}}{[g_r(\theta)]^n},$$

where
$$c_m(n; r+1) = \left[\Delta_u^n c_r^{n-u} c_{m-rn+ru}(u;r)\right]_{u=0}, \ r = 0, 1, \ldots,$$
with $c_m(n;0) = c_m(n) = [D_\theta^m (g_0(\theta))^n]_{\theta=0}/m!$. Therefore,
$$\hat{\theta}(m;r) = \frac{c_{m-1}(n;r)}{c_m(n;r)}, \ m \geq rn+1,$$
and $\hat{\theta}(m;r) = 0, m < rn+1$. The UMVU estimator $\hat{v}(T_n; r)$ of the variance
$$v \equiv V[\hat{\theta}(T_n; r)] = E\{[\hat{\theta}(T_n;r)]^2\} - \theta^2$$
of $\hat{\theta}(T_n; r)$ may similarly be derived as
$$\hat{v}(T_n; r) = \hat{\theta}(T_n; r)\{\hat{\theta}(T_n; r) - \hat{\theta}(T_n - 1; r)\}.$$

Further, when the truncation point is unknown, the unbiasedness condition $E[\hat{r}(Z_n, T_n)] = r$, on using (6.32) and (6.24), may be expressed as
$$\sum_{m=rn}^{\infty} \sum_{z=r}^{[m/n]} \hat{r}(z,m) \frac{[c_m(n;z) - c_m(n; z+1)]\theta^m}{[g_r(\theta)]^n}$$
$$= \sum_{m=rn}^{\infty} \sum_{z=r}^{[m/n]} r \frac{[c_m(n;z) - c_m(n; z+1)]\theta^m}{[g_r(\theta)]^n}.$$

Equating the coefficient of θ^m on both sides of this expression, it follows that
$$\sum_{z=r}^{[m/n]} \hat{r}(z,m)[c_m(n;z) - c_m(n;z+1)] = \sum_{z=r}^{[m/n]} r[c_m(n;z) - c_m(n;z+1)]$$
or equivalently that
$$\sum_{z=r}^{[m/n]} \hat{r}(z,m)[c_m(n;z) - c_m(n;z+1)] = rc_m(n;r).$$

Taking the difference of the last expression, with respect to r, we get the UMVU estimator r as
$$\hat{r}(z,m) = z - \frac{c_m(n; z+1)}{c_m(n; z) - c_m(n; z+1)}.$$
The UMVU estimator $\hat{u}(Z_n, T_n)$ of the variance
$$u \equiv V[\hat{r}(Z_n, T_n)] = E\{[\hat{r}(Z_n, T_n)]^2\} - r^2$$
of $\hat{r}(Z_n, T_n)$ may easily be derived as
$$\hat{u}(z,m) = \frac{c_m(n;z)c_m(n;z+1)}{c_m(n;z) - c_m(n;z+1)}.$$

Also, from the unbiasedness condition $E[\hat{\theta}(Z_n, T_n)] = \theta$, on using (6.32) and (6.24), we get the UMVU estimator of θ as

$$\hat{\theta}(z, m) = \frac{c_{m-1}(n; z) - c_{m-1}(n; z+1)}{c_m(n; z) - c_m(n; z+1)}.$$

Moreover, the UMVU estimator $\hat{v}(Z_n, T_n)$ of the variance

$$v \equiv V[\hat{\theta}(Z_n, T_n)] = E\{[\hat{\theta}(Z_n, T_n)]^2\} - \theta^2$$

of $\hat{\theta}(Z_n, T_n)$ may also be derived as

$$\hat{v}(Z_n, T_n) = \hat{\theta}(Z_n, T_n)\{\hat{\theta}(Z_n, T_n) - \hat{\theta}(Z_n, T_n - 1)\}.$$

Summarizing, the following theorem is shown.

Theorem 6.6 *Let $Y_1, Y_2, \ldots, Y_n$ be a random sample from the population with the left truncated power series distribution*

$$P(Y = y) = [g_r(\theta)]^{-1} c_y \theta^y, \quad y = r, r+1, \ldots, \tag{6.33}$$

where $g_r(\theta) = \sum_{x=r}^{\infty} c_x \theta^x$, $r = 0, 1, \ldots$. Also, let $Z_n = \min\{Y_1, Y_2, \ldots, Y_n\}$ and $T_n = \sum_{j=1}^{n} Y_j$.

(a) *If the truncation point is known, the UMVU estimator of the parameter θ is given by $\hat{\theta}(T_n; r)$ and the UMVU estimator of the variance of $\hat{\theta}(T_n; r)$ is given by $\hat{v}(T_n; r) = \hat{\theta}(T_n; r)\{\hat{\theta}(T_n; r) - \hat{\theta}(T_n - 1; r)\}$, where*

$$\hat{\theta}(m; r) = \frac{c_{m-1}(n; r)}{c_m(n; r)}, \quad m \geq rn + 1, \tag{6.34}$$

and $\hat{\theta}(m; r) = 0$, $m < rn + 1$, with

$$c_m(n; r+1) = \left[\Delta_u^n c_r^{n-u} c_{m-rn+ru}(u; r)\right]_{u=0}, \quad r = 0, 1, \ldots, \tag{6.35}$$

and $c_m(n; 0) = c_m(n) = [D_\theta^m (g_0(\theta))^n]_{\theta=0}/m!$.

(b) *If the truncation point is unknown, the UMVU estimator of the parameter r is given by $\hat{r}(Z_n, T_n)$ and the UMVU estimator of the variance of $\hat{r}(Z_n, T_n)$ is given by $\hat{u}(Z_n, T_n)$, where*

$$\hat{r}(z, m) = z - \frac{c_m(n; z+1)}{c_m(n; z) - c_m(n; z+1)} \tag{6.36}$$

and

$$\hat{u}(z, m) = \frac{c_m(n; z) c_m(n; z+1)}{c_m(n; z) - c_m(n; z+1)}. \tag{6.37}$$

Further, the UMVU estimator of the parameter θ is given by $\hat{\theta}(Z_n, T_n)$ and the UMVU estimator of the variance of $\hat{\theta}(Z_n, T_n)$ is given by

$$\hat{v}(Z_n, T_n) = \hat{\theta}(Z_n, T_n)\{\hat{\theta}(Z_n, T_n) - \hat{\theta}(Z_n, T_n - 1)\},$$

where

$$\hat{\theta}(z,m) = \frac{c_{m-1}(n;z) - c_{m-1}(n;z+1)}{c_m(n;z) - c_m(n;z+1)}. \tag{6.38}$$

Example 6.8 *UMVU estimation for a left truncated logarithmic distribution.* Consider a random sample $Y_1, Y_2, \ldots, Y_n$ from the population with the left truncated logarithmic probability function

$$P(Y = y) = \left(-\log(1-\theta) - \sum_{j=1}^{r-1} \frac{\theta^j}{j}\right)^{-1} \cdot \frac{\theta^y}{y}, \quad y = r, r+1, \ldots.$$

The probability function of $T_n = \sum_{j=1}^{n} Y_j$ is given by (cf. Example 6.3)

$$p_m(n;r) = \left(-\log(1-\theta) - \sum_{j=1}^{r-1} \frac{\theta^j}{j}\right)^{-n} \cdot \frac{n!|s_r(m,n)|\theta^m}{m!},$$

$m = rn, rn+1, \ldots$, where $|s_r(m,n)|$ is the r-associated signless Stirling number of the first kind. Note that $c_m(n;r) = n!|s_r(m,n)|/m!$. Therefore, by Theorem 6.6, (a) if the truncation point is known, the UMVU estimator of the parameter θ is given by $\hat{\theta}(T_n;r)$ and the UMVU estimator of the variance of $\hat{\theta}(T_n;r)$ is given by $\hat{v}(T_n;r) = \hat{\theta}(T_n;r)\{\hat{\theta}(T_n;r) - \hat{\theta}(T_n - 1;r)\}$, where

$$\hat{\theta}(m;r) = \frac{m|s_r(m-1,n)|}{|s_r(m,n)|}, \quad m \geq rn+1,$$

and $\hat{\theta}(m;r) = 0$, $m < rn+1$. (b) If the truncation point is unknown, the UMVU estimator of the parameter r is given by $\hat{r}(Z_n, T_n)$ and the UMVU estimator of the variance of $\hat{r}(Z_n, T_n)$ is given by $\hat{u}(Z_n, T_n)$, where

$$\hat{r}(z,m) = z - \frac{|s_{z+1}(m,n)|}{|s_z(m,n)| - |s_{z+1}(m,n)|}$$

and

$$\hat{u}(z,m) = \frac{|s_z(m,n)| \cdot |s_{z+1}(m,n)|}{|s_z(m,n)| - |s_{z+1}(m,n)|}.$$

Further, the UMVU estimator of the parameter θ is given by $\hat{\theta}(Z_n, T_n)$ and the UMVU estimator of the variance of $\hat{\theta}(Z_n, T_n)$ is given by

$$\hat{v}(Z_n, T_n) = \hat{\theta}(Z_n, T_n)\{\hat{\theta}(Z_n, T_n) - \hat{\theta}(Z_n, T_n - 1)\},$$

where

$$\hat{\theta}(z,m) = \frac{m[|s_z(m-1,n)| - |s_{z+1}(m-1,n)|]}{|s_z(m,n)| - |s_{z+1}(m,n)|}.$$

Example 6.9 *UMVU estimator for a left truncated Poisson distribution.* Consider a random sample $Y_1, Y_2, \ldots, Y_n$ from the population with the left truncated Poisson probability function

$$P(Y = y) = \left(e^\lambda - \sum_{j=0}^{r-1} \frac{\lambda^j}{j!}\right)^{-1} \cdot \frac{\lambda^y}{y!}, \quad y = r, r+1, \ldots.$$

The probability function of $T_n = \sum_{j=1}^n Y_j$ is given by (cf. Example 6.4)

$$p_m(n;r) = \left(e^\lambda - \sum_{j=0}^{r-1} \frac{\lambda^j}{j!}\right)^{-n} \cdot \frac{n! S_r(m,n) \lambda^m}{m!}, \quad m = rn, rn+1, \ldots,$$

where $S_r(m,n)$ is the r-associated Stirling number of the second kind. Note that $c_m(n;r) = n! S_r(m,n)/m!$. Therefore, from Theorem 6.6, we readily deduce the following.

(a) If the truncation point is known, the UMVU estimator of the parameter λ is given by $\hat{\lambda}(T_n; r)$ and the UMVU estimator of the variance of $\hat{\lambda}(T_n; r)$ is given by

$$\hat{v}(T_n; r) = \hat{\lambda}(T_n; r)\{\hat{\lambda}(T_n; r) - \hat{\lambda}(T_n - 1; r)\},$$

where

$$\hat{\lambda}(m; r) = \frac{m S_r(m-1, n)}{S_r(m, n)}, \quad m \geq rn + 1,$$

and $\hat{\lambda}(m; r) = 0$, $m < rn + 1$.

(b) If the truncation point is unknown, the UMVU estimator of the parameter r is given by $\hat{r}(Z_n, T_n)$ and the UMVU estimator of the variance of $\hat{r}(Z_n, T_n)$ is given by $\hat{u}(Z_n, T_n)$, where

$$\hat{r}(z, m) = z - \frac{S_{z+1}(m, n)}{S_z(m, n) - S_{z+1}(m, n)}$$

and

$$\hat{u}(z, m) = \frac{S_z(m, n) S_{z+1}(m, n)}{S_z(m, n) - S_{z+1}(m, n)}.$$

Further, the UMVU estimator of the parameter λ is given by $\hat{\lambda}(Z_n, T_n)$ and the UMVU estimator of the variance of $\hat{\lambda}(Z_n, T_n)$ is given by

$$\hat{v}(Z_n, T_n) = \hat{\lambda}(Z_n, T_n)\{\hat{\lambda}(Z_n, T_n) - \hat{\lambda}(Z_n, T_n - 1)\},$$

where

$$\hat{\lambda}(z, m) = \frac{m[S_z(m-1, n) - S_{z+1}(m-1, n)]}{S_z(m, n) - S_{z+1}(m, n)}.$$

6.6 REFERENCE NOTES

As noted in the introduction the sum of a random sample from a population with a zero truncated power series distribution is a complete sufficient statistic. Its distribution is used in the derivation of UMVU estimators of estimable functions of the parameter θ. In this connection, J. Roy and S. K. Mitra (1957) provided a general expression of the n-fold convolution of left truncated power series distributions. It merely expresses the fact that the distribution of this statistic is also a power series. In the particular case of the n-fold convolution of a zero truncated Poisson, the coefficient function was expressed essentially in terms of the Stirling numbers of the second kind. The UMVU estimator of θ^r was derived as a ratio of the coefficient function. R. F. Tate and R. L. Goen (1958) deduced the probability function of the n-fold convolution of a zero truncated Poisson distribution in the form (6.13) by inverting the corresponding characteristic function. Similarly, they derived the n-fold convolution of a left truncated Poisson distribution away from a point $c > 0$. A combinatorial derivation of the probability function of the n-fold convolution of a zero truncated Poisson distribution in terms of the classical occupancy model was provided by T. Cacoullos (1961). G. P. Patil (1963) investigated the problem of the existence of a UMVU estimator of the parameter θ of a truncated power series distribution, in the framework of a generalized power series distribution, and provided a general technique of obtaining it when it exists. Further results on this topic were derived in G. P. Patil and S. W. Joshi (1970).

The n-fold convolution of a logarithmic distribution, discussed in Section 6.3.2, was obtained by G. P. Patil and S. Bildikar (1966) by expanding the corresponding series function. Investing the characteristic function, H. J. Malik (1969) derived the probability function of the sum of zero truncated binomial random variables. J. C. Ahuja (1970b, 1971b) deduced the probability functions of the n-fold convolutions of zero truncated binomial and negative binomial distributions by considering them as generalized power series and expanding their series functions. T. Cacoullos and Ch. A. Charalambides (1975) provided combinatorial derivations of these distributions and, as in Corollaries 6.3 and 6.4, used the generalized factorial coefficients in their expressions. The general Theorem 6.1 was derived in Ch. A. Charalambides (1986a). The distribution of the sum of independent but not necessarily identically distributed zero truncated Poisson random variables, discussed in Example 6.1, was derived by A. J. Gross (1970), T. Cacoullos (1972), J. C. Ahuja (1972a) and J. Medhi (1973).

Sections 6.4 and 6.5 on convolutions of the left truncated power series distributions and their applications in deriving UMVU estimators of estimable parametric functions are based on the papers of J. Roy and S. K. Mitra (1957), C. J. Park (1973), Ch. A. Charalambides (1974b), S. W. Joshi and C. J. Park (1974), J. C. Ahuja and E. A. Enneking (1974), E. A. Enneking and J. C. Ahuja (1978) and A. Kumar (1980). Recurrence relations for the UMVU estimators of θ of left truncated discrete distributions, appearing in several exercises, were derived by J. C. Ahuja (1970a, 1972b), J. C. Ahuja and E. A. Enneking (1972b) and Ch. A. Charalambides (1974b).

Structural properties of the distributions derived in Theorem 6.2 and Corollary 6.2, under the name Stirling distributions of the first and second kind, were studied by

G. P. Patil and J. K. Wani (1965), J. C. Ahuja (1971a), J. C. Ahuja and E. Enneking (1972a), J. J. Gart (1974, 1975) and J. Singh (1975, 1978).

The n-fold convolutions of a generalized logarithmic distribution and the zero truncated generalized Poisson and generalized negative binomial distributions, given in Exercises 6.12 - 6.15, were first derived by R. C. Gupta (1974a,b, 1977) and K. G. Janardan (1978). The convolutions of left truncated modified power series distributions and their applications in deriving UMVU estimators of estimable parametric functions were discussed by Ch. A. Charalambides (1974c), R. C. Gupta (1974c), J. Medhi (1975), P. N. Jani (1977, 1978), S. R. Patel and P. N. Jani (1977), A. Kumar and P. C. Consul (1980), R. C. Gupta and J. Singh (1982) and V. G. Voinov (1986).

6.7 EXERCISES AND COMPLEMENTS

6.1 *Convolutions of zero truncated negative binomial distributions.* Let $Y_1, Y_2, \ldots, Y_n$ be independent random variables with

$$P(Y_j = x) = (p^{-s} - 1)^{-1} \binom{s_j + x - 1}{x} q^x, \quad x = 1, 2, \ldots,$$

where $q = 1 - p$ and $0 < p < 1$. Show that the probability function of the sum $T_n = \sum_{j=1}^{n} Y_j$ is given by

$$P(T_n = m) = \prod_{j=1}^{n}(p^{-s_j} - 1)^{-1} \frac{q^m}{m!} \sum_{r=0}^{n}(-1)^{n-r} S_{n,n-r},$$

for $m = n, n+1, \ldots$, with

$$S_{n,n-r} = \sum (s_{j_1} + s_{j_2} + \cdots + s_{j_r} + m - 1)_m,$$

where the summation is extended over all r-combinations $\{j_1, j_2, \ldots, j_r\}$ of the n indices $\{1, 2, \ldots, n\}$.

6.2 Under the random occupancy model with logarithmic probabilities

$$q_x = c\frac{\theta^x}{x}, \quad x = 1, 2, \ldots, \quad c = [-\log(1-\theta)]^{-1}, \quad 0 < \theta < 1,$$

let $p_m(n; \theta)$ be the probability that m balls are distributed into the n urns, for $m = n, n+1, \ldots$ (cf. Theorem 6.2).

(a) Show that

$$p_{m+1}(n; \theta) = \frac{\theta}{m+1}\{mp_m(n; \theta) + cnp_m(n-1; \theta)\}.$$

(b) Further, show that $p_m(n; \theta)$ has the following modal properties: If $\theta < 2/n$ it is unimodal with a mode at n and if $\theta = 2/n$ it is bimodal, the modes being at n and

$n+1$ with equal probabilities. Also, if $\theta > 2/n$ it first increases and then decreases with m, except perhaps when it assumes equal values at two consecutive m's at the time of its change from an increasing to a decreasing function.

6.3 *(Continuation).* Let $\hat{\theta}(m;n)$ be the UMVU estimate of θ for $S_n = m$, where $S_n = \sum_{j=1}^n X_j$ with $X_1, X_2, \ldots, X_n$ a random sample from the logarithmic distribution. (a) Show that

$$\hat{\theta}(m+1;n) = \frac{(m+1)\hat{\theta}(m;n-1)}{m\hat{\theta}(m;n-1) - (m-1)\hat{\theta}(m;n) + m},$$

for $m \geq n+1$, with $\hat{\theta}(m;n) = m/(m-1)$ for $n = 1$, and $\hat{\theta}(m;n) = 0$ for $m = n$.

(b) Also, show that $\hat{\theta}(m;n)$ for sample sizes $n = 1$ and $n = 2$ yields unreasonable estimates for all possible values m.

6.4 *(Continuation).* Let $\hat{\theta}(m;n,r)$ be the UMVU estimate of θ for $T_n = m$, where $T_n = \sum_{j=1}^n Y_j$, with $Y_1, Y_2, \ldots, Y_n$ a random sample from a left truncated logarithmic distribution away from $y = r - 1$. Show that

$$\hat{\theta}(m+1;n,r) = \frac{(m+1)\hat{\theta}(m-r+1;n-1,r)}{m\hat{\theta}(m-r+1;n-1,r) - (m-1)\hat{\theta}(m;n,r) + m},$$

for $m \geq rn + 1$, with $\hat{\theta}(m;n,r) = m/(m-1)$ for $n = 1$, and $\hat{\theta}(m;n,r) = 0$ for $m = rn$.

6.5 Under the random occupancy model with Poisson probabilities

$$q_x = e^{-\lambda}\lambda^x/x!, \; x = 0, 1, \ldots, \; 0 < \lambda < 1,$$

let $p_m(n;\lambda)$ be the conditional probability that m balls are distributed into the n urns, given that no urn remains empty, $m = n, n+1, \ldots$ (cf. Corollary 6.2).
(a) Show that

$$p_{m+1}(n;\lambda) = \frac{n\lambda}{m+1}\{p_m(n;\lambda) + cp_m(n-1;\lambda)\},$$

where $c = (e^\lambda - 1)^{-1}$.

(b) Further, show that $p_m(n;\lambda)$ has the following modal properties: If $\lambda < 2/n$ it is unimodal with a mode at n and if $\lambda = 2/n$ it is bimodal, the modes being at n and $n+1$ with equal probabilities. Also, if $\lambda > 2/n$ it first increases and then decreases with m, except perhaps when it assumes equal values at the two consecutive m's at the time of its change from an increasing to a decreasing function.

6.6 *(Continuation).* Let $\hat{\lambda}(m;n)$ be the UMVU estimate of λ for $T_n = m$, where $T_n = \sum_{j=1}^n Y_j$, with $Y_1, Y_2, \ldots, Y_n$ a random sample from the zero truncated Poisson distribution. Show that

$$\hat{\lambda}(m+1;n) = \frac{(m+1)\hat{\lambda}(m,n)}{n\hat{\lambda}(m;n-1) - n\hat{\lambda}(m;n) + m},$$

for $m \geq n+1$, with $\hat{\lambda}(m;n) = m$ for $n = 1$, and $\hat{\lambda}(m;n) = 0$ for $m = n$.

6.7 (*Continuation*). Let $\hat{\lambda}(m;n,r)$ be the UMVU estimate of λ for $T_n = m$, where $T_n = \sum_{j=1}^{n} Y_j$, with $Y_1, Y_2, \ldots, Y_n$ a random sample from a left truncated Poisson distribution away from $y = r - 1$. Show that

$$\hat{\lambda}(m+1;n,r) = \frac{(m+1)\hat{\lambda}(m-r+1;n-1,r)}{n\hat{\lambda}(m-r+1;n-1,r) - n\hat{\lambda}(m;n,r) + m},$$

for $m \geq rn + 1$, with $\hat{\lambda}(m;n,r) = m$ for $n = 1$, and $\hat{\lambda}(m;n,r) = 0$ for $m = rn$.

6.8 (a) Under the random occupancy model with negative binomial probabilities,

$$q_x = \binom{s+x-1}{x} p^s q^x, \quad x = 0, 1, \ldots, \quad q = 1-p, \ 0 < p < 1,$$

let $p_m(n;s,q)$ be the conditional probability that m balls are distributed into the n urns, given that no urn remains empty, $m = n, n+1, \ldots$ (cf. Corollary 6.4). Show that

$$p_{m+1}(n;s,q) = \frac{q}{m+1}\{(sn+m)p_m(n;s,q) + csnp_m(n-1;s,q)\},$$

where $c = (p^{-s} - 1)^{-1}$.

(b) Let $\hat{q}(m;n)$ be the UMVU estimate of q for $T_n = m$, where $T_n = \sum_{j=1}^{n} Y_j$ with $Y_1, Y_2, \ldots, Y_n$ a random sample from the zero truncated negative binomial distribution. Show that

$$\hat{q}(m+1;n) = \frac{(m+1)\hat{q}(m;n-1)}{(sn+m)\hat{q}(m;n-1) - (sn+m-1)\hat{q}(m;n) + m},$$

for $m \geq n+1$, with $\hat{q}(m;n) = m/(s+m-1)$ for $n = 1$, and $\hat{q}(m;n) = 0$ for $m = n$.

6.9 (*Continuation*). Let $\hat{q}(m;n,r)$ be the UMVU estimate of q for $T_n = m$, where $T_n = \sum_{j=1}^{n} Y_j$, with $Y_1, Y_2, \ldots, Y_n$ a random sample from the left truncated negative binomial distribution away from $y = r - 1$. Show that

$$\hat{q}(m+1;n,r) = \frac{(m+1)\hat{q}(m-r+1;n-1,r)}{(sn+m)\hat{q}(m-r+1;n-1,r) - (sn+m-1)\hat{q}(m;n,r) + m},$$

for $m \geq rn + 1$, with $\hat{q}(m;n,r) = m/(s+m-1)$ for $n = 1$, and $\hat{q}(m;n,r) = 0$ for $m = rn$.

6.10 Let $Y_1, Y_2, \ldots, Y_n$ be a random sample from the left truncated geometric distribution with

$$P(Y = y) = pq^{y-r}, \quad y = r, r+1, \ldots,$$

where $q = 1 - p$, $0 < p < 1$ and $r \geq 0$.

(a) Show that the probability function of $T_n = \sum_{j=1}^{n} X_j$ is given by

$$P(T_n = m) = \binom{n+m-rn-1}{n-1} p^n q^{m-rn}, \quad m = rn, rn+1, \ldots.$$

(b) Also, for r known, show that the UMVU estimate of q, for $T_n = m$, is

$$\hat{q}(m;n,r) = \frac{m-rn}{n+m-rn-1},$$

for $m \geq rn+1$ and $\hat{q}(m;n,r) = 0$ for $m \leq rn$.

6.11 *(Continuation).* Suppose that both parameters r and q are unknown and let $Z_n = \min\{Y_1, Y_2, \ldots, Y_n\}$. Show that the UMVU estimates of r and q, for $Z_n = z$ and $T_n = m$, are given by

$$\hat{r}(z,m;n) = z - \frac{\binom{m-nz-1}{n-1}}{\binom{n+m-nz-1}{n-1} - \binom{m-nz-1}{n-1}}$$

and

$$\hat{q}(z,m;n) = \frac{\binom{n+m-nz-2}{n-1} - \binom{m-nz-2}{n-1}}{\binom{n+m-nz-1}{n-1} - \binom{m-nz-1}{n-1}},$$

respectively.

6.12 Let $X_1, X_2, \ldots, X_n$ be a random sample from a generalized logarithmic distribution

$$q_x = \frac{1}{\beta x}\binom{\beta x}{x}\frac{\theta^x(1-\theta)^{\beta x - x}}{[-\log(1-\theta)]}, \quad x = 1, 2, \ldots,$$

with $0 < \theta < 1$, $0 < \theta\beta < 1$ and $\beta \geq 1$. Show that the probability function of $S_n = \sum_{j=1}^{n} X_j$ is given by

$$p_m(n;\theta,\beta) = \frac{\theta^m(1-\theta)^{\beta m - m}}{[-\log(1-\theta)]^n} \cdot \frac{n!}{m!}[|s(m,n;\beta m - m)|$$
$$- (\beta m - m)|s(m-1,n;\beta m - m + 1)|],$$

for $m = n, n+1, \ldots$, where $|s(m,n;r)|$ is the noncentral signless Stirling number of the first kind.

6.13 Under the random occupancy model with generalized Poisson probabilities

$$q_x = e^{-\lambda(1+\theta x)}\frac{\lambda^x(1+\theta x)^{x-1}}{x!}, \quad x = 0, 1, \ldots,$$

with $0 < \lambda < \infty$ and $|\lambda\theta| < 1$, let $p_m(n;\lambda,\theta)$ be the conditional probability that m balls are distributed into the n urns, given that no urn remains empty, $m = n, n+1, \ldots$. Show that

$$p_m(n;\lambda,\theta) = \frac{e^{-\lambda\theta m}\lambda^m}{(e^\lambda - 1)^n} \cdot \frac{n!}{m!}[S(m,n;\theta m) - \theta m S(m-1,n;\theta m)],$$

where $S(m,n;r)$ is the noncentral Stirling number of the second kind.

6.14 Under the random occupancy model with linear function Poisson probabilities

$$q_x = (1 - \lambda\theta)e^{-\lambda(1+\theta x)}\frac{\lambda^x(1+\theta x)^x}{x!}, \quad x = 0, 1, \ldots,$$

with $0 < \lambda < 1$ and $|\lambda\theta| < 1$, let $p_m(n; \lambda, \theta)$ be the conditional probability that m balls are distributed into the n urns, given that no urn remains empty, $m = n, n+1, \ldots$. Show that

$$p_m(n; \lambda, \theta) = \frac{(1-\lambda\theta)e^{-\lambda\theta m}\lambda^m}{(e^\lambda - 1 + \lambda\theta)^n}\frac{n!}{m!}S(m, n; \beta m),$$

where $S(m, n; r)$ is the noncentral Stirling number of the second kind.

6.15 Under the random occupancy model with generalized negative binomial probabilities

$$q_x = \frac{s}{s+\beta x}\binom{s+\beta x}{x}\theta^x(1-\theta)^{s+\beta x-x}, \quad x = 0, 1, \ldots,$$

with $s > 0$, $0 < \theta < 1$ and $|\theta\beta| < 1$, let $p_m(n; s, \beta, \theta)$ be the conditional probability that m balls are distributed into the n urns, given that no urn remains empty, $m = n, n+1, \ldots$. Show that

$$p_m(n; s, \beta, \theta) = \frac{\theta^m(1-\theta)^{\beta m - m}}{[(1-\theta)^{-s} - 1]^n} \cdot \frac{n!}{m!}[C(m, n; s, \beta m)$$
$$- \beta m C(m-1, n; s, \beta m - 1)],$$

where $C(m, n; s, r)$ is the noncentral generalized factorial coefficient.

7
COMPOUND AND MIXTURE DISTRIBUTIONS

7.1 INTRODUCTION

Consider a population of a random number N of elements (individuals) and assume that the jth element, independently of the others, generates (produces) a random number X_j of elements, $j = 1, 2, \ldots$. Then $S_N = \sum_{j=1}^{N} X_j$ is the number of elements of the generated population. It is reasonable to assume that the random variables X_j, $j = 1, 2, \ldots$, are identically distributed. According to Theorem 1.12, the probability generating function $P_{S_N}(t)$ of the sum S_N is given by the composite function $P_{S_N}(t) = P_N(P_X(t))$, where $P_N(u)$ is the probability generating function of N and $P_X(t)$ is the common probability generating function of the random variables X_j, $j = 1, 2, \ldots$. The distribution of S_N is called a *compound distribution*. In this respect, the distribution of N is the *compounded distribution*, while that of X is the *compounding distribution*. The term compound distribution is used by W. Feller (1968) in his classic book on discrete probability and subsequently by several other authors. These distributions are also known as *generalized distributions*, a term used by W. Feller (1943), J. Gurland (1957) and others. The term *stopped sum distribution* is also used.

Another important class of distributions may be constructed as follows. Consider an initial population consisting of a countable number of subpopulations (clusters, categories). Assume that the nth subpopulation generates (produces) a random number S_n of elements, with $q_m(n) = P(S_n = m)$, $m = 0, 1, \ldots$, for $n = 0, 1, \ldots$. Also, let S be the number of elements of the population generated by the initial population,

---

Combinatorial Methods in Discrete Distributions. By Charalambos A. Charalambides
ISBN 0-471-68027-3 Copyright © 2005 John Wiley & Sons, Inc.

with $p_m = P(S = m)$, $m = 0, 1, \ldots$. If c_n is the relative contribution of the nth subpopulation to the generated population, $n = 0, 1, \ldots$, then $p_m = \sum_{n=0}^{\infty} c_n q_m(n)$, $m = 0, 1, \ldots$. The distribution with probability function p_m, $m = 0, 1, \ldots$, is called a *mixture distribution*. In this framework, the distribution with probability function $q_n(m)$, $m = 0, 1, \ldots$, is the *mixed distribution*, while that with probability function c_n, $n = 0, 1, \ldots$, is the *mixing distribution*. A mixture distribution may be interpreted as a compound distribution in the particular case the probability generating function of the mixed distribution is of the form $[G(t)]^n$. Indeed, if $F(u)$ is the probability generating function of the mixing distribution, then the probability generating function $H(t)$ of the mixture distribution is the composite function $H(t) = F(G(t))$.

The discrete stochastic models of compounding and countable mixing are treated in this chapter in terms of urn occupancy models.

7.2 COMPOUND DISCRETE DISTRIBUTIONS

Consider a supply of balls randomly distributed (simultaneously or sequentially) into distinguishable urns. Assume that the number of urns is a random variable with probability function
$$c_n = P(N = n), \; n = 0, 1, \ldots,$$
and let X_j be the number of balls distributed into the jth urn, $j = 1, 2, \ldots$. In the random occupancy model, introduced in Chapter 4, the random variables X_j, $j = 1, 2, \ldots$, are assumed to be independently and identically distributed with a known probability function
$$P(X = x) = q_x, \; x = 0, 1, \ldots.$$
Let $S_N = \sum_{j=1}^{N} X_j$, be the total number of balls distributed into the urns, with
$$P(S_N = m) = p_m, \; m = 0, 1, \ldots,$$
and
$$b_{(r)} = E\left[\binom{S_N}{r}\right] = \sum_{m=r}^{\infty} \binom{m}{r} p_m, \; r = 0, 1, \ldots.$$
Also, consider the probability function of the n-fold convolution of q_x, $x = 0, 1, \ldots$,
$$P(S_n = m) = q_m(n), \; m = 0, 1, \ldots,$$
and its rth binomial moment
$$a_{(r)}(n) = E\left[\binom{S_n}{r}\right] = \sum_{m=r}^{\infty} \binom{m}{r} q_m(n), \; r = 0, 1, \ldots.$$

The probability function of S_N, on using the total probability theorem and the independence of S_n from N, may be expressed as
$$p_m = \sum_{n=0}^{\infty} c_n q_m(n), \; m = 0, 1, \ldots. \tag{7.1}$$

Also,

$$b_{(r)} = \sum_{n=0}^{\infty} c_n a_{(r)}(n), \quad r = 0, 1, \ldots. \tag{7.2}$$

If $q_0 > 0$, let $Y_1, Y_2, \ldots, Y_n$ be independent and identically distributed random variables with a zero truncated discrete distribution

$$P(Y = x) = P(X = x | X > 0) = (1 - q_0)^{-1} q_x, \quad x = 1, 2, \ldots.$$

Also, let $T_n = \sum_{j=1}^{n} Y_j$ and

$$p_m(n) = P(T_n = m) = P(S_n = m | X_1 > 0, X_2 > 0, \ldots, X_n > 0),$$

for $m = n, n+1, \ldots$. Further, consider the power series distribution

$$c_n(q_0) = [f(q_0)]^{-1} c_n q_0^n, \quad n = 0, 1, \ldots, \tag{7.3}$$

with series function

$$f(q_0) = \sum_{n=0}^{\infty} c_n q_0^n,$$

and let

$$\beta_{(k)}(q_0) = [f(q_0)]^{-1} \sum_{n=k}^{\infty} \binom{n}{k} c_n q_0^n, \quad k = 0, 1, \ldots, \tag{7.4}$$

be its kth binomial moment. Also, consider the kth binomial moment of N,

$$\beta_{(k)} = E\left[\binom{N}{k}\right] = \sum_{n=k}^{\infty} \binom{n}{k} c_n, \quad k = 0, 1, \ldots. \tag{7.5}$$

The probability function p_m, $m = 0, 1, \ldots$, of the compound distribution is derived, in the next theorem, in terms of $\beta_{(k)}(q_0)$, $k = 0, 1, \ldots, m$, and the kth difference of $q_m(u)$, $k = 0, 1, \ldots, m$, as a sum of a finite number of terms. Also, its rth binomial moment $b_{(r)}$, $r = 0, 1, \ldots$, is expressed in terms of $\beta_{(k)}$, $k = 0, 1, \ldots, r$, and the kth difference of $a_{(r)}(u)$, $k = 0, 1, \ldots, r$.

Theorem 7.1 (a) *Under the random occupancy model with a random number of urns and assuming that $q_0 > 0$, the probability p_m, $m = 0, 1, \ldots$, that m balls are distributed into the urns is given by*

$$p_m = f(q_0) \sum_{k=0}^{m} \beta_{(k)}(q_0) \left[\Delta_u^k q_0^{-u} q_m(u)\right]_{u=0}. \tag{7.6}$$

If $q_0 = 0$, then

$$p_m = \sum_{n=0}^{m} c_n q_m(n). \tag{7.7}$$

(b) *The rth binomial moment $b_{(r)}$, $r = 0, 1, \ldots$, of the probability function p_m, $m = 0, 1, \ldots$, is given by*

$$b_{(r)} = \sum_{k=0}^{r} \beta_{(k)} \left[\Delta_u^k a_{(r)}(u)\right]_{u=0}. \tag{7.8}$$

Proof (a) The probability function $q_m(n)$, $m = 0, 1, \ldots$ ($n = 0, 1, \ldots$), may be expressed as

$$q_m(n) = \left[(1 + \Delta_u)^n q_0^{n-u} q_m(u)\right]_{u=0} = \sum_{k=0}^{n} \binom{n}{k} \left[\Delta_u^k q_0^{n-u} q_m(u)\right]_{u=0}.$$

Note that $p_m(k) = (1 - q_0)^{-k} [\Delta_u^k q_0^{k-u} q_m(u)]_{u=0}$, $m = k, k+1, \ldots$ ($k = 0, 1, \ldots$), is the conditional probability that m balls are distributed into k urns, given that no urn remains empty (cf. Theorem 6.1). Since $p_m(k) = 0$ for $k > m$, it follows that

$$q_m(n) = \sum_{k=0}^{m} \binom{n}{k} \left[\Delta_u^k q_0^{n-u} q_m(u)\right]_{u=0}.$$

Expression (7.1), upon introducing the last expression of $q_m(n)$, is transformed to

$$p_m = \sum_{n=0}^{\infty} c_n \sum_{k=0}^{m} \binom{n}{k} \left[\Delta_u^k q_0^{n-u} q_m(u)\right]_{u=0}$$

$$= \sum_{k=0}^{m} \left\{\sum_{n=k}^{\infty} \binom{n}{k} c_n q_0^n\right\} \left[\Delta_u^k q_0^{-u} q_m(u)\right]_{u=0},$$

which, by virtue of (7.4), implies (7.6). If $q_0 = 0$, then $q_m(n) = 0$ for $n > m$ and so the series in (7.1) reduces to (7.7).

(b) The rth binomial moment $a_{(r)}(n)$, $r = 0, 1, \ldots$ ($n = 0, 1, \ldots$), may be expressed as

$$a_{(r)}(n) = \left[(1 + \Delta_u)^n a_{(r)}(u)\right]_{u=0} = \sum_{k=0}^{n} \binom{n}{k} \left[\Delta_u^k a_{(r)}(u)\right]_{u=0}.$$

Since $\Delta_u^k a_{(r)}(u) = 0$ for $k > r$, it reduces to

$$a_{(r)}(n) = \sum_{k=0}^{r} \binom{n}{k} \left[\Delta_u^k a_{(r)}(u)\right]_{u=0}.$$

Introducing this expression into (7.2), we find

$$b_{(r)} = \sum_{n=r}^{\infty} c_n \sum_{k=0}^{r} \binom{n}{k} \left[\Delta_u^k a_{(r)}(u)\right]_{u=0} = \sum_{k=0}^{r} \left\{\sum_{n=k}^{\infty} \binom{n}{k} c_n\right\} \left[\Delta_u^k a_{(r)}(u)\right]_{u=0}.$$

The last expression, by virtue of (7.5), implies (7.8).

Example 7.1 *Aggregate insurance claims.* In a collective risk model the number N of insurance claims in a unit time period (usually an economic year) is a random variable with
$$c_n = P(N = n), \; n = 0, 1, \ldots.$$
Let X_j be the amount of the jth claim, $j = 1, 2, \ldots$. It is assumed that the random variables X_j, $j = 1, 2, \ldots$, are independently and identically distributed with a common probability function
$$q_x = P(X = x), \; x = 0, 1, \ldots.$$
Then, according to Theorem 7.1, the probability function $p_m = P(S_N = m)$, $m = 0, 1, \ldots$, of the aggregate insurance claims $S_N = \sum_{j=1}^{N} X_j$ if $q_0 > 0$ is given by (7.6), while if $q_0 = 0$ it is given by (7.7). Its rth binomial moment $b_{(r)} = E\left[\binom{S_N}{r}\right]$, $r = 0, 1, \ldots$, is given by (7.8).

Example 7.2 *Branching processes.* Consider a population of individuals (organisms, particles), which are able to generate (give rise, produce) individuals of like kind. Assume that the probability of any individual to generate x new individuals is given by
$$q_x = P(X = x), \; x = 0, 1, \ldots.$$
An individual (organism, particle) forms the initial (or zero) generation. The direct descendants of the ith generation form the $(i+1)$st generation. In each generation, the individuals generate independently of each other. The distribution of the size of a given generation is of interest.

Let N be the number of individuals of the ith generation, with
$$c_n = P(N = n), \; n = 0, 1, \ldots.$$
If X_j is the number of individuals generated by the jth individual, $j = 1, 2, \ldots$, then $S_N = \sum_{j=1}^{N} X_j$ is the size of the $(i+1)$st generation. Its probability function, by Theorem 7.1, is given by (7.6) or (7.7) according to whether $q_0 > 0$ or $q_0 = 0$. The binomial moments of this distribution are given by (7.8).

7.3 MIXTURE DISCRETE DISTRIBUTIONS

Consider a supply of balls randomly distributed into a countable set of urns $\{0, 1, \ldots, n, \ldots\}$. Assume that the number S_n of balls distributed into the nth urn is a random variable with probability function
$$q_m(n) = P(S_n = m), \; m = 0, 1, \ldots \; (n = 0, 1, \ldots).$$
Let S be the total number of balls distributed into the urns, with
$$P(S = m) = p_m, \; m = 0, 1, \ldots.$$

Further, assume that the relative contribution of balls of the nth urn to the total number of balls distributed into the urns is given by the *weight function*

$$c_n, \quad n = 0, 1, \ldots, \quad \sum_{n=0}^{\infty} c_n = 1.$$

Note that this function may depend on the size of the nth urn. More generally c_n, $n = 0, 1, \ldots$, may be the probability function of a discrete stochastic *attraction parameter* of the distribution $q_m(n)$, $m = 0, 1, \ldots$. Then, the probability function of S is given by

$$p_m = \sum_{n=0}^{\infty} c_n q_m(n), \quad m = 0, 1, \ldots . \tag{7.9}$$

This model is a *countable mixture of discrete distributions*. The distribution with probability function p_m, $m = 0, 1, \ldots$, is called a *mixture distribution*. Also, the distribution with probability function $q_m(n)$, $m = 0, 1, \ldots$, is the *mixed distribution*, while that with probability function c_n, $n = 0, 1, \ldots$, is the *mixing distribution*.

A particular case of this model, in which the dependence of the weight function on the size of each urn becomes transparent, may be obtained by assuming that the nth urn is divided into n cells (compartments), $n = 1, 2, \ldots$. Let $X_{i,n}$ be the number of balls distributed into the ith cell of the nth urn, $i = 1, 2, \ldots, n$, $n = 1, 2, \ldots$. Further, suppose that for each $n = 2, 3, \ldots$, the random variables $X_{1,n}, X_{2,n}, \ldots, X_{n,n}$ are independently and identically distributed with a common probability function $q_x = P(X = x)$, $x = 0, 1, \ldots$. Also, let $P(S_1 = x) = q_x$, $x = 0, 1, \ldots$, and $P(S_0 = 0) = 1$. In this case the probability function $q_m(n)$, $m = 0, 1, \ldots$, of the mixed distribution is the n-fold convolution of q_x, $x = 0, 1, \ldots$. Clearly, if the probability generating function of q_x, $x = 0, 1, \ldots$ is $G(t)$, then the probability generating function of $q_m(n)$, $m = 0, 1, \ldots$, $G_n(t)$, according to Theorem 1.12, is given by $G_n(t) = [G(t)]^n$. Consequently, if the probability generating function of the mixing distribution c_n, $n = 0, 1, \ldots$ is $F(u)$, then from (7.9) the probability generating function $H(t)$ of the mixture distribution, p_m, $m = 0, 1, \ldots$, is deduced as $H(t) = F(G(t))$.

Note that expression (7.9) of countable mixtures of discrete distributions is mathematically the same as expression (7.1) of compound discrete distributions, only the interpretation of the components involved being different. Specifically, c_n, $n = 0, 1, \ldots$, in (7.9) is the mixing distribution, while in (7.1) it is not the compounding but the compounded distribution. Also, $q_m(n)$, $m = 0, 1, \ldots$, in (7.9) is the mixed distribution, while in (7.1) it is the n-fold convolution of the compounding distribution; the compounding distribution is the common distribution of the components of the mixed distribution. Therefore, using this correspondence, any compound discrete distribution can be interpreted as countable mixture discrete distribution. The inverse is also true in the particular case the mixed distribution $q_m(n)$, $m = 0, 1, \ldots$, is the n-fold convolution of a distribution q_x, $x = 0, 1, \ldots$. The most important mixture distributions will be dealt with, in the following sections of this chapter, as compound distributions.

7.4 PARTICULAR COMPOUNDING DISTRIBUTIONS

7.4.1 Poisson Compounding Distribution

Assume that the number X_j of balls distributed into the jth urn, $j = 0, 1, \ldots$, obeys a *Poisson distribution* with

$$q_x = P(X = x) = e^{-\lambda} \frac{\lambda^x}{x!}, \quad x = 0, 1, \ldots, \quad 0 < \lambda < \infty. \tag{7.10}$$

Then, its u-fold convolution obeys also a Poisson distribution with

$$q_m(u) = P(S_u = m) = e^{-u\lambda} \frac{(u\lambda)^m}{m!}, \quad m = 0, 1, \ldots. \tag{7.11}$$

Clearly, the rth binomial moment of this distribution is

$$a_{(r)}(u) = \frac{(u\lambda)^r}{r!}, \quad r = 0, 1, \ldots. \tag{7.12}$$

Introducing expressions (7.10), (7.11) and (7.12) into Theorem 7.1 and using (2.12), the following corollary is deduced.

Corollary 7.1 (a) *Under the random occupancy model with a random number of urns and Poisson probabilities, the probability p_m, $m = 0, 1, \ldots$, that m balls are distributed into the urns is given by*

$$p_m = \frac{\lambda^m}{m!} f(e^{-\lambda}) \sum_{k=0}^{m} k! S(m, k) \beta_{(k)}(e^{-\lambda}), \tag{7.13}$$

where $S(m, k) = [\Delta_u^k u^m]_{u=0}/k!$ is the Stirling number of the second kind.
(b) *The rth binomial moment $b_{(r)}$, $r = 0, 1, \ldots$, of the probability function p_m, $m = 0, 1, \ldots$, is given by*

$$b_{(r)} = \frac{\lambda^r}{r!} \sum_{k=0}^{r} k! S(r, k) \beta_{(k)}. \tag{7.14}$$

Example 7.3 *Neyman type A distribution.* Assume that the number N of clusters of eggs per unit area is a Poisson random variable with probability function

$$c_n = P(N = n) = e^{-\theta} \frac{\theta^n}{n!}, \quad n = 0, 1, \ldots, \quad 0 < \theta < \infty,$$

and kth binomial moment

$$\beta_{(k)} = E\left[\binom{N}{k}\right] = \frac{\theta^k}{k!}, \quad k = 0, 1, \ldots.$$

Further, assume that the number X of larvae developing from any cluster of eggs also obeys a Poisson distribution with

$$q_x = e^{-\lambda} \frac{\lambda^x}{x!}, \quad x = 0, 1, \ldots, \quad 0 < \lambda < \infty.$$

The distribution of the number S_N of larvae per unit area may be deduced from Corollary 7.1 as follows. The power series distribution (7.3), since

$$f(e^{-\lambda}) = \sum_{n=0}^{\infty} c_n (e^{-\lambda})^n = e^{-\theta} \sum_{n=0}^{\infty} \frac{(\theta e^{-\lambda})^n}{n!} = e^{-\theta(1-e^{-\lambda})},$$

reduces to a Poisson distribution with probability function

$$c_n(e^{-\lambda}) = e^{-\theta e^{-\lambda}} \frac{(\theta e^{-\lambda})^n}{n!}, \; n = 0, 1, \ldots.$$

Clearly, its kth binomial moment is

$$\beta_{(k)}(e^{-\lambda}) = \frac{(\theta e^{-\lambda})^k}{k!}.$$

Therefore, the probability function of the *Neyman type A distribution* is given by

$$p_m = e^{-\theta(1-e^{-\lambda})} \frac{\lambda^m}{m!} \sum_{k=0}^{m} S(m,k)(\theta e^{-\lambda})^k, \; m = 0, 1, \ldots.$$

Also, its rth binomial moment is

$$b_{(r)} = \frac{\lambda^r}{r!} \sum_{k=0}^{r} S(r,k) \theta^k, \; r = 0, 1, \ldots.$$

In particular, for $r = 1, 2$ and since $S(r, 0) = 0$, $S(1, 1) = S(2, 1) = S(2, 2) = 1$, we deduce the mean and variance of this distribution as

$$\mu = \theta\lambda, \; \sigma^2 = \theta\lambda(1 + \lambda).$$

Further, using the vertical recurrence relation of the Stirling numbers of the second kind, we deduce for p_m, $m = 0, 1, \ldots$, and $b_{(r)}$, $r = 0, 1, \ldots$, the recurrence relations

$$p_{m+1} = \frac{\theta \lambda e^{-\lambda}}{m+1} \sum_{j=0}^{m} \frac{\lambda^{m-j}}{(m-j)!} p_j, \; m = 0, 1, \ldots,$$

with $p_0 = e^{-\theta(1-e^{-\lambda})}$, and

$$b_{(r+1)} = \frac{\theta \lambda}{r+1} \sum_{j=0}^{r} \frac{\lambda^{r-j}}{(r-j)!} b_{(j)}, \; r = 0, 1, \ldots,$$

with $b_{(0)} = 1$.

Another interesting application of the Neyman type A distribution is discussed in the following example.

Example 7.4 *Bus driver accidents.* Assume that every driver is liable to randomly occurring spells and his/her performance during a spell randomly deviates from the standard. Also, suppose that all drivers behave independently of each other and no accident can occur outside a spell. A stochastic model for the number of bus driver accidents can be constructed by assuming that, in a unit time period, the number N of spells per driver obeys a Poisson distribution with

$$c_n = P(N = n) = e^{-\theta}\frac{\theta^n}{n!}, \ n = 0, 1, \ldots, \ 0 < \theta < \infty.$$

Further, assume that during a spell the number X of accidents a bus driver may have follows a Poisson distribution with

$$q_x = P(X = x) = e^{-\lambda}\frac{\lambda^x}{x!}, \ x = 0, 1, \ldots, \ 0 < \lambda < \infty.$$

Then, as in Example 7.3, the distribution of the number S_N of accidents of a bus driver, during the given unit time period, is a Neyman type A distribution.

Let us now lift the restriction that no accident can occur outside a spell and assume that the number Y of accidents a bus driver may have outside a spell is independently distributed of the number S_N of spell accidents with

$$d_y = P(Y = y) = e^{-\rho}\frac{\rho^y}{y!}, \ y = 0, 1, \ldots, \ 0 < \rho < \infty.$$

Then, the probability function P_m, $m = 0, 1, \ldots$, of the total number $S = S_N + Y$ of accidents of a bus driver, on using the convolution formula, may be obtained as

$$P_m = \sum_{j=0}^{m} p_j d_{m-j} = e^{-\theta(1-e^{-\lambda})-\rho} \sum_{j=0}^{m}\sum_{k=0}^{j} \frac{\lambda^j}{j!} S(j,k)(\theta e^{-\lambda})^k \frac{\rho^{m-j}}{(m-j)!}$$

$$= e^{-\theta(1-e^{-\lambda})-\rho}\frac{\lambda^m}{m!} \sum_{k=0}^{m}\left\{\sum_{j=k}^{m}\binom{m}{j}(\rho/\lambda)^{m-j}S(j,k)\right\}(\theta e^{-\lambda})^k,$$

which by (2.16) reduces to

$$P_m = e^{-\theta(1-e^{-\lambda})-\rho}\frac{\lambda^m}{m!}\sum_{k=0}^{m} S(m,k;\rho/\lambda)(\theta e^{-\lambda})^k, \ m = 0, 1, \ldots,$$

where $S(m, k; \rho/\lambda) = [\Delta_u^k u^m]_{u=\rho/\lambda}/k!$ is the noncentral Stirling number of the second kind. Also, since

$$\gamma_{(r)} = E\left[\binom{Y}{r}\right] = \frac{\rho^r}{r!}, \ r = 0, 1, \ldots,$$

the binomial moments $B_{(r)}$, $r = 0, 1, \ldots$, of this distribution are obtained as

$$B_{(r)} = \sum_{j=0}^{r} b_{(j)} \gamma_{(r-j)} = \sum_{j=0}^{r} \sum_{k=0}^{j} \frac{\lambda^j}{j!} S(j,k) \theta^k \frac{\rho^{r-j}}{(r-j)!}$$

$$= \frac{\lambda^r}{r!} \sum_{k=0}^{r} \left\{ \sum_{j=k}^{r} \binom{r}{j} (\rho/\lambda)^{r-j} S(j,k) \right\} \theta^k,$$

which by (2.16) reduces to

$$B_{(r)} = \frac{\lambda^r}{r!} \sum_{k=0}^{r} S(r, k; \rho/\lambda) \theta^k, \quad r = 0, 1, \ldots.$$

This extended Neyman type A distribution was termed as a *short* distribution to express the reduction of its tail length.

7.4.2 Binomial Compounding Distribution

Assume that the number X_j of balls distributed into the jth urn, $j = 0, 1, \ldots$, obeys a *binomial distribution* with

$$q_x = P(X = x) = \binom{s}{x} p^x q^{s-x}, \quad x = 0, 1, \ldots, s, \tag{7.15}$$

where $0 < p < 1$ and $q = 1 - p$. Then, its u-fold convolution obeys a binomial distribution with

$$q_m(u) = P(S_u = m) = \binom{su}{m} p^m q^{su-m}, \quad m = 0, 1, \ldots, su. \tag{7.16}$$

Clearly, the rth binomial moment of this distribution is

$$a_{(r)}(u) = \binom{su}{r} p^r, \quad r = 0, 1, \ldots, su. \tag{7.17}$$

Introducing expressions (7.15), (7.16) and (7.17) into Theorem 7.1 and using (2.52), the following corollary is deduced.

Corollary 7.2 (a) *Under the random occupancy model with a random number of urns and binomial probabilities, the probability p_m, $m = 0, 1, \ldots$, that m balls are distributed into the urns is given by*

$$p_m = \frac{(p/q)^m}{m!} f(q^s) \sum_{k=0}^{m} k! C(m, k; s) \beta_{(k)}(q^s), \tag{7.18}$$

where $C(m, k; s) = [\Delta_u^k (su)_m]_{u=0}/k!$ is the generalized factorial coefficient.

(b) *The rth binomial moment $b_{(r)}$, $r = 0, 1, \ldots$, of the probability function p_m, $m = 0, 1, \ldots$, is given by*

$$b_{(r)} = \frac{p^r}{r!} \sum_{k=0}^{r} k! C(r, k; s) \beta_{(k)}. \tag{7.19}$$

Example 7.5 *Bernoulli damage process.* The exposure to radiation produces chromosome breakages in cells. For a given dosage and time of exposure, the number N of breakages in individual cells can be assumed to have a Poisson distribution with

$$c_n = P(N = n) = e^{-\theta}\frac{\theta^n}{n!}, \; n = 0, 1, \ldots, \; 0 < \theta < \infty.$$

Also, each breakage independent of others has probability p of persisting and probability $q = 1 - p$ of healing. Then,

$$q_x = P(X = x) = p^x q^{1-x}, \; x = 0, 1, \; 0 < p < 1.$$

The distribution of the number S_N of observable breakages in individual cells is obtained from Corollary 7.2. Clearly,

$$f(q) = \sum_{n=0}^{\infty} c_n q^n = e^{-\theta p}, \; \beta_{(k)}(q) = \frac{(\theta q)^k}{k!},$$

and since $C(m, k; 1) = \delta_{m,k}$,

$$p_m = e^{-\theta p}\frac{(\theta p)^m}{m!}, \; m = 0, 1, \ldots;$$

that is, the distribution of S_N is also a Poisson with parameter θp.

Example 7.6 *Logarithmic distribution with zeros.* Assume that the number N of animals of a particular species per unit of habitat is a random variable with probability function

$$c_n = P(N = n), \; n = 0, 1, \ldots.$$

Also, in an animal-trapping experiment, assume that each animal may be trapped, independently of the others, with probability p. Then,

$$q_x = P(X = x) = p^x q^{1-x}, \; x = 0, 1,$$

where $q = 1 - p$ and $0 < p < 1$. The probability function of the number S_N of animals trapped per unit of habitat, since $C(m, k; 1) = \delta_{m,k}$, is deduced from Corollary 7.2 as

$$p_m = P(S_N = m) = f(q)\beta_{(m)}(q)(p/q)^m, \; m = 0, 1, \ldots.$$

Also, its rth binomial moment is

$$b_{(r)} = \beta_{(r)} p^r, \; r = 0, 1, \ldots.$$

Entomological data sets support the assumption that the random variable N obeys a logarithmic distribution with

$$c_n = P(N = n) = [-\log(1-\theta)]^{-1}\frac{\theta^n}{n}, \quad n = 1, 2, \ldots.$$

In this case,

$$f(q) = \sum_{n=1}^{\infty} c_n q^n = [-\log(1-\theta)]^{-1}[-\log(1-\theta q)]$$

and

$$\beta_{(m)}(q) = [-\log(1-\theta q)]^{-1}\frac{[\theta q/(1-\theta q)]^m}{m}, \quad m = 1, 2, \ldots.$$

Thus, setting

$$w = \frac{-\log(1-\theta q)}{-\log(1-\theta)}, \quad \lambda = \frac{\theta p}{1-\theta q},$$

and since

$$\frac{-\log(1-\lambda)}{-\log(1-\theta)} = \frac{-\log(1-\theta) + \log(1-\theta q)}{-\log(1-\theta)} = 1 - w,$$

the probability function of S_N is deduced as

$$p_0 = w, \quad p_m = (1-w)[-\log(1-\lambda)]^{-1}\frac{\lambda^m}{m}, \quad m = 1, 2, \ldots,$$

where $0 < w < 1$ and $0 < \lambda < 1$. This is a logarithmic distribution with a zero term added. Clearly, the rth binomial moment of this distribution is given by

$$b_{(r)} = [-\log(1-\theta)]^{-1}\frac{[\theta/(1-\theta)]^r p^r}{r}, \quad r = 1, 2, \ldots.$$

In particular, the mean and variance of this distribution are obtained as

$$\mu = \frac{cp\theta}{1-\theta}, \quad \sigma^2 = \frac{cp\theta[(1-c)p\theta + (1-\theta)]}{(1-\theta)^2},$$

with $c = [-\log(1-\theta)]^{-1}$.

The logarithmic distribution with zeros may also be derived as a mixture distribution. Specifically, the mixed distribution is a binomial with parameters N and p, where N is a random variable, and the distribution of N, which is the mixing distribution, is a logarithmic distribution with parameter θ (see Exercise 7.21).

7.4.3 Negative Binomial Compounding Distribution

Assume that the number X_j of balls distributed into the jth urn, $j = 0, 1, \ldots$, obeys a *negative binomial distribution* with

$$q_x = P(X = x) = \binom{s+x-1}{x} p^s q^x, \quad x = 0, 1, \ldots, \tag{7.20}$$

PARTICULAR COMPOUNDING DISTRIBUTIONS   293

where $0 < p < 1$ and $q = 1 - p$. Then, its u-fold convolution obeys a negative binomial distribution with

$$q_m(u) = P(S_u = m) = \binom{su + m - 1}{m} p^{su} q^m, \ m = 0, 1, \ldots. \tag{7.21}$$

Clearly, the rth binomial moment of this distribution is

$$a_{(r)}(u) = \binom{su + r - 1}{r} \frac{q^r}{p^r}, \ r = 0, 1, \ldots. \tag{7.22}$$

Introducing expressions (7.20), (7.21) and (7.22) into Theorem 7.1 and using (2.70), the following corollary is deduced.

Corollary 7.3 (a) *Under the random occupancy model with a random number of urns and negative binomial probabilities, the probability p_m, $m = 0, 1, \ldots$, that m balls are distributed into the urns is given by*

$$p_m = \frac{q^m}{m!} f(p^s) \sum_{k=0}^{m} k! |C(m, k; -s)| \beta_{(k)}(p^s), \tag{7.23}$$

where $|C(m, k; -s)| = [\Delta_u^k (su + m - 1)_m]_{u=0}/k!$ is the absolute generalized factorial coefficient.

(b) *The rth binomial moment $b_{(r)}$, $r = 0, 1, \ldots$, of the probability function p_m, $m = 0, 1, \ldots$, is given by*

$$b_{(r)} = \frac{(q/p)^r}{r!} \sum_{k=0}^{r} k! |C(r, k; -s)| \beta_{(k)}. \tag{7.24}$$

Example 7.7 *An entomological stochastic model.* Assume that the number N of egg masses per plot obeys a Poisson distribution with parameter θ. Data sets indicate that the number X of surviving larvae per egg mass has a more heterogeneous distribution than a Poisson or a binomial. Thus, assuming that the number X of surviving larvae per egg mass obeys a Poisson distribution, with an expected value $E(X) = \lambda$ that varies according to a gamma distribution, it follows that X obeys a negative binomial distribution (cf. Exercise 7.24):

$$q_x = \binom{s + x - 1}{x} p^s q^x, \ x = 0, 1, \ldots.$$

The distribution of number S_N of surviving larvae per plot may be obtained from Corollary 7.3. Indeed, for

$$c_n = e^{-\theta} \frac{\theta^n}{n!}, \ n = 0, 1, \ldots, \ 0 < \theta < \infty,$$

we have $\beta_{(k)} = E[\binom{N}{k}] = \theta^k/k!$, $k = 0, 1, \ldots$. Also, we get

$$f(p^s) = \sum_{n=0}^{\infty} c_n p^{sn} = e^{-\theta(1-p^s)}$$

and $\beta_{(k)}(p^s) = \theta^k p^{sk}/k!$, $k = 0, 1, \ldots$. Therefore,

$$p_m = e^{-\theta(1-p^s)} \frac{q^m}{m!} \sum_{k=0}^{m} |C(m, k; -s)| \theta^k p^{sk}, \quad m = 0, 1, \ldots,$$

and

$$b_{(r)} = \frac{(q/p)^r}{r!} \sum_{k=0}^{r} |C(r, k; -s)| \theta^k, \quad r = 0, 1, \ldots.$$

In particular, for $r = 1$ and $r = 2$, and since $|C(r, 0; -s)| = 0$, $|C(1, 1; -s)| = s$, $|C(2, 1; -s)| = s(s+1)$ and $|C(2, 2; -s)| = s^2$, we deduce the mean and the variance of this distribution as

$$\mu = \frac{s\theta q}{p}, \quad \sigma^2 = \frac{s\theta q}{p} \cdot \frac{1+sq}{p}.$$

Also, using the vertical recurrence relation of the generalized factorial coefficients, we deduce for the probability function p_m, $m = 0, 1, \ldots$, and its binomial moments $b_{(r)}$, $r = 0, 1, \ldots$, the recurrence relations

$$p_{m+1} = \frac{s\theta q p^s}{m+1} \sum_{j=0}^{m} \binom{s+m-j}{m-j} q^{m-j} p_j, \quad m = 0, 1, \ldots,$$

with $p_0 = e^{-\theta(1-p^s)}$, and

$$b_{(r+1)} = \frac{s\theta(q/p)}{r+1} \sum_{j=0}^{r} \binom{s+r-j}{r-j} \left(\frac{q}{p}\right)^{r-j} b_{(j)}, \quad r = 0, 1, \ldots,$$

with $b_{(0)} = 1$.

7.4.4 Logarithmic Compounding Distribution

Assume that the number X_j of balls distributed into the jth urn, $j = 0, 1, \ldots$, obeys a *logarithmic distribution* with

$$q_x = P(X = x) = [-\log(1-\lambda)]^{-1} \frac{\lambda^x}{x}, \quad x = 1, 2, \ldots,$$

where $0 < \lambda < 1$. Note that $P(X > 0) = 1$ and so $q_0 = 0$. Therefore, Theorem 7.1 cannot be applied. Further, since the logarithmic distribution is a power series distribution with series function $g(\lambda) = -\log(1-\lambda)$, the distribution of the sum $S_n = \sum_{j=1}^{n} X_j$ is again a power series with series function $g_n(\lambda) = [-\log(1-\lambda)]^n$. Thus, by virtue of the generating function of the signless Stirling numbers of the first kind (cf. Corollary 2.1),

$$\sum_{m=n}^{\infty} |s(m, n)| \frac{u^m}{m!} = \frac{[-\log(1-u)]^n}{n!},$$

it follows that

$$q_m(n) = [-\log(1-\lambda)]^{-n} n! |s(m,n)| \frac{\lambda^m}{m!}, \quad m = n, n+1, \ldots.$$

Further, the binomial moment generating function of this distribution is

$$\sum_{r=0}^{\infty} a_{(r)}(n) t^r = \left[\frac{-\log(1-\lambda-\lambda t)}{-\log(1-\lambda)} \right]^n$$

and so

$$\sum_{r=k}^{\infty} [\Delta_u^k a_{(r)}(u)]_{u=0} t^r = \left[\frac{-\log(1-\lambda-\lambda t)}{-\log(1-\lambda)} - 1 \right]^k$$

$$= \left[\frac{-\log(1-\lambda t/(1-\lambda))}{-\log(1-\lambda)} \right]^k.$$

Hence,

$$[\Delta_u^k a_{(r)}(u)]_{u=0} = [-\log(1-\lambda)]^{-k} k! |s(r,k)| \frac{[\lambda/(1-\lambda)]^r}{r!}, \quad r = k, k+1, \ldots.$$

Consequently, from (7.7) and (7.8) the following corollary is deduced.

Corollary 7.4 (a) *Under the random occupancy model with a random number of urns and logarithmic probabilities, the probability* p_m, $m = 0, 1, \ldots$, *that* m *balls are distributed into the urns is given by*

$$p_m = \frac{\lambda^m}{m!} \sum_{n=0}^{m} n! |s(m,n)| [-\log(1-\lambda)]^{-n} c_n, \tag{7.25}$$

where $|s(m,n)| = [D_u^n (u+m-1)_m]_{u=0}/n!$ *is the signless Stirling number of the first kind.*

(b) *The* r*th binomial moment* $b_{(r)}$, $r = 0, 1, \ldots$, *of the probability function* p_m, $m = 0, 1, \ldots$, *is given by*

$$b_{(r)} = \frac{[\lambda/(1-\lambda)]^r}{r!} \sum_{k=0}^{r} k! |s(r,k)| [-\log(1-\lambda)]^{-k} \beta_{(k)}. \tag{7.26}$$

Example 7.8 *A negative binomial as a compound distribution.* Returning to the collective risk model presented in Example 7.1, assume that the number N of insurance claims in a unit time period has a Poisson distribution with

$$c_n = P(N = n) = e^{-\theta} \frac{\theta^n}{n!}, \quad n = 0, 1, \ldots, \quad 0 < \theta < \infty.$$

Also, assume that the common distribution of the amount X of any specific claim is the logarithmic with probability function

$$q_x = P(X = x) = [-\log(1-\lambda)]^{-1} \frac{\lambda^x}{x}, \quad x = 1, 2, \ldots, \quad 0 < \lambda < 1.$$

The distribution of the aggregate insurance claims S_N can readily be deduced from Corollary 7.4. Indeed, from (7.25) and setting $a = [-\log(1-\lambda)]^{-1}\theta > 0$, we get the expression

$$p_m = \frac{\lambda^m}{m!} e^{-\theta} \sum_{n=0}^{m} |s(m,n)| a^n,$$

which, by virtue of (2.4), entails

$$p_m = \frac{(a+m-1)_m}{m!} \lambda^m e^{-\theta}, \ m = 0, 1, \ldots.$$

Since

$$e^{-\theta} = e^{a \log(1-\lambda)} = (1-\lambda)^a,$$

this probability function may be written, more suggestively, as

$$p_m = \binom{a+m-1}{m} \lambda^m (1-\lambda)^a, \ m = 0, 1, \ldots,$$

with $a > 0$ and $0 < \lambda < 1$, which is the probability function of a negative binomial distribution.

7.5 COMPOUND POISSON DISTRIBUTIONS

Assume that the number N of urns is a Poisson random variable with probability function

$$c_n = P(N=n) = e^{-\theta} \frac{\theta^n}{n!}, \ n = 0, 1, \ldots, \ 0 < \theta < \infty.$$

Its kth binomial moment is

$$\beta_{(k)} = E\left[\binom{N}{k}\right] = \frac{\theta^k}{k!}, \ k = 0, 1, \ldots.$$

Also, for $q_0 > 0$,

$$f(q_0) = \sum_{n=0}^{\infty} c_n q_0^n = e^{-\theta} \sum_{n=0}^{\infty} \frac{(\theta q_0)^n}{n!} = e^{-\theta(1-q_0)},$$

and the power series distribution (7.3) reduces to a Poisson distribution with

$$c_n(q_0) = e^{-\theta q_0} \frac{(\theta q_0)^n}{n!}, \ n = 0, 1, \ldots.$$

Clearly, its kth binomial moment is

$$\beta_{(k)}(q_0) = \frac{(\theta q_0)^k}{k!}, \ k = 0, 1, \ldots.$$

Therefore, by Theorem 7.1, the probability function of a *compound Poisson distribution* is given by

$$p_m = e^{-\theta(1-q_0)} \sum_{k=0}^{m} \frac{(\theta q_0)^k}{k!} \left[\Delta_u^k q_0^{-u} q_m(u)\right]_{u=0}, \quad m = 0, 1, \ldots, \quad (7.27)$$

if $q_0 > 0$ and by

$$p_m = e^{-\theta} \sum_{n=0}^{m} \frac{\theta^n}{n!} q_m(n), \quad m = 0, 1, \ldots, \quad (7.28)$$

if $q_0 = 0$. In either case the rth binomial moment of a compound Poisson distribution is given by

$$b_{(r)} = \sum_{k=0}^{r} \frac{\theta^k}{k!} \left[\Delta_u^k a_{(r)}(u)\right]_{u=0}, \quad r = 0, 1, \ldots, \quad (7.29)$$

where $a_{(r)}(u)$ is the rth binomial moment of $q_m(u)$, $m = 0, 1, \ldots$.

Further, specifying the distribution $q_x = P(X = x)$, $x = 0, 1, \ldots$, several particular compound Poisson distributions may be obtained.

Example 7.9 *An automobile accidents model.* The number N of automobile accidents in a specified area during a given time period obeys a Poisson distribution with probability function

$$c_n = P(N = n) = e^{-\theta} \frac{\theta^n}{n!}, \quad n = 0, 1, \ldots, \quad 0 < \theta < \infty.$$

Let X_j be the number of cars involved in the jth accident, $j = 1, 2, \ldots$. It is reasonable to assume that X_j, $j = 1, 2, \ldots$, are independently and identically distributed random variables with common probability function

$$q_x = P(X = x), \quad x = 0, 1, \ldots.$$

Then, the number $S_N = \sum_{j=1}^{N} X_j$ of cars involved in the accidents has a compound Poisson distribution with probability function (7.27) or (7.28).

7.5.1 Hermite Distribution

The simplest compound Poisson distribution is constructed with compounding distribution the zero-one Bernoulli. This distribution, which was derived under a Bernoulli damage process in Example 7.5, is also a Poisson distribution with mean given by the product of the means of the component distributions. Proceeding one step further, assume that the number X_j of balls distributed into the jth urn, $j = 1, 2, \ldots$, obeys a binomial distribution with

$$q_x = P(X = x) = \binom{2}{x} p^x q^{2-x}, \quad x = 0, 1, 2,$$

where $q = 1 - p$ and $0 < p < 1$. Its u-fold convolution also obeys a binomial distribution with probability function

$$q_m(u) = P(S_u = m) = \binom{2u}{m} p^m q^{2u-m}, \quad m = 0, 1, \ldots, 2u,$$

and binomial moments

$$a_{(r)}(u) = \binom{2u}{r} p^r, \quad r = 0, 1, \ldots, 2u, \quad a_{(r)}(u) = 0, \quad r > 2u.$$

Hence,

$$q_0^{-u} q_m(u) = \binom{2u}{m} \left(\frac{p}{q}\right)^m, \quad m = 0, 1, \ldots,$$

and since

$$\left[\Delta_u^k \binom{2u}{m}\right]_{u=0} = \frac{k!}{m!} C(m, k; 2),$$

where $C(m, k; s)$ is the generalized factorial coefficient, it follows from (7.27) and (7.29) that

$$p_m = e^{-\theta(1-q^2)} \frac{p^m}{m!} \sum_{k=0}^{m} C(m, k; 2) \theta^k q^{2k-m}, \quad m = 0, 1, \ldots,$$

and

$$b_{(r)} = \frac{p^r}{r!} \sum_{k=0}^{r} C(r, k; 2) \theta^k, \quad r = 0, 1, \ldots.$$

The generalized factorial coefficient $C(m, k; s)$ in the particular case $s = 2$, and since

$$\sum_{m=k}^{\infty} C(m, k; 2) \frac{u^m}{m!} = \frac{[(1+u)^2 - 1]^k}{k!} = \frac{u^k(u+2)^k}{k!}$$

$$= \sum_{m=k}^{2k} \frac{m! 2^{2k-m}}{(m-k)!(2k-m)!} \cdot \frac{u^m}{m!},$$

reduces to

$$C(m, k; 2) = \frac{m! 2^{2k-m}}{(m-k)!(2k-m)!}, \quad m = k, k+1, \ldots, 2k,$$

with $C(m, k; 2) = 0$ for $m < k$ or $m > 2k$. Thus, the probability function p_m, $m = 0, 1, \ldots$, on introducing the parameters

$$\alpha = p\sqrt{2\theta}, \quad \beta = q\sqrt{2\theta},$$

and changing the variable k to $j = m - k$, may be transformed to

$$p_m = e^{-\alpha\beta - \alpha^2/2} \frac{\alpha^m}{m!} \sum_{j=0}^{[m/2]} \frac{m!\beta^{m-2j}}{(m-2j)!j!2^j}, \quad m = 0, 1, \ldots.$$

Similarly, its rth binomial moment may be written as

$$b_{(r)} = \frac{\alpha^r}{r!} \sum_{j=0}^{[r/2]} \frac{r!(\alpha+\beta)^{r-2j}}{(r-2j)!j!2^j}, \quad r = 0, 1, \ldots.$$

Further, these sums may be expressed by modified Hermite polynomials as follows. The *Hermite polynomial* in x of degree n, denoted by $h_n(x)$, is defined by

$$h_n(x) = (-1)^n e^{x^2/2} D_x^n e^{-x^2/2},$$

where D_x is the derivative operator with respect to x. Since $(-1)^n D_x^n f(x) = [D_t^n f(x-t)]_{t=0}$, the generating function of $h_n(x)$, $n = 0, 1, \ldots$, is deduced as

$$\sum_{n=0}^{\infty} h_n(x) \frac{t^n}{n!} = e^{xt - t^2/2}.$$

A *modified Hermite polynomial* in x of degree n is defined by

$$H_n(x) = i^{-n} h_n(ix), \quad n = 0, 1, \ldots, \quad i = \sqrt{-1}$$

and so

$$\sum_{n=0}^{\infty} H_n(x) \frac{t^n}{n!} = e^{xt + t^2/2}.$$

Expanding this generating function into powers of t and equating the coefficients of t^n on both sides of the resulting expression, we get

$$H_n(x) = \sum_{j=0}^{[n/2]} \frac{n! x^{n-2j}}{(n-2j)!j!2^j}, \quad n = 0, 1, \ldots,$$

where $[n/2]$ is the integer part of $n/2$. In particular,

$$H_0(x) = 1, \quad H_1(x) = x, \quad H_2(x) = x^2 + 1.$$

Also, differentiating the generating function, with respect to t, and equating the coefficients of t^{n-1} in the resulting expression, we deduce the recurrence relation

$$H_{n+1}(x) = xH_n(x) + nH_{n-1}(x), \quad n = 1, 2, \ldots,$$

with initial conditions $H_0(x) = 1$ and $H_1(x) = x$.

Clearly, the probability function p_m, $m = 0, 1, \ldots$, and its binomial moments $b_{(r)}$, $r = 0, 1, \ldots$, on introducing the modified Hermite polynomial, are expressed as

$$p_m = e^{\alpha\beta - \alpha^2/2} H_m(\beta) \frac{\alpha^m}{m!}, \quad m = 0, 1, \ldots,$$

and

$$b_{(r)} = H_r(\alpha + \beta) \frac{\alpha^r}{r!}, \quad r = 0, 1, \ldots.$$

In particular, for $r = 1, 2$, we deduce the mean and variance as

$$\mu = \alpha(\alpha + \beta) = 2\theta p, \quad \sigma^2 = \alpha(2\alpha + \beta) = 2\theta p(1 + p).$$

This distribution is called a *Hermite distribution*.

The recurrence relation of the modified Hermite polynomials entails for the probability function p_m, $m = 0, 1, \ldots$, and its binomial moments $b_{(r)}$, $r = 0, 1, \ldots$, the following recurrence relations:

$$(m + 1)p_{m+1} = \alpha\beta p_m + \alpha^2 p_{m-1}, \quad m = 1, 2, \ldots,$$

with $p_0 = e^{-\alpha\beta - \alpha^2/2}$, $p_1 = \alpha\beta e^{-\alpha\beta - \alpha^2/2}$, and

$$(r + 1)b_{(r+1)} = \alpha(\alpha + \beta)b_{(r)} + \alpha^2 b_{(r-1)}, \quad r = 1, 2, \ldots,$$

with $b_{(0)} = 1$, $b_{(1)} = \alpha(\alpha + \beta)$.

7.5.2 Generalized Hermite Distribution

A generalization of the Hermite distribution is constructed by assuming that the number X_j of balls distributed into the jth urn, $j = 1, 2, \ldots$, obeys a binomial distribution with

$$q_x = P(X = x) = \binom{s}{x} p^x q^{s-x}, \quad x = 0, 1, \ldots, s,$$

where $q = 1 - p$, $0 < p < 1$ and s is a positive integer. The u-fold convolution of this distribution is also a binomial distribution with probability function

$$q_m(u) = P(S_u = m) = \binom{su}{m} p^m q^{su-m}, \quad m = 0, 1, \ldots, su,$$

and binomial moments

$$a_{(r)}(u) = \binom{su}{r} p^r, \quad r = 0, 1, \ldots, su, \quad a_{(r)}(u) = 0, \quad r > su.$$

Thus,

$$q_0^{-u} q_m(u) = \binom{su}{m} \left(\frac{p}{q}\right)^m, \quad m = 0, 1, \ldots,$$

and since
$$\left[\Delta_u^k \binom{su}{m}\right]_{u=0} = \frac{k!}{m!}C(m,k;s),$$
with $C(m,k;s)$ the generalized factorial coefficient, it follows from (7.27) and (7.29) that
$$p_m = e^{-\theta(1-q^s)}\frac{p^m}{m!}\sum_{k=0}^{m}C(m,k;s)\theta^k q^{sk-m}, \quad m=0,1,\ldots,$$
and
$$b_{(r)} = \frac{p^r}{r!}\sum_{k=0}^{r}C(r,k;s)\theta^k, \quad r=0,1,\ldots.$$

Further, these sums may be expressed by generalized Hermite polynomials as follows. A *generalized Hermite polynomial* in x, studied by Bell (1934a), is defined by
$$H_n(x;a,s) = e^{-ax^s}D_x^n e^{ax^s},$$
where s is a positive integer and a is a real number. Since $D_x^n f(x) = [D_t^n f(x+t)]_{t=0}$,
$$\sum_{n=0}^{\infty} H_n(x;a,s)\frac{t^n}{n!} = e^{a[(x+t)^s - x^s]}.$$

Expanding this generating function into powers of t and using the exponential generating function of the generalized factorial coefficients,
$$\sum_{n=k}^{\infty} C(n,k;s)\frac{u^n}{n!} = \frac{[(1+u)^s - 1]^k}{k!},$$
we deduce the expression
$$H_n(x;a,s) = \sum_{k=0}^{n} C(n,k;s)a^k x^{sk-n}.$$

In particular,
$$H_0(x;a,s) = 1, \quad H_1(x;a,s) = asx^{s-1}, \quad H_2(x;a,s) = a(s)_2 x^{s-2} + a^2 s^2 x^{2s-2}.$$

Also, differentiating the generating function of the generalized Hermite polynomials with respect to t and equating the coefficients of t^{n-1} in the resulting expression, we get the recurrence relation
$$H_{n+1}(x;a,s) = as\sum_{j=0}^{n}\binom{s-1}{j}x^{s-1-j}(n)_j H_{n-j}(x;a,s),$$
for $n=1,2,\ldots$, with initial condition $H_0(x;a,s) = 1$.

Clearly, the probability function p_m, $m = 0, 1, \ldots$, and its binomial moments $b_{(r)}$, $r = 0, 1, \ldots$, on introducing the generalized Hermite polynomial, are expressed as

$$p_m = e^{-\theta(1-q^s)} H_m(q; \theta, s) \frac{p^m}{m!}, \quad m = 0, 1, \ldots,$$

and

$$b_{(r)} = H_r(1; \theta, s) \frac{p^r}{r!}, \quad r = 0, 1, \ldots.$$

In particular, for $r = 1, 2$, we deduce the mean and variance as

$$\mu = \theta s p, \quad \sigma^2 = \theta s p (1 + (s-1)p).$$

This distribution, which is a compound Poisson-binomial, may be called a *generalized Hermite distribution*.

Recurrence relations for the probability function p_m, $m = 0, 1, \ldots$, and its binomial moments $b_{(r)}$, $r = 0, 1, \ldots$, may be deduced from the recurrence relation of the generalized Hermite polynomials. Thus,

$$(m+1)p_{m+1} = \theta s p \sum_{j=0}^{m} \binom{s-1}{j} p^j q^{s-j-1} p_{m-j}, \quad m = 0, 1, \ldots,$$

with $p_0 = e^{-\theta(1-q^s)}$, and

$$(r+1)b_{(r+1)} = \theta s p \sum_{j=0}^{r} \binom{s-1}{j} p^j b_{(r-j)}, \quad r = 0, 1, \ldots,$$

with $b_{(0)} = 1$.

7.5.3 Pólya-Aeppli Distribution

Assume that the number X_j of balls distributed into the jth urn, $j = 1, 2, \ldots$, obeys a geometric distribution with

$$q_x = P(X = x) = pq^{x-1}, \quad x = 1, 2, \ldots,$$

where $q = 1 - p$ and $0 < p < 1$. Its n-fold convolution is a Pascal distribution with probability function

$$q_m(n) = P(S_n = m) = \binom{m-1}{n-1} p^n q^{m-n}, \quad m = n, n+1, \ldots.$$

Further, the generating function of the geometric distribution is $g(t) = pt/(1-qt)$ and so the binomial moment generating function of its u-fold convolution is

$$\sum_{r=0}^{\infty} a_{(r)}(u) t^r = [g(t+1)]^u.$$

Taking the kth difference, with respect to u, and since

$$\left[\Delta_u^k [g(t+1)]^u\right]_{u=0} = [g(t+1) - 1]^k = \left[\frac{t/p}{1-qt/p}\right]^k,$$

we get

$$\sum_{r=0}^{\infty} [\Delta_u^k a_{(r)}(u)]_{u=0} t^r = \sum_{r=k}^{\infty} \binom{r-1}{k-1} \frac{q^{r-k}}{p^r} t^r$$

and so

$$[\Delta_u^k a_{(r)}(u)]_{u=0} = \binom{r-1}{k-1} \frac{q^{r-k}}{p^r}.$$

The probability function p_m, $m = 0, 1, \ldots$, since $q_0 = 0$, is deduced from (7.28) as

$$p_0 = e^{-\theta}, \; p_m = e^{-\theta} q^m \sum_{n=1}^{m} \binom{m-1}{n-1} \frac{(\theta p/q)^n}{n!}, \; m = 1, 2, \ldots.$$

Also, the rth binomial moment of this distribution is deduced from (7.29) as

$$b_{(r)} = \left(\frac{q}{p}\right)^r \sum_{k=1}^{r} \binom{r-1}{k-1} \frac{(\theta/q)^k}{k!}, \; r = 1, 2, \ldots.$$

Further, these sums may be expressed by Laguerre polynomials as follows. The *Laguerre polynomial* in x of degree n is defined by

$$L_n^{(1)}(x) = \sum_{k=0}^{n} (-1)^k \binom{n+1}{n-k} \frac{x^k}{k!}, \; n = 0, 1, \ldots.$$

Its generating function is readily deduced as

$$\sum_{n=0}^{\infty} L_n^{(1)}(x) t^n = (1-t)^{-2} e^{-xt/(1-t)}.$$

Differentiating it, with respect to t, and equating the coefficients of t^{n-1} in the resulting expression, we get the recurrence relation

$$n L_n^{(1)}(x) = (2n - x) L_{n-1}^{(1)}(x) - n L_{n-2}^{(1)}(x), \; n = 2, 3, \ldots,$$

with $L_1^{(1)}(x) = 1$ and $L_2^{(1)}(x) = 2 - x$. Also, the explicit expression of the Laguerre polynomial may be transformed to

$$L_n^{(1)}(x) = -\frac{n+1}{x} \sum_{k=0}^{n} (-1)^{k+1} \binom{n}{k} \frac{x^{k+1}}{(k+1)!}$$

$$= -\frac{n+1}{x} \sum_{j=1}^{n+1} (-1)^j \binom{n}{j-1} \frac{x^j}{j!}.$$

Then, the probability function p_m, $m = 0, 1, \ldots$, and its binomial moments $b_{(r)}$, $r = 1, 2, \ldots$, on introducing the last expression of the Laguerre polynomials, can be written as

$$p_0 = e^{-\theta}, \quad p_m = e^{-\theta} L_{m-1}^{(1)}(-\theta p/q) \frac{\theta p q^{m-1}}{m}, \quad m = 1, 2, \ldots,$$

and

$$b_{(r)} = L_{r-1}^{(1)}(-\theta/q) \frac{\theta q^{r-1}}{rp^r}, \quad r = 1, 2, \ldots.$$

In particular, the mean and variance of this distribution are given by

$$\mu = \frac{\theta}{p}, \quad \sigma^2 = \frac{\theta(1+q)}{p^2}.$$

This distribution is called a *Pólya-Aeppli distribution*.

The recurrence relation for the Laguerre polynomials entails for the probability function p_m, $m = 0, 1, \ldots$, and its binomial moments $b_{(r)}$, $r = 0, 1, \ldots$, the recurrence relations

$$(m+1)p_{m+1} = (2mq + \theta p)p_m - (m-1)q^2 p_{m-1}, \quad m = 1, 2, \ldots,$$

with $p_0 = e^{-\theta}$, $p_1 = e^{-\theta}\theta p$, and

$$(r+1)b_{(r+1)} = \frac{2rq + \theta}{p} b_{(r)} + \frac{(r-1)q^2}{p^2} b_{(r-1)}, \quad r = 1, 2, \ldots,$$

with $b_{(0)} = 1$, $b_{(1)} = \theta/p$.

7.5.4 Poisson Distribution of Order k

Let the number N of urns be a Poisson random variable with probability function

$$c_n = P(N = n) = e^{-k\lambda} \frac{(k\lambda)^n}{n!}, \quad n = 0, 1, \ldots, \quad 0 < \lambda < \infty,$$

where k is a positive integer. Further, assume that the number X_j of balls distributed into the jth urn, $j = 1, 2, \ldots$, obeys a discrete uniform distribution, with probability function

$$q_x = P(X = x) = \frac{1}{k}, \quad x = 1, 2, \ldots, k.$$

Since

$$g(t) = \sum_{x=1}^{k} q_x t^x = \frac{t(1-t^k)}{k(1-t)},$$

the n-fold convolution of this distribution has generating function

$$g_n(t) = \sum_{m=n}^{kn} q_m(n) t^m = \frac{t^n(1-t^k)^n}{k^n(1-t)^n}.$$

Therefore,
$$q_m(n) = \frac{Q(m,n;k)}{k^n}, \quad m = n, n+1, \ldots, kn,$$
where
$$Q(m,n;k) = \sum_{j=0}^{n}(-1)^j \binom{n}{j}\binom{m-jk-1}{n-1}$$
is the number of ways of distributing m like balls into n distinguishable urns so that each urn contains at least one and no more than k balls. Also, the binomial moment generating function of this distribution is
$$\sum_{r=0}^{\infty} a_{(r)}(n)t^r = [g(t+1)]^n = \left[\frac{(1+t)^{k+1}-1-t}{kt}\right]^n,$$
and so
$$\sum_{r=j}^{\infty}\left[\Delta_u^j a_{(r)}(u)\right]_{u=0} t^r = [g(t+1)-1]^j = \left[\frac{(1+t)^{k+1}-1-(k+1)t}{kt}\right]^j.$$

Comparing the last generating function with the generating function of the associated generalized factorial coefficient $C_2(r,j;s)$, with scale parameter $s = k+1$ (see Section 3.3),
$$\sum_{r=2j}^{\infty} C_2(r,j;k+1)\frac{t^r}{r!} = \frac{[(1+t)^{k+1}-1-(k+1)t]^j}{j!},$$
we get
$$\left[\Delta_u^j a_{(r)}(u)\right]_{u=0} = \frac{j!}{(r+j)!k^j} C_2(r+j,j;k+1), \quad r = j, j+1, \ldots.$$

Then, the probability function and the binomial moments of the compound distribution are deduced from (7.28) and (7.29) as
$$p_m = e^{-k\lambda} \sum_{n=0}^{m} Q(m,n;k)\frac{\lambda^n}{n!}, \quad m = 0, 1, \ldots,$$
and
$$b_{(r)} = \sum_{j=0}^{r} C_2(r+j,j;k+1)\frac{\lambda^j}{(r+j)!}, \quad r = 0, 1, \ldots.$$

In particular, for $r = 1, 2$, and since (from Table 3.3)
$$C_2(n,0;k+1) = 0, \quad n > 0,$$
$$C_2(n,1;k+1) = (k+1)_n, \quad n = 2, 3, \ldots, \quad C_2(4,2;k+1) = 3(k+1)_2^2,$$

the mean and variance of this distribution are obtained as

$$\mu = \frac{k(k+1)\lambda}{2}, \quad \sigma^2 = \frac{k(k+1)(2k+1)\lambda}{6}.$$

This distribution is called a *Poisson distribution of order k*.

Some more light into the structure of the Poisson distribution of order k can be shed by the following stochastic process of order k. Consider a nonnegative integer valued homogeneous stochastic process $\{S(t), t \geq 0\}$ with independent increments and let

$$p_m(t) = P[S(t) = m], \quad m = 0, 1, \ldots.$$

Assume that, for small $\Delta t > 0$,

$$p_j(\Delta t) = \lambda \Delta t + o(\Delta t), \quad j = 1, 2, \ldots, k,$$

$$p_0(\Delta t) = 1 - k\lambda \Delta t + o(\Delta t),$$

with

$$\lim_{\Delta t \to 0} \frac{o(\Delta t)}{\Delta t} = 0.$$

Note that these conditions imply $P[S(\Delta t) > k] = o(\Delta t)$. Then, the following differential equations for the probabilities $p_m(t)$, $m = 0, 1, \ldots$, are readily deduced:

$$\frac{dp_0(t)}{dt} = -k\lambda p_0(t),$$

$$\frac{dp_m(t)}{dt} = \lambda \sum_{j=1}^{\min\{m,k\}} p_{m-j}(t) - k\lambda p_m(t), \quad m = 1, 2, \ldots.$$

Multiplying the mth equation by u^m and summing for all $m = 0, 1, \ldots$, we get for the generating function

$$h(u, t) = \sum_{m=0}^{\infty} p_m(t) u^m$$

the differential equation

$$\frac{\partial h(u, t)}{\partial t} = -k\lambda[1 - g(u)]h(u, t), \quad g(u) = \frac{u(1 - u^k)}{k(1 - u)},$$

which, on using the initial condition $p_0(0) = 1$, implies

$$h(u, t) = \exp\{-k\lambda t[1 - g(u)]\}.$$

In the unit time interval, $t = 1$, this is the probability generating function of the Poisson distribution of order k.

7.6 COMPOUND LOGARITHMIC DISTRIBUTIONS

Assume that the number N of urns has a logarithmic distribution with probability function

$$c_n = P(N = n) = [-\log(1-\theta)]^{-1}\frac{\theta^n}{n}, \quad n = 1, 2, \ldots, \quad 0 < \theta < 1.$$

Its kth binomial moment is

$$\beta_{(k)} = [-\log(1-\theta)]^{-1}\frac{[\theta/(1-\theta)]^k}{k}, \quad k = 1, 2, \ldots.$$

Further, for $q_0 > 0$,

$$f(q_0) = \sum_{n=1}^{\infty} c_n q_0^n = [-\log(1-\theta)]^{-1} \sum_{n=1}^{\infty} \frac{(\theta q_0)^n}{n}$$
$$= [-\log(1-\theta)]^{-1}[-\log(1-\theta q_0)]$$

and the power series distribution (7.3) reduces to a logarithmic distribution with probability function

$$c_n(q_0) = [-\log(1-\theta q_0)]^{-1}\frac{(\theta q_0)^n}{n}, \quad n = 1, 2, \ldots,$$

and binomial moments

$$\beta_{(k)}(q_0) = [-\log(1-\theta q_0)]^{-1}\frac{[\theta q_0/(1-\theta q_0)]^k}{k}, \quad k = 1, 2, \ldots.$$

Therefore, by Theorem 7.1, the probability function of a *compound logarithmic distribution* is given by

$$p_0 = [-\log(1-\theta)]^{-1}[-\log(1-\theta q_0)],$$

$$p_m = [-\log(1-\theta)]^{-1}\sum_{k=1}^{m}\frac{[\theta q_0/(1-\theta q_0)]^k}{k}\left[\Delta_u^k q_0^{-u} q_m(u)\right]_{u=0}, \quad (7.30)$$

for $m = 1, 2, \ldots$. If $q_0 = 0$, then

$$p_m = [-\log(1-\theta)]^{-1}\sum_{n=1}^{m}\frac{\theta^n}{n} q_m(n), \quad m = 1, 2, \ldots. \quad (7.31)$$

In either case the rth binomial moment of the compound logarithmic distribution is given by

$$b_{(r)} = [-\log(1-\theta)]^{-1}\sum_{k=1}^{r}\frac{[\theta/(1-\theta)]^k}{k}\left[\Delta_u^k a_{(r)}(u)\right]_{u=0}, \quad (7.32)$$

for $r = 1, 2, \ldots$, where $a_{(r)}(u)$ is the rth binomial moment of $q_m(u)$, $m = 0, 1, \ldots$.

7.6.1 Chebyshev Distribution of the First Kind

The simplest compound logarithmic distribution arises when the compounding is the zero-one Bernoulli distribution. This distribution, which was constructed in Example 7.6, is a logarithmic distribution with a zero term added. Taking one step further, assume that the number X_j of balls distributed into the jth urn, $j = 1, 2, \ldots$, follows a binomial distribution with probability function

$$q_x = P(X = x) = \binom{2}{x} p^x q^{2-x}, \; x = 0, 1, 2,$$

where $q = 1 - p$ and $0 < p < 1$. Its u-fold convolution is also a binomial distribution with probability function

$$q_m(u) = P(S_u = m) = \binom{2u}{m} p^m q^{2u-m}, \; m = 0, 1, \ldots, 2u,$$

and binomial moments

$$a_{(r)}(u) = \binom{2u}{r} p^r, \; r = 0, 1, \ldots, 2u, \; a_{(r)}(u) = 0, \; r > 2u.$$

Hence,

$$q_0^{-u} q_m(u) = \binom{2u}{m} \left(\frac{p}{q}\right)^m, \; m = 0, 1, \ldots,$$

and since

$$\left[\Delta_u^k \binom{2u}{m}\right]_{u=0} = \frac{k!}{m!} C(m, k; 2),$$

where $C(m, k; 2)$ is the generalized factorial coefficient, it follows from (7.30) and (7.32) that

$$p_0 = [-\log(1-\theta)]^{-1}[-\log(1-\theta q^2)],$$

$$p_m = [-\log(1-\theta)]^{-1} \frac{(p/q)^m}{m!} \sum_{k=1}^{m} (k-1)! C(m, k; 2) [\theta q^2/(1-\theta q^2)]^k,$$

for $m = 1, 2, \ldots$, and

$$b_{(r)} = [-\log(1-\theta)]^{-1} \frac{p^r}{r!} \sum_{k=1}^{r} (k-1)! C(r, k; 2) [\theta/(1-\theta)]^k,$$

for $r = 1, 2, \ldots$. Since (see Section 7.5.1)

$$C(m, k; 2) = \frac{m! 2^{2k-m}}{(m-k)!(2k-m)!}, \; m = k, k+1, \ldots, 2k,$$

with $C(m, k; 2) = 0$ for $m < k$ or $m > 2k$, the probability function p_m, $m = 0, 1, \ldots$, on introducing the parameters

$$\alpha = p\sqrt{2\theta/(1-\theta q^2)}, \; \beta = q\sqrt{2\theta/(1-\theta q^2)},$$

and changing the variable k to $j = m - k$, may be transformed to

$$p_0 = [-\log(1-\theta)]^{-1}[-\log(1-\theta q^2)],$$

$$p_m = [-\log(1-\theta)]^{-1} a^m \sum_{j=0}^{[m/2]} \frac{1}{m-j} \binom{m-j}{j} \frac{\beta^{m-2j}}{2^j}, \quad m = 1, 2, \ldots.$$

Similarly, its binomial moments may be written as

$$b_{(r)} = [-\log(1-\theta)]^{-1}(\gamma p)^r \sum_{j=0}^{[r/2]} \frac{1}{r-j} \binom{r-j}{j} \frac{\gamma^{m-2j}}{2^j}, \quad r = 1, 2, \ldots,$$

where

$$\gamma = \sqrt{2\theta/(1-\theta)}.$$

Further, these sums may be expressed by modified Chebyshev polynomials of the first kind as follows. The *Chebyshev polynomial of the first kind* of degree n, denoted by $t_n(x)$, may be defined through its generating function by

$$\sum_{n=1}^{\infty} t_n(x) \frac{t^n}{n!} = -\log(1 - xt + t^2/2).$$

Expanding it into powers of t and equating the coefficients of t^n on both sides of the resulting expression, we find

$$t_n(x) = n! \sum_{j=0}^{[n/2]} (-1)^j \frac{1}{n-j} \binom{n-j}{j} \frac{x^{n-2j}}{2^j}, \quad n = 1, 2, \ldots.$$

A *modified Chebyshev polynomial of the first kind*, with positive coefficients, is defined by

$$T_n(x) = i^{-n} t_n(ix), \quad n = 1, 2, \ldots, \quad i = \sqrt{-1}.$$

Then,

$$T(t; x) = \sum_{n=1}^{\infty} T_n(x) \frac{t^n}{n!} = -\log(1 - xt - t^2/2)$$

and

$$T_n(x) = n! \sum_{j=0}^{[n/2]} \frac{1}{n-j} \binom{n-j}{j} \frac{x^{n-2j}}{2^j},$$

where $[n/2]$ is the integer part of $n/2$. In particular,

$$T_1(x) = x, \quad T_2(x) = x^2 + 1.$$

Differentiating the generating function $T(t; x)$, with respect to t, and multiplying the resulting expression by $(1 - xt - t^2/2)$, we get the relation

$$\frac{dT(t;x)}{dt} = x + t + (xt + t^2/2) \frac{dT(t;x)}{dt},$$

which entails for the modified Chebyshev polynomials of the first kind the recurrence relation
$$T_{n+1}(x) = nxT_n(x) + \binom{n}{2}T_{n-1}(x), \; n = 2, 3, \ldots,$$
with initial conditions $T_1(x) = x$ and $T_2(x) = x^2 + 1$.

Clearly, the probability function p_m, $m = 0, 1, \ldots$, and its binomial moments $b_{(r)}$, $r = 1, 2, \ldots$, on introducing the modified Chebyshev polynomial of the first kind, are expressed as
$$p_0 = [-\log(1-\theta)]^{-1}[-\log(1-\theta q^2)],$$
$$p_m = [-\log(1-\theta)]^{-1} T_m(\beta) \frac{\alpha^m}{m!}, \; m = 1, 2, \ldots,$$
and
$$b_{(r)} = [-\log(1-\theta)]^{-1} T_r(\gamma) \frac{(\gamma p)^r}{r!}, \; r = 1, 2, \ldots.$$

In particular, for $r = 1, 2$, the mean and variance are obtained as
$$\mu = c\frac{2p\theta}{1-\theta}, \; \sigma^2 = c(1+p)\frac{2p\theta}{(1-\theta)} + c(1-c)\left(\frac{2p\theta}{1-\theta}\right)^2,$$
with $c = [-\log(1-\theta)]^{-1}$. This distribution may be called a *Chebyshev distribution of the first kind*.

The recurrence relation of the modified Chebyshev polynomials of the first kind yields for p_m, $m = 0, 1, \ldots$, and $b_{(r)}$, $r = 1, 2, \ldots$, the following recurrence relations
$$(m+1)p_{m+1} = \frac{2m\theta pq p_m + (m-1)\theta p^2 p_{m-1}}{1-\theta q^2}, \; m = 2, 3, \ldots$$
with $p_1 = 2c\theta pq/(1-\theta q^2)$, $p_2 = c\theta p(1+\theta q^2)/(1-\theta q^2)^2$, and
$$(r+1)b_{(r+1)} = \frac{2r\theta p b_{(r)} + (r-1)\theta p^2 b_{(r-1)}}{1-\theta}, \; r = 2, 3, \ldots,$$
with $b_{(1)} = 2c\theta p/(1-\theta)$, $b_{(2)} = c\theta p(1+\theta)/(1-\theta)^2$.

7.6.2 Logarithmic Distribution of Order k

Suppose that the number N of urns obeys a logarithmic distribution with probability function
$$c_n = P(N = n) = [-k\log p]^{-1}\frac{(1-p^k)^n}{n}, \; n = 1, 2, \ldots, \; 0 < p < 1,$$
where k is a positive integer. Also, assume that the number X_j of balls distributed into the jth urn, $j = 1, 2, \ldots$, has a right truncated geometric distribution with probability function
$$q_x = P(X = x) = (1-p^k)^{-1} q p^{x-1}, \; x = 1, 2, \ldots, k,$$

where $q = 1 - p$. Clearly,

$$g(t) = \sum_{x=1}^{k} q_x t^x = t \left(\frac{1 - p^k t^k}{1 - pt} \right) \bigg/ \left(\frac{1 - p^k}{1 - p} \right)$$

and so the generating function of the n-fold convolution of q_x, $x = 1, 2, \ldots, k$, is given by

$$g_n(t) = \sum_{m=n}^{kn} q_m(n) t^m = t^n \left(\frac{1 - p^k t^k}{1 - pt} \right)^n \bigg/ \left(\frac{1 - p^k}{1 - p} \right)^n.$$

Hence,

$$q_m(n) = (1 - p^k)^{-n} Q(m, n; k) q^n p^{m-n}, \quad m = n, n+1, \ldots, kn,$$

where

$$Q(m, n; k) = \sum_{j=0}^{n} (-1)^j \binom{n}{j} \binom{m - jk - 1}{n - 1}.$$

Then, the probability function of the compound distribution, on using (7.31) with $\theta = 1 - p^k$, is deduced as

$$p_m = [-k \log p]^{-1} \sum_{n=1}^{m} \frac{1}{n} Q(m, n; k) q^n p^{m-n}, \quad m = 1, 2, \ldots.$$

This distribution is called a *logarithmic distribution of order k*.

The general expression of the binomial moments of this distribution is not manageable and so for the calculation of the mean and variance it is preferable to use the expressions

$$E(S_N) = E(N) E(X), \quad V(S_N) = E(N) V(X) + V(N) [E(X)]^2.$$

Clearly, the mean $E(N)$ and the variance $V(N)$ of the logarithmic distribution, with parameter $\theta = 1 - p^k$, are given by

$$E(N) = \frac{1 - p^k}{p^k [-k \log p]}, \quad V(N) = \frac{1 - p^k}{p^{2k} [-k \log p]} - \frac{(1 - p^k)^2}{p^{2k} [-k \log p]^2}.$$

Further, the mean $E(X)$ and the variance $V(X)$ of the right truncated geometric distribution may be obtained, by differentiating its probability generating function, as

$$E(X) = \frac{1}{q} - \frac{k p^k}{1 - p^k}, \quad V(X) = \frac{p}{q^2} - \frac{k^2 p^k}{(1 - p^k)^2}.$$

Then, after some algebraic manipulations, the mean $\mu = E(S_N)$ and the variance $\sigma^2 = V(S_N)$ of the logarithmic distribution of order k are deduced as

$$\mu = \frac{1 - p^k - k q p^k}{q p^k [-k \log p]}, \quad \sigma^2 = \frac{1 - (2k+1) q p^k - p^{2k+1}}{q p^{2k} [-k \log p]} - \frac{(1 - p^k - k q p^k)^2}{q^2 p^{2k} [-k \log p]^2}.$$

7.7 COMPOUND NEGATIVE BINOMIAL DISTRIBUTIONS

Assume that the number N of urns has a negative binomial distribution with probability function

$$c_n = P(N = n) = \binom{s+n-1}{n}(1-\theta)^s \theta^n, \quad n = 0, 1, \ldots,$$

where $0 < \theta < 1$ and $s > 0$. Its kth binomial moment is

$$\beta_{(k)} = \binom{s+k-1}{k}\left(\frac{\theta}{1-\theta}\right)^k, \quad k = 0, 1, \ldots.$$

Further, for $q_0 > 0$,

$$f(q_0) = \sum_{n=0}^{\infty} c_n q_0^n = (1-\theta)^s \sum_{n=0}^{\infty} \binom{s+n-1}{n}(\theta q_0)^n = \frac{(1-\theta)^s}{(1-\theta q_0)^s}$$

and the power series distribution (7.3) reduces to a negative binomial distribution with probability function

$$c_n(q_0) = \binom{s+n-1}{n}(1-\theta q_0)^s(\theta q_0)^n, \quad n = 0, 1, \ldots,$$

and binomial moments

$$\beta_{(k)}(q_0) = \binom{s+k-1}{k}\left(\frac{\theta q_0}{1-\theta q_0}\right)^k, \quad k = 0, 1, \ldots.$$

Therefore, according to Theorem 7.1, the probability function of a *compound negative binomial distribution* is given by

$$p_m = \left(\frac{1-\theta}{1-\theta q_0}\right)^s \sum_{k=0}^{m} \binom{s+k-1}{k}\left(\frac{\theta q_0}{1-\theta q_0}\right)^k \left[\Delta_u^k q_0^{-u} q_m(u)\right]_{u=0}, \quad (7.33)$$

for $m = 0, 1, \ldots$. If $q_0 = 0$, then

$$p_m = (1-\theta)^s \sum_{n=0}^{m} \binom{s+n-1}{n} \theta^n q_m(n), \quad m = 0, 1, \ldots. \quad (7.34)$$

The rth binomial moment of the compound negative binomial distribution is also deduced from Theorem 7.1, as

$$b_{(r)} = \sum_{k=0}^{r} \binom{s+k-1}{k}\left(\frac{\theta}{1-\theta}\right)^k \left[\Delta_u^k a_{(r)}(u)\right]_{u=0}, \quad r = 0, 1, \ldots, \quad (7.35)$$

where $a_{(r)}(u)$ is the rth binomial moment of $q_m(u)$, $m = 0, 1, \ldots$.

7.7.1 Chebyshev Distribution of the Second Kind

Consider the compound geometric distribution with compounding a binomial distribution with probability function

$$q_x = P(X = x) = \binom{2}{x} p^x q^{2-x}, \quad x = 0, 1, 2,$$

where $q = 1 - p$ and $0 < p < 1$. Then,

$$q_m(u) = P(S_u = m) = \binom{2u}{m} p^m q^{2u-m}, \quad m = 0, 1, \ldots, 2u,$$

and

$$a_{(r)}(u) = E\left[\binom{S_u}{r}\right] = \binom{2u}{r} p^r, \quad r = 0, 1, \ldots, 2u, \quad a_{(r)} = 0, \; r > 2u.$$

Also,

$$q_0^{-u} q_m(u) = \binom{2u}{m} \left(\frac{p}{q}\right)^m, \quad m = 0, 1, \ldots$$

and

$$\left[\Delta_u^k \binom{2u}{m}\right]_{u=0} = \frac{k!}{m!} C(m, k; 2),$$

where $C(m, k; 2)$ is the generalized factorial coefficient. Therefore, from (7.33) and (7.35) with $s = 1$, it follows that the probability function and binomial moments of the compound distribution are given by

$$p_m = \left(\frac{1-\theta}{1-\theta q^2}\right) \frac{(p/q)^m}{m!} \sum_{k=0}^{m} k! C(m, k; 2) \left(\frac{\theta q^2}{1-\theta q^2}\right)^k, \quad m = 0, 1, \ldots,$$

and

$$b_{(r)} = \frac{p^r}{r!} \sum_{k=0}^{r} k! C(r, k; 2) \left(\frac{\theta}{1-\theta}\right)^k, \quad r = 0, 1, \ldots,$$

where (see Section 7.5.1)

$$C(m, k; 2) = \frac{m! 2^{2k-m}}{(m-k)!(2k-m)!}, \quad m = k, k+1, \ldots, 2k,$$

with $C(m, k; 2) = 0$ for $m < k$ or $m > 2k$. Thus, the probability function p_m, $m = 0, 1, \ldots$, on introducing the parameters

$$\alpha = p\sqrt{2\theta/(1-\theta q^2)}, \quad \beta = q\sqrt{2\theta/(1-\theta q^2)},$$

and changing the variable k to $j = m - k$, is transformed to

$$p_m = (1 - \alpha\beta - \alpha^2/2) \alpha^m \sum_{j=0}^{[m/2]} \binom{m-j}{j} \frac{\beta^{m-2j}}{2^j}, \quad m = 0, 1, \ldots.$$

Similarly, its binomial moments may be written as

$$b_{(r)} = (\gamma p)^r \sum_{j=0}^{[r/2]} \binom{r-j}{j} \frac{\gamma^{m-2j}}{2^j}, \quad r = 0, 1, \ldots,$$

where
$$\gamma = \sqrt{2\theta/(1-\theta)}.$$

Further, these sums may be expressed by modified Chebyshev polynomials of the second kind. Indeed, the *Chebyshev polynomial of the second kind* of degree n, denoted by $u_n(x)$, may be defined through its generating function by

$$\sum_{n=0}^{\infty} u_n(x) \frac{t^n}{n!} = (1 - xt + t^2/2)^{-1}.$$

Expanding it into powers of t and equating the coefficients of t^n on both sides of the resulting expression, we get

$$u_n(x) = n! \sum_{j=0}^{[n/2]} (-1)^j \binom{n-j}{j} \frac{x^{n-2j}}{2^j}, \quad n = 0, 1, \ldots.$$

A *modified Chebyshev polynomial of the second kind*, with positive coefficients, is defined by
$$U_n(x) = i^{-n} u_n(ix), \quad n = 0, 1, \ldots, \quad i = \sqrt{-1}.$$

Then,
$$U(t; x) = \sum_{n=0}^{\infty} U_n(x) \frac{t^n}{n!} = (1 - xt - t^2/2)^{-1}$$

and
$$U_n(x) = n! \sum_{j=0}^{[n/2]} \binom{n-j}{j} \frac{x^{n-2j}}{2^j}, \quad n = 0, 1, \ldots.$$

In particular,
$$U_0(x) = 1, \quad U_1(x) = x, \quad U_2(x) = 2x^2 + 1.$$

Differentiating the generating function $U(t; x)$, with respect to t, and multiplying the resulting expression by $(1 - xt - t^2/2)$, we get the relation

$$\frac{dU(t; x)}{dt} = (x+t)U(t; x) + (xt + t^2/2)\frac{dU(t; x)}{dt},$$

which entails for the modified Chebyshev polynomials of the second kind the recurrence relation

$$U_{n+1}(x) = (n+1)xU_n(x) + \binom{n+1}{2} U_{n-1}(x), \quad n = 1, 2, \ldots,$$

with initial conditions $U_0(x) = 1$ and $U_1(x) = x$.

Introducing the modified Chebyshev polynomials of the second kind into the expressions of the probability function p_m, $m = 0, 1, \ldots$, and its binomial moments $b_{(r)}$, $r = 0, 1, \ldots$, we find

$$p_m = (1 - \alpha\beta - \alpha^2/2)U_m(\beta)\frac{\alpha^m}{m!}, \quad m = 0, 1, \ldots,$$

and

$$b_{(r)} = U_r(\gamma)\frac{(\gamma p)^r}{r!}, \quad r = 0, 1, \ldots.$$

In particular, for $r = 1, 2$, the mean and variance are deduced as

$$\mu = \frac{2\theta p}{1 - \theta}, \quad \sigma^2 = \frac{2\theta p[2p + (1 - \theta)(1 - p)]}{(1 - \theta)^2}.$$

This distribution may be called a *Chebyshev distribution of the second kind*.

The recurrence relation of the modified Chebyshev polynomials of the second kind entails for the probability function p_m, $m = 0, 1, \ldots$, and its binomial moments $b_{(r)}$, $r = 0, 1, \ldots$, the recurrence relations

$$p_{m+1} = \frac{2\theta pq p_m + \theta p^2 p_{m-1}}{1 - \theta q^2}, \quad m = 1, 2, \ldots,$$

with $p_0 = (1 - \theta)/(1 - \theta q^2)$, $p_1 = 2\theta(1 - \theta)pq/(1 - \theta q^2)^2$, and

$$b_{(r+1)} = \frac{2\theta p b_{(r)} + \theta p^2 b_{(r-1)}}{1 - \theta}, \quad r = 1, 2, \ldots,$$

with $b_{(0)} = 1$, $b_{(1)} = 2\theta p/(1 - \theta)$.

7.7.2 Gegenbauer Distribution

The Gegenbauer distribution is a compound negative binomial distribution with compounding a binomial distribution with probability function

$$q_x = P(X = x) = \binom{2}{x} p^x q^{2-x}, \quad x = 0, 1, 2,$$

where $q = 1 - p$ and $0 < p < 1$. The probability function p_m, $m = 0, 1, \ldots$, and it binomial moments $b_{(r)}$, $r = 0, 1, \ldots$, following the steps of derivation of the Chebyshev distribution of the second kind, are obtained as

$$p_m = (1 - \alpha\beta - \alpha^2/2)^s \alpha^m \sum_{j=0}^{[m/2]} \binom{m-j}{j}\binom{s+m-j-1}{m-j}\frac{\beta^{m-2j}}{2^j},$$

for $m = 0, 1, \ldots$, and

$$b_{(r)} = (\gamma p)^r \sum_{j=0}^{[r/2]} \binom{r-j}{j}\binom{s+r-j-1}{r-j}\frac{\gamma^{m-2j}}{2^j},$$

for $r = 0, 1, \ldots$, where
$$\alpha = p\sqrt{2\theta/(1-\theta q^2)}, \ \beta = q\sqrt{2\theta/(1-\theta q^2)}, \ \gamma = \sqrt{2\theta/(1-\theta)}.$$

Further, consider the *Gegenbauer polynomials* $g_n^{(s)}(x)$, $n = 0, 1, \ldots$, with $-1/2 < s < 0$ or $0 < s < \infty$, which may be defined by their generating function as
$$\sum_{n=0}^{\infty} g_n^{(s)}(x) \frac{t^n}{n!} = (1 - xt + t^2/2)^{-s}.$$

Then,
$$g_n^{(s)}(x) = n! \sum_{j=0}^{[n/2]} (-1)^j \binom{n-j}{j} \binom{s+n-j-1}{n-j} \frac{x^{n-2j}}{2^j}, \ n = 0, 1, \ldots.$$

A *modified Gegenbauer polynomial*, with positive coefficients, is defined by
$$G_n^{(s)}(x) = i^{-n} g_n^{(s)}(ix), \ n = 0, 1, \ldots, \ i = \sqrt{-1},$$
so that
$$\sum_{n=0}^{\infty} G_n^{(s)}(x) \frac{t^n}{n!} = (1 - xt - t^2/2)^{-s}$$
and
$$G_n^{(s)}(x) = n! \sum_{j=0}^{[n/2]} \binom{n-j}{j} \binom{s+n-j-1}{n-1} \frac{x^{n-2j}}{2^j}, \ n = 0, 1, \ldots.$$

In particular,
$$G_0^{(s)}(x) = 1, \ G_1^{(s)}(x) = sx, \ G_2^{(s)}(x) = s(s+1)x^2 + s.$$

Also, differentiating the generating function, with respect to t, we find the following recurrence relation:
$$G_{n+1}^{(s)}(x) = (n+s)xG_n^{(s)}(x) + \left\{\binom{n}{2} + sn\right\} G_{n-1}^{(s)}(x), \ n = 1, 2, \ldots,$$
with initial conditions $G_0^{(s)}(x) = 1$ and $G_1^{(s)}(x) = sx$.

Introducing the modified Gegenbauer polynomials into the expressions of the probability function p_m, $m = 0, 1, \ldots$, and its binomial moments $b_{(r)}$, $r = 0, 1, \ldots$, we get
$$p_m = (1 - \alpha\beta - \alpha^2/2)^s G_m^{(s)}(\beta) \frac{\alpha^m}{m!}, \ m = 0, 1, \ldots,$$
and
$$b_{(r)} = G_r^{(s)}(\gamma) \frac{(\gamma p)^r}{r!}, \ r = 0, 1, \ldots.$$

The mean and variance are deduced from the last expression as

$$\mu = \frac{2s\theta p}{1-\theta}, \quad \sigma^2 = \frac{2s\theta p[2p + (1-\theta)(1-p)]}{(1-\theta)^2}.$$

This distribution may be called a *Gegenbauer distribution*.

The recurrence relation of the modified Gegenbauer polynomials entails for the probability function p_m, $m = 0, 1, \ldots$, and its binomials moments $b_{(r)}$, $r = 0, 1, \ldots$, the recurrence relations

$$p_{m+1} = \frac{2(m+s)\theta p q p_m + (m+2s-1)\theta p^2 p_{m-1}}{(m+1)(1-\theta q^2)}, \quad m = 1, 2, \ldots,$$

with $p_0 = (1-\theta)^s/(1-\theta q^2)^s$, $p_1 = 2s\theta pq(1-\theta)^s/(1-\theta q^2)^{s+1}$, and

$$b_{(r+1)} = \frac{2(r+s)\theta p b_{(r)} + (r+2s-1)\theta p^2 b_{(r-1)}}{(r+1)(1-\theta)}, \quad r = 1, 2, \ldots,$$

with $b_{(0)} = 1$, $b_{(1)} = 2s\theta p/(1-\theta)$.

7.7.3 Geometric Distribution of Order k

Suppose that the number N of urns obeys a geometric distribution with probability function

$$c_n = P(N = n) = p^k(1-p^k)^n, \quad n = 0, 1, \ldots,$$

where $0 < p < 1$ and k is a positive integer. Also, assume that the number X_j of balls distributed into the jth urn, $j = 1, 2, \ldots$, has a right truncated geometric distribution with probability function

$$q_x = (1-p^k)^{-1} q p^{x-1}, \quad x = 1, 2, \ldots, k,$$

where $q = 1 - p$. The n-fold convolution of this distribution was derived in Section 7.6.2 as

$$q_m(n) = (1-p^k)^{-n} Q(m, n; k) q^n p^{m-n}, \quad m = n, n+1, \ldots, kn,$$

where

$$Q(m, n; k) = \sum_{j=0}^{n}(-1)^j \binom{n}{j}\binom{m - jk - 1}{n - 1}.$$

The probability function of the compound distribution, on using (7.34) with $s = 1$ and $\theta = 1 - p^k$, is obtained as

$$p_m = p^k \sum_{n=0}^{m} Q(m, n; k) q^n p^{m-n}, \quad m = 0, 1, \ldots.$$

This distribution is called a *geometric distribution of order k*.

The expression of the binomial moments, which can be deduced from (7.35) with $s = 1$ and $\theta = 1 - p^k$, due to the involved expression of the binomial moments of $q_m(n)$, $m = n, n+1, \ldots, kn$, is not manageable. Nevertheless, the mean and variance of this compound distribution are easily obtained by using the expressions

$$\mu \equiv E(S_N) = E(N)E(X),$$

$$\sigma^2 \equiv V(S_N) = E(N)V(X) + V(N)[E(X)]^2.$$

Indeed, the mean $E(N)$ and the variance $V(N)$ of the geometric distribution, with probability of success p^k, are given by

$$E(N) = \frac{1 - p^k}{p^k}, \quad V(N) = \frac{1 - p^k}{p^{2k}}.$$

Also, the mean $E(X)$ and the variance $V(X)$ of the right truncated geometric distribution are easily obtained, by differentiating its probability generating function, as

$$E(X) = \frac{1}{q} - \frac{kp^k}{1 - p^k}, \quad V(X) = \frac{p}{q^2} - \frac{k^2 p^k}{(1 - p^k)^2}.$$

Then, a simple calculation yields

$$\mu = \frac{1 - p^k}{qp^k} - k, \quad \sigma^2 = \frac{1 - (2k+1)qp^k - p^{2k+1}}{q^2 p^{2k}}.$$

It is noteworthy that in a sequence of independent Bernoulli trials, each with probability of success p, the probability function of the number of trials that precede the occurrence of a success run of length k is given by p_m, $m = 0, 1, \ldots$. W. Feller (1968) deduced the probability function p_{z-k}, $z = k, k+1, \ldots$, as an application of the theory of recurrent events.

7.8 PARTITION POLYNOMIALS AND COMPOUND DISTRIBUTIONS

In this section, the use of partition polynomials in expressing the probability function and binomial moments of compound discrete distributions in terms of the component probability functions and binomial moments is briefly presented.

The partition polynomials, introduced by E. T. Bell (1927, 1934a,b), are multivariable polynomials that are defined by a sum extended over all partitions of their index. Such a sum had previously been used by F. Faà di Bruno (1855a,b) to express the derivatives of a composite function in terms of the derivatives of the component functions. The later partition polynomials have found many applications in combinatorics, probability theory and statistics. Since they are quite general polynomials, they include as particular cases the exponential polynomials and their inverses, the logarithmic polynomials, as well as the potential polynomials, owing their particular name to the form of their generating functions.

A *partition of a positive integer n into k parts* is an unordered collection of k positive integers $\{r_1, r_2, \ldots, r_k\}$, $r_1 \geq r_2 \geq \cdots \geq r_k \geq 1$, $k = 1, 2, \ldots, n$, whose sum equals n. In the case of a partition of n into an unspecified number of parts, the term *partition of n* is used. Since in a partition the order of its parts does not count, they are registered in decreasing order of magnitude. Thus, a partition of a positive integer n into k parts is a solution in positive integers of the linear equation:

$$x_1 + x_2 + \cdots + x_k = n, \quad x_1 \geq x_2 \geq \cdots \geq x_k \geq 1.$$

In a partition of n, let $k_i \geq 0$ be the number of parts that are equal to i, $i = 1, 2, \ldots, n$. Then,

$$k_1 + 2k_2 + \cdots + nk_n = n, \quad k_i \geq 0, \quad i = 1, 2, \ldots, n. \tag{7.36}$$

In the case of a partition of n into k parts, in addition, it holds that

$$k_1 + k_2 + \cdots + k_n = k. \tag{7.37}$$

If $\{k_{i_1}, k_{i_2}, \ldots, k_{i_{r-1}}, k_{i_r}\}$, with $i_1 < i_2 < \cdots < i_{r-1} < i_r$, is the set of the positive k_i in (7.36), the corresponding partition of n is denoted by

$$i_r^{k_{i_r}} i_{r-1}^{k_{i_{r-1}}} \cdots i_1^{k_{i_1}},$$

omitting the exponents that are equal to 1.

The polynomial $B_n \equiv B_n(x_1, x_2, \ldots, x_n)$ in the variables $x_1, x_2, \ldots, x_n$, defined by the sum

$$B_n = \sum \frac{n!}{k_1!(1!)^{k_1} k_2!(2!)^{k_2} \cdots k_n!(n!)^{k_n}} x_1^{k_1} x_2^{k_2} \cdots x_n^{k_n}, \tag{7.38}$$

where the summation is extended over all partitions of n, is called the *exponential Bell partition polynomial*.

The polynomial $B_{n,k} \equiv B_{n,k}(x_1, x_2 \ldots, x_n)$ in the variables $x_1, x_2, \ldots, x_n$ of degree k, defined by the sum

$$B_{n,k} = \sum \frac{n!}{k_1!(1!)^{k_1} k_2!(2!)^{k_2} \cdots k_n!(n!)^{k_n}} x_1^{k_1} x_2^{k_2} \cdots x_n^{k_n}, \tag{7.39}$$

where the summation is extended over all partitions of n into k parts, is called the *partial Bell partition polynomial*.

Clearly,

$$B_n(ax_1, a^2 x_2, \ldots, a^n x_n) = a^n B_n(x_1, x_2, \ldots, x_n) \tag{7.40}$$

and

$$B_{n,k}(abx_1, a^2 bx_2, \ldots, a^n bx_n) = a^n b^k B_{n,k}(x_1, x_2, \ldots, x_n). \tag{7.41}$$

Also,
$$B_n(x_1, x_2, \ldots, x_n) = \sum_{k=0}^{n} B_{n,k}(x_1, x_2, \ldots, x_n), \quad n = 0, 1, \ldots. \quad (7.42)$$

Multiplying (7.39) by $u^k t^n/n!$ and summing the resulting expression for $k = 0, 1, \ldots, n$ and $n = 0, 1, \ldots$, we get the bivariate generating function of the partial Bell partition polynomials as

$$B(t, u) = \sum_{n=0}^{\infty} \sum_{k=0}^{n} B_{n,k}(x_1, x_2, \ldots, x_n) u^k \frac{t^n}{n!} = \exp\{u[g(t) - x_0]\}, \quad (7.43)$$

where $g(t) = \sum_{r=0}^{\infty} x_r t^r/r!$. Interchanging the order of summation on the left-hand side, we deduce the vertical generating function

$$B_k(t) = \sum_{n=k}^{\infty} B_{n,k}(x_1, x_2, \ldots, x_n) \frac{t^n}{n!} = \frac{[g(t) - x_0]^k}{k!}, \quad k = 0, 1, \ldots. \quad (7.44)$$

Setting $u = 1$ in (7.43), we find, by virtue of (7.42), the generating function of the exponential Bell partition polynomials as

$$B(t) = \sum_{n=0}^{\infty} B_n(x_1, x_2, \ldots, x_n) \frac{t^n}{n!} = \exp[g(t) - x_0]. \quad (7.45)$$

Introducing a sequence of parameters c_k, $k = 0, 1, \ldots$, the exponential Bell partition polynomial $B_n(x_1, x_2, \ldots, x_n)$ is extended to the general partition polynomial. More precisely, using the symbolic calculus, the polynomial

$$B_n(cx_1, cx_2, \ldots, cx_n) = \sum_{k=0}^{n} c_k B_{n,k}(x_1, x_2, \ldots, x_n), \quad c^k \equiv c_k, \quad (7.46)$$

for $n = 0, 1, \ldots$, is called the *partition polynomial*. These polynomials were used by F. Faà di Bruno to express the derivatives $h_n = [d^n h(t)/dt^n]_{t=a}$, $n = 0, 1, \ldots$, of a composite function $h(t) = f(g(t))$ in terms of the derivatives $g_r = [d^r g(t)/dt^r]_{t=a}$, $r = 1, 2, \ldots$, and $f_k = [d^k f(u)/du^k]_{u=g(a)}$, $k = 0, 1, \ldots$, of the component functions as

$$h_n = \sum_{k=0}^{n} f_k B_{n,k}(g_1, g_2, \ldots, g_n)$$
$$= B_n(fg_1, fg_2, \ldots, fg_n), \quad f^k \equiv f_k. \quad (7.47)$$

The partition polynomials include as particular cases, in addition to the exponential partition polynomials, the logarithmic and the potential partition polynomials. Specifically,

$$L_n(x_1, x_2, \ldots, x_n) = B_n(cx_1, cx_2, \ldots, cx_n), \quad c^k \equiv c_k, \quad (7.48)$$

for $n = 0, 1, \ldots$, with $c_0 = 0$, $c_k = (-1)^{k-1}(k-1)!$, $k = 1, 2, \ldots, n$, is the *logarithmic partition polynomial*. Note that $B_0 = c_0$ and so $L_0 = 0$. Also,

$$C_{n,s}(x_1, x_2, \ldots, x_n) = B_n(cx_1, cx_2, \ldots, cx_n), \quad c^k \equiv c_k, \tag{7.49}$$

for $n = 0, 1, \ldots$, with $c_k = (s)_k$, $k = 0, 1, \ldots, n$ and s a real number, is the *potential partition polynomial*. From the vertical generating function (7.44) and expression (7.46), with $c_0 = 0$, $c_k = (-1)^{k-1}(k-1)!$, $k = 1, 2, \ldots, n$ and $c_k = (s)_k$, $k = 0, 1, \ldots, n$, the generating functions of these polynomials are readily deduced as

$$L(t) = \sum_{n=1}^{\infty} L_n(x_1, x_2, \ldots, x_n) \frac{t^n}{n!} = \log\{1 + [g(t) - x_0]\} \tag{7.50}$$

and

$$C_s(t) = \sum_{n=0}^{\infty} C_{n,s}(x_1, x_2, \ldots, x_n) \frac{t^n}{n!} = [1 + (g(t) - x_0)]^s, \tag{7.51}$$

respectively. A derivation of the Faà di Bruno formula, together with recurrence relations and other properties of the partition polynomials can be found in L. Comtet (1974), J. Riordan (1958, 1968) and Ch. A. Charalambides (2002).

Finally, it is worth noting that a comparison of definition (7.39) of the partial Bell partition polynomials with expressions (2.29) and (2.30) of the Stirling numbers and expression (2.62) of the generalized factorial coefficients entails

$$B_{n,k}(0!, 1!, \ldots, (n-1)!) = |s(n, k)|,$$

$$B_{n,k}(1, 1, \ldots, 1) = S(n, k)$$

and

$$B_{n,k}((s)_1, (s)_2, \ldots, (s)_n) = C(n, k; s).$$

The derivation of partition polynomial expressions for the probability function and binomial moments of compound discrete distribution is facilitated by the following lemma.

Lemma 7.1 (a) *Under the random occupancy model with $q_0 > 0$, the probability $q_m(n)$, $m = 0, 1, \ldots$, that m balls are distributed into the urns is given by*

$$q_m(n) = \frac{q_0^n}{m!} \sum_{k=0}^{m} (n)_k B_{m,k}(1!q_0^{-1}q_1, 2!q_0^{-1}q_2, \ldots, m!q_0^{-1}q_m)$$

$$= \frac{q_0^n}{m!} C_{m,n}(1!q_0^{-1}q_1, 2!q_0^{-1}q_2, \ldots, m!q_0^{-1}q_m). \tag{7.52}$$

If $q_0 = 0$, then

$$q_m(n) = \frac{n!}{m!} B_{m,n}(1!q_1, 2!q_2, \ldots, m!q_m). \tag{7.53}$$

322 COMPOUND AND MIXTURE DISTRIBUTIONS

(b) *The rth binomial moment $a_{(r)}(n)$, $r = 0, 1, \ldots$, of the probability function $q_m(n)$, $m = 0, 1, \ldots$, is given by*

$$a_{(r)}(n) = \frac{1}{r!} \sum_{k=0}^{r} (n)_k B_{r,k}(1!a_{(1)}, 2!a_{(2)}, \ldots, r!a_{(r)})$$

$$= \frac{1}{r!} C_{r,n}(1!a_{(1)}, 2!a_{(2)}, \ldots, r!a_{(r)}), \qquad (7.54)$$

where $a_{(j)} = E\bigl[\binom{X}{j}\bigr]$, $j = 0, 1 \ldots$.

Proof (a) The probability function $q_m(n) = P(S_n = m)$, $m = 0, 1, \ldots$, is given in terms of the probabilities $q_x = P(X = x)$, $x = 0, 1, \ldots$, by the sum

$$q_m(n) = \sum q_{x_1} q_{x_2} \cdots q_{x_n},$$

where the summation is extended over all $x_i \geq 0$, $i = 1, 2, \ldots, n$, such that $x_1 + x_2 + \cdots + x_n = m$. Letting $k_j \geq 0$ be the number of urns containing j balls each, $j = 0, 1, \ldots, m$, this sum may be transformed to the sum

$$q_m(n) = \sum \frac{n!}{k_0! k_1! \cdots k_m!} q_0^{k_0} q_1^{k_1} \cdots q_m^{k_m},$$

where the summation is extended over all nonnegative integer solutions $(k_0, k_1, \ldots, k_m)$ of the equations $k_1 + 2k_2 + \cdots + mk_m = m$ and $k_0 + k_1 + \cdots + k_m = n$. Introducing the variable $k = n - k_0$ and first taking the sum over all partitions of m into k parts and then summing over all $k = 0, 1, \ldots, m$, we get the expression

$$q_m(n) = \frac{q_0^n}{m!} \sum_{k=0}^{m} (n)_k \sum \frac{m! q_0^{-k}}{k_1! k_2! \cdots k_m!} q_1^{k_1} q_2^{k_2} \cdots q_m^{k_m},$$

which, using (7.39) and (7.49), implies (7.52). If $q_0 = 0$, then

$$q_m(n) = \frac{n!}{m!} \sum \frac{m!}{k_1! k_2! \cdots k_m!} q_1^{k_1} q_2^{k_2} \cdots q_m^{k_m},$$

which, using (7.39), implies (7.53).

(b) The rth binomial moment $a_{(r)}(n) = E\bigl[\binom{S_n}{r}\bigr]$, $r = 0, 1, \ldots$, is given in terms of the binomial moments $a_{(j)} = E\bigl[\binom{X}{j}\bigr]$, $j = 0, 1, \ldots$, by the sum

$$a_{(r)}(n) = \sum a_{(j_1)} a_{(j_2)} \cdots a_{(j_n)},$$

where the summation is extended over all $j_i \geq 0$, $i = 1, 2, \ldots, n$, such that $j_1 + j_2 + \cdots + j_n = m$. Letting $k_j \geq 0$ be the number of indices among $j_1, j_2, \ldots, j_n$ that are equal to j, $j = 0, 1, \ldots, r$, and since $a_{(0)} = 1$, this sum is transformed to the sum

$$a_{(r)}(n) = \sum \frac{n!}{k_0! k_1! \cdots k_r!} a_{(1)}^{k_1} a_{(2)}^{k_2} \cdots a_{(r)}^{k_r},$$

where the summation is extended over all nonnegative integer solutions $(k_0, k_1, \ldots, k_r)$ of the equations $k_1 + 2k_2 + \cdots + rk_r = r$ and $k_0 + k_1 + \cdots + k_r = n$. Then, proceeding as above, we get (7.54).

Theorem 7.2 (a) *Under the random occupancy model with a random number of urns and assuming that $q_0 > 0$, the probability p_m, $m = 0, 1, \ldots$, that m balls are distributed into the urns is given by*

$$p_m = \frac{1}{m!} f(q_0) \sum_{k=0}^{m} k! \beta_{(k)}(q_0) B_{m,k}(1! q_0^{-1} q_1, 2! q_0^{-1} q_2, \ldots, m! q_0^{-1} q_m). \quad (7.55)$$

If $q_0 = 0$, then

$$p_m = \frac{1}{m!} \sum_{k=0}^{m} k! c_k B_{m,k}(1! q_1, 2! q_2, \ldots, m! q_m). \quad (7.56)$$

(b) *The rth binomial moment $b_{(r)}$, $r = 0, 1, \ldots$, of the probability function p_m, $m = 0, 1, \ldots$, is given by*

$$b_{(r)} = \frac{1}{r!} \sum_{k=0}^{r} k! \beta_{(k)} B_{r,k}(1! a_{(1)}, 2! a_{(2)}, \ldots, r! a_{(r)}). \quad (7.57)$$

Proof (a) Introducing expression (7.52) into (7.1), the probability function p_m, $m = 0, 1, \ldots$, is obtained as

$$p_m = \frac{1}{m!} \sum_{n=0}^{\infty} c_n q_0^n \sum_{k=0}^{m} (n)_k B_{m,k}(1! q_0^{-1} q_1, 2! q_0^{-1} q_2, \ldots, m! q_0^{-1} q_m)$$

$$= \frac{1}{m!} \sum_{k=0}^{m} \left\{ \sum_{n=k}^{\infty} (n)_k c_n q_0^n \right\} B_{m,k}(1! q_0^{-1} q_1, 2! q_0^{-1} q_2, \ldots, m! q_0^{-1} q_m)$$

and since, by (7.4),

$$\sum_{n=k}^{\infty} (n)_k c_n q_0^n = f(q_0) k! \beta_{(k)}(q_0),$$

(7.55) is established. If $q_0 = 0$, then introducing (7.53) into (7.7), expression (7.56) of the probability function p_m, $m = 0, 1, \ldots$, is deduced.

(b) Introducing expression (7.54) into (7.2), the rth binomial moment $b_{(r)}$, $r = 0, 1, \ldots$, is obtained as

$$b_{(r)} = \frac{1}{r!} \sum_{n=0}^{\infty} c_n \sum_{k=0}^{r} (n)_k B_{r,k}(1! a_{(1)}, 2! a_{(2)}, \ldots, r! a_{(r)})$$

$$= \frac{1}{r!} \sum_{k=0}^{r} \left\{ \sum_{k=0}^{\infty} (n)_k c_n \right\} B_{r,k}(1! a_{(1)}, 2! a_{(2)}, \ldots, r! a_{(r)}),$$

and since, by (7.5),
$$\sum_{n=k}^{\infty} (n)_k c_n = k! \beta_{(k)},$$
(7.57) is established.

Theorem 7.2, by specifying the distribution $c_n = P(N = n)$, $n = 0, 1, \ldots$, of the number N of urns, yields the probability function and binomial moments of the compound Poisson, logarithmic, binomial and negative binomial distributions.

Assume that
$$c_n = P(N = n) = e^{-\theta} \frac{\theta^n}{n!}, \quad n = 0, 1, \ldots, \quad 0 < \theta < \infty.$$

Then (cf. Section 7.5),
$$\beta_{(k)} = \frac{\theta^k}{k!}, \quad k = 0, 1, \ldots,$$
and
$$f(q_0) = e^{-\theta(1-q_0)}, \quad \beta_{(k)}(q_0) = \frac{(\theta q_0)^k}{k!}, \quad k = 0, 1, \ldots.$$

Thus, from Theorem 7.2, and using (7.41) and (7.42), we deduce the probability function
$$P(S_N = m) = p_m, \quad m = 0, 1, \ldots,$$
and the binomial moments
$$b_{(r)} = E\left[\binom{S_N}{r}\right], \quad r = 0, 1, \ldots,$$
of the *compound Poisson distribution* as
$$p_m = e^{-\theta(1-q_0)} \frac{1}{m!} \sum_{k=0}^{m} \theta^k B_{m,k}(1!q_1, 2!q_2, \ldots, m!q_m)$$
$$= e^{-\theta(1-q_0)} \frac{1}{m!} B_m(1!\theta q_1, 2!\theta q_2, \ldots, m!\theta q_m), \quad m = 0, 1, \ldots,$$
and
$$b_{(r)} = \frac{1}{r!} \sum_{k=0}^{r} \theta^k B_{r,k}(1!a_{(1)}, 2!a_{(2)}, \ldots, k!a_{(r)})$$
$$= \frac{1}{r!} B_r(1!\theta a_{(1)}, 2!\theta a_{(2)}, \ldots, r!\theta a_{(r)}), \quad r = 1, 2, \ldots.$$

If
$$c_n = P(N = n) = [-\log(1-\theta)]^{-1} \frac{\theta^n}{n!}, \quad n = 1, 2, \ldots, \quad 0 < \theta < 1,$$

then (cf. Section 7.6)

$$\beta_{(k)} = [-\log(1-\theta)]^{-1}\frac{[\theta/(1-\theta)]^k}{k}, \quad k = 1, 2, \ldots,$$

and

$$f(q_0) = [-\log(1-\theta)]^{-1}[-\log(1-\theta q_0)],$$

$$\beta_{(k)}(q_0) = [-\log(1-\theta q_0)]^{-1}\frac{[\theta q_0/(1-\theta q_0)]^k}{k}, \quad k = 1, 2, \ldots.$$

Thus, from Theorem 7.2 on using (7.41) and (7.48) together with (7.46), we obtain the probability function and binomial moments of the *compound logarithmic distribution* as

$$p_0 = [-\log(1-\theta)]^{-1}[-\log(1-\theta q_0)],$$

$$p_m = \frac{[-\log(1-\theta)]^{-1}}{m!} \sum_{k=1}^{m}(k-1)!\lambda^k B_{m,k}(1!q_1, 2!q_2, \ldots, m!q_m)$$

$$= \frac{[\log(1-\theta)]^{-1}}{m!} L_m(-1!\lambda q_1, -2!\lambda q_2, \ldots, -m!\lambda q_m),$$

for $m = 1, 2, \ldots$, with $\lambda = \theta(1-\theta q_0)^{-1}$, and

$$b_{(r)} = \frac{[-\log(1-\theta)]^{-1}}{r!} \sum_{k=1}^{r}(k-1)!\mu^k B_{r,k}(1!a_{(1)}, 2!a_{(2)}, \ldots, r!a_{(r)})$$

$$= \frac{[\log(1-\theta)]^{-1}}{r!} L_r(-1!\mu a_{(1)}, -2!\mu a_{(2)}, \ldots, -r!\mu a_{(r)}),$$

for $r = 1, 2, \ldots$, with $\mu = \theta(1-\theta)^{-1}$.

Suppose that

$$c_n = P(N = n) = \binom{s}{n}\theta^n(1-\theta)^{s-n}, \quad n = 0, 1, \ldots,$$

where $0 < \theta < 1$ and s is a positive integer. Then,

$$\beta_{(k)} = \binom{s}{k}\theta^k, \quad k = 0, 1, \ldots,$$

and

$$f(q_0) = [1 - \theta(1-q_0)]^s,$$

$$\beta_{(k)}(q_0) = \binom{s}{k}\left(\frac{\theta q_0}{1-\theta(1-q_0)}\right)^k, \quad k = 0, 1, \ldots.$$

Thus, from Theorem 7.2, on using (7.41), (7.46) and (7.49), we deduce the probability function and binomial moments of the *compound binomial distribution* as

$$p_m = \frac{[1 - \theta(1 - q_0)]^s}{m!} \sum_{k=0}^{m} (s)_k \lambda^k B_{m,k}(1!q_1, 2!q_2, \ldots, m!q_m)$$

$$= \frac{[1 - \theta(1 - q_0)]^s}{m!} C_{m,s}(1!\lambda q_1, 2!\lambda q_2, \ldots, m!\lambda q_m), \quad m = 0, 1, \ldots,$$

with $\lambda = \theta[1 - \theta(1 - q_0)]^{-1}$, and

$$b_{(r)} = \frac{1}{r!} \sum_{k=0}^{r} (s)_k \theta^k B_{r,k}(1!a_{(1)}, 2!a_{(2)}, \ldots, r!a_{(r)})$$

$$= \frac{1}{r!} C_{r,s}(1!\theta a_{(1)}, 2!\theta a_{(2)}, \ldots, r!\theta a_{(r)}), \quad r = 1, 2, \ldots.$$

Finally, if

$$c_n = P(N = n) = \binom{s + n - 1}{n}(1 - \theta)^s \theta^n, \quad n = 0, 1, \ldots,$$

where $0 < \theta < 1$ and $s > 0$, then (cf. Section 7.7)

$$\beta_{(k)} = \binom{s + k - 1}{k}\left(\frac{\theta}{1 - \theta}\right)^k, \quad k = 0, 1, \ldots$$

and

$$f(q_0) = \left(\frac{1 - \theta}{1 - \theta q_0}\right)^s, \quad \beta_{(k)}(q_0) = \binom{s + k - 1}{k}\left(\frac{\theta q_0}{1 - \theta q_0}\right)^k, \quad k = 0, 1, \ldots.$$

The probability function and binomial moments of the *compound negative binomial distribution* are deduced from Theorem 7.2, on using (7.41) and (7.49) together with (7.46), as

$$p_m = \left(\frac{1 - \theta}{1 - \theta q_0}\right)^s \frac{1}{m!} \sum_{k=0}^{m} (-s)_k (-\lambda)^k B_{m,k}(1!q_1, 2!q_2, \ldots, m!q_m)$$

$$= \left(\frac{1 - \theta}{1 - \theta q_0}\right)^s \frac{1}{m!} C_{m,-s}(-1!\lambda q_1, -2!\lambda q_2, \ldots, -m!\lambda q_m),$$

for $m = 0, 1, \ldots$, with $\lambda = \theta(1 - \theta q_0)^{-1}$, and

$$b_{(r)} = \frac{1}{r!} \sum_{k=0}^{r} (-s)_k (-\mu)^k B_{r,k}(1!a_{(1)}, 2!a_2, \ldots, r!a_{(r)})$$

$$= \frac{1}{r!} C_{r,-s}(-1!\mu a_{(1)} - 2!\mu a_{(2)}, \ldots, -m!\mu a_{(r)}),$$

for $r = 1, 2, \ldots$, with $\mu = \theta(1 - \theta)^{-1}$.

7.9 REFERENCE NOTES

The treatment of the discrete compound and mixture distributions in terms of the random occupancy model, with a random number of urns, is based on the paper of Ch. A. Charalambides (1986a), where the basic Theorem 7.1 was derived. It seems that H. W. Watson and F. Galton (1874) expressed, for the first time, the probability generating function of the size of a generation in a branching process as a composite function; the component functions being the probability generating functions of the size of the parental population and the number of individuals a parent may generate (see Example 7.2). The brief discussion of the countable mixture of discrete distributions in Section 7.3 and the connection of the particular mixture distributions to compound distributions is based on the papers of W. Feller (1943) and J. Gurland (1957). In these papers instead of mixture and compound distributions the terms compound and generalized distributions, respectively, were used. While W. Feller (1968) in his classical book adopted the names mixtures and compound distributions, J. Gurland (1958, 1965) consistently used the terms compound and generalized for these distributions.

W. Feller (1943) introduced a general class of contagious distributions and made a distinction between *true contagion*, where each favorable event increases (or decreases) the probability of future favorable events, and *apparent contagion*, where the events are independent and contagion is due to population heterogeneity. As particular cases he treated the compound (generalized) and the mixture (compound) Poisson distributions. As he quoted, F. E. Satterthwaite (1942) had previously derived the compound Poisson distribution in connection with insurance problems. L. Jánossy, A. Rényi and J. Aczél (1950) and A. Rényi (1951) thoroughly studied the compound Poisson distributions under the name *composed* Poisson distributions. H. P. Galliher, P. M. Morse and M. Simond (1959) and C. D. Kemp (1967) treated the compound Poisson distributions under the name *stuttering* Poisson distributions. J. B. Douglas (1970, 1971) systematically studied the mixture and compound (stopped sum) distributions and exemplified the use of the Stirling numbers in expressing probabilities and moments of these and other discrete distributions. A wealth of information on the statistical analysis with contagious distributions is provided in the book of J. B. Douglas (1980). J. G. Skellam (1952), in the context of spatial distribution of plants, discussed by means of generating functions the compound Poisson distribution and the related mixture distributions. Particular attention was given to Poisson-binomial and Poisson-Pascal distributions. H. R. Thompson (1954), departing from Darwin's model, provided another spatial mode of genesis of the compound Poisson distribution.

C. G. Khatri and I. R. Patel (1961) introduced through probability generating functions three wide classes of discrete distributions corresponding to the compound Poisson, general binomial (binomial and negative binomial) and logarithmic distributions; most of the mixture distributions are also included in these classes. Recurrence relations for the probability function and the factorial cumulants for each of these classes together with methods of estimation were given.

The Neyman type A distribution, which may be viewed either as a compound or as a mixture Poisson distribution (see Example 7.3 and Exercise 7.17), was developed by J. Neyman (1939). The expression of the probability function of this distribution in terms of the Stirling numbers of the second kind was deduced by F. Cernuschi and L. Castagnetto (1946). W. L. Cresswell and P. Froggatt (1963) derived the distribution of the number of spell accidents of a bus driver, during a given time period, as a Neyman type A distribution. Also, they deduced the total number of accidents (within and outside spells) of a bus driver as a *short* Neyman type A distribution. A recurrence relation for the probabilities of this distribution was derived by C. D. Kemp (1967) (see Example 7.4 and Exercise 7.6). J. B. Douglas (1955) pointed out that the probability function of the Neyman type A distribution could be expressed in terms of the power moments of a specific Poisson distribution. The parameter of this Poisson distribution is the product of the constant multiplier of the parameter of the mixed Poisson and the probability at zero of the mixing Poisson distribution (see Exercises 7.16 and 7.17). A lucid exposition of the properties, estimation and applications, together with a wealth of bibliographic information, is provided in the book of N. L. Johnson, S. Kotz and A. W. Kemp (1992). A compound Poisson distribution with compounding a shifted Poisson distribution was introduced by M. Thomas (1949) (see Exercise 7.5). A generalization of the Neyman's contagious distributions of types A, B and C was studied by G. Beall and R. R. Rescia (1953). A further generalization was discussed by J. Gurland (1958).

The Bernoulli damage process of Example 7.5 was treated by D. G. Catcheside (1948). The logarithmic distribution with zeros was introduced and studied by C. B. Williams (1947) and D. G. Kendall (1948). This distribution was also derived by C. G. Khatri (1962) and G. P. Patil (1964) as a mixture of a binomial with varying number of trials and mixing a logarithmic distribution (see Exercise 7.21). The negative binomial distribution as a compound Poisson distribution with compounding a logarithmic distribution, presented in Example 7.8, was derived by R. Lüders (1934) and independently by M. H. Quenouille (1949).

Starting from the probability generating function of a compound (generalized) Poisson distribution, C. D. Kemp and A. W. Kemp (1965) considered the case in which the probability generating function of the compounding (generalizing) distribution is a polynomial of order two. The probabilities of this particular compound Poisson distribution were expressed in terms of modified Hermite polynomials. This definition of the Hermite distribution allowed the authors to provide three models leading to it. Thus, this distribution is (i) a particular case ($s=2$) of the compound Poisson-binomial, (ii) a convolution of an ordinary Poisson with a doublet Poisson and (iii) a sum of two correlated Poisson. As J. O. Irwin (1963) quoted, A. G. McKendrich (1926) derived the distribution of the sum of the two correlated Poisson random variables in terms of a polynomial of the parameters, without noticing that is was a modified Hermite polynomial. A. W. Kemp and C. D. Kemp (1966) provided yet another derivation of the Hermite distribution, as a Poisson mixture with mixing a normal distribution. Besides C. D. Kemp and A. W. Kemp (1965), A. G. McKendrich (1926), Y. C. Patel, L. R. Shenton and K. O. Bowman (1974) and Y. C. Patel (1976, 1977) discussed various parameter estimation methods for the Hermite distribution.

R. P. Gupta and G. C. Jain (1974), replacing the exponent 2 in the generating function of the Hermite distribution by an integer $m \geq 2$, studied the resulting distribution under the name generalized Hermite distribution.

The compound Poisson-binomial distribution was first discussed by J. G. Skellam (1952), who fitted it, for $s = 3$, to spatial data on plants. J. U. McGuire, T. A. Brindley and T. A. Bancroft (1957) also used the compound Poisson-binomial distribution to incorporate the heterogeneity that is present in the distribution of the larval populations of corn borer and fitted it, for the particular case $s = 2$ (Hermite distribution), to a collected set of data. The expressions of the probabilities and binomial moments of this distribution in terms of the generalized factorial coefficients, in view of the expression connecting these numbers to the Stirling numbers of the first and second kind, coincide with the corresponding expression derived by R. Shumway and J. Gurland (1960a,b). Also, in these papers, estimation methods simplifying the computations were explored. S. K. Katti and J. Gurland (1962) discussed yet another estimation method, which seems to yield estimators with high efficiency in practical problems.

G. Pólya (1930) presented the Pólya-Aeppli distribution and ascribed its derivation to his student Aeppli, who discussed it in his thesis at Zurich in 1924. D. A. Evans (1953) provided an explicit expression and a recurrence relation for the probabilities of this distribution. An alternative recurrence relation was given by C. D. Kemp (1967). In H. P. Galliher, P. M. Morse and M. Simond (1959) the probabilities of the Pólya-Aeppli distribution were expressed in terms of Laguerre polynomials. The more general compound Poisson-Pascal (negative binomial) distribution was introduced by J. G. Skellam (1952), who proposed to call it the generalized Pólya-Aeppli distribution. The derivation of this distribution in the entomological stochastic model of Example 7.7 is based on the paper of S. K. Katti and J. Gurland (1961); the recurrence relation for the probabilities was derived in this paper.

I. G. Plunkett and G. C. Jain (1975) derived the Gegenbauer distribution as a Hermite mixture with mixing a gamma distribution (see Exercise 7.25). A. W. Kemp (1979) studied it as a convolution of a binomial with a pseudo binomial distribution. J. Medhi and M. Borah (1984), replacing the exponent 2 in the generating function of the Gegenbauer distribution by an integer $m \geq 2$, studied the resulting distribution under the name generalized Gegenbauer distribution. For further mathematical properties and information on the Hermite, Chebyshev and Gegenbauer polynomials the interested reader is referred to T. S. Chihara (1978).

A. N. Philippou and A. A. Muwafi (1982) considered the problem of finding the probability function of the number W_k of trials required until the occurrence of the kth consecutive success in a sequence of independent Bernoulli trials with constant success probability. The probability function $p_m = P(Z_k = m)$, $m = 0, 1, \ldots$, of $Z_k = W_k - k$, was combinatorially obtained as a multiple sum over all partitions of m, with parts not greater than k. The probability generating function of this distribution was derived by W. Feller (1968) as an application of the theory of recurrent events but the origin of this problem was attributed by I. Todhunter (1865) to A. De Moivre (1718). V. R. R. Uppuluri and G. P. Patil (1983) provided another derivation of the probability function using generating functions. A. N. Philippou, C. Georgiou and

G. N. Philippou (1983) named the distribution of W_k geometric distribution of order k and further studied it. Also, they treated the r-fold convolution Y_k of W_k and called it negative binomial distribution of order k. The Poisson distribution of order k was derived as the limiting distribution of $Y_k - rk$ as r tends to infinity. S. Aki, H. Kuboki and K. Hirano (1984) and K. Hirano (1986) derived the logarithmic distribution of order k as the limit of the zero truncated distribution of $Y_k - rk$ as r tends to zero. Inspired by the first of these papers, Ch. A. Charalambides (1986b) pointed out that these distributions of order k are simply compound (generalized) power series distributions. Also, the Poisson process of order k, discussed at the end of Section 7.5.4, was studied in that paper. An overview of the distributions of order k and their applications to the reliability of consecutive-k-out-of-n failure systems is provided by A. N. Philippou (1986). It should be noted that the shifted distributions of $W_k - k$ and $Y_k - rk$ and not the distributions of the random variables W_k and Y_k are treated in this chapter as geometric and negative binomial distributions of order k.

Finally, Section 7.8, on the use of partition polynomials in expressing the probability function and binomial moments of the class of compound (generalized) discrete distributions in terms of the component probability functions and binomial moments, is based on the papers of Ch. A. Charalambides (1977b, 1986a).

7.10 EXERCISES AND COMPLEMENTS

7.1 The rth binomial moment of a compound discrete distribution is given in Theorem 7.1 as

$$E\left[\binom{S_N}{r}\right] = \sum_{k=0}^{r} E\left[\binom{N}{k}\right] \left[\Delta_u^k E\left\{\binom{S_u}{r}\right\}\right]_{u=0}.$$

Deduce for $r = 1$ and $r = 2$ the expressions

$$E(S_N) = E(N)E(X)$$

and

$$E\left[\binom{S_N}{2}\right] = E(N)E\left[\binom{X}{2}\right] + E\left[\binom{N}{2}\right][E(X)]^2,$$

respectively. Further, conclude that

$$V(S_N) = E(N)V(X) + V(N)[E(X)]^2.$$

7.2 Let the number N of urns be a Poisson random variable with probability function

$$c_n = P(N = n) = e^{-\theta}\frac{\theta^n}{n!},\ n = 0, 1, \ldots,\ 0 < \theta < \infty,$$

and assume that the number X_j of balls distributed into the jth urn, $j = 1, 2, \ldots$, is geometrically distributed with

$$q_x = P(X = x) = pq^x,\ x = 0, 1, \ldots,$$

where $q = 1 - p$ and $0 < p < 1$. Show that the probability function and binomial moments of the number S_N of balls distributed into the urns are given by

$$p_0 = e^{-\theta q}, \quad p_m = e^{-\theta q} q^m \sum_{k=1}^{m} \binom{m-1}{k-1} \frac{(\theta p)^k}{k!}, \quad m = 1, 2, \ldots,$$

and

$$b_{(r)} = \left(\frac{q}{p}\right)^r \sum_{k=1}^{r} \binom{r-1}{k-1} \frac{\theta^k}{k!}, \quad r = 1, 2, \ldots,$$

or equivalently by

$$p_0 = e^{-\theta q}, \quad p_m = e^{-\theta q} L_{m-1}^{(1)}(-\theta p) \frac{\theta p q^m}{m}, \quad m = 1, 2, \ldots,$$

and

$$b_{(r)} = L_{r-1}^{(1)}(-\theta) \frac{\theta q^r}{r p^r}, \quad r = 1, 2, \ldots,$$

where

$$L_n^{(1)}(x) = \sum_{k=0}^{n} (-1)^k \binom{n+1}{n-k} \frac{x^k}{k!}, \quad n = 0, 1, \ldots,$$

is the Laguerre polynomial. This is a Pólya-Aeppli distribution (cf. Section 7.5.3).

7.3 *A compound Poisson process.* Consider a nonnegative integer valued homogeneous stochastic process $\{S(t), t \geq 0\}$ with independent increments and let

$$p_m(t) = P[S(t) = m], \quad m = 0, 1, \ldots.$$

Also, let q_x, $x = 0, 1, \ldots$, be a probability function and $\theta > 0$ a constant. Assume that, for small $\Delta t > 0$,

$$p_j(\Delta t) = \theta q_j \Delta t + o(\Delta t), \quad j = 1, 2, \ldots,$$

$$p_0(\Delta t) = 1 - \theta(1 - q_0)\Delta t + o(\Delta t),$$

with $\lim_{\Delta t \to 0} o(\Delta t)/\Delta t = 0$. Derive for the probability generating function

$$h(u, t) = \sum_{m=0}^{\infty} p_m(t) u^m$$

the differential equation

$$\frac{\partial h(u, t)}{\partial t} = -\theta[1 - g(u)] h(u, t),$$

where

$$g(u) = \sum_{x=0}^{\infty} q_x u^x,$$

and using the initial condition $p_0(0) = P[S(0) = 0] = 1$, conclude that

$$h(u,t) = e^{-\theta t[1-g(u)]}.$$

Expanding it into powers of u, deduce the probability function $p_m(t)$, $m = 0, 1, \ldots$, in the form (7.27) or (7.28) depending on whether $q_0 > 0$ or $q_0 = 0$. Further, expanding the binomial moment generating function $h(u+1, t)$ into powers of u, find the binomial moments $b_{(r)}$, $r = 0, 1, \ldots$, in the form (7.29).

7.4 Show that the probability function p_m, $m = 0, 1, \ldots$, and the binomial moments $b_{(r)}$, $r = 0, 1, \ldots$, of the Poisson distribution of order k, introduced in Section 7.5.4, satisfy the following recurrence relations

$$p_m = \frac{\lambda}{m} \sum_{j=1}^{\min\{m,k\}} j p_{m-j}, \quad m = 1, 2, \ldots,$$

with $p_0 = e^{-k\lambda}$, and

$$b_{(r)} = \frac{\lambda}{r} \sum_{j=1}^{\min\{r,k\}} j \binom{k+1}{j+1} b_{(r-j)}, \quad r = 1, 2, \ldots,$$

with $b_{(0)} = 1$.

7.5 *Thomas distribution.* A stochastic model for the distribution of plants of a given species in randomly dropped quadrants (square lattices) was constructed as follows. Consider an area over which a parent population of plants is randomly distributed. With each of these parent plants a random number of offspring is associated. In an experiment the area is divided into squares and the probability that a square contains $m = 0, 1, \ldots$, plants is estimated by the corresponding frequency. The parent plants may be taken to represent cluster centers, and the offspring as additional plants (after the parent). A model suited to situations in which the parent and the offspring plants are included in the count for each cluster is built by assuming that the number N of parent plants (clusters) obeys a Poisson distribution with

$$c_n = P(N = n) = e^{-\theta} \frac{\theta^n}{n!}, \quad n = 0, 1, \ldots, \quad 0 < \theta < \infty,$$

while the number X of plants in a cluster has a shifted Poisson distribution with

$$q_x = P(X = x) = e^{-\lambda} \frac{\lambda^{x-1}}{(x-1)!}, \quad x = 1, 2, \ldots, \quad 0 < \lambda < \infty.$$

(a) Show that the probability function p_m, $m = 0, 1, \ldots$, and binomial moments $b_{(r)}$, $r = 0, 1, \ldots$, of the compound distribution are given by

$$p_m = \frac{e^{-\theta}}{m!} \sum_{n=0}^{m} \binom{m}{n} (\theta e^{-\lambda})^n (n\lambda)^{m-n}, \quad m = 0, 1, \ldots,$$

and

$$b_{(r)} = \frac{1}{r!} \sum_{j=0}^{r} \sum_{i=0}^{j} \binom{r}{j} S(j,i;r-j) \lambda^j \theta^{r-j+i}, \quad r = 0, 1, \ldots,$$

where $S(j,i;r)$ is the noncentral Stirling number of the second kind.

(b) In particular, deduce the mean and variance.

7.6 *Extended Neyman type A distribution.* Show that the probability function P_m, $m = 0, 1, \ldots$, and the binomial moments $B_{(r)}$, $r = 0, 1, \ldots$, of the number of bus driver accidents, discussed in Example 7.4, satisfy the recurrence relations

$$P_{m+1} = \frac{\rho}{m+1} P_m + \frac{\theta \lambda e^{-\lambda}}{m+1} \sum_{j=0}^{m} \frac{\lambda^{m-j}}{(m-j)!} P_j, \quad m = 0, 1, \ldots,$$

with $P_0 = e^{-\theta(1-e^{-\lambda})-\rho}$, and

$$B_{(r+1)} = \frac{\rho}{r+1} B_{(r)} + \frac{\theta \lambda}{r+1} \sum_{j=0}^{r} \frac{\lambda^{r-j}}{(r-j)!} B_{(j)}, \quad r = 0, 1, \ldots,$$

with $B_{(0)} = 1$. In particular, for $r = 0$ and $r = 1$, deduce the mean and variance of this distribution as

$$\mu = \theta\lambda + \rho, \quad \sigma^2 = \theta\lambda(1+\lambda) + \rho.$$

7.7 *Generalized Chebyshev distribution of the first kind.* The Chebychev distribution of the first kind, discussed in Section 7.6.1, may be extended by assuming that the number X_j of balls distributed into the jth urn, $j = 1, 2, \ldots$, follows a binomial distribution with probability function

$$q_x = P(X = x) = \binom{s}{x} p^x q^{s-x}, \quad x = 0, 1, \ldots, s,$$

where $q = 1 - p$, $0 < p < 1$ and s is a positive integer. Show that the probability function and the binomial moments of the compound distribution are given by

$$p_0 = [-\log(1-\theta)]^{-1}[-\log(1-\theta q^s)],$$

$$p_m = [-\log(1-\theta)]^{-1} \frac{(p/q)^m}{m!} \sum_{k=1}^{m} (k-1)! C(m,k;s)[\theta q^s/(1-\theta q^s)]^k,$$

for $m = 1, 2, \ldots$, and

$$b_{(r)} = [-\log(1-\theta)]^{-1} \frac{p^r}{r!} \sum_{k=1}^{r} (k-1)! C(r,k;s)[\theta/(1-\theta)]^k,$$

for $r = 1, 2, \ldots$, where $C(m,k;s)$ is the generalized factorial coefficient. In particular, deduce the mean and variance of this distribution.

7.8 (*Continuation*). A generalized Chebyshev polynomial in x of the first kind $T_n(x; a, s)$, $n = 1, 2, \ldots$, may be defined through its generating function by

$$T(t; x) = \sum_{n=1}^{\infty} T_n(x; a, s) \frac{t^n}{n!} = -\log\{1 - a[(x+t)^s - x^s]\},$$

where s is a positive integer and a is a real number.

(a) Show that

$$T_n(x; a, s) = \sum_{k=1}^{n} (k-1)! C(n, k; s) a^k x^{sk-n}.$$

In particular,

$$T_1(x; a, s) = asx^{s-1}, \quad T_2(x; a, s) = a(s)_2 x^{s-2} + a^2 s^2 x^{2s-2}.$$

(b) Derive the recurrence relation

$$T_{n+1}(x; a, s) = a(s)_{n+1} + a \sum_{k=1}^{n} \binom{s}{k}(n)_k x^{s-k} T_{n-k+1}(x; a, s),$$

for $n = 2, 3, \ldots$, with initial conditions $T_1(x; a, s) = asx^{s-1}$ and $T_2(x; a, s) = a(s)_2 x^{s-2} + a^2 s^2 x^{2s-2}$.

(c) Express the probability function and binomial moments of the compound logarithmic-binomial distribution in terms of the generalized Chebyshev polynomials of the first kind and deduce recurrence relations for these sequences.

7.9 Show that the probability function p_m, $m = 1, 2, \ldots$, of the logarithmic distribution of order k, examined in Section 7.6.2, satisfies the following recurrence relation

$$p_m = [-k \log p]^{-1} q p^{m-1} \zeta(m, k) + \frac{q}{m} \sum_{j=1}^{\min\{m,k\}} (m-j) p^{j-1} p_{m-j},$$

for $m = 2, 3, \ldots$, with $p_0 = [-k \log p]^{-1} q$, where the zeta function $\zeta(m, k) = 1$, for $m \leq k$ and $\zeta(m, k) = 0$, for $m > k$.

7.10 *Compound binomial distributions.* Assume that the number N of urns has a binomial distribution with probability function

$$c_n = P(N = n) = \binom{s}{n} \theta^n (1-\theta)^{s-n}, \quad n = 0, 1, \ldots, s,$$

where $0 < \theta < 1$ and s is a positive integer. Let

$$q_x = P(X = x), \quad x = 0, 1, \ldots,$$

be the probability function of the number X of balls distributed into any specific urn. Show that, if $q_0 > 0$, the probability function of the compound binomial distribution is given by

$$p_m = (1 - \theta + \theta q_0)^s \sum_{k=0}^{m} \binom{s}{k} (\theta q_0)^k \left[\Delta_u^k q_0^{-u} q_m(u)\right]_{u=0}, \quad m = 0, 1, \ldots,$$

while if $q_0 = 0$, it is given by

$$p_m = (1 - \theta)^s \sum_{n=0}^{m} \binom{s}{n} [\theta/(1-\theta)]^n q_m(n), \quad m = 0, 1, \ldots.$$

Also, show that the rth binomial moment of this distribution is given by

$$b_{(r)} = \sum_{k=0}^{r} \binom{s}{k} \theta^k \left[\Delta_u^k a_{(r)}(u)\right]_{u=0}, \quad r = 0, 1, \ldots,$$

where $a_{(r)}(u)$ is the rth binomial moment of $q_m(u)$, $m = 0, 1, \ldots$.

7.11 (*Continuation*). If the compounding distribution is also a binomial with probability function

$$q_x = \binom{r}{x} p^x q^{r-x}, \quad x = 0, 1, \ldots, r,$$

where $q = 1-p$, $0 < p < 1$ and r is a positive integer, (a) find the probability function p_m, $m = 0, 1, \ldots$, and binomial moments $b_{(j)}$, $j = 0, 1, \ldots$, of the compound distribution. (b) In particular, deduce the mean and variance. (c) Derive recurrence relations for the probability function and binomial moments.

7.12 *Generalized Gegenbauer distribution.* The Gegenbauer distribution, examined in Section 7.7.2, may be extended by assuming that the number X_j of balls distributed into the jth urn, $j = 1, 2, \ldots$, follows a binomial distribution with probability function

$$q_x = P(X = x) = \binom{r}{x} p^x q^{r-x}, \quad x = 0, 1, \ldots, r,$$

where $q = 1 - p$, $0 < p < 1$ and r is a positive integer. Show that the probability function and the binomial moments of the compound distribution are given by

$$p_m = \left(\frac{1-\theta}{1-\theta q^r}\right)^s \frac{(p/q)^m}{m!} \sum_{k=0}^{m} (s+k-1)_k C(m, k; r) \left(\frac{\theta q^r}{1-\theta q^r}\right)^k,$$

for $m = 0, 1, \ldots,$ and

$$b_{(j)} = \frac{p^j}{j!} \sum_{k=0}^{j} (s+j-1)_j C(j, k; r) \left(\frac{\theta}{1-\theta}\right)^k,$$

for $j = 0, 1, \ldots$, where $C(m, k; r)$ is the generalized factorial coefficient. In particular, find the mean and variance of this distribution.

7.13 (*Continuation*). A generalized Gegenbauer polynomial in x, denoted by $G_n^{(s)}(s; a, r)$, $n = 1, 2, \ldots$, may be defined through its generating function by

$$G(t; x) = \sum_{n=0}^{\infty} G_n^{(s)}(x; a, r) \frac{t^n}{n!} = (1 - a[(x + t)^r - x^r])^{-s},$$

where s and r are positive integers and a is a real number.

(a) Show that

$$G_n^{(s)}(x; a, r) = \sum_{k=0}^{n} (s + k - 1)_k C(n, k; r) a^k x^{rk-n}.$$

In particular,

$$G_0^{(s)}(x; a, r) = 1, \quad G_1^{(s)}(x; a, r) = asr x^{r-1}.$$

(b) Derive the recurrence relation

$$G_{n+1}^{(s)}(x; a, r) = a \sum_{k=0}^{n} \binom{n}{k} \left(s + \frac{n-k}{k+1}\right) (r)_{k+1} x^{r-k-1} G_{n-k}^{(s)}(x; a, r),$$

for $n = 0, 1, \ldots$, with initial condition $G_0^{(s)}(x; a, r) = 1$.

(c) Express the probability function and binomial moments of the compound negative binomial-binomial distribution in terms of the generalized Gegenbauer polynomials and deduce recurrence relations for these sequences.

7.14 Show that the probability function p_m, $m = 0, 1, \ldots$, of the geometric distribution of order k, discussed in Section 7.7.3, satisfies the recurrence relation

$$p_m = \sum_{j=1}^{\min\{m,k\}} qp^{j-1} p_{m-j}, \quad m = 1, 2, \ldots,$$

with $p_0 = p^k$.

7.15 *Negative binomial Poisson distribution.* This is a compound distribution with compounded a negative binomial distribution

$$c_n = P(N = n) = \binom{s + n - 1}{n} (1 - \theta)^s \theta^n, \quad n = 0, 1, \ldots,$$

where $0 < \theta < 1$, $s > 0$, and compounding a Poisson distribution

$$q_x = e^{-\lambda} \frac{\lambda^x}{x!}, \quad x = 0, 1, \ldots, \quad 0 < \lambda < \infty.$$

(a) Find the probability function p_m, $m = 0, 1, \ldots$, and binomial moments $b_{(r)}$, $r = 0, 1, \ldots$, of the compound distribution. (b) In particular, deduce the mean and variance. (c) Derive recurrence relations for the sequences p_m, $m = 0, 1, \ldots$, and $b_{(r)}$, $r = 0, 1, \ldots$.

7.16 *Poisson mixtures.* Consider a population consisting of a countable number of clusters. Assume that each cluster generates a random number of individuals with a Poisson probability function

$$q_m(N; \lambda) = e^{-N\lambda} \frac{(N\lambda)^m}{m!}, \ m = 0, 1, \ldots, \ 0 < \lambda < \infty,$$

where N is a nonnegative integer valued random variable with

$$c_n = P(N = n), \ n = 0, 1, \ldots .$$

Show that the probability function of the mixture distribution is given by

$$p_m = \frac{\lambda^m}{m!} f(e^{-\lambda}) \mu'_m(e^{-\lambda}), \ m = 0, 1, \ldots,$$

where $\mu'_m(e^{-\lambda})$ is the mth (power) moment of the power series distribution

$$c_n(e^{-\lambda}) = [f(e^{-\lambda})]^{-1} c_n (e^{-\lambda})^n, \ n = 0, 1, \ldots,$$

with series function $f(e^{-\lambda}) = \sum_{n=0}^{\infty} c_n (e^{-\lambda})^n$. Further, expressing the power moments $\mu'_m(e^{-\lambda})$, $m = 0, 1, \ldots$, in terms of the binomial moments $\beta_{(k)}(e^{-\lambda})$, $k = 0, 1, \ldots$, verify that this countable Poisson mixture coincides with a compound distribution with compounding a Poisson distribution (cf. Corollary 7.1).

7.17 *(Continuation).* If the mixing distribution is Poisson with

$$c_n = P(N = n) = e^{-\theta} \frac{\theta^n}{n!}, \ n = 0, 1, \ldots, \ 0 < \theta < \infty,$$

show that the mixture is the Neyman type A distribution.

7.18 *(Continuation).* (a) Assume that the mixing distribution is a zero-one Bernoulli distribution with

$$c_n = P(N = n) = p^n q^{1-n}, \ n = 0, 1,$$

where $q = 1 - p$ and $0 < p < 1$. Show that the mixture distribution has probability function

$$p_0 = q + pe^{-\lambda}, \ p_m = pe^{-\lambda} \frac{\lambda^m}{m!}, \ m = 1, 2, \ldots,$$

which is a modified Poisson distribution.

(b) More generally, let the mixing distribution be a binomial with probability function

$$c_n = P(N = n) = \binom{s}{n} p^n q^{s-n}, \ n = 0, 1, \ldots, s,$$

where $q = 1 - p$, $0 < p < 1$ and s is a positive integer. Express the probability function of the mixture distribution as a finite sum with coefficients being the Stirling numbers of the second kind.

7.19 (*Continuation*). Assume that the mixing distribution is a geometric with

$$c_n = P(N = n) = pq^n, \ n = 0, 1, \ldots,$$

or more generally a negative binomial with

$$c_n = P(N = n) = \binom{s+n-1}{n} p^s q^n, \ n = 0, 1, \ldots,$$

where $q = 1 - p$, $0 < p < 1$ and $s > 0$. Express the probability function of the mixture distribution as a finite sum with coefficients being the Stirling numbers of the second kind.

7.20 (*Continuation*). Let the mixing distribution be a logarithmic with

$$c_n = P(N = n) = [-\log(1-\theta)]^{-1} \frac{\theta^n}{n}, \ n = 1, 2, \ldots, \ 0 < \theta < 1.$$

Derive the probability function of the mixture distribution in terms of the Stirling numbers of the second kind.

7.21 *Bernoulli mixtures*. Consider a population consisting of a countable number of clusters. Assume that the size N of each cluster is a random variable with probability function

$$c_n = P(N = n), \ n = 0, 1, \ldots.$$

Also, assume that each element of a cluster, during its lifetime, may or may not generate, independently of the others, an element with probabilities p and $q = 1 - p$, respectively. Thus, each cluster generates a random number of individuals with a binomial probability function

$$q_m(N) = \binom{N}{m} p^m q^{N-m}, \ m = 0, 1, \ldots, N.$$

Show that the probability function of the mixture distribution is given by

$$p_m = (p/q)^m f(q) \beta_{(m)}(q), \ m = 0, 1, \ldots,$$

where $\beta_{(m)}(q)$ is the mth binomial moment of the power series distribution

$$c_n(q) = [f(q)]^{-1} c_n q^n, \ n = 0, 1, \ldots,$$

with series function $f(q) = \sum_{n=0}^{\infty} c_n q^n$. In particular, show that:
 (a) If the mixing distribution is a Poisson with mean θ, then the mixture distribution is also Poisson with mean θp.
 (b) If the mixing distribution is a logarithmic with parameter θ, then the mixture distribution is a logarithmic with a zero term added (cf. Example 7.6.)
 (c) If the mixing distribution is a binomial with probability of success θ, then the mixture distribution is also a binomial with probability of success θp.

7.22 *Binomial mixtures.* Consider a population consisting of a countable number of subpopulations. Assume that each subpopulation generates a random number of elements with a binomial probability function

$$q_m(N;s) = \binom{sN}{m} p^m q^{sN-m}, \quad m = 0, 1, \ldots, sN,$$

where $q = 1 - p$, $0 < p < 1$, s is a positive integer and N is a nonnegative integer valued random variable with

$$c_n = P(N = n), \quad n = 0, 1, \ldots.$$

(a) Show that the probability function of the mixture distribution is given by

$$p_m = (p/q)^m f(q^s) \beta_{(m;s)}(q^s), \quad m = 0, 1, \ldots,$$

where $\beta_{(m;s)}(q^s)$ is the mth binomial moment of sY, where Y is a random variable having the power series distribution

$$c_n(q^s) = [f(q^s)]^{-1} c_n q^{sn}, \quad n = 0, 1, \ldots,$$

with series function $f(q^s) = \sum_{n=0}^{\infty} c_n q^{sn}$.

(b) Further, expressing the binomial moments $\beta_{(m;s)}(q^s)$, $m = 0, 1, \ldots$, of sY in terms of the binomial moments $\beta_{(k)}(q^s)$, $k = 0, 1, \ldots$, of Y, verify that this countable binomial mixture coincides with a compound distribution with compounding a binomial distribution (cf. Corollary 7.2).

(c) In particular, if the mixing distribution is a Poisson with

$$c_n = e^{-\theta} \frac{\theta^n}{n!}, \quad n = 0, 1, \ldots, \quad 0 < \theta < \infty,$$

find the probability function of the mixture distribution.

7.23 *Negative binomial mixtures.* In a population consisting of a countable number of clusters, assume that each cluster generates a random number of individuals with a negative binomial probability function

$$q_m(N;s) = \binom{sN + m - 1}{m} p^{sN} q^m, \quad m = 0, 1, \ldots,$$

where $q = 1 - p$, $0 < p < 1$, $s > 0$ and N is a nonnegative integer valued random variable with

$$c_n = P(N = n), \quad n = 0, 1, \ldots.$$

(a) Show that the probability function of the mixture distribution is given by

$$p_m = q^m f(p^s) \beta_{[m;s]}(p^s), \quad m = 0, 1, \ldots,$$

where $\beta_{[m;s]}(p^s)$ is the mth ascending binomial moment of sY, where Y is a random variable obeying the power series distribution

$$c_n(p^s) = [f(p^s)]^{-1} c_n p^{sn}, \quad n = 0, 1, \ldots,$$

with series function $f(p^s) = \sum_{n=0}^{\infty} c_n p^{sn}$.

340 COMPOUND AND MIXTURE DISTRIBUTIONS

(b) Also, expressing the ascending binomial moments $\beta_{[m;s]}(p^s)$, $m = 0, 1, \ldots$, of sY in terms of usual binomial moments $\beta_{(k)}(p^s)$, $k = 0, 1, \ldots$, of Y verify that this countable negative binomial mixture coincides with a compound distribution with compounding a negative binomial distribution (cf. Corollary 7.3).

(c) In particular, if the mixing distribution is a logarithmic distribution with

$$c_n = [-\log(1-\theta)]^{-1} \frac{\theta^n}{n}, \quad n = 1, 2, \ldots, \quad 0 < \theta < 1,$$

deduce the probability function of the mixture distribution.

7.24 *A negative binomial as a mixture distribution.* Let the conditional distribution of a random variable X, given that $Y = y$, be a Poisson with probability function

$$q_{x|y} = P(X = x | Y = y) = e^{-\theta y} \frac{(\theta y)^x}{x!}, \quad x = 0, 1, \ldots,$$

where $0 < \theta < \infty, 0 < y < \infty$, and assume that the random variable Y has a gamma distribution with probability density function

$$f(y) = \frac{\lambda^s}{\Gamma(s)} y^{s-1} e^{-\lambda y}, \quad 0 < y < \infty,$$

where $0 < \lambda < \infty, 0 < s < \infty$. Show that the random variable X obeys a negative binomial distribution with probability function

$$q_x = \binom{s + x - 1}{x} p^s q^x, \quad x = 0, 1, \ldots,$$

where $p = \lambda/(\theta + \lambda)$ and $q = \theta/(\theta + \lambda)$.

7.25 *A Gegenbauer as a mixture distribution.* Let the conditional distribution of a random variable X, given $Y = y$, be a Hermite with probability generating function

$$g(t; y) = e^{\theta y [2pq(t-1) + p^2(t^2-1)]},$$

where $q = 1 - p$, $0 < p < 1$, $0 < \theta < \infty$, and assume that the random variable Y has a gamma distribution with probability function

$$f(y) = \frac{\lambda^s}{\Gamma(s)} y^{s-1} e^{-\lambda y}, \quad 0 < y < \infty,$$

where $0 < \lambda < \infty$, $0 < s < \infty$. Show that the random variable X obeys a Gegenbauer distribution with probability generating function

$$h(t) = (1 - \theta[2pq(t-1) + p^2(t^2-1)]/\lambda)^{-s}.$$

7.26 *Alternative interpretation of a compound distribution as a mixture.* Consider the compound distribution $p_m = P(S_N = m)$, $m = 0, 1, \ldots$, with compounded distribution $c_n = P(N = n)$, $n = 0, 1, \ldots$, and compounding distribution

$q_x = P(X = x)$, $x = 0, 1, \ldots$. If $q_0 > 0$, show that this compound distribution may be interpreted as a mixture distribution with mixed distribution

$$p_m(k) = (1 - q_0)^{-k}\left[\Delta_u^k q_0^{k-u} q_m(u)\right]_{u=0}, \quad m = k, k+1, \ldots \; (k = 0, 1, \ldots),$$

the k-fold convolution of a zero truncated distribution $P(X = x | X > 0) = (1 - q_0)^{-1} q_x$, $x = 0, 1, \ldots$, and mixing distribution

$$d_k = f(q_0)\beta_{(k)}(q_0)[(1 - q_0)/q_0]^k, \quad k = 0, 1, \ldots,$$

where $\beta_{(k)}(q_0)$, $k = 0, 1, \ldots$, is the kth binomial moment of the power series distribution

$$c_n(q_0) = [f(q_0)]^{-1} c_n q_0^n, \quad n = 0, 1, \ldots$$

with series function $f(q_0) = \sum_{n=0}^{\infty} c_n q_0^n$. In particular show that:

(a) The compound Poisson distribution may be interpreted as a mixture distribution with mixing distribution a Poisson with mean $\theta(1 - q_0)$.

(b) The Hermite distribution may be interpreted as a mixture distribution with mixed distribution

$$d_k = \binom{k}{m-k}\left(\frac{p}{2q}\right)^m \left(\frac{4q^2}{1-q^2}\right)^k, \quad k = 0, 1, \ldots, [m/2]$$

and mixing distribution a Poisson with mean $\theta(1 - q^2)$.

(c) The generalized Hermite distribution may be interpreted as a mixture distribution with mixed distribution

$$d_k = (q^{-s} - 1)^{-k} k! C(m, k; s)\frac{(p/q)^m}{m!}, \quad m = k, k+1, \ldots,$$

the k-fold convolution of a zero truncated binomial distribution (cf. Section 6.3.3) and mixing distribution a Poisson with mean $\theta(1 - q^s)$.

7.27 (*Continuation*). Show that:

(a) The compound logarithmic distribution may be interpreted as a mixture distribution with mixing distribution a logarithmic with parameter $\lambda = \theta q_0/(1 - \theta q_0)$.

(b) The Chebyshev distribution of the first kind may be interpreted as a mixture distribution with mixed distribution

$$d_k = \binom{k}{m-k}\left(\frac{p}{2q}\right)^m \left(\frac{4q^2}{1-q^2}\right)^k, \quad k = 0, 1, \ldots, [m/2]$$

and mixing distribution a logarithmic with parameter $\lambda = \theta q^2/(1 - \theta q^2)$.

(c) The generalized Chebyshev distribution of the first kind (cf. Exercise 7.7) may be interpreted as a mixture distribution with mixed distribution

$$d_k = (q^{-s} - 1)^{-k} k! C(m, k; s)\frac{(p/q)^m}{m!}, \quad m = k, k+1, \ldots,$$

the k-fold convolution of a zero truncated binomial distribution (cf. Section 6.3.3) and mixing distribution a logarithmic with parameter $\lambda = \theta q^s/(1 - \theta q^s)$.

7.28 (*Continuation*). Show that:

(a) The compound negative binomial distribution may be interpreted as a mixture distribution with mixing distribution a negative binomial with failure probability $\theta q_0/(1 - \theta q_0)$.

(b) The Gegenbauer distribution may be interpreted as a mixture distribution with mixed distribution

$$d_k = \binom{k}{m-k}\left(\frac{p}{2q}\right)^m \left(\frac{4q^2}{1-q^2}\right)^k, \quad k = 0, 1, \ldots, [m/2]$$

and mixing distribution a negative binomial with failure probability $\theta q^2/(1 - \theta q^2)$.

(c) The generalized Gegenbauer distribution (cf. Exercise 7.12) may be interpreted as a mixture distribution with mixed distribution

$$d_k = (q^{-r} - 1)^{-k} k! C(m, k; r) \frac{(p/q)^m}{m!}, \quad m = k, k+1, \ldots,$$

the k-fold convolution of a zero truncated binomial distribution (cf. Section 6.3.3) and mixing distribution a negative binomial with failure probability $\theta q^r/(1 - \theta q^r)$.

7.29 Under the random occupancy model with $q_0 > 0$, show that the conditional probability $p_m(n)$ that m balls are distributed into the n distinguishable urns, given that no urn remains empty, $m = n, n+1, \ldots$, is given by

$$p_m(n) = \frac{n!}{m!}(1 - q_0)^{-n} B_{m,n}(1!q_1, 2!q_2, \ldots, m!q_m),$$

where $B_{m,n}(x_1, x_2, \ldots, x_m)$ is the partial Bell partition polynomial.

7.30 Under the random occupancy model with $q_0 > 0$, show that the conditional probability $p_k(m, n)$ that k urns are occupied, given that m balls are distributed into the n distinguishable urns, $k = 0, 1, \ldots, n$, is given by

$$p_k(m, n) = \frac{(n)_k q_0^{n-k}}{m! q_m(n)} B_{m,k}(1!q_1, 2!q_2, \ldots, m!q_m),$$

where $B_{m,k}(x_1, x_2, \ldots, x_m)$ is the partial Bell partition polynomial.

Appendix
HINTS AND ANSWERS TO EXERCISES

CHAPTER 1

1.1 (a_1) If the order in which the balls are placed into the urns does not count, establish a one-to-one correspondence between the set of distributions of k distinguishable balls into n distinguishable urns and the set of k-permutations of the n urns with repetition. Then, applying Lemma 1.1, conclude the required number.

(b_1) Establish a one-to-one correspondence between the set of distributions of k indistinguishable balls into n distinguishable urns and the set of k-combinations of the n urns with repetition and conclude the required number.

(a_2) The required number equals the number of distributions of k indistinguishable balls into n distinguishable urns of unlimited capacity, derived in (b_1), multiplied by the number $k!$ of permutations of the k distinguishable balls.

(b_2) Establish a one-to-one correspondence between the set of distributions of k indistinguishable balls into n distinguishable urns, each of capacity limited to one ball, and the set of k-combinations of the n urns without repetition and conclude the required number.

Combinatorial Methods in Discrete Distributions. By Charalambos A. Charalambides
ISBN 0-471-68027-3 Copyright © 2005 John Wiley & Sons, Inc.

1.2 The number of distributions of i like balls into n distinguishable urns, according to the second part of Exercise 1.1 is equal to

$$\binom{n+i-1}{i}$$

if the urns are of unlimited capacity, while it is equal to

$$\binom{n}{i}$$

if the capacity of each urn is limited to one ball. Then, apply the multiplication principle to deduce the number of distributions of $r_i \geq 0$ distinguishable kinds of balls, each including i like balls, into n distinguishable urns. Finally, apply once more the multiplication principle to establish the required formulae.

1.3 Let

$$\Omega = \{(j_1, j_2, \ldots, j_{2k+1}) : j_r = 1, 2, \ldots, 2n,\ r = 1, 2, \ldots, 2k+1\}$$

be the set of all possible results. Then,

$$A = \{(j_1, j_2, \ldots, j_{2k+1}) \in \Omega : j_1 + j_2 + \cdots + j_{2k+1} > (2n+1)k + n\}$$

and

$$A' = \{(i_1, i_2, \ldots, i_{2k+1}) \in \Omega : i_1 + i_2 + \cdots + i_{2k+1} \leq (2n+1)k + n\}.$$

The element $(j_1, j_2, \ldots, j_{2k+1})$ in the set A uniquely corresponds to the element $(i_1, i_2, \ldots, i_{2k+1})$ in the set A', with $i_r = (2n+1) - j_r,\ r = 1, 2, \ldots, 2k+1$. Indeed,

$$i_1 + i_2 + \cdots + i_{2k+1} = (2n+1)(2k+1) - (j_1 + j_2 + \cdots + j_{2k+1})$$
$$\leq (2n+1)k + n$$

if and only if

$$j_1 + j_2 + \cdots + j_{2k+1} > (2n+1)k + n.$$

1.4 Let $A_{m,j}$ be the event that j balls among the m balls drawn from the first urn are white and $B_{n,k}$ be the event that k balls among the n balls drawn from the second urn are white. Then,

$$P(A_{m,j}) = \binom{r}{j}\binom{s}{m-j} \bigg/ \binom{r+s}{m},\quad j = 0, 1, \ldots, m,$$

and

$$P(B_{n,k} | A_{m,j}) = \binom{j}{k}\binom{m-j}{n-k} \bigg/ \binom{m}{n},\quad k = 0, 1, \ldots, n.$$

Using the total probability theorem,

$$P(B_{n,k}) = \sum_{j=0}^{m} P(A_{m,j}) P(B_{n,k}|A_{m,j}),$$

deduce the required probability $p_{n,k} = P(B_{n,k})$.

1.5 Using the total probability theorem and since

$$P(B_{i_1} B_{i_2} \cdots B_{i_r} | A_k) = \frac{(k)_r (n-r)_{m-r}}{(n)_m} = \frac{(k)_r}{(n)_r},$$

for $\{i_1, i_2, \ldots, i_r\} \subseteq \{1, 2, \ldots, m\}$, deduce the expression

$$P(B_{i_1} B_{i_2} \cdots B_{i_r}) = \sum_{k=r}^{n} p_k \frac{(k)_r}{(n)_r},$$

which implies the exchangeability of the events $B_1, B_2, \ldots, B_m$.

1.6 Let C_m be the event that all m drawn balls are white. Then, the required conditional probability $P(A_n|C_m)$, on using Bayes' theorem, is obtained as

$$P(A_n|C_m) = \binom{n}{m} \Big/ \sum_{k=m}^{n} \binom{k}{m}.$$

Since

$$\sum_{k=m}^{n} \binom{k}{m} = \binom{n+1}{m+1},$$

this expression is simplified to

$$P(A_n|C_m) = (m+1)/(n+1).$$

1.7 Use the expression $AB = \sum_{k=1}^{\infty} A A_k B$ and successively the additivity and multiplicative properties of probability to get

$$P(AB) = \sum_{k=1}^{\infty} P(AA_k B) = \sum_{k=1}^{\infty} P(A) P(A_k|A) P(B|AA_k),$$

which implies the required formula.

1.8 Let A_j be the event that exactly j white balls are drawn in n trials. Then, $A = A_k + A_{k+1} + \cdots + A_n$ and using the total conditional probability formula of Exercise 1.7, deduce for $P(B|A)$ the required expression.

1.9 Use the conditional probability expression

$$P(A_r|B_1 B_2) = \frac{P(B_2 A_r|B_1)}{P(B_2|B_1)}$$

and the multiplicative conditional probability formula

$$P(B_2 A_r | B_1) = P(A_r | B_1) P(B_2 | B_1 A_r)$$

together with the total conditional probability formula

$$P(B_2 | B_1) = \sum_{k=1}^{\infty} P(A_k | B_1) P(B_2 | B_1 A_k).$$

The last formula, which is a variation of the total conditional probability formula of Exercise 1.7, may be similarly shown.

1.10 In the last formula of Exercise 1.9, express the general term of the sum as

$$P(A_k) P(B_1 B_2 \cdots B_m | A_k) / P(B_1 B_2 \cdots B_{m-1}),$$

for all $k = 1, 2, \ldots, r, \ldots$, and then, in both the numerator and denominator, use the conditional independence of the events $B_1, B_2, \ldots, B_m$, given A_k, to deduce the first expression. Also, in the same formula express the term $P(B_m | B_1 B_2 \cdots B_{m-1} A_k)$ as

$$P(B_1 B_2 \cdots B_m | A_k) / P(B_1 B_2 \cdots B_{m-1} | A_k),$$

for all $k = 1, 2, \ldots, r, \ldots$, and then, in both the numerator and denominator, use the conditional independence of the events $B_1, B_2, \ldots, B_m$, given A_k, to deduce the second expression.

1.11 For any r-combination $\{i_1, i_2, \ldots, i_r\}$ of the n indices $\{1, 2, \ldots, n\}$, use the total probability theorem to express the probability $P(B_{i_1} B_{i_2} \cdots B_{i_r})$ as

$$P(B_{i_1} B_{i_2} \cdots B_{i_r}) = \sum_{k=1}^{\infty} P(A_k) P(B_{i_1} B_{i_2} \cdots B_{i_r} | A_k).$$

Then, utilize the conditional independence of the events $B_1, B_2, \ldots, B_n$, given the event A_k, and the assumption $P(B_{i_m} | A_k) = a_k$, for all $k = 1, 2, \ldots$, to conclude that the probability $P(B_{i_1} B_{i_2} \cdots B_{i_r})$ depends only on r.

1.12 Express the conditional probability as

$$P(A_{i_1} A_{i_2} \cdots A_{i_r} | B_{n,k}) = \frac{P(A_{i_1} A_{i_2} \cdots A_{i_r} B_{n,k})}{P(B_{n,k})},$$

where $P(A_{i_1} A_{i_2} \cdots A_{i_r} B_{n,k})$ is the probability $P_{n-r,k-r}$ that exactly $k - r$ among the $n-r$ exchangeable events $A_{i_{r+1}}, A_{i_{r+2}}, \ldots, A_{i_n}$ occur, with $\{i_{r+1}, i_{r+2}, \ldots, i_n\}$ $= \{1, 2, \ldots, n\} - \{i_1, i_2, \ldots, i_r\}$ and $P(B_{n,k})$ is the probability $P_{n,k}$ that exactly k among the n exchangeable events $A_1, A_2, \ldots, A_n$ occur. Then, use Corollary 1.5 to deduce the required expression for $P(A_{i_1} A_{i_2} \cdots A_{i_r} | B_{n,k})$. Further, by the total probability theorem,

$$P(B_{m,r}) = \sum_{k=0}^{n} P(B_{n,k}) P(B_{m,r} | B_{n,k}),$$

where the conditional probability $P(B_{m,r}|B_{n,k})$ may be expressed as

$$P(B_{m,r}|B_{n,k}) = \binom{m}{r} \sum_{j=0}^{m-r} (-1)^j \binom{m-r}{j} P(A_{i_1} A_{i_2} \cdots A_{i_{r+j}} | B_{n,k}).$$

Then, use the first part of this exercise to find

$$P(B_{m,r}|B_{n,k}) = \binom{k}{r} \sum_{j=0}^{m-r} (-1)^j \binom{n-r-j}{n-m} \binom{k-r}{j} \bigg/ \binom{n}{m}$$

and show the combinatorial identity

$$\sum_{j=0}^{m-r} (-1)^j \binom{n-r-j}{n-m} \binom{k-r}{j} = \binom{n-k}{m-r}$$

to establish the required expression.

1.13 For any r-combination $\{i_1, i_2, \ldots, i_r\}$ of the n indices $\{1, 2, \ldots, n\}$ and $\{i_{r+1}, i_{r+2}, \ldots, i_n\} = \{1, 2, \ldots, n\} - \{i_1, i_2, \ldots, i_r\}$,

$$P(A_{i_1} A_{i_2} \cdots A_{i_r} A'_{i_{r+1}} A'_{i_{r+2}} \cdots A'_{i_n}) = \sum_{j=r}^{n} (-1)^{j-r} S_{n,r,j}(i_1, i_2, \ldots, i_r),$$

with

$$S_{n,r,j}(i_1, i_2, \ldots, i_r) = \sum P(A_{i_1} A_{i_2} \cdots A_{i_r} A_{h_1} A_{h_2} \cdots A_{h_{j-r}}),$$

where the summation is extended over all $(j-r)$-combinations $\{h_1, h_2, \ldots, h_{j-r}\}$ of the $n-r$ indices $\{i_{r+1}, i_{r+2}, \ldots, i_n\}$. Utilize the exchangeability of the events $A_1, A_2, \ldots, A_n$, $P(A_{i_1} A_{i_2} \cdots A_{i_r}) = p_r$, to get the expression

$$P(A_{i_1} A_{i_2} \cdots A_{i_r} A'_{i_{r+1}} A'_{i_{r+2}} \cdots A'_{j_n}) = \sum_{j=r}^{n} (-1)^{j-r} \binom{n-r}{j-r} p_j,$$

which implies that the probability

$$P(A_{i_1} A_{i_2} \cdots A_{i_r} A'_{i_{r+1}} A'_{i_{r+2}} \cdots A'_{i_n}) = P_r, \quad r = 1, 2, \ldots, n,$$

depends only on r and not on the specific collection of indices. Conversely, for any r-combination $\{i_1, i_2, \ldots, i_r\}$ of the n indices $\{1, 2, \ldots, n\}$ and $\{i_{r+1}, i_{r+2}, \ldots, i_n\}$ $= \{1, 2, \ldots, n\} - \{i_1, i_2, \ldots, i_r\}$,

$$P(A_{i_1} A_{i_2} \cdots A_{i_r}) = \sum_{j=r}^{n} Q_{n,r,j}(i_1, i_2, \ldots, i_r),$$

with

$$Q_{n,r,j}(i_1, i_2, \ldots, i_r) = \sum P(A_{i_1} A_{i_2} \cdots A_{i_r} A_{h_1} \cdots A_{h_{j-r}} A'_{h_{j-r+1}} \cdots A'_{h_{n-r}}),$$

where the summation is extended over all $(j - r)$-combinations $\{h_1, h_2, \ldots, h_{j-r}\}$ of the $n - r$ indices $\{i_{r+1}, i_{r+2}, \ldots, i_n\}$ and $\{h_{j-r+1}, h_{j-r+2}, \ldots, h_{n-r}\} = \{i_{r+1}, i_{r+2}, \ldots, i_n\} - \{h_1, h_2, \ldots, h_{j-r}\}$. Use the hypothesis

$$P(A_{i_1} A_{i_2} \cdots A_{i_r} A'_{i_{r+1}} A'_{i_{r+2}} \cdots A'_{i_n}) = P_r, \quad r = 1, 2, \ldots, n,$$

to get the expression

$$P(A_{i_1} A_{i_2} \cdots A_{i_r}) = \sum_{j=r}^{n} \binom{n-r}{j-r} P_j,$$

which implies the exchangeability of the events $A_1, A_2, \ldots, A_n$.

1.14 Use the probabilities

$$P(B_j) = \binom{s_j}{k_j} p^{k_j} q^{s_j - k_j}, \quad k_j = 0, 1, \ldots, s_j, \quad j = 1, 2, \ldots, n,$$

$$P(A) = \binom{s}{k} p^k q^s, \quad k = 0, 1, \ldots, s$$

to find the expressions

$$P(B_1 B_2 \cdots B_n | A) = \binom{s_1}{k_1} \binom{s_2}{k_2} \cdots \binom{s_n}{k_n} \bigg/ \binom{s}{k}, \quad k_j = 0, 1, \ldots, s_j,$$

$$P(B_j | A) = \binom{s_j}{k_j} \binom{s - s_j}{k - k_j} \bigg/ \binom{s}{k}, \quad k_j = 0, 1, \ldots, s_j, \quad j = 1, 2, \ldots, n,$$

which imply

$$P(B_1 B_2 \cdots B_n | A) \neq P(B_1 | A) P(B_2 | A) \cdots P(B_n | A).$$

1.15 Use the probabilities

$$P(B_j) = \binom{s_j + k_j - 1}{k_j} p^{k_j} q^{s_j}, \quad k_j = 0, 1, \ldots, \quad j = 1, 2, \ldots, n,$$

$$P(A) = \binom{s + k - 1}{k} p^k q^s, \quad k = 0, 1, \ldots,$$

to find the expressions

$$P(B_1 B_2 \cdots B_n | A) = \frac{\binom{s_1 + k_1 - 1}{k_1} \binom{s_2 + k_2 - 1}{k_2} \cdots \binom{s_n + k_n - 1}{k_n}}{\binom{s + k - 1}{k}},$$

and

$$P(B_j|A) = \frac{\binom{s_j + k_j - 1}{k_j}\binom{s - s_j + k - k_j - 1}{k - k_j}}{\binom{s + k - 1}{k}}, \quad j = 1, 2, \ldots, n,$$

which imply

$$P(B_1 B_2 \cdots B_n|A) \neq P(B_1|A)P(B_2|A) \cdots P(B_n|A).$$

1.16 Apply the total conditional probability formula of Exercise 1.7 and then use the independence of the events A and B to establish the required expression. Conversely, if the given relation holds, then

$$P(B|A)P(C|AB) + P(B'|A)P(C|AB') = P(B)P(C|AB) + P(B')P(C|AB'),$$

and using the conditional probability formula, find the relation

$$P(ABC)P(AB)P(AB') + P(AB'C)P(AB)P(AB')$$
$$= P(ABC)P(AB')P(A)P(B) + P(AB'C)P(AB)P(A)P(B').$$

Since

$$P(B') = 1 - P(B), \quad P(AB') = P(A) - P(AB), \quad P(AB'C) = P(AC) - P(ABC),$$

the last relation, after some algebraic manipulations, can be factored as

$$[P(ABC)P(A) - P(AB)P(AC)] \cdot [P(AB) - P(A)P(B)] = 0.$$

Then, use the assumption $P(BC|A) \neq P(B|A)P(C|A)$ to conclude that

$$P(AB) = P(A)P(B).$$

1.17 Start with the formulae given in the hint of Exercise 1.13 and utilize the independence or conditional independence probability relations.

1.18 (a) The number of permutations of any subset of balls

$$\{j_1, j_2, \ldots, j_r\}, 1 \leq j_1 < j_2 < \cdots < j_r \leq n,$$

in which the ball j_r occupies the rth position, equals the number of permutations of the set $\{j_1, j_2, \ldots, j_{r-1}\}$ and so

$$P(A_r) = \frac{(r-1)!}{r!} = \frac{1}{r}, \quad r = 1, 2, \ldots, n.$$

(b) The number of permutations of the set

$$\{j_1, j_2, \ldots, j_k, \ldots, j_{r-1}\}, \quad 1 \leq j_1 < j_2 < \cdots < j_k < \cdots < j_{r-1} < j_r \leq n,$$

in which the ball j_s occupies the kth position, for $s = k, k+1, \ldots, r-1$, while the positions $\{1, 2, \ldots, k-1\}$ are occupied by any $k-1$ of the $s-1$ balls $\{j_1, j_2, \ldots, j_{s-1}\}$ and the positions $\{k+1, k+2, \ldots, r-1\}$ are occupied by the remaining $r-k-1$ balls, is given by $(s-1)_{k-1}(r-k-1)!$, $s = k, k+1, \ldots, r-1$. Since

$$\sum_{s=k}^{r-1}(s-1)_{k-1}(r-k-1)! = (k-1)!(r-k-1)!\sum_{s=k}^{r-1}\binom{s-1}{k-1}$$

$$= (k-1)!(r-k-1)!\binom{r-1}{k} = \frac{(r-1)!}{k},$$

conclude that

$$P(A_k A_r) = \frac{(r-1)!/k}{r!} = \frac{1}{kr}, \quad \{k, r\} \subseteq \{1, 2, \ldots, n\}.$$

Similarly, deduce that

$$P(A_{i_1} A_{i_2} \cdots A_{i_r}) = \frac{(i_r - 1)!/(i_1 i_2 \cdots i_{r-1})}{i_r!} = \frac{1}{i_1 i_2 \cdots i_r},$$

for $\{i_1, i_2, \ldots, i_r\} \subseteq \{1, 2, \ldots, n\}$, and conclude that the events $A_1, A_2, \ldots, A_n$ are independent.

1.19 Let A_i be the event that the ith urn remains empty, $i = 1, 2, \ldots, n$. Using the result of Exercise 1.2, show that the events $A_1, A_2, \ldots, A_n$ are exchangeable and apply Corollary 1.5 to get the required expressions.

1.20 Apply Corollary 1.6 to the events $A_1, A_2, \ldots, A_n$ defined in the hint of Exercise 1.19 to get the binomial moments of the number K_0 of empty urns. Then, use Theorem 1.11 to deduce the binomial moments of the number $K = n - K_0$ of occupied urns (cf. Example 1.13).

1.21 Let A_i be the event that a ball bearing the number i is drawn at the ith drawing, $i = 1, 2, \ldots, n$. Show that these events are exchangeable and apply Corollary 1.5 to find the required expressions.

1.22 Apply Theorem 1.10 to the events $A_1, A_2, \ldots, A_n$ defined in the hint of Exercise 1.21 (cf. Example 1.11).

1.23 Apply Corollary 1.8 to the events $A_1, A_2, \ldots, A_n$ defined in the hint of Exercise 1.21 (cf. Example 1.10).

1.24 Consider the event A that the gambler wins one euro on the first game and its complement A'. Using the total probability theorem, express the gambler's ruin probability in terms of his ruin probabilities after the result of the first game.

1.25 Consider the event A that a wins the first game after resuming the series of games and its complement A'. Using the total probability theorem, express the probability $p_{k,r}$ in terms of the probabilities of a to win the series of games after the result of this game.

1.26 Let A_j be the event that an r-tuple coincidence occurs at the jth trial, $j = 1, 2, \ldots, n$. Show that these events are exchangeable and apply Corollary 1.5 to find the required expressions.

1.27 Derive the probability generating function as

$$G(t) = \sum_{k=ns}^{n(s+m-1)} p_k(n, m, s) t^k = \frac{t^{sn}(1-t^m)^n}{m^n(1-t)^n}$$

and, expanding it into powers of t, conclude the required expression for the probability $p_k(n, m, s)$. Further, deduce the generating function

$$H(t) = \sum_{k=ns}^{n(s+m-1)} P_k(n, m, s) t^k$$

from the relation $H(t) = G(t)/(1-t)$ and, expanding it into powers of t, find the required expression for the probability $P_k(n, m, s)$.

1.28 The general inclusion and exclusion formula (1.35) can be generalized to any real or complex valued finitely additive set function. Specifically, let Ω be an arbitrary set, and Q a field of subsets of Ω. Let G be any finitely additive set function defined on Q. Denote by $B_{n,k}$ the set of elements that belong to exactly k among n sets $A_1, A_2, \ldots, A_n$ in Q. Then,

$$G(B_{n,k}) = \sum_{r=k}^{n} (-1)^{r-k} \binom{r}{k} S_{n,r}, \quad k = 0, 1, \ldots, n,$$

with $S_{n,0} = G(\Omega)$, and

$$S_{n,r} = \sum G(A_{i_1} A_{i_2} \cdots A_{i_r}),$$

where the summation is extended over all r-combinations $\{i_1, i_2, \ldots, i_r\}$ of the n indices $\{1, 2, \ldots, n\}$. Apply this formula to the n-dimensional simplex

$$\Omega = \{(u_1, u_2, \ldots, u_n) : u_1 + u_2 + \cdots + u_n \leq x, \; u_i \geq 0, \; i = 1, 2, \ldots, n\}$$

and A_i the subset of Ω for which $u_i > 1$, $i = 1, 2, \ldots, n$, and since

$$G(\Omega) = x^n/n!, \quad G(A_{i_1} A_{i_2} \cdots A_{i_r}) = (x-r)^n/n!, \text{ if } r \leq x$$

and $G(A_{i_1} A_{i_2} \cdots A_{i_r}) = 0$ if $r > x$, deduce the required expression.

1.29 Let A_j be the event that the jth urn remains empty, $j = 1, 2, \ldots, n$. Then, for any r indices $\{i_1, i_2, \ldots, i_r\}$ out of the n indices $\{1, 2, \ldots, n\}$,

$$P(A_{i_1} A_{i_2} \cdots A_{i_r}) = \frac{(s - s_{i_1} - s_{i_2} - \cdots - s_{i_r} + m - 1)_m}{(s + m - 1)_m}$$

$$= \frac{(s_{j_1} + s_{j_2} + \cdots + s_{j_{n-r}} + m - 1)_m}{(s + m - 1)_m},$$

with $\{j_1, j_2, \ldots, j_{n-r}\} = \{1, 2, \ldots, n\} - \{i_1, i_2, \ldots, i_r\}$. Proceeding as in Example 1.14, deduce the expressions

$$p_k(m, n; s) = \sum_{r=0}^{k} (-1)^{k-r} \binom{n-r}{k-r} S_{n,n-r},$$

$$b_{(r)}(m, n; s) = \sum_{k=0}^{r} (-1)^k \binom{n-r}{r-k} S_{n,k},$$

and

$$P_k(m, n; s) = \sum_{r=0}^{k} (-1)^{k-r} \binom{n-r-1}{n-k-1} S_{n,n-r},$$

with

$$S_{n,r} = \sum \frac{(s_{j_1} + s_{j_2} + \cdots + s_{j_{n-r}} + m - 1)_m}{(s + m - 1)_m}.$$

1.30 It can be shown by mathematical induction. Clearly, it holds for $n = 1$ and $k = 0, 1$. Suppose that $P(A_{m,k}) = (m-k)/m$, for $0 \le k \le m \le n-1$, with $n \ge 2$, and show that $P(A_{n,k}) = (n-k)/n$.

1.31 Let A_i be the event that a success occurs at the ith trial, $i = 1, 2, \ldots, n$. The $p_k(n)$ is the probability that exactly k among the n events $A_1, A_2, \ldots, A_n$ occur and so applying Theorem 1.8 deduce the required expression.

1.32 Apply Corollary 1.8 to the events $A_1, A_2, \ldots, A_n$ defined in the hint of Exercise 1.30.

1.33 Work as in the derivation of Theorem 1.8 and Corollary 1.5.

1.34 Work as in the derivation of Corollary 1.6.

1.35 Work as in the derivation of Corollary 1.7.

1.36 Set $m = n$ in Exercise 1.33 and, using the assumption of disjointness of the pairs (A_h, B_h), $h = 1, 2, \ldots, n$, remove the zero terms in the expression of $S_{n;i,j} \equiv S_{n,n;i,j}$. Then, in the expression of $P_{n;k,r} \equiv P_{n,n;k,r}$ the upper limit in the inner sum reduces to $n - j$ and in the outer sum to $n - k$.

1.37 Work as in the derivation of Corollary 1.6.

1.38 Work as in the derivation of Corollary 1.7.

1.39 Let A_h and B_h be the events of the occurrence of A and B, respectively, at the hth trial, $h = 1, 2, \ldots, n$. Apply the bivariate inclusion and exclusion formula to the pairs of disjoint events (A_h, B_h), $h = 1, 2, \ldots, n$.

1.40 Use the relation

$$q_{k,r}(s) = (1 - p_{s+k+r} - q_{s+k+r}) p_{k,r}(s + k + r - 1)$$

and the expression of the probability $p_{k,r}(n)$ given in Exercise 1.38.

CHAPTER 2

2.1 Expand the identity

$$(t)_n = \frac{(t)_{n+1}}{t} \cdot \frac{1}{1 - n/t}$$

into powers of t and equate the coefficients of t^k on both sides of the resulting identity to deduce the first recurrence relation.

Use the triangular recurrence relation for the Stirling numbers of the second kind to show that the generating function

$$\phi_k(u) = \sum_{n=k}^{\infty} S(n,k) u^n$$

satisfies the recurrence relation

$$\phi_k(u) = u(1 - ku)^{-1} \phi_{k-1}(u), \quad k = 1, 2, \ldots.$$

Expand it into powers of u and equate the coefficients of u^n on both sides of the resulting identity to deduce the second recurrence relation.

2.2 Applying the Lagrange inversion formula on the given series, deduce for the Stirling numbers of the first kind the expression

$$s(n,k) = \binom{n-1}{k-1} \left[\frac{d^{n-k}}{dt^{n-k}} \left(\frac{e^t - 1}{t} \right)^{-n} \right]_{t=0}.$$

Then, use the expression

$$\left[\frac{d^m}{dt^m} (h(t))^s \right]_{t=0} = \sum_{r=0}^{m} \binom{s}{r} \binom{m-s}{m-r} \left[\frac{d^m}{dt^m} (h(t))^r \right]_{t=0}$$

to find

$$s(n,k) = \binom{n-1}{k-1} \sum_{k=0}^{n-k} \binom{-n}{r} \binom{2n-k}{n-k-r} \left[\frac{d^{n-k}}{dt^{n-k}} \left(\frac{e^t - 1}{r} \right)^r \right]_{t=0}.$$

The last expression, by virtue of the exponential generating function of the Stirling numbers of the second kind (cf. Theorem 2.3), yields the required formula.

2.3 Applying the Lagrange inversion formula on the power series $u = \phi(t) = \log(1 + t)$, $\varphi(0) = 0$, with inverse $t = \varphi^{-1}(u) = e^u - 1$ and working as in Exercise 2.2, derive the expression

$$S(n,j) = \sum_{r=0}^{n-j} (-1)^r \binom{n+r-1}{j-1} \binom{2n-j}{n-j-r} s(n-j+r, r),$$

which, for $j = n - k$, reduces to the required expression.

2.4 Expand the function $(t+u)_n$ into powers of $t+u$ and into factorials of t and u. Then, expand both sides of the resulting identity into powers of t and u and deduce the required formula.

2.5 Expand the function $(t+u)^n$ into factorials of $t+u$ and into powers of t and u. Then, expanding both sides of the resulting identity into factorials of t and u, deduce the required formula.

2.6 Let $Y_j = 1$, if X_j is a record, and $Y_j = 0$, if X_j is not a record. Show that $Y_1, Y_2, \ldots, Y_j, \ldots$, are independent random variables, with

$$P(Y_j = 1) = \frac{1}{j}, \quad P(Y_j = 0) = \frac{j-1}{j}, \quad j = 1, 2, \ldots,$$

and derive the probability function of N_n. Then, use the relation

$$P(L_k = n) = P(N_{n-1} = k-1, Y_n = 1)$$

to deduce the probability function of L_k.

2.7 Applying the Lagrange inversion formula on the power series $u = \phi(t) = (1+t)^s - 1$, $\phi(0) = 0$, with inverse $t = \phi^{-1}(u) = (1+u)^{1/s} - 1$ and working as in Exercise 2.2, derive the expression

$$C(n, j; s^{-1}) = \sum_{r=0}^{n-j} (-1)^r \binom{n+r-1}{j-1} \binom{2n-j}{n-j-r} C(n-j+r, r; s),$$

which, for $j = n - k$, reduces to the required expression.

2.8 Expand the function $(st + su)_n$ into factorials of $t+u$ and into factorials of st and su. Further, expand both sides of the resulting identity into factorials of t and u and deduce the required expression.

2.9 Use the expansion that defines the signless Stirling numbers of the first kind to derive the relation

$$\sum_{k=0}^{n} |s(n+1, k+1)| t^k = n! \exp\left\{\sum_{r=1}^{\infty} (-1)^{r-1} \zeta_n(r) t^r / r \right\}.$$

Expanding the right-hand side into powers of t, deduce the required expression.

2.10 Use the expression of Exercise 2.9 and

$$\lim_{n \to \infty} \zeta_n(1)/\log(1+n) = 1, \quad \lim_{n \to \infty} [\log(1+n)]^j/n = 0,$$

$$\zeta_n(s) \leq \pi^2/6, \quad s = 2, 3, \ldots,$$

to show that

$$\lim_{n \to \infty} |s(n+1, k+1)| [\log(1+n)]^{-k}/n! = 1/k!.$$

Use the explicit expressions of $S(n,k)$ and $C(n,k;s)$ to derive the limits

$$\lim_{n\to\infty} S(n,k)/k^n = 1/k!, \quad \lim_{n\to\infty} C(n,k;s)/(sn)_n = 1/k!.$$

2.11 Expand both members of the relation

$$t[g(t) - 1] = -(e^t - 1 - t)g(t)$$

into powers of t to derive the recurrence relation.

Rewrite the generating function as

$$g(t) = \log[1 + (e^t - 1)]/(e^t - 1)$$

and expand it, first into powers of $u = e^t - 1$ and then into powers of t to get the expression of the Bernoulli numbers in terms of the Stirling numbers of the second kind. Multiply it by $s(r,n)$ and sum for $n = 0, 1, \ldots, r$.

2.12 Differentiate the generating function $g_r(t)$ to get the differential equation

$$g_{r+1}(t) = (1-t)g_r(t) - (t/r)g'_r(t),$$

which, expanded into powers of t, yields the recurrence relation.

2.13 Express the ascending factorial $(t + r + n - 1)_n$ into a sum of ascending factorials $(t + r - m + j - 1)_j$, $j = 0, 1, \ldots, n$, using Vandermode's formula, and then expand both members of the resulting expression into powers of t, using (2.8), to find the required expression.

2.14 Express the power $(t+r)^n$ into a sum of powers $(t+r-m)^j$, $j = 0, 1, \ldots, n$, using Newton's binomial formula, and then expand both members of the resulting expression into factorials of t, using (2.7), to find the required expression.

2.15 Express the noncentral factorial $(t + r_2 - r_1)_n$ into powers of $t + r_2$ and then expand the powers $(t+r_2)^j$, $j = 0, 1, \ldots, n$, into factorials of t, using (2.6) and (2.7). In addition, express the noncentral factorial $(t + r_2 - r_1)_n$ directly into factorials of t, using Vandermonde's formula, and equate the coefficients of $(t)_k$ on both sides of the resulting expression. Also, express the power $(t - r_2 + r_1)^n$ into factorials of $t - r_2$ and then expand the factorials $(t - r_2)_j$, $j = 0, 1, \ldots, n$, into powers of t, using (2.6) and (2.7). Further, express the power $(t + r_1 - r_2)^n$ directly into powers of t, using Newton's binomial formula, and equate the coefficients of t^k in the resulting expression.

2.16 Work with the function $(t + u - r_1 - r_2)_n$ as in Exercise 2.4.

2.17 Work with the function $(t + u + r_1 + r_2)^n$ as in Exercise 2.5.

2.18 Work with the function $(st + su + r_1 + r_2)^n$ as in Exercise 2.8.

2.19 Work as in Example 2.4.

2.20 Work as in Example 2.7.

CHAPTER 3

3.1 Start with the generating function of the r-associated signless Stirling numbers of the first kind (3.9) and follow the steps of the derivation of Theorem 2.5.

3.2 Start with the generating function of the r-associated Stirling numbers of the second kind (3.8) and follow the steps of the derivation of Theorem 2.5.

3.3 Start with the generating function of the r-associated generalized factorial coefficients (3.17) and follow the steps of the derivation of Theorem 2.15.

3.4 The enumerator for the occupancy of the jth urn, when it may contain at most r balls, is

$$E_r(u; x_j) = 1 + x_j u + \frac{x_j^2 u^2}{2!} + \cdots + \frac{x_j^r u^r}{r!}, \quad j = 1, 2, \ldots, k,$$

and so the enumerator for the occupancy of the k distinguishable urns, when each urn may contain at most r balls, is

$$G_{k,r}(u; x_1, x_2, \ldots, x_k) = \prod_{j=1}^{k} \left(1 + x_j u + \frac{x_j^2 u^2}{2!} + \cdots + \frac{x_j^r u^r}{r!} \right).$$

Setting $x_j = 1$, $j = 1, 2, \ldots, k$, deduce the generating function of $B_r(n, k)$, $n = 0, 1, \ldots, rk$. Further, differentiate $F_{k,r}(u)$ to get the differential equation

$$F'_{k,r}(u) = k F_{k,r}(u) - \frac{u^r}{r!} F_{k-1,r}(u),$$

which, expanded into powers of u, yields the first recurrence relation. Finally, derive the expression

$$k F_{k,r}(u) = \sum_{j=0}^{r} F_{k-1,r}(u) \frac{u^j}{j!},$$

which implies the second recurrence relation.

3.5 The enumerator for the occupancy of the jth urn, when it may contain at most r balls, is

$$E_r(u; x_j) = 1 + s x_j u + \binom{s}{2} x_j^2 u^2 + \cdots + \binom{s}{r} x_j^r u^r, \quad j = 1, 2, \ldots, k,$$

and so the enumerator for the occupancy of the k distinguishable urns, when each urn may contain at most r balls, is

$$G_{k,r}(u; x_1, x_2, \ldots, x_k) = \prod_{j=1}^{k} \left[\sum_{i=0}^{r} \binom{s}{i} x_j^i u^i \right].$$

Setting $x_j = 1$, $j = 1, 2, \ldots, k$, deduce the generating function of $B_r(n, k; s)$, $n = 0, 1, \ldots, rk$. Then, continue as in Exercise 3.4.

3.6 The enumerator for the occupancy of the jth urn, when it may contain at most r balls, is

$$E_r(u; x_j) = 1 + sx_ju + \binom{s+1}{2}x_j^2 u^2 + \cdots + \binom{s+r-1}{r}x_j^r u^r,$$

for $j = 1, 2, \ldots, k$, and so the enumerator for the occupancy of the k distinguishable urns, when each urn may contain at most r balls, is

$$G_{k,r}(u; x_1, x_2, \ldots, x_k) = \prod_{j=1}^{k}\left[\sum_{i=0}^{r}\binom{s+i-1}{i}x_j^i u^i\right].$$

Setting $x_j = 1$, $j = 1, 2, \ldots, k$, deduce the generating function of $|B_r(n, k; -s)|$, $n = 0, 1, \ldots, rk$. Then, continue as in Exercise 3.4.

3.7 Using the triangular recurrence relation (3.36), derive the expression

$$|s(n+1, k; \boldsymbol{a}_0)|^2 - |s(n+1, k-1; \boldsymbol{a}_0)| \cdot |s(n+1, k+1; \boldsymbol{a}_0)|$$
$$= \{|s(n, k-1; \boldsymbol{a}_0)|^2 - |s(n, k-2; \boldsymbol{a}_0)| \cdot |s(n, k; \boldsymbol{a}_0)|\}$$
$$+ a_m^2\{|s(n, k; \boldsymbol{a}_0)|^2 - |s(n, k-1; \boldsymbol{a}_0)| \cdot |s(n, k+1; \boldsymbol{a}_0)|\}$$
$$+ a_m\{|s(n, k; \boldsymbol{a}_0)| \cdot |s(n, k-1; \boldsymbol{a}_0)| - |s(n, k-2; \boldsymbol{a}_0)| \cdot |s(n, k+1; \boldsymbol{a}_0)|\}.$$

Then, by induction on n, deduce the required inequality.

3.8 Using the triangular recurrence relation (3.38), derive the expression

$$[S(n+1, k; \boldsymbol{a}_0)]^2 - S(n+1, k-1; \boldsymbol{a}_0)S(n+1, k+1; \boldsymbol{a}_0)$$
$$= \{[S(n, k-1; \boldsymbol{a}_0)]^2 - S(n, k-2; \boldsymbol{a}_0)S(n, k; \boldsymbol{a}_0)\}$$
$$+ (a_k^2 - a_{k-1}a_{k+1})[S(n, k; \boldsymbol{a}_0)]^2$$
$$+ a_{k-1}a_{k+1}\{[S(n, k; \boldsymbol{a}_0)]^2 - S(n, k-1; \boldsymbol{a}_0)S(n, k+1; \boldsymbol{a}_0)\}$$
$$+ (2a_k - a_{k-1} - a_{k+1})S(n, k; \boldsymbol{a}_0)S(n, k-1; \boldsymbol{a}_0)$$
$$+ a_{k+1}\{S(n, k; \boldsymbol{a}_0)S(n, k-1; \boldsymbol{a}_0) - S(n, k-2; \boldsymbol{a}_0)S(n, k+1, \boldsymbol{a}_0)\}.$$

Then, by induction on n, deduce the required inequality.

3.9 Consider the function $H_n(q, x) = \prod_{i=1}^{n}(1 + xq^{i-1})$ and its expansion into powers of x,

$$H_n(q, x) = \sum_{k=0}^{n} H_{n,k}(q)x^k.$$

Expand both members of the identity

$$(1+x)H_n(q, xq) = (1 + xq^n)H_n(q, x)$$

into powers of x to get the recurrence relation

$$H_{n,k}(q) = q^{k-1}\frac{[n-k+1]_q}{[k]_q}H_{n,k-1}(q), \quad k = 1, 2, \ldots, n,$$

with $H_{n,0}(q) = 1$. Iterating it, deduce that

$$H_{n,k}(q) = q^{\binom{k}{2}} \begin{bmatrix} n \\ k \end{bmatrix}_q, \quad k = 0, 1, \ldots, n.$$

Similarly, considering the function $G_n(q,x) = \prod_{i=1}^{n}(1-xq^{i-1})^{-1}$ and its expansion into powers of x,

$$G_n(q,x) = \sum_{k=0}^{\infty} G_{n,k}(q) x^k,$$

derive the recurrence relation

$$G_{n,k}(q) = \frac{[n+k-1]_q}{[k]_q} G_{n,k-1}(q), \quad k = 1, 2, \ldots, n,$$

with $G_{n,0}(q) = 1$, which entails

$$G_{n,k}(q) = \begin{bmatrix} n+k-1 \\ k \end{bmatrix}_q, \quad k = 0, 1, \ldots, n.$$

For the derivation of the third formula, consider the function $F_n(q,x) = q^{xn}$ and its expansion into q-factorials of x,

$$F_n(q,x) = \sum_{k=0}^{n} F_{n,k}(q) [x]_{k,q},$$

and derive the recurrence relation

$$F_{n,k}(q) = (q-1)q^{k-1} \frac{[n-k+1]_q}{[k]_q} F_{n,k-1}(q), \quad k = 1, 2, \ldots, n,$$

with $F_{n,0}(q) = 1$, which implies

$$F_{n,k}(q) = q^{\binom{k}{2}} (q-1)^k \begin{bmatrix} n \\ k \end{bmatrix}_q, \quad k = 0, 1, \ldots, n.$$

3.10 Expand the identities

$$\prod_{i=1}^{r+s} (1+xq^{i-1}) = \prod_{i=1}^{r} (1+xq^{i-1}) \prod_{j=1}^{s} (1+xq^{r+j-1})$$

and

$$\prod_{i=1}^{r+s} (1-xq^{i-1})^{-1} = \prod_{i=1}^{r} (1-xq^{i-1})^{-1} \prod_{j=1}^{s} (1-xq^{r+j-1})^{-1}$$

into powers of x, using the q-binomial and the q-negative binomial formulae of Exercise 3.9, respectively, to derive the first two formulae. Show the third formula by induction on n.

3.11 (a) Expand the noncentral q-factorial $[t-r]_{n,q}$ into powers of $[t-r]_q$, and then expand the powers of $[t-r]_q = q^{-r}([t]_q - [r]_q)$ into powers of $[t]_q$. Comparing the resulting expansion with the definition of the noncentral q-Stirling numbers of the first kind, deduce the first expression. Further, expand the noncentral q-factorial $[t-r]_{n,q}$ into q-factorials $[t]_{j,q}$, $j = 0, 1, \ldots, n$, using the q-Vandermonde's formula given in Exercise 3.10, and then expand the q-factorials $[t]_{j,q}$, $j = 0, 1, \ldots, n$, into powers of $[t]_q$. Comparing the resulting expansion with the definition of the noncentral q-Stirling numbers of the first kind, deduce the second expression.

(b) Rewrite the definition of the noncentral q-Stirling numbers of the first kind as

$$([t]_q - [r]_q)([t]_q - [r+1]_q) \cdots ([t]_q - [r+n-1]_q) = \sum_{k=0}^{n} s_q(n,k;r)[t]_q^k,$$

and execute the multiplications to get the required expression. Similarly, for $r = 0$, the formula

$$([t]_q - [1]_q)([t]_q - [2]_q) \cdots ([t]_q - [n-1]_q) = \sum_{k=0}^{n} s_q(n,k)[t]_q^{k-1}$$

implies the required expression.

(c) Expand both members of $[t-r]_{n+1,q} = [t-r-n]_q[t-r]_{n,q}$ into powers of $[t]_q$, using the definition of the noncentral q-Stirling numbers of the first kind, and conclude the triangular recurrence relation.

3.12 (a) Expand the power $[t+r]_q^n$ into q-factorials $[t+r]_{j,q}$, $j = 0, 1, \ldots, n$, and then expand the q-factorial $[t+r]_{j,q}$ into q-factorials $[t]_{k,q}$, $k = 0, 1, \ldots, j$, using the q-Vandermonde's formula given in Exercise 3.10. Comparing the resulting expansion with the definition of the noncentral q-Stirling numbers of the second kind, deduce the first expression. Further, expand the power $[t+r]_q^n = (q^r[t]_q + [r]_q)^n$ into powers $[t]_q^j$, $j = 0, 1, \ldots, n$, and then expand the power $[t]_q^j$ into q-factorials $[t]_{k,q}$, $k = 0, 1, \ldots, j$. Comparing the resulting expansion with the definition of the noncentral q-Stirling numbers of the second kind, deduce the second expression.

(b) Expand both members of $[t+r]_q^{n+1} = [t+r]_q[t+r]_q^n$ into q-factorials $[t]_{k,q}$, $k = 0, 1, \ldots, n+1$ and using the relation $[t+r]_q[t]_{k,q} = q^{k+r}[t]_{k+1,q} + [k+r]_q[t]_{k,q}$, conclude the triangular recurrence relation.

(c) Expand the noncentral q-factorial $[t-r]_{n,q}$ into powers $[t]_q^j$, $j = 0, 1, \ldots, n$, and then expand these powers into noncentral q-factorials $[t-r]_{k,q}$, $k = 0, 1, \ldots, j$. Finally, equating the coefficients of $[t-r]_{k,q}$ in both sides of the resulting expression, deduce the first orthogonal relation. The second relation may be similarly derived.

3.13 (a) Using the triangular recurrence relation for the noncentral q-Stirling numbers of the second kind, given in Exercise 3.12, deduce the recurrence relation

$$\phi_k(u;r,q) = u(1-[r+k]_q u)^{-1}\phi_{k-1}(u;r,q), \quad k = 1, 2, \ldots,$$

with $\phi_0(u;r,q) = (1-[r]_q u)^{-1}$. Iterate it to get the required expression.

(b) Expand both members of

$$\phi_k(u;r,q) = u(1 - [r+k]_q u)^{-1} \phi_{k-1}(u;r,q), \quad k = 1, 2, \ldots$$

into powers of u and equate the coefficients of u^n on both sides of the resulting expression to get the vertical recurrence relation. Further, expand each factor in the expression of the generating function $\phi_k(u;r,q)$ into powers of u and equate coefficients of u^n on both sides of the resulting expansion to find

$$S_q(n,k;r) = \sum [r]_q^{j_0} [r+1]_q^{j_1} \cdots [r+k]_q^{j_k},$$

where the summation is extended over all $j_i \geq 0$, $i = 0, 1, \ldots, k$, with $j_0 + j_1 + \cdots + j_k = n - k$, which is equivalent to the required expression.

3.14 The generalized Stirling number of the second kind $S(n, k; a_0)$ for $a_i = [i+r]_q$, $i = 0, 1, \ldots$, reduces to the noncentral q-Stirling number of the second kind $S_q(n, k; r)$ (see Remark 3.2). Thus, setting $a_i = [i+r]_q$, $i = 0, 1, \ldots$, into the explicit expression (3.35) and since

$$(a_j)_{j,a_0} = q^{\binom{j}{2}+rj}[j]_q!, \quad (a_j)_{k-j,a_{j+1}} = (-1)^{k-j} q^{(j+r)(k-j)}[k-j]_q!,$$

deduce the first expression. Also, the generating function $\phi_k(u;r,q)$, given in Exercise 3.13, upon using the formula

$$\prod_{j=0}^{k}(t - [r+j]_q)^{-1} = \sum_{j=0}^{k} c_j \cdot (t - [r+j]_q)^{-1},$$

where

$$c_j^{-1} = \frac{d}{dt}\left[\prod_{i=0}^{k}(t - [r+i]_q)\right]_{t=[r+j]_q} = (-1)^{k-j} q^{-\binom{j+1}{2}+(j+r)k}[j]_q![k-j]_q!,$$

may be expanded into powers of u yielding the same expression. Further, starting from

$$\sum_{k=0}^{n} q^{\binom{k}{2}+rk} S_q(n,k;r)[t]_{k,q} = [t+r]_q^n,$$

expand the right-hand side $[t+r]_q^n = (1 - q^{t+r})^n/(1-q)^n$ into powers of q^{t+r}, using Newton's binomial formula, and then, expand the power $(q^t)^j$, $j = 0, 1, \ldots, n$, into q-factorials $[t]_{k,q}$, $k = 0, 1, \ldots, j$, using the second q-binomial formula given in Exercise 3.9, deduce the expansion

$$[t+r]_q^n = \sum_{k=0}^{n} q^{\binom{k}{2}} \left\{ \sum_{j=k}^{n} (-1)^{j-k} q^{rj} \begin{bmatrix} j \\ k \end{bmatrix}_q \binom{n}{j} \right\} [t]_{k,q},$$

which implies the second expression.

3.15 Multiplying the expression defining the q-Stirling numbers of the first kind (see Remark 3.2) by $q^{\binom{n}{2}+rn}u^n/[n]_q!$ and summing it for $n = 0, 1, \ldots$, deduce, by virtue of the general q-binomial formula

$$\prod_{i=1}^{\infty} \frac{1+uq^{i-1}}{1+uq^{t+i-1}} = \sum_{k=0}^{\infty} q^{\binom{k}{2}} \begin{bmatrix} t \\ k \end{bmatrix}_q u^k,$$

the generating function

$$\sum_{n=0}^{\infty}\sum_{k=0}^{n} s_q(n,k;r)[t]_q^k \frac{u^n}{[n]_q!} = \prod_{i=1}^{\infty} \frac{1+uq^{r+i-1}}{1+uq^{t+i-1}},$$

which, on using the relation $q^t = 1 - (1-q)[t]_q$ and then replacing $[t]_q$ by t, implies the first generating function. The second generating function may be derived by using the first explicit expression of the q-Stirling numbers of the second kind given in Exercise 3.14.

3.16 (a) Expand the q-factorial $[st + r]_{n,q} = [s(t + r/s)]_{n,q}$ into q-factorials $[t + r/s]_{j,q^s}$, $j = 0, 1, \ldots, n$ and then expand the q-factorial $[t + r/s]_{j,q^s}$ into q-factorials $[t]_{k,q^s}$, $k = 0, 1, \ldots, j$, using the q-Vandermonde's formula given in Exercise 3.10. Comparing the resulting expansion with the definition of $C_q(n, k; s, r)$, deduce the first expression. Also, expand the q-factorial $[st + r]_{n,q}$ into q-factorials $[st]_{j,q}$, $j = 0, 1, \ldots, n$, and then expand these q-factorials into q-factorials $[t]_{k,q^s}$, $k = 0, 1, \ldots, j$. Comparing the resulting expansion with the definition of $C_q(n, k; s, r)$, conclude the second expression.

(b) Expand both members of the recurrence relation

$$[st + r]_{n+1,q} = [st + r - n]_q [st + r]_{n,q}$$

into q-factorials $[t]_{k,q^s}$, $k = 0, 1, \ldots, n+1$, and using the relations

$$[st - n + r]_q = q^{-n+r}([s]_q[t]_{q^s} - [n-r]_q),$$

and

$$[t]_{q^s}[t]_{k,q^s} = [k]_{q^s}[t]_{k,q^s} + q^{sk}[t]_{k+1,q^s},$$

deduce the triangular recurrence relation.

3.17 The generalized Lah number $C(n, k; a_0, b_0)$ for $a_i = [i-r]_q$ and $b_i = [si]_q$, $i = 0, 1, \ldots$, reduces to $[s]_q^{-k}C_q(n, k; s, r)$ (see Remark 3.4). Thus, setting $a_i = [i-r]_q$ and $b_i = [si]_q$, $i = 0, 1, \ldots$, into the explicit expression (3.45) and since

$$(b_j)_{n,a_0} = q^{\binom{n}{2}-rn}[sj+r]_{n,q}, \quad (b_j)_{j,b_0} = q^{s\binom{j}{2}}[s]_q^j[j]_{q^s}!$$

and

$$(b_j)_{k-j,b_{j+1}} = (-1)^{k-j}q^{sj(k-j)}[s]_q^{k-j}[k-j]_{q^s}!,$$

deduce the first expression. Further, starting from

$$\sum_{k=0}^{n} q^{s\binom{k}{2}} C_q(n,k;s,r)[t]_{k,q^s} = q^{\binom{n}{2}-rn}[st+r]_{n,q},$$

and using the first q-binomial formula given in Exercise 3.9, expand the q-factorial

$$[st+r]_{n,q} = (1-q)^{-n} \prod_{i=1}^{n} (1 - q^{st+r-n+i})$$

into powers of $q^{st+r-n+1}$ and then, using the second q-binomial formula in Exercise 3.9, expand the power q^{stj}, $j = 0, 1, \ldots, n$, into q-factorials $[t]_{k,q^s}$, $k = 0, 1, \ldots, j$, to deduce the second expression.

3.18 Starting from

$$\sum_{k=0}^{n} q^{s\binom{k}{2}} C_q(n,k;s,s\rho-r)[t]_{k,q^s} = q^{\binom{n}{2}+rn-s\rho n}[s(t+\rho)-r]_{n,q},$$

expand the q-factorial $[s(t+\rho)-r]_{n,q}$ into powers of $[s(t+\rho)]_q = [s]_q[t+\rho]_{q^s}$ and then expand the power $[t+\rho]_{q^s}^j$, $j = 0, 1, \ldots, n$, into q-factorials $[t]_{k,q^s}$, $k = 0, 1, \ldots, j$. The resulting expansion entails the required expression. The limiting formulae may be deduced from the corresponding defining expressions of the noncentral generalized q-factorial coefficients and the noncentral q-Stirling numbers.

3.19 Let

$$C_{k;q}(t) = \sum_{n=k}^{\infty} q^{-\binom{n+1}{2}+r(n+1)} C_q(n,k;s,r) \frac{1}{[st+r]_{n+1,q}}.$$

Use the triangular recurrence relation for the noncentral generalized q-factorial coefficients $C_q(n,k;s,r)$ together with the relation

$$\frac{q^{-\binom{n}{2}+rn}}{[st+r]_{n,q}} = \frac{q^{-\binom{n+1}{2}+r(n+1)}([st]_q - [n-r]_q)}{[st+r]_{n+1,q}},$$

to derive the recurrence relation

$$C_{k;q}(t) = \frac{q^{-sk}}{[t-k]_{q^s}} C_{k-1;q}(t), \quad k = 1, 2, \ldots, \quad C_{0;q}(t) = \frac{1}{[s]_q[t]_{q^s}}.$$

Iterating it, deduce that

$$C_{k;q}(t) = \frac{q^{-s\binom{k+1}{2}}}{[s]_q[t]_{k+1,q^s}}.$$

3.20 Expand the q-factorial $[t+n-r-1]_{n,q}$ into q-factorials $[t]_{k,q}$, $k = 0, 1, \ldots, n$, using the q-Vandermonde's formula given in Exercise 3.10, and comparing this expansion with the definition of the noncentral signless q-Lah numbers, conclude the required expression.

CHAPTER 4

4.1 Let A_j be the event that the jth urn contains i balls, $j = 1, 2, \ldots, n$, given that m balls are distributed into the n urns, and $\{j_1, j_2, \ldots, j_r\} \subseteq \{1, 2, \ldots, n\}$. Also, let B_j be the event of placing any ball into the jth urn, $j = 1, 2, \ldots, n$, when m balls are allocated into the n urns. Then, $P(A_{j_1} A_{j_2} \cdots A_{j_r})$ is the probability that, in a multinomial model of m independent trials with $P(B_{j_1}) = p_{j_1}$, $P(B_{j_2}) = p_{j_2}, \ldots$, $P(B_{j_r}) = p_{j_r}$ and $P(B'_{j_1} B'_{j_2} \cdots B'_{j_r}) = 1 - p_{j_1} - p_{j_2} - \cdots - p_{j_r}$, each of the events $B_{j_1}, B_{j_2}, \ldots, B_{j_r}$ appears i times, and so, applying Corollary 1.6, deduce the required expression.

4.2 Work as in Exercise 4.1 and replace the multinomial model by the multivariate hypergeometric model with parameters $s_{j_1}, s_{j_2}, \ldots, s_{j_r}$ and $s - s_{j_1} - s_{j_2} - \cdots - s_{j_r}$.

4.3 Work as in Exercise 4.1 and replace the multinomial model by the multivariate negative hypergeometric model with parameters $s_{j_1}, s_{j_2}, \ldots, s_{j_r}$ and $s - s_{j_1} - s_{j_2} - \cdots - s_{j_r}$.

4.4 In the expression of the binomial moments of K_i, given that $S_n = m$, which is derived in Exercise 4.1, put $p_j = 1/n$, $j = 1, 2, \ldots, n$, and $r = 1, 2$ to get

$$E(K_i | S_n = m) = n \binom{m}{i} \left(\frac{1}{n}\right)^i \left(1 - \frac{1}{n}\right)^{m-i} = n p_{m,i}(1/n)$$

and

$$E\left[\binom{K_i}{2} \Big| S_n = m\right] = \binom{n}{2}\binom{m}{i}\binom{m-i}{i}\left(\frac{1}{n}\right)^{2i}\left(1 - \frac{2}{n}\right)^{m-2i}$$
$$= \binom{n}{2} p_{m,i}(1/n) p_{m-i,i}(1/(n-1)).$$

4.5 In the expression of the binomial moments of K_i, given $S_n = m$, which is derived in Exercise 4.2, put $s_j = s$, $j = 1, 2, \ldots, n$, and $r = 1, 2$ to find

$$E(K_i | S_n = m) = n \binom{m}{i} \frac{(s)_i (s(n-1))_{m-i}}{(sn)_m} = n p_{m,i}(sn, s)$$

and

$$E\left[\binom{K_i}{2} \Big| S_n = m\right] = \binom{n}{2}\binom{m}{i}\binom{m-i}{i}\frac{(s)_i^2 (s(n-2))_{m-2i}}{(sn)_m}$$
$$= \binom{n}{2} p_{m,i}(sn, s) p_{m-i}(s(n-1), s).$$

4.6 The relation $p_{k,m}(n, i) = p_k(m, n, i) q_m(n)$ together with the expression

$$\sum_{k=0}^{\min\{n, [m/i]\}} p_k(m, n, i) t^k = \sum_{r=0}^{\min\{n, [m/i]\}} b_{(r)}(m, n, i)(t-1)^r,$$

on using (4.7) in Theorem 4.1, gives

$$\sum_{k=0}^{\min\{n,[m/i]\}} p_{k,m}(n,i) t^k = \sum_{r=0}^{\min\{n,[m/i]\}} \binom{n}{r} q_i^r q_{m-ir}(n-r).$$

Multiply it by u^m and sum for $m = 0, 1, \ldots$, to derive the bivariate generating function $g_{n,i}(t,u)$.

4.7 Let A_h be the event that the hth urn contains at least i balls, $h = 1, 2, \ldots, n$, given that $S_n = m$, and $\{h_1, h_2, \ldots, h_r\} \subseteq \{1, 2, \ldots, n\}$. Then, the probability that j among m distinguishable balls are placed into r specified urns so that each urn is occupied by at least i balls, on using Example 3.1, is given by

$$\binom{m}{j}(n-r)^{m-j} r! S_i(j,r), \quad j = ir, ir+1, \ldots, m.$$

Summing these probabilities, find the probability $P(A_{h_1} A_{h_2} \cdots A_{h_r})$ and using Corollary 1.6, deduce the required expression.

4.8 Work as in Exercise 4.7 and use Example 3.2 to conclude that the probability that j among m like balls are placed into r specified urns, with s cells, so that each urn is occupied by at least i balls, is given by

$$\frac{r! C_i(j,r;s)}{j!} \cdot \frac{\binom{s(n-r)}{m-j}}{\binom{sn}{m}}, \quad j = ir, ir+1, \ldots, m.$$

4.9 The event $\{Z_{(1)} > z\}$ is equivalent to the event $\{Z_1 > z, Z_2 > z, \ldots, Z_n > z\}$ and so (cf. Example 4.1)

$$P(z;m,n) = 1 - \sum \frac{n!}{z_1! z_2! \cdots z_n!} p_1^{z_1} p_2^{z_2} \cdots p_n^{z_n},$$

where the summation is extended over all $z_j > z$, $j = 1, 2, \ldots, n$, such that $z_1 + z_2 + \cdots + z_n = m$. Multiplying it by $u^m/m!$ and summing for $m = 0, 1, \ldots$, derive the expression of the generating function $G(u; z, n)$. Also, the event $\{Z_{(n)} \leq z\}$ is equivalent to the event $\{Z_1 \leq z, Z_2 \leq z, \ldots, Z_n \leq z\}$ and so

$$Q(z;m,n) = \sum \frac{m!}{z_1! z_2! \cdots z_n!} p_1^{z_1} p_2^{z_2} \cdots p_n^{z_n},$$

where the summation is extended over all $z_j \leq z$, $j = 1, 2, \ldots, n$, such that $z_1 + z_2 + \cdots + z_n = m$. Multiplying it by $u^m/m!$ and summing for $m = 0, 1, \ldots$, deduce the expression of the generating function $H(u; z, n)$.

4.10 As in Exercise 4.9 show first that

$$P(z;m,n) = 1 - \sum \binom{s_1}{z_1}\binom{s_2}{z_2}\cdots\binom{s_n}{z_n} \bigg/ \binom{c}{m},$$

where the summation is extended over all $z_j > z$, $j = 1, 2, \ldots, n$, such that $z_1 + z_2 + \cdots + z_n = m$, and

$$Q(z; m, n) = \sum \binom{s_1}{z_1}\binom{s_2}{z_2}\cdots\binom{s_n}{z_n} \bigg/ \binom{c}{m},$$

where the summation is extended over all $z_j \leq z$, $j = 1, 2, \ldots, n$, such that $z_1 + z_2 + \cdots + z_n = m$. Then, multiplying each expression by $u^m/m!$ and summing for $m = 0, 1, \ldots$, deduce the expressions of the generating functions $G(u; z, n)$ and $H(u; z, n)$, respectively.

4.11 As in Exercise 4.9 show first that

$$P(z; m, n) = 1 - \sum \frac{\binom{s_1 + z_1 - 1}{z_1}\binom{s_2 + z_2 - 1}{z_2}\cdots\binom{s_n + z_n - 1}{z_n}}{\binom{c + m - 1}{m}},$$

where the summation is extended over all $z_i > z$, $j = 1, 2, \ldots, n$, such that $z_1 + z_2 + \cdots + z_n = m$, and

$$Q(z; m, n) = \sum \frac{\binom{s_1 + z_1 - 1}{z_1}\binom{s_2 + z_2 - 1}{z_2}\cdots\binom{s_n + z_n - 1}{z_n}}{\binom{c + m - 1}{m}},$$

where the summation is extended over all $z_i \leq z$, $j = 1, 2, \ldots, n$, such that $z_1 + z_2 + \cdots + z_n = m$. Then, multiplying each expression by $u^m/m!$ and summing for $m = 0, 1, \ldots$, deduce the expressions of the generating functions $G(u; z, n)$ and $H(u; z, n)$, respectively.

4.12 Let A_j be the event that the jth urn contains i balls, $j = 1, 2, \ldots, n$, and $\{j_1, j_2, \ldots, j_r\} \subseteq \{1, 2, \ldots, n\}$. Then,

$$P(A_{j_1} A_{j_2} \cdots A_{j_r}) = \binom{n - r + m - ir - 1}{m - ir} \bigg/ \binom{n + m - 1}{m},$$

which implies that the events $A_1, A_2, \ldots, A_n$ are exchangeable. Thus, applying Corollaries 1.5 and 1.6, establish the expressions of $p_k(m, n, i)$ and $b_{(r)}(m, n, i)$, respectively.

4.13 Use the fact that the number of distributions of m indistinguishable balls into k distinguishable urns so that no urn remains empty is given by $\binom{m-1}{k-1}$ to establish the expression of the probability $p_k(m, n)$. Also, assume that one ball is placed into each of r specified urns and then use the fact that the number of distributions of the remaining $m - r$ balls into n distinguishable urns without any restriction is given by $\binom{n+m-r-1}{m-r}$ to establish the expression of the binomial moment $b_{(r)}(m, n)$.

Note that these expressions may also be deduced from the corresponding expressions of Exercise 4.12 by putting $i = 0$.

4.14 Let A_j be the event that the jth urn contains at least i balls, $j = 1, 2, \ldots, n$. Show that the events $A_1, A_2, \ldots, A_n$ are exchangeable with

$$p_r = P(A_{j_1} A_{j_2} \cdots A_{j_r}) = \binom{n + m - ir - 1}{m - ir} \bigg/ \binom{n + m - 1}{m}.$$

Since $P(Z_{(k)} \geq i)$ is the probability that at least $n - k + 1$ among the n events $A_1, A_2, \ldots, A_n$ occur, on using Corollary 1.7, establish the required expression.

4.15 Let A_j be the event that the jth urn contains at least $i+1$ balls, $j = 1, 2, \ldots, n$. Show that the events $A_1, A_2, \ldots, A_n$ are exchangeable with

$$p_r = P(A_{j_1} A_{j_2} \cdots A_{j_r}) = \binom{m - ir - 1}{n - 1} \bigg/ \binom{m - 1}{r - 1}.$$

Since $P(Z_{(k)} > i)$ is the probability that at least $n - k + 1$ among the n events $A_1, A_2, \ldots, A_n$ occur, on using Corollary 1.7, deduce the required expression.

4.16 Let A_j be the event that the jth urn contains i balls, $j = 1, 2, \ldots, n$. Show that the events $A_1, A_2, \ldots, A_n$ are exchangeable with

$$p_r = P(A_{j_1} A_{j_2} \cdots A_{j_r}) = L(m - ir, n - r; s)/L(m, n; s)$$

and then apply Corollaries 1.5 and 1.6 to establish the expressions of the probability $p_k(m, n, i)$ and the binomial moment $b_{(r)}(m, n, i)$, respectively.

4.17 The n pupae prey may be considered as n distinguishable urns. An attack on the jth pupae prey or on the jth prey corpse corresponds to the placement of a ball into the jth urn, $j = 1, 2, \ldots, n$. Since all pupae prey and prey corpses are equally vulnerable, the probability of placing a ball into the jth urn is $p_j = 1/n$, $j = 1, 2, \ldots, n$. This is equivalent to the assumption that the number $X_j(t)$ of attacks suffered by the jth pupae prey or the jth prey corpse obeys a Poisson distribution with $E[X_j(t)] = \lambda t$, $j = 1, 2, \ldots, n$. Thus, apply Corollary 4.4 to deduce the probability function $P[K_0(t) = k|S(t) = m]$, $k = 0, 1, \ldots, n$. Further, since all adults are assumed to act independently of one another, the probability generating function of $S(t) = \sum_{i=1}^{r} S_i(t)$ is $[g(u;t)]^r$. Therefore, using the expression $S(m, n-k) = [\Delta_u^{n-k} u^m]_{u=0}/(n-k)!$, derived the probability function $P[K_0(t) = k]$, $k = 0, 1, \ldots$, and its generating function $f(v; t)$.

4.18 Let Y_n be the total number of balls staying in the n urns. Then,

$$P(Y_n = y|S_n = m) = \binom{m}{y} p^y (1-p)^{m-y}, \quad y = 0, 1, \ldots, m,$$

and

$$P(K = k|Y_n = y) = \binom{n}{k} \left[\Delta_u^k \left(\frac{u}{n}\right)^y\right]_{u=0}, \quad k = 0, 1, \ldots, y.$$

Use the relation

$$P(K = k|S_n = m) = \sum_{y=0}^{m} P(Y_n = y|S_n = m) P(K = k|Y_n = y)$$

to derive the required expression.

4.19 Follow the steps of derivation of parts (a) and (b) of Theorem 4.1 to obtain the expressions of the probability function $p_k(m,n,r,i)$ and the binomial moment $b_{(j)}(m,n,i)$. Working as in the derivation of the generating function of Exercise 4.6, establish the expression of the generating function $g_{n,r,i}(t,u)$.

4.20 Setting $i=0$ in the expressions of the probability function and its binomial moments in Exercise 4.19 and working as in the derivation of parts (a) and (b) of Corollary 4.1, deduce the required expressions.

4.21 Work as in the derivation of Corollary 4.4.

4.22 Work as in the derivation of Corollary 4.6.

4.23 The probability P_k that at most k urns are occupied equals the probability that at least $n-k$ urns are empty, and so using Corollary 1.7 together with Lemma 4.1, establish the required expression.

4.24 Let A_h and B_h be the events that the hth urn contains i and j balls, respectively, $i \neq j, h = 1, 2, \ldots, n$. Then, (A_h, B_h) is a pair of disjoint events and $p_{k,r}(m,n;i,j)$ is the probability that k among the events $A_1, A_2, \ldots, A_n$ and r among the events $B_1, B_2, \ldots, B_n$ occur. Thus, apply the bivariate inclusion and exclusion formula given in Exercise 1.35 to derive the probability $p_{k,r}(m,n;i,j)$. Also, apply the expression of the joint binomial moments to find $b_{(k,r)}(m,n;i,j)$.

4.25 Since $q_x = e^{-\lambda}\lambda^x/x!$ and $q_m(n) = e^{-n\lambda}(n\lambda)^m/m!$, the general expressions given in Exercise 4.24 reduce to the required formulae.

CHAPTER 5

5.1 The probability function $q_m(k,n) = P(W_k = m)$, $m = k, k+1, \ldots$, by using the expression $q_m(k,n) = P_{k-1}(m-1,n) - P_{k-1}(m,n)$, where $P_k(m,n) = P(K \leq k|S_n = m)$ is given by (cf. Exercise 1.29 and the hint to its solution)

$$P_k(m,n) = \sum_{r=0}^{k}(-1)^{k-r}\binom{n-r-1}{n-k-1}S_{n,n-r},$$

with

$$S_{n,n-r} = \sum \frac{(s_{j_1} + s_{j_2} + \cdots + s_{j_r} + m - 1)_m}{(s+m-1)_m},$$

where the summation is extended over all r-combinations $\{j_1, j_2, \ldots, j_r\}$ of the n indices $\{1, 2, \ldots, \}$. Thus,

$$q_m(k,n) = \sum_{r=0}^{k-1}(-1)^{k-r-1}\binom{n-r-1}{k-r-1}T_{n,n-r},$$

with

$$T_{n,n-r} = \sum \frac{(s - s_{j_1} - s_{j_2} - \cdots - s_{j_r})(s_{j_1} + s_{j_2} + \cdots + s_{j_r} + m - 2)_{m-1}}{(s+m-1)_m},$$

where the summation is extended over all r-combinations $\{j_1, j_2, \ldots, j_r\}$ of the n indices $\{1, 2, \ldots, n\}$. Further, working as in Example 5.3, the ith ascending binomial moment $b_{[i]}(k, n) = E\left[\binom{W_k+i-1}{i}\right]$ is obtained as

$$b_{[i]}(k, n) = \sum_{r=0}^{k-1}(-1)^{k-r-1}\binom{n-r-1}{k-r-1}U_{n,n-r},$$

with

$$U_{n,n-r} = \sum \frac{(s-1)_i}{(s-s_{j_1}-s_{j_2}-\cdots-s_{j_r}-1)_i},$$

where the summation is extended over all r-combinations $\{j_1, j_2, \ldots, j_r\}$ of the n indices $\{1, 2, \ldots, n\}$.

5.2 Express the probability function $c_k(r, n) = P(L_r = k)$, $k = 1, 2, \ldots, n$, as $c_k(r, n) = P_k(r+k, n) - P_{k-1}(r+k-1, n)$, where $P_k(m, n) = P(K \leq k | S_n = m)$ is given by (cf. Exercise 1.29 and the hint to its solution)

$$P_k(m, n) = \sum_{i=0}^{k}(-1)^{k-i}\binom{n-i-1}{k-i}S_{n,n-i}(m),$$

with

$$S_{n,n-i}(m) = \sum \frac{(s_{j_1}+s_{j_2}+\cdots+s_{j_i}+m-1)_m}{(s+m-1)_m},$$

where the summation is extended over all i-combinations $\{j_1, j_2, \ldots, j_i\}$ of the n indices $\{1, 2, \ldots, n\}$. Then, proceeding as in Example 5.2, deduce the expression

$$c_k(r, n) = \sum_{i=0}^{k}(-1)^{k-i}\binom{n-i-1}{k-i}T_{n,n-i},$$

with

$$T_{n,n-i} = \sum\left[s_{j_1}+s_{j_2}+\cdots+s_{j_i}-s+\frac{n-i}{n-k}(s+r+k-1)\right]$$
$$\times (s_{j_1}+s_{j_2}+\cdots+s_{j_i}+r+k-2)_{r+k-1}/(s+r+k-1)_{r+k},$$

where the summation is extended over all i-combinations $\{j_1, j_2, \ldots, j_i\}$ of the n indices $\{1, 2, \ldots, n\}$.

5.3 Note that the Bose-Einstein occupancy model may be considered as a random occupancy model with geometric probabilities: $q_x = pq^x$, $x = 0, 1, \ldots$. Then, $q_m(u) = \binom{u+m-1}{m}p^u q^m$, $m = 0, 1, \ldots$, and using Theorem 5.1, deduce for the probability function $q_m(k, n)$, $m = k, k+1, \ldots$, the required expression and for its ascending binomial moments the expression

$$b_{[r]}(k, n) = \binom{n-1}{k-1}\left[\Delta_u^{k-1}\frac{(n)_{r+1}}{(n-u)_{r+1}}\right]_{u=0},$$

which, after evaluating the indicated differences, reduces to the required expression.

5.4 Use Theorems 5.2 and 5.3 with $q_x = pq^x$, $x = 0, 1, \ldots$, to derive the expressions for $q_{r;m}(k,n)$ and $c_k(r,n)$, respectively.

5.5 (a) The probability $q_m(k, n; p)$ is equal the product of the probability $p_{k-1}(m-1, n; p)$ that under the extended random occupancy model $k-1$ urns are occupied, given that $m-1$ balls are distributed into the n urns, and the probability $(n-k+1)p/n$ that the mth ball is placed into any of the $n-k+1$ empty urns and stays there. Thus, using the expression of $p_{k-1}(m-1, n; p)$ given in Exercise 4.18, deduce the expression of $q_m(k, n; p)$.

(b) Express the binomial moment $b_{[r]}(k, n; p)$ as

$$b_{[r]}(k,n;p) = p\binom{n-1}{k-1} \sum_{m=1}^{\infty} \binom{r+m-1}{m-1} \left[\Delta_u^{k-1}\left(\frac{u+nq/p}{n/p}\right)^{m-1}\right]_{u=0},$$

which, after executing the summation, reduces to

$$b_{[r]}(k,n;p) = p\binom{n-1}{k-1}\left[\Delta_u^{k-1}\left(1 - \frac{u+nq/p}{n/p}\right)^{-r-1}\right]_{u=0}$$

$$= p^{-r}\binom{n-1}{k-1}\left[\Delta_u^{k-1}\left(1 - \frac{u}{n}\right)^{-r-1}\right]_{u=0}.$$

5.6 (a) Under the extended sequential occupancy model, let $Y_{j;p}$ be the number of balls required to be distributed into the n urns after the $(j-1)$st urn, empty before, is occupied by one ball and until the jth urn, empty before, is occupied by one ball, $j = 1, 2, \ldots, k$. These random variables are independent geometric with varying success probability.

(b) The expected value of $Y_{j;p}$ is

$$E(Y_{j;p}) = \frac{n}{(n-j+1)p}.$$

5.7 Follow the steps of the derivation of Theorem 5.1.

5.8 Work as in the derivation of Corollary 5.1.

5.9 Work as in the derivation of Corollary 5.4.

5.10 Follow the steps of the derivation of Theorem 5.2.

5.11 (a) Under the modified sequential occupancy model, let Y_i be the number of balls required to be distributed into the $n+r$ urns after the $(i-1)$st target urn, empty before, receives one ball and until the ith target urn, empty before, receives one ball, $i = 1, 2, \ldots, k$. These random variables are independent geometric with varying success probability.

(b) The expected value of Y_i is

$$E(Y_i) = \frac{n+r}{n+i-1}.$$

5.12 Introduce the same random variables as in Exercise 5.11. In this case these random variables are not independent. The required conditional probability function can be deduced from that of Exercise 5.10.

5.13 Follow the steps of the derivation of Theorem 5.3. For the particular complementary waiting time distributions work as in the derivation of Corollaries 5.3 and 5.6.

5.14 The (sequential) random occupancy model, with the number of balls distributed into any specific urn obeying a Poisson distribution and given the total number of balls distributed into the n urns, is equivalent to the assumption that distinguishable balls are allocated into the n urns with probability $1/n$ for each ball to fall into any specific urn. Thus, for $m \leq r$,

$$P(M_r > m) = \frac{(n)_m}{n^m},$$

while for $m > r$,

$$P(M_r = m | M_r > m - 1) = \frac{r-1}{n}$$

and

$$P(M_r > m | M_r > m - 1) = 1 - \frac{r-1}{n}.$$

Apply repeatedly the last expression to find

$$P(M_r > m) = P(M_r > r)\left(1 - \frac{r-1}{n}\right)^{m-r} = \frac{(n)_r}{n^r}\left(1 - \frac{r-1}{n}\right)^{m-r}.$$

From the expressions of $P(M_r > m)$ deduce the required probability function.

5.15 Let A_j be the event that the jth number is not drawn in r drawings, $j = 1, 2, \ldots, n$. Then,

$$P(T_{k;s} > r) = P(A_{j_1} \cup A_{j_2} \cup \cdots \cup A_{j_k})$$

and applying the inclusion and exclusion formula (1.32), derive the expression

$$P(T_{k;s} > r) = \sum_{i=1}^{k}(-1)^{i-1}\binom{k}{i}\binom{n-i}{s}^r\bigg/\binom{n}{s}^r,$$

which implies

$$P(T_{k;s} = r) = \sum_{i=1}^{k}(-1)^{i-1}\binom{k}{i}\left(1 - \frac{(n-i)_s}{(n)_s}\right)\left(\frac{(n-i)_s}{(n)_s}\right)^{r-1}$$

$$= \frac{k}{n}\left[\Delta_u^{k-1}\frac{n}{n-u}\left(1 - \frac{(u)_s}{(n)_s}\right)\left(\frac{(u)_s}{(n)_s}\right)^{r-1}\right]_{u=n-k}.$$

Note that this probability function is a linear combination of k geometric probability functions. Utilize it to find

$$E\left[\binom{T_{k;s}+j-1}{j}\right] = \sum_{i=1}^{k}(-1)^{i-1}\binom{k}{i}\left(1-\frac{(n-i)_s}{(n)_s}\right)^{-j}$$

$$= \frac{k}{n}\left[\Delta_u^{k-1}\frac{n}{n-u}\left(\frac{(u)_s}{(n)_s}\right)^{-j}\right]_{u=n-k}.$$

CHAPTER 6

6.1 Work as in Example 6.2 and use the result of Example 4.3.

6.2 (a) The recurrence relation may be deduced from (6.16) by using the triangular recurrence relation of the signless Stirling numbers of the first kind (2.35).
(b) From expression (6.16) deduce that

$$\frac{p_{m-1}(n;\theta)}{p_m(n;\theta)} = \frac{m|s(m-1,n)|}{|s(m,n)|} \cdot \frac{1}{\theta}$$

and, in particular, for $m = n+1$,

$$\frac{p_n(n;\theta)}{p_{n+1}(n;\theta)} = \frac{2}{n} \cdot \frac{1}{\theta}.$$

Using these expressions, prove the modal property.

6.3 (a) The UMVU estimate $\hat{\theta}(m;n)$ is given by (cf. Example 6.6)

$$\hat{\theta}(m;n) = \frac{m|s(m-1,n)|}{|s(m,n)|}, \quad m \geq n+1.$$

Use the triangular recurrence relation of the signless Stirling number of the first kind, (2.35), to derive the required recurrence relation.
(b) Use $|s(m,1)| = 1/m$ and $|s(m,2)| = \sum_{j=1}^{m}1/j$.

6.4 The UMVU estimate $\hat{\theta}(m;n,r)$ is given by (cf. Example 6.9)

$$\hat{\theta}(m;n,r) = \frac{m|s_r(m-1,n)|}{|s_r(m,n)|}, \quad m \geq rn+1.$$

Use recurrence relation (3.10) for the r-associated signless Stirling numbers of the first kind to derive the required recurrence relation.

6.5 Work as in Exercise 6.2.

6.6 Work as in Exercise 6.3.

6.7 Work as in Exercise 6.4.

6.8 For (a) work as in Exercise 6.2 (a), and for (b) work as in Exercise 6.3 (a).

6.9 Work as in Exercise 6.4.

6.10 (a) Apply Corollary 6.5. (b) Apply Theorem 6.6 (a).

6.11 Apply Theorem 6.6 (b).

6.12 Expand the series function $[-\log(1-\theta)]^n$ into powers of $\theta(1-\theta)^{\beta-1}$ using the Lagrange formula:

$$f(t) = f(0) + \sum_{m=1}^{\infty} \left[\frac{d^{m-1}}{dt^{n-1}} \left(g^m(t) \frac{df(t)}{dt} \right) \right]_{t=0} \cdot \frac{u^m}{m!}, \quad u = \frac{t}{g(t)}.$$

6.13 Expand the series function $(e^\lambda - 1)^n$ into powers of $\lambda e^{-\lambda\theta}$ using the Lagrange formula given in the hint of Exercise 6.12.

6.14 Expand the series function $(e^\lambda - 1 + \lambda\theta)^n$ into powers of $\lambda e^{-\lambda\theta}$ using the Lagrange formula:

$$f(t)\left(1 - u\frac{dg\theta)}{dt}\right)^{-1} = \sum_{m=0}^{\infty} \left[\frac{d^m}{dt^m} (g^m(t)f(t)) \right]_{t=0} \frac{u^m}{m!}, \quad u = \frac{t}{g(t)}.$$

6.15 Expand the series function $[(1-\theta)^{-s} - 1]^n$ into powers of $\theta(1-\theta)^{\beta-1}$ using the Lagrange formula given in the hint of Exercise 6.12.

CHAPTER 7

7.1 The required expressions are deduced by showing that

$$[\Delta_u E(S_u)]_{u=0} = E(X_1) = E(X)$$

and

$$\left[\Delta_u E\left\{\binom{S_u}{2}\right\}\right]_{u=0} = E\left[\binom{X_1}{2}\right] = E\left[\binom{X}{2}\right],$$

$$\left[\Delta_u^2 E\left\{\binom{S_u}{2}\right\}\right]_{u=0} = E\left[\binom{S_2}{2} - 2\binom{S_1}{2}\right]$$

$$= E\left[\binom{X_2}{2} - \binom{X_1}{2} + X_1 X_2\right] = [E(X)]^2.$$

7.2 Use (7.27) and (7.29).

7.3 Express the probability $p_m(t + \Delta t)$ in terms of the probabilities $p_j(t)$, $j = 0, 1, \ldots, m$, and then derive the differential equations

$$\frac{dp_0(t)}{dt} = -\theta(1 - q_0)p_0(t),$$

and
$$\frac{dp_m(t)}{dt} = \theta \sum_{j=1}^{m} q_j p_{m-j}(t) - \theta(1-q_0)p_m(t), \quad m = 1, 2, \ldots.$$

Multiplying the mth equation by u^m and summing for all $m = 0, 1, \ldots$, derive the required partial differential equation for the probability generating function $h(u,t)$.

7.4 Differentiating the probability generating function
$$h(u) = \sum_{m=0}^{\infty} p_m u^m = \exp\left\{-k\lambda\left(1 - \frac{1}{k}\sum_{j=1}^{k} u^j\right)\right\},$$

derive the expression
$$\frac{dh(u)}{du} = \lambda h(u) \sum_{j=1}^{k} j u^{j-1},$$

which implies the recurrence relation for the probability function p_m, $m = 0, 1, \ldots$. The recurrence relation for the binomial moments can be similarly derived.

7.5 (a) The n-fold convolution of the probability function $q_x = e^{-\lambda}\lambda^{x-1}/(x-1)!$, $x = 1, 2, \ldots$, is
$$q_m(n) = e^{-n\lambda}\frac{(n\lambda)^{m-n}}{(m-n)!}, \quad m = n, n+1, \ldots,$$

and so from (7.28) deduce the probability function p_m, $m = n, n+1, \ldots$. The rth binomial moment $a_{(r)}(n)$ of $q_m(n)$, $m = n, n+1, \ldots$, upon using the expression
$$\binom{m}{r} = \sum_{j=0}^{r}\binom{m-n}{j}\binom{n}{r-j},$$

is obtained as
$$a_{(r)}(n) = \sum_{j=0}^{r}\binom{n}{r-j}\frac{(n\lambda)^j}{j!}.$$

Then, the rth binomial moment $b_{(r)}$ may be deduced from (7.29), by using the expansion
$$u^j = \sum_{i=0}^{j} S(j, i; r-j)(u-r+j)_i.$$

(b) The mean μ and the variance σ^2 are deduced as
$$\mu = \theta(1+\lambda), \quad \sigma^2 = \theta(1+3\lambda+\lambda^2).$$

7.6 Differentiating the probability generating function
$$G(t) = \exp\{\theta[e^{\lambda(t-1)} - 1] + \rho(t-1)\},$$

derive the expression
$$\frac{dG(\theta)}{dt} = [\theta\lambda e^{\lambda(t-1)} + \rho]G(t),$$
which implies the recurrence relation for the probability function P_m, $m = 0, 1, \ldots$. The recurrence relation for the binomial moments can be similarly derived.

7.7 The probability function p_m, $m = 0, 1, \ldots$, and its binomial moments $b_{(r)}$, $r = 0, 1, \ldots$, may be deduced from (7.30) and (7.32), respectively. The mean μ and the variance σ^2 are deduced as
$$\mu = c\frac{sp\theta}{1-\theta}, \quad \sigma^2 = c[1 + (s-1)p]\frac{sp\theta}{1-\theta} + c(1-c)\left(\frac{sp\theta}{1-\theta}\right)^2,$$
where $c = [-\log(1-\theta)]^{-1}$.

7.8 (a) Using the exponential generating function of the generalized factorial coefficients (see Theorem 2.13), expand the generating function $T(t;x)$ into powers of t to establish the required expression.

(b) Differentiating the generating function $T(t;x)$, with respect to t, and multiplying the resulting expression by $1 - a[(x+t)^s - x^s]$, deduce the relation
$$\frac{dT(t;x)}{dt} = as(x+t)^{s-1} + a[(x+t)^s - x^s]\frac{dT(t;x)}{dt},$$
which entails for the generalized Chebyshev polynomials the required recurrence relation.

(c) Expand the probability generating function of the compound logarithmic-binomial distribution
$$\sum_{m=0}^{\infty} p_m t^m = [-\log(1-\theta)]^{-1}[-\log\{1 - \theta(pt+q)^s\}]$$
$$= [-\log(1-\theta)]^{-1}[-\log(1-\theta q^s)]$$
$$+ [-\log(1-\theta)]^{-1}[-\log\{1 - \theta(1-\theta q^s)^{-1}[(pt+q)^s - q^s]\}],$$
using the generalized Chebyshev polynomials of the first kind, to find
$$p_0 = [-\log(1-\theta)]^{-1}[-\log(1-\theta q^s)],$$
$$p_m = [-\log(1-\theta)]^{-1}T_m(q;\theta(1-\theta q^s)^{-1}, s)\frac{p^m}{m!}, \quad m = 1, 2, \ldots.$$
Similarly, find
$$b_{(r)} = [-\log(1-\theta)]^{-1}T_r(1;\theta(1-\theta)^{-1}, s)\frac{p^r}{r!}, \quad r = 1, 2, \ldots.$$
Using the recurrence relation for the generalized Chebyshev polynomials, deduce the recurrence relations
$$p_m = [-\log(1-\theta)]^{-1}\frac{\theta}{1-\theta q^s}\binom{s}{m}p^m + \frac{\theta}{1-\theta q^s}\sum_{k=1}^{m}\binom{s}{k}p^k q^{s-k}\frac{m-k}{m}p_{m-k},$$

for $m = 1, 2, \ldots$, with $p_0 = [-\log(1-\theta)]^{-1}[-\log(1-\theta q^s)]$, and

$$b_{(r)} = [-\log(1-\theta)]^{-1} \frac{\theta}{1-\theta} \binom{s}{r} p^r + \frac{\theta}{1-\theta} \sum_{k=1}^{r} \binom{s}{k} p^k \frac{r-k}{r} b_{(r-k)},$$

for $r = 1, 2, \ldots$, with $b_{(0)} = 1$.

7.9 Differentiating the probability generating function

$$h(t) = \sum_{m=1}^{\infty} p_m t^m = [-k \log p]^{-1} \left[-\log \left(1 - \sum_{j=1}^{k} qp^{j-1} t^j \right) \right],$$

derive the expression

$$\frac{dh(t)}{dt} = [-k \log p]^{-1} \sum_{j=1}^{k} jqp^{j-1} t^{j-1} + \frac{dh(t)}{dt} \sum_{j=1}^{k} qp^{j-1} t^j,$$

which implies the required recurrence relation.

7.10 The probability function and binomial moments of the compound binomial distribution can be deduced from Theorem 7.1.

7.11 (a) The probability function and binomial moments of the compound distribution may be obtained from the corresponding expressions of Exercise 7.10 as

$$p_m = (1 - \theta + \theta q^r)^s \frac{(p/q)^m}{m!} \sum_{k=0}^{m} (s)_k C(m, k; r) (\theta q^r)^k, \quad m = 0, 1, \ldots,$$

and

$$b_{(j)} = \frac{p^j}{j!} \sum_{k=0}^{j} (s)_k C(j, k; r) \theta^k, \quad j = 0, 1, \ldots.$$

(b) The mean μ and variance σ^2 are deduced as

$$\mu = sr\theta p, \quad \sigma^2 = sr\theta p[q + r(1-\theta)p].$$

(c) Differentiating the probability generating function

$$h(t) = \sum_{m=0}^{\infty} p_m t^m = [1 + \theta\{(pt+q)^r - 1\}]^s,$$

derive the expression

$$\frac{dh(t)}{dt} = sr\theta p(pt+q)^{r-1} h(t) - \theta\{(pt+q)^r - 1\} \frac{dh(t)}{dt},$$

which expanded into powers of t entails the recurrence relation

$$p_m = \frac{\theta}{m} \sum_{j=1}^{m} \binom{r}{j} p^j q^{r-j} (sj + j - m) p_{m-j}, \quad m = 1, 2, \ldots,$$

with $p_0 = (1-\theta+\theta q^r)^s$. Similarly, differentiating the binomial moment generating function $h(t+1)$ establish the recurrence relation

$$b_{(j)} = \frac{\theta}{j}\sum_{i=1}^{j}\binom{r}{i}p^i(si+i-j)b_{(j-i)}, \quad j=1,2,\ldots,$$

with $b_{(0)} = 1$.

7.12 The probability function p_m, $m = 0, 1, \ldots$, and its binomial moments $b_{(j)}$, $j = 0, 1, \ldots$, may be deduced from (7.33) and (7.35), respectively. The mean μ and variance σ^2 are deduced as

$$\mu = srp\theta/(1-\theta), \quad \sigma^2 = srp\theta[q(1-\theta)+rp]/(1-\theta)^2.$$

7.13 (a) Using the exponential generating function of the generalized factorial coefficients (see Theorem 2.13), expand the generating function $G(t;x)$ into powers of t to get the required expression.

(b) Differentiating the generating function $G(t;x)$, with respect to t, and multiplying the resulting expression by $1 - a[(x+t)^r - x^r]$, deduce the relation

$$\frac{dG(t;x)}{dt} = asr(x+t)^{r-1}G(t;x) + a[(x+t)^r - x^r]\frac{dG(t;x)}{dt},$$

which, expanded into powers of t, entails the required recurrence relation.

(c) Expand the probability generating function of the compound negative binomial distribution

$$\sum_{m=0}^{\infty}p_m t^m = \left(\frac{1-\theta}{1-\theta q^r}\right)^s \left(1 - \frac{\theta}{1-\theta q^r}[(pt+q)^r - q^r]\right)^{-s},$$

using the generalized Gegenbauer polynomials, to get

$$p_m = \left(\frac{1-\theta}{1-\theta q^r}\right)^s G_m^{(s)}(q;\theta(1-\theta q^r)^{-1},r)\frac{p^m}{m!}, \quad m=0,1,\ldots.$$

Similarly, find

$$b_{(j)} = G_j^{(s)}(1;\theta(1-\theta)^{-1},r)\frac{p^j}{j!}, \quad j=0,1,\ldots.$$

Using the recurrence relation for the generalized Gegenbauer polynomials, deduce the recurrence relations

$$p_m = \frac{\theta}{m(1-\theta q^r)}\sum_{k=1}^{m}\binom{r}{k}p^k q^{r-k}(m+sk-k)p_{m-k}, \quad m=1,2,\ldots,$$

with $p_0 = (1-\theta)(1-\theta q^r)^{-1}$, and

$$b_{(j)} = \frac{\theta}{j(1-\theta)}\sum_{k=1}^{j}\binom{r}{k}p^k(j+sk-k)b_{(j-k)}, \quad j=1,2,\ldots,$$

with $b_{(0)} = 1$.

7.14 Differentiating the probability generating function

$$h(t) = \sum_{m=0}^{\infty} p_m t^m = p^k \left(1 - \sum_{j=1}^{k} qp^{j-1} t^j\right)^{-1},$$

derive the expression

$$\frac{dh(t)}{dt} = \left(\sum_{j=1}^{k} jqp^{j-1} t^{j-1}\right) h(t) + \left(\sum_{j=1}^{k} qp^{j-1} t^j\right) \frac{dh(t)}{dt},$$

which, expanded into powers of t, implies the required recurrence relation.

7.15 (a) Use (7.33) and (7.35) to find

$$p_m = \left(\frac{1-\theta}{1-\theta e^{-\lambda}}\right)^s \frac{\lambda^m}{m!} \sum_{k=0}^{m} (s+k-1)_k S(m,k) \left(\frac{\theta e^{-\lambda}}{1-\theta e^{-\lambda}}\right)^k, \quad m = 0, 1, \ldots,$$

and

$$b_{(r)} = \frac{\lambda^r}{r!} \sum_{k=0}^{r} (s+k-1)_k S(r,k) \left(\frac{\theta}{1-\theta}\right)^k, \quad r = 0, 1, \ldots.$$

(b) The mean μ and variance σ^2 are deduced as

$$\mu = \frac{s\theta\lambda}{1-\theta}, \quad \sigma^2 = \frac{s\theta\lambda(1-\theta+\lambda)}{(1-\theta)^2}.$$

(c) Differentiating the probability generating function

$$h(t) = \sum_{m=0}^{\infty} p_m t^m = (1-\theta)^s [1 - \theta e^{\lambda(t-1)}]^{-s}$$

$$= \left(\frac{1-\theta}{1-\theta e^{-\lambda}}\right)^s \left[1 - \frac{\theta e^{-\lambda}}{1-\theta e^{-\lambda}}(e^{\lambda t} - 1)\right]^{-s},$$

derive the expression

$$\frac{dh(t)}{dt} = \frac{s\theta\lambda e^{-\lambda}}{1-\theta e^{-\lambda}} e^{\lambda t} h(t) + \frac{\theta e^{-\lambda}}{1-\theta e^{-\lambda}}(e^{\lambda t} - 1) \frac{dh(t)}{dt},$$

which entails the recurrence relation

$$p_m = \frac{\theta e^{-\lambda}}{m(1-\theta e^{-\lambda})} \sum_{j=1}^{m} (m+sj-j) \frac{\lambda^j}{j!} p_{m-j}, \quad m = 1, 2, \ldots,$$

with $p_0 = (1-\theta)^s(1-\theta e^{-\lambda})^{-s}$. Similarly, differentiating the binomial moment generating function $h(t+1)$, derive the recurrence relation

$$b_{(r)} = \frac{\theta}{r(1-\theta)} \sum_{j=1}^{r} (r+sj-j) \frac{\lambda^j}{j!} b_{(r-j)}, \quad r = 1, 2, \ldots,$$

with $b_{(0)} = 1$.

7.16 Use (7.9) to establish the required expression for p_m, $m = 0, 1, \ldots$. Then, by

$$\mu'_m(e^{-\lambda}) = \sum_{k=0}^{m} k! S(m,k) \beta_{(k)}(e^{-\lambda}),$$

deduce that

$$p_m = \frac{\lambda^m}{m!} f(e^{-\lambda}) \sum_{k=0}^{m} k! S(m,k) \beta_{(k)}(e^{-\lambda}), \quad m = 0, 1, \ldots.$$

7.17 In this case

$$f(e^{-\lambda}) = e^{-\theta(1-e^{-\lambda})}, \quad \beta_{(k)}(e^{-\lambda}) = \frac{(\theta e^{-\lambda})^k}{k!}, \quad k = 0, 1, \ldots,$$

and from the second expression of the probabilities p_m, $m = 0, 1, \ldots$, in Exercise 7.16, deduce the expression

$$p_m = e^{-\theta(1-e^{-\lambda})} \frac{\lambda^m}{m!} \sum_{k=0}^{m} S(m,k)(\theta e^{-\lambda})^k, \quad m = 0, 1, \ldots.$$

7.18 (a) In this case
$$f(e^{-\lambda}) = q + pe^{-\lambda},$$
$$\mu_0(e^{-\lambda}) = 1, \quad \mu'_m(e^{-\lambda}) = (q + pe^{-\lambda})^{-1} pe^{-\lambda}, \quad m = 1, 2, \ldots,$$

and from the first expression of the probabilities p_m, $m = 0, 1, \ldots$, in Exercise 7.16, deduce the required expression.

(b) In the more general case

$$f(e^{-\lambda}) = (q + pe^{-\lambda})^s, \quad \beta_{(k)}(e^{-\lambda}) = \binom{s}{k} \left(\frac{pe^{-\lambda}}{q + pe^{-\lambda}}\right)^k, \quad k = 0, 1, \ldots,$$

and from the second expression of the probabilities p_m, $m = 0, 1, \ldots$, in Exercise 7.16, deduce the expression

$$p_m = \frac{\lambda^m}{m!} \sum_{k=0}^{m} (s)_k S(m,k) (pe^{-\lambda})^k (q + pe^{-\lambda})^{s-k}, \quad m = 0, 1, \ldots.$$

7.19 In the general case

$$f(e^{-\lambda}) = \left(\frac{p}{1 - qe^{-\lambda}}\right)^s,$$

$$\beta_{(k)}(e^{-\lambda}) = \binom{s+k-1}{k} \left(\frac{qe^{-\lambda}}{1 - qe^{-\lambda}}\right)^k, \quad k = 0, 1, \ldots,$$

and from the second expression of the probabilities p_m, $m = 0, 1, \ldots$, in Exercise 7.16, deduce the expression

$$p_m = \left(\frac{p}{1 - qe^{-\lambda}}\right)^s \frac{\lambda^m}{m!} \sum_{k=0}^{m} (s + k - 1)_k S(m, k) \left(\frac{qe^{-\lambda}}{1 - qe^{\lambda}}\right)^k, \quad m = 0, 1, \ldots.$$

7.20 In this case

$$f(e^{-\lambda}) = [-\log(1 - \theta)]^{-1}[-\log(1 - \theta e^{-\lambda})],$$

$$\beta_{(k)}(e^{-\lambda}) = [-\log(1 - \theta e^{-\lambda})]^{-1} \frac{[\theta e^{-\lambda}/(1 - \theta e^{-\lambda})]^k}{k}, \quad k = 1, 2, \ldots,$$

and from the second expression of the probabilities p_m, $m = 0, 1, \ldots$, in Exercise 7.16, deduce the expression

$$p_0 = [-\log(1 - \theta)]^{-1}[-\log(1 - \theta e^{-\lambda})],$$

$$p_m = [-\log(1 - \theta)]^{-1} \frac{\lambda^m}{m!} \sum_{k=1}^{m} (k - 1)! S(m, k) \left(\frac{\theta e^{-\lambda}}{1 - \theta e^{-\lambda}}\right)^k, \quad m = 1, 2, \ldots.$$

7.21 Use (7.9) to establish the required expression for p_m, $m = 0, 1, \ldots$.
(a) Show that

$$f(q) = e^{\theta p}, \quad \beta_{(m)}(q) = \frac{(\theta q)^m}{m!}, \quad m = 0, 1, \ldots,$$

and from the expression of the probabilities p_m, $m = 0, 1, \ldots$, deduce that

$$p_m = e^{-\theta p} \frac{(\theta p)^m}{m!}, \quad m = 0, 1, \ldots.$$

(b) Show that

$$f(q) = [-\log(1 - \theta)]^{-1}[-\log(1 - \theta q)],$$

$$\beta_{(m)}(q) = [-\log(1 - \theta q)]^{-1} \frac{[\theta q/(1 - \theta q)]^m}{m}, \quad m = 1, 2, \ldots,$$

and from the expression of the probabilities p_m, $m = 0, 1, \ldots$, deduce that

$$p_0 = w, \quad p_m = (1 - w)[-\log(1 - \lambda)]^{-1} \frac{\lambda^m}{m!}, \quad m = 1, 2, \ldots,$$

where

$$w = [-\log(1 - \theta)]^{-1}[\log(1 - \theta q)], \quad \lambda = \theta p/(1 - \theta q).$$

(c) Show that

$$f(q) = (1 - \theta p)^s, \quad \beta_{(m)}(q) = \binom{s}{m} \left(\frac{\theta q}{1 - \theta p}\right)^m, \quad m = 0, 1, \ldots,$$

and from the expression of the probabilities p_m, $m = 0, 1, \ldots$, deduce that

$$p_m = \binom{s}{m}(\theta p)^m(1-\theta p)^{s-m}, \quad m = 0, 1, \ldots, s.$$

7.22 (a) Use (7.9) to establish the required expression for the probability function of the mixture distribution.

(b) Using the expansion

$$(sn)_m = \sum_{k=0}^{m} C(m, k; s)(n)_k,$$

deduce the expression

$$\beta_{(m;s)}(q^s) = \frac{1}{m!}\sum_{k=0}^{m} k! C(m, k; s)\beta_{(k)}(q^s),$$

and conclude that

$$p_m = \frac{(p/q)^m}{m!} f(q^s) \sum_{k=0}^{m} k! C(m, k; s)\beta_{(k)}(q^s), \quad m = 0, 1, \ldots.$$

(c) Show that

$$f(q^s) = e^{-\theta(1-q^s)}, \quad \beta_{(k)}(q^s) = \frac{(\theta q^s)^k}{k!}, \quad k = 0, 1, \ldots,$$

and conclude that

$$p_m = e^{-\theta(1-q^s)}\frac{(p/q)^m}{m!}\sum_{k=0}^{m} C(m, k; s)(\theta q^s)^k, \quad m = 0, 1, \ldots.$$

7.23 (a) Use (7.9) to establish the required expression for the probability function of the mixture distribution.

(b) Using the expansion

$$(sn + m - 1)_m = \sum_{k=0}^{m} |C(m, k; -s)|(n)_k,$$

deduce the expression

$$\beta_{[m;s]}(p^s) = \frac{1}{m!}\sum_{k=0}^{m} k! |C(m, k; -s)|\beta_{(k)}(p^s),$$

and conclude that

$$p_m = \frac{q^m}{m!} f(p^s) \sum_{k=0}^{m} k! |C(m, k; -s)|\beta_{(k)}(p^s), \quad m = 0, 1, \ldots.$$

(c) Show that

$$f(p^s) = [-\log(1-\theta)]^{-1}[-\log(1-\theta p^s)],$$

$$\beta_{(k)}(p^s) = [-\log(1-\theta p^s)]^{-1}\frac{[\theta p^s/(1-\theta p^s)]^k}{k}, \quad k = 1, 2, \ldots,$$

and conclude that

$$p_0 = [-\log(1-\theta)]^{-1}[-\log(1-\theta p^s)],$$

$$p_m = [-\log(1-\theta)]^{-1}\frac{q^m}{m!}\sum_{k=1}^{m}(k-1)!|C(m,k;-s)|\left(\frac{\theta p^s}{1-\theta p^s}\right)^k, \quad m = 1, 2, \ldots.$$

7.24 The probability function of the random variable X is obtained by evaluating the integral

$$q_x = \frac{\theta^x \lambda^s}{x!\Gamma(s)}\int_0^\infty y^{s+x-1}e^{-(\theta+\lambda)y}dy.$$

7.25 The probability generating function of the random variable X is obtained by evaluating the integral

$$h(t) = \frac{\lambda^s}{\Gamma(s)}\int_0^\infty y^{s-1}e^{-(\lambda-\theta[2pq(t-1)+p^2(t^2-1)])y}dy.$$

7.26 The probability function of a compound discrete distribution for $q_0 > 0$, by Theorem 7.1, may be expressed as

$$p_m = \sum_{k=0}^{m} d_k p_m(k), \quad m = 0, 1, \ldots,$$

where

$$p_m(k) = (1-q_0)^{-k}\left[\Delta_u^k q_0^{k-u} q_m(u)\right]_{u=0}, \quad m = k, k+1, \ldots \quad (k = 0, 1, \ldots),$$

according to Theorem 6.1, is the k-fold convolution of the zero truncated distribution $P(X = x|X > 0) = (1-q_0)^{-1}q_x$, $x = 0, 1, \ldots$, and

$$d_k = f(q_0)\beta_{(k)}(q_0)[(1-q_0)/q_0]^k, \quad k = 0, 1, \ldots,$$

is clearly a legitimate probability function. This expression entails the required interpretation. Further, use this general interpretation in the particular cases of compound Poisson, Hermite and generalized Hermite distributions.

7.27 Use the interpretation of a compound discrete distribution as a mixture given in Exercise 7.26 in the particular case of the compound logarithmic distribution and in the also particular cases of the Chebyshev distribution of the first kind and generalized Chebyshev distribution of the second kind.

7.28 Use the interpretation of a compound discrete distribution as a mixture given in Exercise 7.26 in the particular cases of the compound negative binomial, Gegenbauer and generalized Gegenbauer distributions.

7.29 Note that under the random occupancy model, the conditional probability $p_x = P(X = x | X > 0)$ that x balls are distributed into any specific urn, given that is not empty, is given by $p_x = (1 - q_0)^{-1} q_x$, $x = 1, 2, \ldots$, and $p_0 = 0$. Thus, using (7.53), deduce that

$$p_m(n) = \frac{n!}{m!} B_{m,n}(1! p_1, 2! p_2, \ldots, m! p_m)$$
$$= \frac{n!}{m!} B_{m,n}(1!(1 - q_0)^{-1} q_1, 2!(1 - q_0)^{-1} q_2, \ldots, m!(1 - q_0)^{-1} q_m),$$

which, by (7.41), entails the required expression.

7.30 Under the random occupancy model, the conditional probability that $k_j \geq 0$ urns are occupied by j balls each for $j = 0, 1, \ldots, m$, given that m balls are distributed into the n urns, is

$$\frac{n!}{k_0! k_1! \cdots k_m!} \frac{q_0^{k_0} q_1^{k_1} \cdots q_m^{k_m}}{q_m(n)}, \quad k_0 + k_1 + \cdots + k_m = n.$$

For fixed $n - k_0 = k$, write this probability in the form

$$\frac{(n)_k q_0^{n-k}}{m! q_m(n)} \cdot \frac{m!}{k_1! k_2! \cdots k_m!} q_1^{k_1} q_2^{k_2} \cdots q_m^{k_m}$$

and sum over all nonnegative integer solutions of the equations

$$k_1 + 2k_2 + \cdots + m k_m = m, \quad k_1 + k_2 + \cdots + k_m = k$$

to deduce the required expression.

References

Abramowitz, M. and Stegun, I. A. (1965) *Handbook of Mathematical Functions*, Dover, New York.

Ahuja, J. C. (1970a) On a certain recurrence relation, *Annals of Mathematical Statistics*, **41**, 665 - 666.

Ahuja, J. C. (1970b) On the distribution of sum of independent positive binomial variables, *Canadian Mathematical Bulletin*, **13**, 151 - 152.

Ahuja, J. C. (1971a) Certain properties of the Stirling distribution of the second kind, *Australian Journal of Statistics*, **13**, 133 - 136.

Ahuja, J. C. (1971b) Distribution of the sum of independent decapitated negative binomial variables, *Annals of Mathematical Statistics*, **42**, 383 - 384.

Ahuja, J. C. (1972a) Convolution of the truncated Poisson distribution, *American Statistician*, **26** (5), 46.

Ahuja, J. C. (1972b) Recurrence relation for minimum variance unbiased estimation of certain left-truncated Poisson distributions, *Journal of the Royal Statistical Society, Series C: Applied Statistics*, **21**, 81 - 86.

Ahuja, J. C. and Enneking, E. A. (1972a) Certain properties of the distribution of the sum of independent zero-one-truncated Poisson variables, *Australian Journal of Statistics*, **14**, 50-53.

Ahuja, J. C. and Enneking, E. A. (1972b) Recurrence relation for the minimum variance unbiased estimator of a parameter of a left-truncated Poisson distribution, *Journal of the American Statistical Association*, **67**, 232-233.

Ahuja, J. C. and Enneking, E. A. (1974) Convolution of independent left-truncated negative binomial variables and limiting distributions, *Annals of the Institute of Statistical Mathematics*, **26**, 265-270.

Aki, S., Kuboki, H. and Hirano, K. (1984) On discrete distributions of order k, *Annals of the Institute of Statistical Mathematics*, **36**, 431-440.

Arfwedson, G. (1951) A probability distribution connected with Stirling's second class numbers, *Skandinavisk Aktuarietidskrift*, **34**, 121-132.

Arnold, B. C. (1972) The waiting time until duplication, *Journal of Applied Probability*, **9**, 841-846.

Bartholomew, D. J. (1982) *Stochastic Models in Social Processes*, 3rd ed., John Wiley & Sons, New York.

Balakrishnan, N. and Nevzorov, V. B. (1997) Stirling numbers and records, in *Advances in Combinatorial Methods and Applications to Probability and Statistics* (Ed. N. Balakrishnan), pp. 189-200, Birkhäuser, Boston.

Barton, D. E. and David, F. N. (1959a) Contagious occupancy, *Journal of the Royal Statistical Society, Series B*, **21**, 120-133.

Barton, D. E. and David, F. N. (1959b) Haemacytometer counts and occupancy theory, *Trabajos de Estadistica*, **10**, 13-18.

Barton, D. E. and David, F. N. (1959c) Sequential occupancy, *Biometrika*, **46**, 218-223.

Beall, G. and Rescia, R. R. (1953) A generalization of the Neyman's contagious distributions, *Biometrics*, **9**, 354-386.

Bell, E. T. (1927) Partition polynomials, *Annals of Mathematics*, **29**, 38-46.

Bell, E. T. (1934a) Exponential numbers, *American Mathematical Monthly*, **41**, 411-419.

Bell, E. T. (1934b) Exponential polynomials, *Annals of Mathematics*, **35**, 258-277.

Berg, S. (1974) Factorial series distributions with applications to capture-recapture problems. *Scandinavian Journal of Statistics*, **1**, 145-152.

Berg, S. (1975) Some properties and applications of a ratio of Stirling numbers of the second kind, *Scandinavian Journal of Statistics*, **2**, 91 - 94.

Berg, S. (1976) A note on the UMVU estimate in a multiple-recapture census, *Scandinavian Journal of Statistics*, **3**, 86 - 88.

Berg, S. (1987) Associated Lah numbers, factorial series distributions and unbiased estimation of a size parameter, *Biometrical Journal*, **29**, 163 - 171.

Bernstein, L. (1965) Rational approximations of algebraic irrationals by means of a modified Jacobi-Perron algorithm, *Duke Mathematical Journal*, **32**, 161 - 176.

Blissard, J. (1867) On the properties of $\Delta^n 0^n$ class numbers and of other analogous to them, as investigated by means of representative notation, *Quarterly Journal of Pure and Applied Mathematics*, **8**, 85 - 110.

Bonferroni, C. E. (1936) Teorie statistica delle classi e calcolo delle probabilità, *Publicazioni del R. Instituto Superiore di Scienze Economiche e Commerciali di Firenze*, **8**, 1 - 62.

Boole, G. (1860) *A Treatise on the Calculus of Finite Differences*, Stechert, New York. (Reprinted by Chelsea, New York, 1950.)

Broder, A. Z. (1984) The r-Stirling numbers, *Discrete Mathematics*, **49**, 241 - 259.

Cacoullos, T. (1961) A combinatorial derivation of the distribution of the truncated Poisson sufficient statistic, *Annals of Mathematical Statistics*, **32**, 904 - 905.

Cacoullos, T. (1972) On the convolution of Poisson distributions with the zero class missing, *American Statistician*, **26** (2), 44.

Cacoullos, T. and Charalambides, Ch. A. (1975) On minimum variance unbiased estimation for truncated binomial and negative binomial distributions, *Annals of the Institute of Statistical Mathematics*, **27**, 235 - 244.

Carlitz, L. (1932) On arrays of numbers, *American Journal of Mathematics*, **54**, 739 - 752.

Carlitz, L. (1933) On Abelian fields, *Transactions of the American Mathematical Society*, **35**, 122 - 136.

Carlitz, L. (1948) q-Bernoulli numbers and polynomials, *Duke Mathematical Journal*, **15**, 987 - 1000.

Carlitz, L. (1960) Note on Norlund's polynomials $B_n^{(z)}$, *Proceedings of the American Mathematical Society*, **11**, 452 - 455.

Carlitz, L. (1965) Note on a paper of L. Bernstein, *Duke Mathematical Journal*, **32**, 177 - 180.

Carlitz, L. (1971) Note on the numbers of Jordan and Ward, *Duke Mathematical Journal*, **38**, 783 - 790.

Carlitz, L. (1979) Degenerate Stirling, Bernoulli and Eulerian numbers, *Utilitas Mathematica*, **15**, 51 - 88.

Carlitz, L. (1980a) Weighted Stirling numbers of the first and second kind-I, *Fibonacci Quarterly*, **18**, 147 - 162.

Carlitz, L. (1980b) Weighted Stirling numbers of the first and second kind-II, *Fibonacci Quarterly*, **18**, 242 - 257.

Catcheside, D. G. (1948) Genetic effects of radiation, in *Advances in Genetics*, Vol. 2 (Ed. M. Demerec), pp. 271 - 358, Academic Press, New York.

Cauchy, A. (1833) *Resumes Analytiques*, Turin.

Cayley, A. (1887) Note on a formula for $\Delta^n 0^i / n^i$ when n and i are very large numbers, *Proceedings of the Royal Society of Edinburgh*, **14**, 149 - 153.

Cerasoli, M. (1983) Poisson randomization in occupancy problems, *Journal of Mathematical Analysis and Applications*, **94**, 150 - 165.

Cernuschi, F. and Castagnetto, L. (1946) Chains of rare events, *Annals of Mathematical Statistics*, **17**, 53 - 61.

Charalambides, Ch. A. (1974a) The generalized Stirling and C-numbers, *Sankhyā, Series A*, **36**, 419 - 436.

Charalambides, Ch. A. (1974b) Minimum variance unbiased estimation for a class of left-truncated discrete distributions, *Sankhyā, Series A*, **36**, 397 - 418.

Charalambides, Ch. A. (1974c) The distributions of sufficient statistics of truncated generalized logarithmic series, Poisson and negative binomial distributions, *Canadian Journal of Statistics*, **2**, 261 - 275.

Charalambides, Ch. A. (1976) The asymptotic normality of certain combinatorial distributions, *Annals of the Institute of Statistical Mathematics*, **28**, 499 - 506.

Charalambides, Ch. A. (1977a) A new kind of numbers appearing in the n-fold convolution of truncated binomial and negative binomial distributions, *SIAM Journal on Applied Mathematics*, **33**, 279 - 288.

Charalambides, Ch. A. (1977b) On the generalized discrete distributions and the Bell polynomials, *Sankhyā, Series B*, **39**, 36 - 44.

Charalambides, Ch. A. (1979) Some properties and applications of the differences of the generalized factorials, *SIAM Journal on Applied Mathematics*, **36**, 273 - 280.

Charalambides, Ch. A. (1981) On a restricted occupancy model and its applications, *Biometrical Journal*, **23**, 601 - 610.

Charalambides, Ch. A. (1983) On restricted and pseudo-contagious occupancy distributions, *Journal of Applied Probability*, **20**, 872 - 876.

Charalambides, Ch. A. (1984a) On the weighted Stirling and other related numbers and some combinatorial applications, *Fibonacci Quarterly*, **22**, 297 - 309.

Charalambides, Ch. A. (1984b) Probabilities and moments of generalized discrete distributions using finite differences operators, *Communications in Statistics - Theory and Methods*, **13**, 3225 - 3241.

Charalambides, Ch. A. (1986a) Derivation of probabilities and moments of certain generalized discrete distributions via urn models, *Communications in Statistics - Theory and Methods*, **15**, 677 - 696.

Charalambides, Ch. A. (1986b) On discrete distributions of order k, *Annals of the Institute of Statistical Mathematics*, **38**, 557 - 568.

Charalambides, Ch. A. (1996) On the q-differences of the generalized q-factorials, *Journal of Statistical Planning and Inference*, **54**, 31 - 43.

Charalambides, Ch. A. (1997) A unified derivation of occupancy and sequential occupancy distributions, in *Advances in Combinatorial Methods and Applications to Probability and Statistics* (Ed. N. Balakrishnan), pp. 259 - 273, Birkhäuser, Boston.

Charalambides, Ch. A. (2001) General occupancy distributions, in *Probability and Statistical Models with Applications* (Eds. Ch. A. Charalambides, M. V. Koutras and N. Balakrishnan), pp. 253 - 268, Chapman & Hall/CRC, Boca Raton.

Charalambides, Ch. A. (2002) *Enumerative Combinatorics*, Chapman & Hall/CRC, Boca Raton.

Charalambides, Ch. A. (2004) Non-central generalized q-factorial coefficients and q-Stirling numbers, *Discrete Mathematics*, **275**, 67 - 85.

Charalambides, Ch. A. (2005a) Derivation of a joint occupancy distribution via a bivariate inclusion and exclusion formula, *Metrika* (to appear).

Charalambides, Ch. A. (2005b) A unified derivation of the complementary waiting time distribution in sequential occupancy, *Methodology and Computing in Applied Probability* (to appear).

Charalambides, Ch. A. and Koutras, M. (1983) On the differences of the generalized factorials at an arbitrary point and their combinatorial applications, *Discrete Mathematics*, **47**, 183 - 201.

Charalambides, Ch. A. and Singh, J. (1988) A review of the Stirling numbers, their generalizations and statistical applications, *Communications in Statistics - Theory and Methods*, **17**, 2533 - 2595.

Chihara, T. S. (1978) *An Introduction to Orthogonal Polynomials*, Gordon and Breach, New York.

Chung, K. L. (1941) On the probability of the occurrence of at least m events among n arbitrary events, *Annals of Mathematical Statistics*, **12**, 328-338.

Chung, K. L. (1943a) Generalization of Poincaré's formula in the theory of probability, *Annals of Mathematical Statistics*, **14**, 63-65.

Chung, K. L. (1943b) On fundamental systems of probabilities of a finite number of events, *Annals of Mathematical Statistics*, **14**, 123-133.

Chung, K. L. (1943c) Further results on probabilities of a finite number of events, *Annals of Mathematical Statistics*, **14**, 234-237.

Comtet, L. (1972) Nombres de Stirling généraux et fonctions symmetriques, *Comptes Rendus de l' Académie des Sciences, Paris*, **275**, 747-750.

Comtet, L. (1973) Une formule explicite pour les puissances successives de l' opérateur de derivation de Lie, *Comptes Rendus de l' Académie des Sciences, Paris*, **276**, 165-168.

Comtet, L. (1974) *Advanced Combinatorics*, Reidel, Dordrecht, Holland.

Craig, C. C. (1953) On the utilization of marked specimens in estimating populations of flying insects, *Biometrika*, **40**, 170-176.

Cresswell, W. L. and Froggatt, P. (1963) *The Causation of Bus Driver Accidents*, Oxford University Press, London.

Crippa, D. and Simon, K. (1997) q-Distributions and Markov processes, *Discrete Mathematics*, **170**, 81-98.

Crippa, D., Simon, K. and Trunz, P. (1997) Markov processes involving q-Stirling numbers, *Combinatorics, Probability and Computing*, **6**, 165-178.

Csörgo, M. and Guttman, I. (1962) On the empty cell test, *Technometrics*, **4**, 235-247.

Darroch, J. N. (1958) The multiple-recapture census I. Estimation of a closed population, *Biometrika*, **45**, 343-359.

David, F. N. (1950) Two combinatorial tests of whether a sample has come from a given population, *Biometrika*, **37**, 97-110.

David, F. N. and Barton, D. E. (1962) *Combinatorial Chance*, Griffins, London.

David, F. N., Kendall, M. G. and Barton, D. E. (1966) *Symmetric Functions and Allied Tables*, Cambridge University Press, Cambridge.

Davies, P. (1983) Some statistical tests for constrained occupancy problems, *Biometrics*, **39**, 719-725.

De Moivre, A. (1718) *The Doctrine of Chances*, Pearson, London (2nd ed. 1738 and 3rd ed. 1756).

Douglas, J. B. (1955) Fitting the Neyman type A (two parameter) contagious distribution, *Biometrics*, **11**, 149-173.

Douglas, J. B. (1970) Statistical models in discrete distributions, in *Random Counts in Scientific Work*, Vol. 3 (Ed. G. P. Patil), pp. 203-232, Pennsylvania State University Press, University Park.

Douglas, J. B. (1971) Stirling numbers in discrete distributions, in *Statistical Ecology*, Vol. 1 (Eds. G. P. Patil, E. C. Pielou and W. E. Waters), pp. 69-98, Pennsylvania State University Press, University Park.

Douglas, J. B. (1980) *Analysis with Standard Contagious Distributions*, International Co-operative Publishing House, Burtonsville, MD.

Enneking, E. A. and Ahuja, J. C. (1978) On minimum variance unbiased estimation for certain left-truncated discrete distributions, *Sankhyā, Series B*, **40**, 105-110.

Euler, L. (1760) Theoremata arithmetica nova methodo demonstrata, *Novi Commentarii Academiae Petropolitanae*, **8**, 74-104 (Opera Omnia I-2, 531-535.)

Evans, D. A. (1953) Experimental evidence concerning contagious distributions in ecology. *Biometrika*, **40**, 186-211.

Faà di Bruno, F. (1855a), Sullo sviluppo delle funzioni, *Annali di Scienze Mathematiche et Fisiche di Tortolini*, **6**, 479-480.

Faà di Bruno, F. (1855b), Note sur un nouvelle formule de calcul différentiel, *Quarterly Journal of Mathematics*, **1**, 359-360.

Fang, K.-T. (1982) A restricted occupancy problem, *Journal of Applied Probability*, **19**, 707-711.

Fang, K.-T. (1985) Occupancy problems, in *Encyclopedia of Statistical Sciences*, Vol. 6 (Eds. S. Kotz, N. L. Johnson and C. B. Read), pp. 402-406, John Wiley & Sons, New York.

Feller, W. (1943) On a general class of "contagious" distributions, *Annals of Mathematical Statistics*, **14**, 389-400.

Feller, W. (1968) *An Introduction to Probability Theory and Its Applications*, Vol. 1, 3rd ed., John Wiley & Sons, New York.

Fisher, R. A. and Yates, F. (1953) *Statistical Tables*, Oliver Boyd, London.

Fréchet, M. (1940) *Le Probabilités associées à un système d' événements compatibles et dépendants, I: Èvénements en nombre fini fixe*, Actualités Scientifiques et Industrielles, No. 859, Hermann, Paris.

Fréchet, M. (1943) *Le Probabilités associées à un système d' événements compatibles et dépendants II: Cas particuliers et applications*, Actualités Scientifiques et Industrielles, No. 942, Hermann, Paris.

Freund, J. E. and Pozner, A. N. (1956) Some results on restricted occupancy theory, *Annals of Mathematical Statistics*, **27**, 537 - 540.

Galliher, H. P., Morse, P. M. and Simond, M. (1959) Dynamics of two classes of continuous review inventory systems, *Operations Research*, **7**, 362 - 384.

Garsia, A. M. and Remmel, J. B. (1986) q-Counting rook configurations and a formula of Frobenius, *Journal of Combinatorial Theory, Series A*, **41**, 246 - 275.

Gart, J. J. (1974) Exact moments of the variance test for left-truncated Poisson distributions, *Sankhyā, Series B*, **36**, 406 - 416.

Gart, J. J. (1975) The Poisson distribution: The theory and application of some conditional tests, in *Statistical Distributions in Scientific Work*, Vol. 2 (Eds. G. P. Patil et al.), pp. 125 - 140, Reidel, Dordrecht, Holland.

Gittelsohn, A. M. (1969) An occupancy problem, *American Statistician*, **23** (2), 11 - 12.

Glasser, G. J. (1963) Random numbers, sample selection and occupancy problems, *Journal of the Royal Statistical Society, Series A*, **126**, 115 - 119.

Glaymann, M. and Djokovic, D. (1962) Sur une relation récurrente concernant les nombres de Stirling, *Univ. Beogrand. Publ. Elekrotehn. Fak. Ser. Mat. Fiz.*, No. 78 - 83, 1 - 5.

Gould, H. W. (1958) A theorem concerning the Bernstein polynomial, *Mathematics Magazine*, **31**, 259 - 264.

Gould, H. W. (1961) The q-Stirling numbers of first and second kinds, *Duke Mathematical Journal*, **28**, 281 - 289.

Gould, H. W. (1964) The operator $(a^x \Delta)^n$ and Stirling numbers of the first kind, *American Mathematical Monthly*, **71**, 850 - 858.

Goodman, L. A. (1949) On the estimation of the number of classes in a population, *Annals of Mathematical Statistics*, **20**, 572 - 579.

Goodman, L. A. (1953) Sequential sampling tagging for population size problems, *Annals of Mathematical Statistics*, **24**, 56 - 59.

Gross, A. J. (1970) A note on the convolution of Poisson distributions with the zero class missing, *American Statistician*, **24** (5), 42 - 43.

Grunert, J. A. (1822) *Mathematische Abhandlungen*, Altona.

Grunert, J. A. (1843) Ueber die Summirung der Reihen von der Form $A_0\phi(0)$, $A_1\phi(1)x$, $A_2\phi(2)x^2, \ldots, A_n\phi(n)x^n, \ldots$ wo A eine beliebige constante Gröfse, A_n eine beliebige und $\phi(n)$ eine ganze rationale algebraische Function der positiven ganzen Zahl n bezeichnet, *Journal für die reine und angewandte Mathematik*, **25**, 240-279.

Gupta, H. (1950) Tables of distributions, *Research Bulletin of East Panjab University*, 13-44.

Gupta, R. C. (1974a) Distribution of the sum of independent decapitated generalized negative binomial variables, *Sankhyā, Series B*, **36**, 67-69.

Gupta, R. C. (1974b) Distribution of the sum of independent decapitated generalized Poisson variables, *Sankhyā, Series B*, **36**, 212-214.

Gupta, R. C. (1974c) Modified power series distribution and some of its applications, *Sankhyā, Series B*, **36**, 288-298.

Gupta, R. C. (1977) Minimum variance unbiased estimation in a modified power series distribution and some of its applications, *Communications in Statistics, A*, **6**, 977-991.

Gupta, R. C. and Singh, J. (1982) Estimation of probabilities in the class of modified power series distributions, *Math. Operationsforsch. Statist., Ser. Statistics*, **13**, 71-77.

Gupta, R. P. and Jain, G. C. (1974) A generalized Hermite distribution and its properties, *SIAM Journal of Applied Mathematics*, **27**, 359-363.

Gurland, J. (1957) Some interrelations among compound and generalized distributions, *Biometrika*, **44**, 265-268.

Gurland, J. (1958) A generalized class of contagious distributions, *Biometrics*, **14**, 229-249.

Gurland, J. (1965) A method of estimation for some generalized Poisson distributions, in *Classical and Contagious Discrete Distributions* (Ed. G. P. Patil) pp. 141-158, Statistical Publishing Society, Calcutta; Pergamon Press, Oxford.

Hahn, W. (1949) Über Orthogonalpolynome, die q-Differenzengleichungen genügen, *Mathematische Nachrichten*, **2**, 4-34.

Harkness, W. L. (1970) The classical occupancy problem revisited, in *Random Counts in Scientific Work*, Vol. 3 (Ed. G. P. Patil), pp. 107-126, Pennsylvania State University Press, University Park.

Harris, B. (1968) Statistical inference in the classical occupancy problem. Unbiased estimation of the number of classes, *Journal of the American Statistical Association*, **63**, 837-847.

Hill, B. M. (1974) The rank-frequency form of Zipf's law, *Journal of the American Statistical Association*, **69**, 1017 - 1026.

Hirano, K. (1986) Some properties of the distributions of order k, in *Fibonacci Numbers and Their Applications* (Eds. A. N. Philippou, G. E. Bergum and A. F. Horadam), pp. 43 - 53, Reidel, Dordrecht, Holland.

Holst, L. (1979) A unified approach to limit theorems for urn models, *Journal of Applied Probability*, **16**, 154 - 162.

Holst, L. (1980) On matrix occupancy, committee and capture-recapture problems, *Scandinavian Journal of Statistics*, **7**, 139 - 146.

Holst, L. (1986) On birthdays, collectors', occupancy and other classical urn problems, *International Statistical Review*, **54**, 15 - 27.

Holst, L. (1988) Rates of Poisson convergence for some coverage and urn problems using coupling, *Journal of Applied Probability*, **25**, 717 - 724.

Holst, L. and Hüsler, J. (1985) Sequential urn schemes and birth processes, *Advances in Applied Probability*, **17**, 257 - 279.

Irwin, J. O. (1963) The place of mathematics in medicine and biological statistics, *Journal of the Royal Statistical Society, Series A*, **126**, 1 - 44.

Janardan, K. G. (1978) Distribution of the sum of independent zero-truncated linear function Poisson random variables, *Biometrical Journal*, **20**, 115 - 117.

Jani, P. N. (1977) Minimum variance unbiased estimation for some left-truncated modified power series distributions, *Sankhyā, Series B*, **39**, 258 - 278.

Jani, P. N. (1978) New numbers appearing in minimum variance unbiased estimation for decapitated generalized, negative binomial and Poisson distributions, *Journal of the Indian Statistical Association*, **16**, 41 - 48.

Jánossy, L., Rényi, A. and Aczél, J. (1950) On composed Poisson distributions, I, *Acta Mathematica Academiae Scientiarum Hungaricae*, **1**, 209 - 224.

Janson, S. (1983) Limit theorems for some sequential occupancy problems, *Journal of Applied Probability*, **20**, 545 - 553.

Johnson, N. L. and Kotz, S. (1977) *Urn Models and Their Application*, John Wiley & Sons, New York.

Johnson, N. L. and Kotz, S. (1982) Developments in discrete distributions, 1969 - 1980, *International Statistical Review*, **50**, 71 - 101.

Johnson, N. L., Kotz, S. and Kemp, A. W. (1992) *Univariate Discrete Distributions*, 2nd ed., John Wiley & Sons, New York.

Jones, H. L. (1959) How many of a group of random numbers will be usable in selecting a particular sample? *Journal of the American Statistical Association*, **54**, 102-122.

Jordan, C. (1867) De quelques formules de probabilité, *Comptes Rendus de l' Académie des Sciences, Paris*, **65**, 993-994.

Jordan, Ch. (1926) Sur la probabilité des épreuves répétées, le théorème de Bernoulli et son inversion, *Bulletin de la Société Mathématique de France*, **54**, 101-137.

Jordan, Ch. (1927a) Sur un cas généralisé de la probabilité des épreuves répétées, *Comptes Rendus de l' Académie des Sciences, Paris*, **184**, 315-317.

Jordan, Ch. (1927b) Sur un cas généralisé de la probabilité des épreuves répétées, *Acta Scientiarum Mathematicarum, Szeged*, **3**, 193-210.

Jordan, Ch. (1933) On Stirling numbers, *Tôhoku Mathematical Journal*, **37**, 254-278.

Jordan, Ch. (1934) Le thèorém de probabilité de Poincaré généralisé au cas de plusiers variables indépendantes, *Acta Scientiarum Mathematicarum, Szeged*, **7**, 103-111.

Jordan, Ch. (1939a) *Calculus of Finite Differences*, Rotting and Romwalter, Sorron, Hungary. (Reprinted by Chelsea, New York, 1947, 1960, 1965.)

Jordan, Ch. (1939b) Problémes de la probabilité des épreuves répétées dans le cas général, *Bulletin de la Société Mathématique de France*, **67**, 223-242.

Joshi, S. W. and Park, C. J. (1974) Minimum variance unbiased estimation for truncated power series distribution, *Sankhyā, Series A*, **36**, 305-314.

Karanicoloff, Chr. (1961) Sur une representation des nombres de Stirling dans une forme explicite, *Univ. Beograd. Publ. Elektotehn. Fak. Ser. Mat. Fiz.*, No. 65-69, 9-10.

Karlin, S. and McGregor, J. (1972) Addendum to a paper of W. Ewens, *Theoretical Population Biology*, **3**, 113-116.

Katti, S. K. and Gurland, J. (1961) The Poisson Pascal distribution, *Biometrics*, **17**, 527-538.

Katti, S. K. and Gurland, J. (1962) Some methods of estimation for the Poisson binomial distribution, *Biometrics*, **18**, 42-51.

Kemp, A. W. (1979) Convolutions involving binomial pseudo variables, *Sankhyā, Series A*, **41**, 232-243.

Kemp, A. W. and Kemp, C. D. (1966) An alternative derivation of the Hermite distribution, *Biometrika*, **53**, 627-628.

Kemp, C. D. (1967) "Stuttering-Poisson" distributions, *Journal of the Statistical and Social Enquiry Society of Ireland*, **21**, 151 - 157.

Kemp, C. D. and Kemp, A. W. (1965) Some properties of the Hermite distribution, *Biometrika*, **52**, 381 - 394.

Kendall, D. G. (1948) On some modes of population growth leading to R. A. Fisher's logarithmic series distribution, *Biometrika*, **35**, 6 - 15.

Khatri, C. G. (1962) On the distributions obtained by varying the number of trials in a binomial distribution, *Annals of the Institute of Statistical Mathematics*, **13**, 47 - 51.

Khatri, C. G. and Patel, I. R. (1961) Three classes of univariate discrete distributions, *Biometrics*, **17**, 567 - 575.

Kolchin, V. F., Sevastyanov, B. A. and Chistyakov, V. P. (1978) *Random Allocations*, Wiston, Washington, DC.

Kotz, S. and Balakrishnan, N. (1997) Advances in urn models during the past two decades, in *Advances in Combinatorial Methods and Applications to Probability and Statistics* (Ed. N. Balakrishnan), pp. 203 - 257, Birkhäuser, Boston.

Koutras, M. (1982) Non-central Stirling numbers and some applications, *Discrete Mathematics*, **42**, 73 - 89.

Kullback, S. (1937) On certain distributions derived from the multinomial distribution, *Annals of Mathematical Statistics*, **8**, 128 - 144.

Kumar, A. (1980) A note on minimum variance unbiased estimators of power series distributions, *Communications in Statistics - Theory and Methods*, **9**, 549 - 556.

Kumar, A. and Consul, P. C. (1980) Minimum variance unbiased estimation for modified power series distribution, *Communications in Statistics - Theory and Methods*, **9**, 1261 - 1275.

Lagrange, L. (1770) Nouvelle methode pour resoudre les equations litterales par le moyen de series, *Mem. Acad. Roy. Sci. Belles-Lettres de Berlin*, **24**.

Lah, I. (1955) Eine neue Art von Zahlen, ihre Eigenschaften und Anwendung in der mathematischen Statistik, *Mitteilungsbl. Math. Statist.*, **7**, 203 - 212.

Laplace, P. S. (1812) *Théorie Analytique des Probabilités*, Courvier, Paris.

Leite, J. G. and Pereira, C. A. de B. (1987) An urn model for multi-sample capture/recapture sequential tagging process, *Sequential Analysis*, **6**, 179 - 186.

Leite, J. G. and Pereira, C. A. de B. (1994) Urn scheme to obtain properties of Stirling numbers of second kind, *International Journal of Mathematical and Statistical Sciences*, **3**, 205 - 216.

Leite, J. G., Pereira, C. A. de B. and Rodrigues, F. W. (1993) Waiting time to exhaust lottery numbers, *Communications in Statistics - Theory and Methods*, **22**, 301-310.

Lüders, R. (1934) Die Statistik der seltenen Ereignisse, *Biometrika*, **26**, 108-128.

MacMahon, P. A. (1915) *Combinatory Analysis*, Vol. 1, Cambridge University Press, London. (Reprinted by Chelsea, New York, 1960.)

MacMahon, P. A. (1916) *Combinatory Analysis*, Vol. 2, Cambridge University Press, London. (Reprinted by Chelsea, New York, 1960.)

Malik, H. J. (1969) Distribution of the sum of truncated binomial variates, *Canadian Mathematical Bulletin*, **12**, 334-336.

Mantel, N. (1974) Approaches to a health research occupancy problem, *Biometrics*, **30**, 355-362.

Mantel, N. and Pasternack, B. S. (1968) A class of occupancy problems, *American Statistician*, **22** (2), 23-24.

McGovern, W. E. (1980) Missing die probabilities, expected die production and the index figure, *The American Numismatic Society Museum Notes*, **25**, 209-223.

McGuire, J. U., Brindley, T. A. and Bancroft, T. A. (1957) The distribution of European corn borer *Pyrausta Nubilalis* (Hbn.) in field corn, *Biometrics*, **13**, 65-78; Errata and extensions, *Biometrics* **14** (1958), 432-434.

McKendrich, A. G. (1926) Applications of mathematics to medical problems, *Proceedings of the Edinburgh Mathematical Society*, **44**, 98-130.

Medhi, J. (1973) On a generalised Stirling distribution, *Metrika*, **20**, 215-218.

Medhi, J. (1975) On the convolutions of left-truncated generalised negative binomial and Poisson variables, *Sankhyā, Series B*, **37**, 293-299.

Medhi, J. and Borah, M. (1984) On generalized Gegenbauer polynomials and associated probabilities, *Sankhyā, Series B*, **46**, 157-165.

Mertz, D. B. and Davies, R. B. (1968) Cannibalism of the pupal stage by adult flour beetles: An experiment and a stochastic model, *Biometrics*, **24**, 247-275.

Milne-Thomson, L. M. (1933) *The Calculus of Finite Differences*, MacMillan, London.

Mitrinovic, D. S. and Djokovic, D. (1960) Sur une relation de recurrence concernant les nombres de Stirling, *Comptes Rendus de l' Académie des Sciences, Paris*, **250**, 2110-2111.

Montmort, P. R. (1708) *Essai d' Analyse sur les Jeux de Hazards*, Paris.

Nath, H. B. (1974) On the collector's sequential sample size, *Trabajos de Estatistica*, **25**, 85 - 88.

Netto, E. (1927) *Lehrbuch der Kombinatorik*, 2nd ed., Teubner, Berlin. (Reprinted by Chelsea, New York, 1958.)

Neyman, J. (1939) On a new class of "contagious" distributions, applicable in entomology and bacteriology, *Annals of Mathematical Statistics*, **10**, 35 - 57.

Nielsen, N. (1906) *Hambuch der Theorie des Gammafunktion*, Leipzig. (Reprinted by Chelsea, New York, 1966.)

Nielsen, N. (1923) *Traite Elementaire des Nombres de Bernoulli*, Gauthier-Villars, Paris.

Nishimura, K. and Sibuya, M. (1997) Extended Stirling family of discrete probability distributions, *Communications in Statistics - Theory and Methods*, **26**, 1727 - 1744.

Noack, A. (1950) A class of random variables with discrete distributions, *Annals of Mathematical Statistics*, **21**, 127 - 132.

Nörlund, N. E. (1924) *Vorlesungen über Differenzenrechnung*, Berlin. (Reprinted by Chelsea, New York, 1954.)

Okamoto, M. (1952) On a non-parametric test, *Osaka Mathematical Journal*, **4**, 77 - 82.

Park, C. J. (1973) The power series distribution with unknown truncation parameter, *Annals of Statistics*, **1**, 395 - 399.

Patel, S. R. and Jani, P. N. (1977) On minimum variance unbiased estimation of generalized Poisson distribution and decapitated generalized Poisson distribution, *Journal of the Indian Statistical Association*, **15**, 157 - 159.

Patel, Y. C. (1976) Even point estimation and moment estimation in Hermite distribution, *Biometrics*, **32**, 865 - 873.

Patel, Y. C. (1977) Higher moments of moment estimators and even point estimators for the parameters of the Hermite distribution, *Annals of the Institute of Statistical Mathematics*, **29**, 119 - 130.

Patel, Y. C., Shenton, L. R. and Bowman, K. O. (1974) Maximum likelihood estimation for the parameters of the Hermite distribution, *Sankhyā, Series B*, **36**, 154 - 162.

Pathak, P. K. (1964) On estimating the size of a population and its inverse by capture mark method, *Sankhyā, Series A*, **26**, 75 - 80.

Patil, G. P. (1963) Minimum variance unbiased estimation and certain problems of additive number theory, *Annals of Mathematical Statistics*, **34**, 1050 - 1056.

Patil, G. P. (1964) On certain compound Poisson and compound binomial distributions, *Sankhyā, Series A*, **26**, 293 - 294.

Patil, G. P. and Bildikar, S. (1966) On minimum variance unbiased estimation for the logarithmic series distribution, *Sankhyā, Series A*, **28**, 239 - 250.

Patil, G. P. and Joshi, S. W. (1970) Further results on minimum variance unbiased estimation and additive number theory, *Annals of Mathematical Statistics*, **41**, 567 - 575.

Patil, G. P. and Wani, J. K. (1965) On certain structural properties of the logarithmic series distribution and the first type Stirling distribution, *Sankhyā, Series A*, **27**, 271 - 280.

Philippou, A. N. (1986) Distributions and Fibonacci polynomials of order k, longest runs, and reliability of consecutive-k-out-of-n: F systems, in *Fibonacci Numbers and Their Applications* (Eds. A. N. Philippou, G. E. Bergum and A. F. Horadam), pp. 203 - 227, Reidel, Dordrecht, Holland.

Philippou, A. N. and Muwafi, A. A. (1982) Waiting for the k-th consecutive success and the Fibonacci sequence of order k, *Fibonacci Quarterly*, **20**, 28 - 32.

Philippou, A. N., Georgiou, C. and Philippou, G. N. (1983) A generalized geometric distribution and some of its properties, *Statistics & Probability Letters*, **1**, 171 - 175.

Platonov, M. L. (1964) On the numbers of a combinatorial structure (in Russian), *Sibirskii Matematicheskii Zhurnal*, **5**, 1317 - 1325.

Platonov, M. L. (1976) Elementary applications of combinatorial numbers in probability theory, *Theory of Probability and Mathematical Statistics*, **11**, 129 - 137.

Plunkett, I. G. and Jain, G. C. (1975) Three generalized negative binomial distributions, *Biometrische Zeitschrift*, **17**, 279 - 302.

Poincaré, H. (1896) *Calcul des Probabilités*, Gauthier-Villars, Paris.

Poisson, S. D. (1837) *Recherchés sur la Probabilité des Jugements en Matière Criminelle et en Matière Civile, Précédées des Regles Générales du calcul des Probabilités*, Bachelier, Imprimeur-Libraine pour des Mathematiques, la Physique, etc., Paris.

Pólya, G. (1930) Sur quelques points de la théorie des probabilités, *Annales de l' Institut H. Poincaré*, **1**, 117 - 161.

Price, G. B. (1946) Distributions derived from the multinomial expansion, *American Mathematical Monthly*, **53**, 59 - 74.

Quenouille, M. H. (1949) A relation between the logarithmic, Poisson, and negative binomial series, *Biometrics*, **5**, 162 - 164.

Raj, D. and Khamis, S. H. (1958) Some remarks on sampling with replacement, *Annals of Mathematical Statistics*, **29**, 550 - 557.

Rao, C. R. and Chakravarti, I. M. (1956) Some small sample tests of significance for a Poisson distribution, *Biometrics*, **12**, 264 - 282.

Rényi, A. (1951) On composed Poisson distributions, II, *Acta Mathematica Academiae Scientiarum Hungaricae*, **2**, 83 - 96.

Rényi, A. (1962) Theorie des elements sailants d' une suite d' observations, in *Collogium on Combinatorial Methods in Probability Theory*, pp. 104-115, Aarchus University, Aarchus, Denmark.

Richards, P. I. (1968) A generating function (Problem 67 - 18), *SIAM Review*, **10**, 455 - 456.

Riordan, J. (1937) Moment recurrence relations for binomial, Poisson and hypergeometric frequency distributions, *Annals of Mathematical Statistics*, **8**, 103 - 111.

Riordan, J. (1958). *An Introduction to Combinatorial Analysis*, John Wiley & Sons, New York.

Riordan, J. (1968) *Combinatorial Identities*, John Wiley & Sons, New York.

Romanovsky, V. (1934) Su due problemi di distribuzione casuale, *Giornale dell' Istituto Italiano degli Attuari*, **5**, 196 - 218.

Roy, J. and Mitra, S. K. (1957) Unbiased minimum variance estimation in a class of discrete distributions, *Sankhyā, Series A*, **18**, 371 - 378.

Satterthwaite, F. E. (1942) Generalized Poisson distribution, *Annals of Mathematical Statistics*, **13**, 410 - 417.

Schläfli, L. (1867) Erganzung der abhandlung uber die entwickelung des products, *Journal für die reine und angewandte Mathematik*, **67**, 179 - 182.

Schlömilch, O. (1852) Recherches sur les coefficients des faculties analytiques, *Journal für die reine und angewandte Mathematik*, **44**, 344 - 355.

Schlömilch, O. (1895) *Compendium der Hoheren Analysis*, Braunschweig.

Sen, A. and Balakrishnan, N. (1999) Convolution of geometrics and a reliability problem, *Statistics & Probability Letters*, **43**, 421 - 426.

Shanmugan, R. (1984) On central versus factorial moments, *South African Statistical Journal*, **18**, 97 - 110.

Shorrock, R. W. (1972) On record values and record times, *Journal of Applied Probability*, **9**, 316 - 326.

Shumway, R. and Gurland, J. (1960a) A fitting procedure for some generalized Poisson distributions, *Skandinavisk Aktuarietidskrift*, **43**, 87 - 108.

Shumway, R. and Gurland, J. (1960b) Fitting the Poisson binomial distribution, *Biometrics*, **16**, 522 - 533.

Singh, J. (1975) A note on the Stirling distribution of the second kind, *Communications in Statistics*, **4**, 753 - 759.

Singh, J. (1978) A characterization of positive Poisson distribution and its statistical application, *SIAM Journal on Applied Mathematics*, **34**, 545 - 548.

Skellam, J. G. (1952) Studies in statistical ecology I. Spatial pattern, *Biometrika*, **39**, 346 - 362.

Sprott, D. A. (1969) A note on a class of occupancy problems, *American Statistician*, **23** (2), 12 - 13.

Steffensen, J. F. (1927) *Interpolation*, Williams and Wilkins, Baltimore. (Reprinted by Chelsea, New York, 1950.)

Stevens, W. L. (1937) Significance of grouping, *Annals of Eugenics*, **9**, 117 - 129.

Stirling, J. (1730) *Methodus Differentialis sine Tractatus de Summatione et Interpolatione Serierum Infinitarum*, Londini (English translation by F. Holliday with title: *The Differential Method*, London, 1749.)

Strait, P. T. (1971) A note on independence and conditional probabilities, *American Statistician*, **25** (2), 17 - 18.

Sylvester, J. (1883) Note sur le théorèm de Legendre cité dans une Note insérée dans les Comptes Rendus, *Comptes Rendus de l' Académie des Sciences, Paris*, **96**, 463 - 465.

Taga, Y. and Isii, K. (1959) On a stochastic model concerning the pattern of communication-diffusion of news in a social group, *Annals of the Institute of Statistical Mathematics*, **11**, 25 - 43.

Takács, L. (1962) A generalization of the ballot problem and its application in the theory of queues, *Journal of the American Statistical Association*, **57**, 327 - 337.

Takács, L. (1967) On the method of inclusion and exclusion, *Journal of the American Statistical Association*, **62**, 102 - 113.

Takács, L. (1991) A moment convergence theorem, *American Mathematical Monthly*, **98**, 742 - 746.

Tate, R. F. and Goen, R. L. (1958) Minimum variance unbiased estimation for the truncated Poisson distribution, *Annals of Mathematical Statistics* **29**, 755 - 765.

Tauber, S. (1962) On quasi-orthogonal numbers, *American Mathematical Monthly*, **69**, 365 - 372.

Tauber, S. (1965) On generalized Lah-numbers, *Proceedings of the Edinburgh Mathematical Society*, **14**, 229 - 232.

Thomas, M. (1949) A generalization of Poisson's binomial limit for use in ecology, *Biometrika*, **36**, 18 - 25.

Thompson, H. R. (1954) A note on contagious distributions, *Biometrika*, **41**, 268 - 271.

Todhunter, I. (1865) *A History of the Mathematical Theory of Probability*, Macmillan, Cambridge. (Reprinted by Chelsea, New York, 1965.)

Turner, M. E. (1957) The distribution of red blood cells in the hemacytometer, *Biometrics*, **13**, 485 - 495.

Uppuluri, V. R. R. and Carpenter, J. A. (1971) A generalization of the classical occupancy problem, *Journal of Mathematical Analysis and Applications*, **34**, 316 - 324.

Uppuluri, V. R. R. and Patil, G. P. (1983) Waiting times and generalized Fibonacci sequences, *Fibonacci Quarterly*, **21**, 342 - 349.

Vandermonde, N. (1772) Mémoire sur des Irrationneles de différens ordres avec application au cercle, *Histoire de l' Acedemie Royale des Sciences*, Part 1, 489 - 498.

Voinov, V. G. (1986) On Jani's paper on the minimum variance unbiased estimation for a left-truncated modified power series distribution, *Sankhyā, Series B*, **48**, 144 - 150.

Ward, M. (1934) The representation of the Stirling's numbers and the Stirling's polynomials as sums of factorials, *American Journal of Mathematics*, **56**, 87 - 95.

Watson, H. W. and Galton, F. (1874) On the probability of extinction of families, *Journal of the Anthropological Institute*, **4**, 138 - 144.

Woodbury, M. A. (1949) On a probability distribution, *Annals of Mathematical Statistics*, **20**, 311 - 313.

Williams, C. B. (1947) The logarithmic series and its application to biological problems, *Journal of Ecology*, **34**, 253 - 272.

White, C. (1971) The committee problem, *American Statistician*, **25**(4), 25 - 26.

Worpitzky, J. (1883) Studien über die Bernoullischen and Eulerschen Zahlen, *Journal für die reine und angewandte Mathematik*, **94**, 203 - 232.

Yamato, H. (1990) A probabilistic approach to Stirling numbers of the first kind, *Communications in Statistics - Theory and Methods*, **19**, 3915 - 3923.

Author Index

Abramowitz, M., 114, 383
Aczél, J., 327, 392
Ahuja, J. C., 275–276, 383–384, 389
Aki, S., 330, 384
Arfwedson, G., 203, 384
Arnold, B. C., 245, 384
Balakrishnan, N., 151, 203, 384, 394, 398
Bancroft, T. A., 329, 395
Bartholomew, D. J., 113, 384
Barton, D. E., 113–114, 202, 244, 384, 388
Beall, G., 328, 384
Bell, E. T., 114, 318, 384
Berg, S., 169, 203–204, 245, 384–385
Bernstein, L., 114, 385
Bildikar, S., 275, 397
Blissard, J., 113, 385
Bonferroni, C. E., 61, 385
Boole, G., 113, 385
Borah, M., 329, 395
Bowman, K. O., 328, 396
Brindley, T. A., 329, 395
Broder, A. Z., 114, 385
Cacoullos, T., 275, 385
Carlitz, L., 113–114, 150, 385–386
Carpenter, J. A., 204, 400
Castagnetto, L., 328, 386
Catcheside, D. G., 328, 386
Cauchy, A., 15, 113, 386
Cayley, A., 113, 386
Cerasoli, M., 203, 244, 386

Cernuschi, F., 328, 386
Chakravarti, I. M., 203, 398
Charalambides, Ch. A., 60–61, 114, 150–151, 202–203, 244, 275–276, 321, 327, 330, 385–387
Chihara, T. S., 329, 388
Chistyakov, V. P., 204, 394
Chung, K. L., 61, 388
Comtet, L., 60, 113–114, 150, 321, 388
Consul, P. C., 276, 394
Craig, C. C., 203, 388
Cresswell, W. L., 328, 388
Crippa, D., 150, 388
Csörgo, M., 183, 388
Darroch, J. N., 245, 388
David, F. N., 113–114, 202–203, 244, 384, 388
Davies, P., 204, 389
Davies, R. B., 204, 395
De Moivre, A., 60, 73, 203, 329, 389
Djokovic, D., 113–114, 390, 395
Douglas, J. B., 327–328, 389
Enneking, E. A., 275–276, 384, 389
Euler, L., 60, 389
Evans, D. A., 329, 389
Faà di Bruno, F., 318, 389
Fang, K.-T., 202–203, 389
Feller, W., 114, 281, 327, 329, 389
Fisher, R. A., 114, 389
Fréchet, M., 61, 390
Freund, J. E., 203, 390

Froggatt, P., 328, 388
Galliher, H. P., 327, 329, 390
Galton, F., 327, 400
Garsia, A. M., 150, 390
Gart, J. J., 276, 390
Georgiou, C., 330, 397
Gittelsohn, A. M., 204, 390
Glasser, G. J., 245, 390
Glaymann, M., 114, 390
Goen, R. L., 275, 399
Goodman, L. A., 203, 245, 390
Gould, H. W., 113–114, 150, 390
Gross, A. J., 275, 390
Grunert, J. A., 113, 391
Gupta, H., 114, 391
Gupta, R. C., 276, 391
Gupta, R. P., 329, 391
Gurland, J., 114, 281, 327–329, 391, 393, 399
Guttman, I., 183, 388
Hüsler, J., 245, 392
Hahn, W., 150, 391
Harkness, W. L., 203, 391
Harris, B., 203, 245, 391
Hill, B. M., 203, 392
Hirano, K., 330, 384, 392
Holst, L., 204, 245, 392
Irwin, J. O., 328, 392
Isii, K., 113, 399
Jánossy, L., 327, 392
Jain, G. C., 329, 391, 397
Janardan, K. G., 276, 392
Jani, P. N., 276, 392, 396
Janson, S., 245, 392
Johnson, N. L., 203–204, 244–245, 328, 392
Jones, H. L., 203, 393
Jordan, C., 60, 393
Jordan, Ch., 60, 73, 113–114, 150, 393
Joshi, S. W., 275, 393, 397
Karanicoloff, Chr., 114, 393
Karlin, S., 113, 393
Katti, S. K., 329, 393
Kemp, A. W., 203, 328–329, 392–394
Kemp, C. D., 327–329, 393–394
Kendall, D. G., 328, 394
Kendall, M. G., 114, 388
Khamis, S. H., 203, 245, 398
Khatri, C. G., 327–328, 394
Kolchin, V. F., 204, 394
Kotz, S., 203–204, 244–245, 328, 392, 394
Koutras, M., 114, 387, 394
Kuboki, H., 330, 384
Kullback, S., 203, 394
Kumar, A., 275–276, 394
Lüders, R., 328, 395
Lagrange, L., 113, 394
Lah, I., 114, 394

Laplace, P. S., 113, 394
Leite, J. G., 114, 204, 245, 394–395
MacMahon, P. A., 60, 159, 203, 395
Malik, H. J., 275, 395
Mantel, N., 204, 395
McGovern, W. E., 203, 395
McGregor, J., 113, 393
McGuire, J. U., 329, 395
McKendrich, A. G., 328, 395
Medhi, J., 275–276, 329, 395
Mertz, D. B., 204, 395
Milne-Thomson, L. M., 113, 395
Mitra, S. K., 275, 398
Mitrinovic, D. S., 113, 395
Montmort, P. R., 60, 73, 395
Morse, P. M., 327, 329, 390
Muwafi, A. A., 329, 397
Nörlund, N. E., 113–114, 396
Nath, H. B., 245, 396
Netto, E., 113, 396
Nevzorov, V. B., 151, 384
Neyman, J., 328, 396
Nielsen, N., 74, 113–114, 396
Nishimura, K., 204, 245, 396
Noack, A., 168, 396
Okamoto, M., 203, 396
Pólya, G., 329, 397
Park, C. J., 275, 393, 396
Pasternack, B. S., 204, 395
Patel, I. R., 327, 394
Patel, S. R., 276, 396
Patel, Y. C., 328, 396
Pathak, P. K., 204, 396
Patil, G. P., 275–276, 328–329, 396–397, 400
Pereira, C. A. de B., 114, 204, 245, 394–395
Philippou, A. N., 329–330, 397
Philippou, G. N., 330, 397
Platonov, M. L., 150–151, 397
Plunkett, I. G., 329, 397
Poincaré, H., 60, 397
Poisson, S. D., 68, 151, 397
Pozner, A. N., 203, 390
Price, G. B., 203, 397
Quenouille, M. H., 328, 397
Rényi, A., 113, 327, 392, 398
Raj, D., 203, 245, 398
Rao, C. R., 203, 398
Remmel, J. B., 150, 390
Rescia, R. R., 328, 384
Richards, P. I., 203, 398
Riordan, J., 60, 113–114, 150, 159, 203, 321, 398
Rodrigues, F. W., 245, 395
Romanovsky, V., 203, 398
Roy, J., 275, 398
Satterthwaite, F. E., 327, 398
Schläfli, L., 113, 398

Schlömilch, O., 86, 113–114, 398
Sen, A., 151, 398
Sevastyanov, B. A., 204, 394
Shanmugan, R., 114, 398
Shenton, L. R., 328, 396
Shorrock, R. W., 113, 398
Shumway, R., 114, 329, 399
Sibuya, M., 204, 245, 396
Simon, K., 150, 388
Simond, M., 327, 329, 390
Singh, J., 114, 151, 276, 387, 391, 399
Skellam, J. G., 327, 329, 399
Sprott, D. A., 204, 399
Steffensen, J. F., 113, 399
Stegun, I. A., 114, 383
Stevens, W. L., 203, 399
Stirling, J., 73, 113–114, 399
Strait, P. T., 64, 399
Sylvester, J., 60, 399
Taga, Y., 113, 399

Takács, L., 61, 68, 399
Tate, R. F., 275, 399
Tauber, S., 150, 400
Thomas, M., 328, 400
Thompson, H. R., 327, 400
Todhunter, I., 329, 400
Trunz, P., 150, 388
Turner, M. E., 203, 400
Uppuluri, V. R. R., 204, 329, 400
Vandermonde, N., 14, 400
Voinov, V. G., 276, 400
Wani, J. K., 276, 397
Ward, M., 150, 400
Watson, H. W., 327, 400
White, C., 204, 400
Williams, C. B., 328, 400
Woodbury, M. A., 151, 400
Worpitzky, J., 113, 400
Yamato, H., 114, 400
Yates, F., 114, 389

Subject Index

Addition principle, 5
Aggregate insurance claims
   binomial moments of, 285
   probability function of, 285
Ascending binomial moments, 48
   expressed by binomial moments, 50
Ascending factorial moments, 48
Associated generalized factorial coefficients, 131
   generating function for, 131
   recurrence relation for, 132
   table of, 133
r-associated generalized factorial coefficients, 131
   absolute, 134
      application of, 134, 209
      generating function for, 134
   application of, 133, 207, 209
   expressed by multiple sum, 152
   generating function for, 131
   recurrence relation for, 131
Associated Stirling numbers of the first kind, 124
   generating function for, 123
   signless, 124
      generating function for, 124
      recurrence relation for, 127
      table of, 128
r-associated Stirling numbers of the first kind, 125
   generating function for, 124
   signless, 125
      application of, 264, 273

      expressed by multiple sum, 151
      generating function for, 125
      in enumeration of permutations by cycles, 138
      recurrence relation for, 125
Associated Stirling numbers of the second kind, 124
   generating function for, 124
   recurrence relation for, 127
   table of, 128
r-associated Stirling numbers of the second kind, 125
   application of, 129, 207–208, 265, 274
   expressed by multiple sum, 151–152
   generating function for, 124
   in enumeration of partitions by subsets, 136
   recurrence relation for, 126
Attraction parameter in mixing, 286
Automobile accidents model, 297
Ballot formula, general, 68
Bayes' theorem, 21
Bell numbers, 88
   in enumeration of partitions by subsets, 110
   table of, 90
Bernoulli damage process, 291
Bernoulli numbers
   expressed by Stirling numbers, 117
   generating function for, 117
   recurrence relation for, 117

405

SUBJECT INDEX

Bernoulli trials
  with varying success probability, 68–69, 81, 145–146, 149
Binomial coefficient, 15
  generalized, 15
q-binomial coefficient, 140
Binomial distribution
  as compounding distribution, 290
  generalized, 68
  zero truncated
    convolution of, 254, 260
    UMVU estimation for, 269
Binomial formula, 12
  general, 12
q-binomial formula, 153–154
Binomial moment generating function, 56
Binomial moments, 31, 47
  expressed by ascending binomial moments, 50
  expressed by sum of probabilities, 31
Bivariate
  distribution function, 41
  generating function, general form of, 55
  inclusion and exclusion formula, 69–70
  probability function, 42
  probability generating function, 56
Bonferroni's inequalities, 36
  classical, 38
Bose-Einstein model, 168, 178
Bus driver accidents
  binomial moments of, 290
  probability function of, 289
Candidates participate, in at least one committee
  probability function of, 202
Capture-recapture sampling, 186
  multiple, 199
  sequential, 238
Cars involved in accidents
  distribution of, 297
Cartesian product, 2
Cauchy's formula, 15
q-Cauchy formula, 154
Chebyshev distribution
  of the first kind, 310
    binomial moments of, 310
    probability function of, 310
  of the first kind, generalized, 334
    binomial moments of, 333
    probability function of, 333
  of the second kind, 315
    binomial moments of, 315
    probability function of, 315
Chebyshev polynomial
  of the first kind, 309
    generalized, 334
    modified, 309
  of the second kind, 314

    modified, 314
Coincidences
  multiple, 67
  problem of, 35
  generalized, 66
Complementary waiting-time occupancy distribution
  classical
    probability function of, 219, 229
  classical, extended
    probability function of, 249
  general probabilities, extended
    probability function of, 249
  general probabilities
    probability function of, 225
  probability function of
    under Bose-Einstein model, 246
  pseudo-contagious
    probability function of, 236, 245
  restricted
    probability function of, 220, 233
  restricted, extended
    probability function of, 249
Compound binomial distribution
  binomial moments of, 326, 335
  probability function of, 326, 335
  with binomial compounding, 335
Compound distribution, 281
  binomial moments of, 284, 323
  connection with mixture distribution, 286, 340
  probability function of, 283, 323
  with binomial compounding
    binomial moments of, 291
    probability function of, 290
  with logarithmic compounding
    binomial moments of, 295
    probability function of, 295
  with negative binomial compounding
    binomial moments of, 293
    probability function of, 293
  with Poisson compounding
    binomial moments of, 287
    probability function of, 287
Compound logarithmic distribution
  binomial moments of, 307, 325
  probability function of, 307, 325
  with Bernoulli compounding, 291
  with binomial compounding, 333–334
Compound negative binomial distribution
  binomial moments of, 312, 326
  probability function of, 312, 326
  with binomial compounding, 336
  with Poisson compounding, 336
Compound Poisson distribution, 59
  binomial moments of, 297, 324
  probability function of, 297, 324

with binomial compounding, 297, 300
with geometric compounding, 302, 330
with logarithmic compounding, 295
with negative binomial compounding, 293
with Poisson compounding, 287
Compound Poisson process, 331
Compounded distribution, 281
Compounding distribution, 281
Conditional
    expected value, 44
    independence, 24
        and exchangeability, 64
    probability, 19
    probability function, 43
    standard deviation, 46
    variance, 46
Convolution of discrete distribution
    binomial moments of, 322
    probability function of, 321
Convolution of generalized logarithmic
        distribution
    probability function of, 279
Convolution of left truncated
    discrete distribution
        probability function of, 262
    geometric distribution
        probability function of, 278
    logarithmic distribution
        probability function of, 264
    Poisson distribution
        probability function of, 265
    power series distribution
        probability function of, 264
Convolution of logarithmic distribution
    binomial moments of, 260
    modal property of, 276
    probability function of, 259, 276
Convolution of right truncated
    discrete distribution
        probability function of, 265
    power series distribution
        probability function of, 266
Convolution of truncated discrete distribution
    connection with occupancy distribution, 253
Convolution of zero truncated
    binomial distribution
        binomial moments of, 260
        probability function of, 255, 260
    discrete distribution
        binomial moments of, 255
        probability function of, 255, 342
    generalized negative binomial distribution
        probability function of, 280
    generalized Poisson distribution
        probability function of, 279
    linear function Poisson distribution
        probability function of, 280
    negative binomial distribution
        binomial moments of, 261
        probability function of, 261, 276
        recurrence relation for, 278
    Poisson distribution
        binomial moments of, 258
        modal property of, 277
        probability function of, 254, 258
        recurrence relation for, 277
    power series distribution
        binomial moments of, 257
        probability function of, 257
Correlation coefficient, 46
Countable mixtures model, 286
Coupon collector's problem, 107, 233
Covariance, 46
De Morgan's formulae, 4
Derivative(s)
    of factorials, 76
    of generalized factorials, 141
    operator, 12
        as a limit of difference operator, 12
        expressed by difference operator, 13
Difference equation, 15
Difference(s), 11
    of generalized factorials, 98, 105
    of powers, 76
    operator, 11
        expressed by derivative operator, 13
Displacement operator, 12
Distribution function, 40
    bivariate, 41
    marginal, 41
Distribution of shares, 67
Divided differences, 141
    of generalized factorials, 148
    of powers, 142
Division(s) of a finite set
    definition of, 108
    enumeration of, 108
Empty squares, in hemacytometer counts
    probability function of, 177
Entomological stochastic model, 293
Equal sets, 2
Equivalent sets, 3
Event, 16
    dependent, 25
Events
    conditionally independent, 24
    dependent but conditionally independent, 24
    exchangeable, 30
    independent, 22
    independent but not conditionally independent, 24
    pairwise but not completely independent, 22

SUBJECT INDEX

Expansion of
  ascending factorial into factorials, 97
  factorial into powers, 74
  generalized factorial into factorials, 96
  generalized factorial into generalized factorials with different increments, 146
  generalized factorial into powers, 139
  noncentral ascending factorial into factorials, 97
  noncentral ascending factorial into powers, 75
  noncentral factorial into powers, 75
  noncentral generalized factorial into factorials, 97
  power into factorials, 74
  power into generalized factorials, 139
  power into noncentral factorials, 75
  reciprocal factorial into reciprocal powers, 94
  reciprocal q-factorial into reciprocal q-factorials, 158
  reciprocal q-factorial into reciprocal q-powers, 156
  reciprocal generalized factorial into reciprocal factorials, 106
  reciprocal generalized factorial into reciprocal generalized factorials with different increments, 149
  reciprocal generalized factorial into reciprocal powers, 145
  reciprocal power into generalized factorials, 145
  reciprocal power into reciprocal factorials, 94
  reciprocal q-power into reciprocal q-factorials, 156
Expected value, 44
  conditional, 44
q-exponential function, 157
Faà di Bruno's formula, 320
Factorial, 13
q-factorial, 140
Factorial
  ascending, 14
    expansion into factorials, 97
  descending, 14
    expansion into powers, 74
  generalized, 14, 96, 139
    expansion into factorials, 96
Factorial moment generating function, 56
Factorial moments, 47
Factorial series distribution, 169
Fermi-Dirac model, 168, 174
Gambler's ruin, 67
Gaussian coefficient, 140
Gegenbauer distribution, 317
  as mixture distribution, 340
  binomial moments of, 316
  generalized, 335–336
  probability function of, 316
Gegenbauer polynomial, 316

generalized, 336
modified, 316
Generalized binomial coefficient, 15
Generalized Chebyshev distribution
  of the first kind, 334
    binomial moments of, 333
    probability function of, 333
Generalized distribution, 281
Generalized factorial, 14, 96, 139
Generalized factorial coefficients, 106
  absolute
    application of, 179, 190, 234, 236, 240, 261, 269, 293–294
  absolute, noncentral, 105
    application of, 120, 179
    as differences of generalized factorials, 105
    recurrence relation for, 105
  application of, 175, 177, 184, 189, 230, 233, 239, 260, 269, 290, 301, 333, 336
  asymptotic expression of, 117
  connection of, 116
  definition of, 97
  expressed by multiple sum, 101
  limiting expression of, 102
  noncentral
    application of, 108, 119–120, 175, 213, 248–249, 280
    as differences of generalized factorials, 98
    definition of, 97
    explicit expression of, 101
    expressed by central, 98–99
    expressed by noncentral Stirling numbers, 99
    generating function for, 100
    orthogonality relation for, 103
    properties of, 119
    recurrence relation for, 104
  recurrence relation for, 105
  table of, 106
Generalized q-factorial coefficients, 147, 157
  noncentral, 147, 157–158
    explicit expression of, 157
    expressed by central, 157
    expressed by noncentral q-Stirling numbers, 157
    limiting expression of, 158
    recurrence relation for, 157
Generalized factorial
  expansion into factorials, 96
Generalized Gegenbauer distribution, 336
  binomial moments of, 335
  probability function of, 335
Generalized Hermite distribution, 302
  binomial moments of, 302
  probability function of, 302
Generalized logarithmic distribution
  convolution of, 279

SUBJECT INDEX   409

Generalized negative binomial distribution
   zero truncated
      convolution of, 280
Generalized Poisson distribution
   convolution of, 279
Generating function
   binomial moment, 56
   bivariate, general form of, 55
   factorial moment, 56
   general form of, 55
   probability, 55
Generation size, in branching process
   binomial moments of, 285
   probability function of, 285
Geometric distribution, left truncated
   convolution of, 278
   UMVU estimation for, 279
Geometric distribution of order k, 317
   mean of, 318
   probability function of, 317, 336
   variance of, 318
Geometric progression, sum of terms of, 8
Groups represented, in at least one committee
   probability function of, 201–202
Hanoi tower, transfer of, 11
Hermite distribution, 300
   binomial moments of, 300
   generalized, 302
   probability function of, 300
Hermite polynomial, 299
   generalized, 301
   modified, 299
Inclusion and exclusion principle, 26
   bivariate, general expression of, 69–70
   general expression of, 28
   limiting expression of, 38
Independence
   and conditional probability, 64
   conditional, 24
   stochastic, 22
Indicator random variable, 51
Individuals caught, in multiple capture-recapture
   binomial moments of, 199
   probability function of, 199
Intersection, 3
Inverse relations, 82
Lagrange inversion formula, 115
Laguerre polynomial, 303
Lah numbers
   definition of, 97
   generalized, 146, 149
      as divided differences of generalized factorials, 148
      explicit expression of, 148
      expressed by generalized Stirling numbers, 148
      in non-identical Bernoulli trials, 150
      orthogonality relation for, 149
      recurrence relation for, 148–149
   noncentral
      definition of, 97
   signless
      definition of, 97
      explicit expression of, 98
   signless, noncentral
      definition of, 97
      explicit expression of, 98
q-Lah numbers, 147
   noncentral, 147, 158
   signless
      explicit expression of, 158
   signless, noncentral, 158
      explicit expression of, 158
Linear function Poisson distribution
   zero truncated
      convolution of, 280
Logarithmic distribution
   as compounding distribution, 294
   convolution of, 259
   left truncated
      convolution of, 264
      UMVU estimation for, 273, 277
   modal property of the convolution of, 276
   recurrence relation for the convolution of, 276
   UMVU estimation for, 268, 277
   with zeros
      as compound distribution, 291
      as mixture Bernoulli distribution, 338
Logarithmic distribution of order k, 311
   mean of, 311
   probability function of, 311, 334
   variance of, 311
Lottery numbers
   drawn in r drawings
      binomial moments of, 201
      probability function of, 201
   in sequential drawings
      waiting-time probability function for, 244
      waiting-time probability function to exhaust, 250
   not drawn in r drawings
      asymptotic probability function of, 39
      probability function of, 39
Map, 2
   bijective, 3
   injective, 3
   surjective, 3
Marginal
   distribution function, 41
   probability function, 42
Maxwell-Boltzmann model, 168, 170
Missing dies, in a series of coins

estimation of, 188
probability function of, 188
Mixed distribution, 282, 286
Mixing distribution, 282, 286
Mixture Bernoulli distribution
probability function of, 338
with logarithmic mixing, 338
Mixture binomial distribution
probability function of, 339
with Poisson mixing, 339
Mixture distribution, 282, 286
connection with compound distribution, 286
Mixture negative binomial distribution
probability function of, 339
with logarithmic mixing, 340
Mixture Poisson distribution
probability function of, 337
with Bernoulli mixing
probability function of, 337
with binomial mixing, 337
with geometric mixing, 338
with logarithmic mixing, 338
with negative binomial mixing, 338
Moments
ascending binomial, 48
ascending factorial, 48
binomial, 47
central, 47
definition of, 47
factorial, 47
connection with ordinary moments, 83
ordinary (power), 47
connection with factorial moments, 83
Montmort-Moivre problem, 58
Multiple coincidences, 67
Multiplication principle, 8
Multiplicative theorem, 19
Nörlund-Bernoulli numbers
expressed by Stirling numbers, 118
generating function for, 117
recurrence relation for, 117
Negative binomial distribution
as compound distribution, 295
as compounding distribution, 292
as mixture distribution, 340
generalized, 69
left truncated
UMVU estimation for, 278
zero truncated
convolution of, 261, 276
recurrence relation for the convolution of, 278
UMVU estimation for, 269, 278
q-negative binomial formula, 154
Negative trinomial distribution
generalized, 72

Newton's binomial formula, 12
general, 12
Neyman type A distribution, 287
as mixture Poisson, 337
binomial moments of, 288
extended (short), 290
binomial moments of, 290, 333
probability function of, 289, 333
probability function of, 288
q-number, 140
Occupancy distribution
Bernoulli probabilities
binomial moments of, 197
probability function of, 197
binomial moments of, 166, 169
binomial moments of the number of urns
occupied by i balls each, 165
under Bose-Einstein model, 209
binomial moments of the number of urns
occupied by at least i balls each, 207
binomial probabilities
binomial moments of, 198
probability function of, 198
bivariate
binomial moments of, 214
probability function of, 214
classical
binomial moments of, 53, 162, 171
binomial moments of the number of urns
occupied by i balls each, 170, 204
binomial moments of the number of urns
occupied by at least i balls each, 207
distribution function of, 53
expected value of, 171
expected value of the number of urns
occupied by i balls each, 205
probability function of, 53, 162, 171
probability function of the number of urns
occupied by i balls each, 170
probability generating function of, 207
variance of, 172
variance of the number of urns occupied by i
balls each, 205
classical, bivariate
binomial moments of, 214
probability function of, 214
classical, extended
binomial moments of, 213
probability function of, 213
connection with convolution of truncated
discrete distribution, 253
distribution function of, 167
extended
binomial moments of, 213
binomial moments of the number of target
urns occupied by i balls each, 212

probability function of, 213
probability function of the number of target
  urns occupied by i balls each, 212
probability generating function for, 212
general probabilities
  binomial moments of, 194–195
  distribution function of, 214
  probability function of, 193, 195
geometric probabilities
  binomial moments of, 196
  probability function of, 196
negative binomial probabilities
  binomial moments of, 198
  probability function of, 198
probability function of, 166, 169, 342
probability function of the number of pupae
  prey surviving from predators attacks, 211
probability function of the number of urns
  occupied by i balls each, 165
  under Bose-Einstein model, 209
probability generating function of, 206
pseudo-contagious
  binomial moments of, 164, 179
  binomial moments of the number of urns
    occupied by i balls each, 179, 205
  expected value of, 179
  probability function of, 164, 179
  probability function of the number of urns
    occupied by i balls each, 178
  variance of, 179
randomized
  probability function of, 212
restricted
  binomial moments of, 54, 163, 175
  binomial moments of the number of urns
    occupied by i balls each, 175, 205
  binomial moments of the number of urns
    occupied by at least i balls each, 207
  distribution function of, 54
  expected value of, 175
  expected value of the number of urns
    occupied by i balls each, 206
  probability function of, 54, 163, 175
  probability function of the number of urns
    occupied by i balls each, 175
  probability generating function of, 207
  variance of, 176
  variance of the number of urns occupied by i
    balls each, 206
restricted, extended
  binomial moments of, 214
  probability function of, 213
restricted Bose-Einstein
  binomial moments of, 181
  binomial moments of the number of urns
    occupied by i balls each, 211

expected value of, 182
probability function of, 181
probability function of the number of urns
  occupied by i balls each, 210
probability generating function of, 207
variance of, 182
tail probability of order statistic of occupancy
  numbers
  under Bose-Einstein model, 210
tail probability of the number of urns occupied
  by i balls each, 165
Occupancy model, 160
  classical, 161, 170
    joint probability function of occupancy
      numbers, 161
  in statistical physics, 167
  pseudo-contagious, 163, 178
    joint probability function of occupancy
      numbers, 163
  restricted, 162, 174
    joint probability function of occupancy
      numbers, 162
Occupancy numbers, 159
  generating function of distribution function of
    the maximum of, 208
    in classical occupancy model, 208
    in pseudo-contagious occupancy model, 209
    in restricted occupancy model, 208
  generating function of distribution function of
    the minimum of, 208
    in classical occupancy model, 208
    in pseudo-contagious occupancy model, 209
    in restricted occupancy model, 208
  joint probability function of
    in classical occupancy model, 161
    in pseudo-contagious occupancy model, 163
    in restricted occupancy model, 162
Operator $\Psi = t\Delta$, powers of
  expressed by powers of Δ, 85
Operator $\Theta = tD$, powers of
  expressed by powers of D, 84
Ordered n-tuple, 2
Ordered pair, 2
Orthogonality relation
  for generalized Lah numbers, 149
  for generalized Stirling numbers, 142
  for noncentral generalized factorial coefficients,
    103
  for noncentral Stirling numbers, 79
  for noncentral q-Stirling numbers, 155
Pólya-Aeppli distribution, 304
  binomial moments of, 304, 331
  probability function of, 304, 331
Partition polynomial, 320
  exponential Bell, 319
    generating function of, 320

logarithmic, 321
  generating function of, 321
partial Bell, 319
  generating function of, 320
potential, 321
  generating function of, 321
Partition(s) of a finite set
  definition of, 108
  enumeration by subsets of, 109
  universal generating function for, 135
Permutations of a finite set
  decomposed into cycles
    universal generating function for, 138
  enumeration by cycles of, 111
Poincaré formula, 26
Poisson distribution of order k, 306
  binomial moments of, 305, 332
  mean of, 306
  probability function of, 305, 332
  variance of, 306
Poisson distribution
  as compounding distribution, 287
  left truncated
    convolution of, 264
    UMVU estimation for, 274, 278
  zero truncated
    convolution of, 253, 258
    modal property of the convolution of, 277
    recurrence relation for the convolution of, 277
    UMVU estimation for, 268, 277
Positive points, assigned in product ratings
  binomial moments of, 182
  probability function of, 182
Power series distribution, 168, 185, 236, 251, 256
  convolution of, 168, 185, 236, 256
  left truncated, 263, 270
    convolution of, 264
    UMVU estimation for, 272
  right truncated, 266
    convolution of, 266
  zero truncated, 251, 257, 267
    convolution of, 257
    UMVU estimation for, 268
Power set, 2
Principle of
  addition, 5
  inclusion and exclusion, 26
    general expression of, 28
    limiting expression of, 38
  multiplication, 8
Probability
  classical, 18
  conditional, 19
  definition of, 17
Probability function, 41

bivariate, 42
conditional, 43
expressed by binomial moments, 49
marginal, 42
Probability generating function, 55
Random experiment(s), 16
  n-dimensional (compound), 25
  independent, 25
Random occupancy model, 160
  classical, 161, 170
    probability function of random occupancy numbers, 161
  extended, 212
  general, 191
  pseudo-contagious, 163, 178
    probability function of random occupancy numbers, 163
  restricted, 162, 174
    probability function of random occupancy numbers, 162
  with random number of urns, 283
Random occupancy numbers, 160
  probability function of
    in classical random occupancy model, 161
    in pseudo-contagious random occupancy model, 163
    in restricted random occupancy model, 162
Random variable
  ascending binomial moments of, 48
  ascending factorial moments of, 48
  binomial moments of, 47
  central moments of, 47
  conditional expected value of, 44
  conditional standard deviation of, 46
  conditional variance of, 46
  definition of, 40
  discrete, 41
  distribution function of, 40
  expected value of, 44
  factorial moments of, 47
  indicator, 51
  marginal distribution function of, 41
  marginal probability function of, 42
  moments of, 47
  ordinary moments of, 47
  probability function of, 41
  probability generating function of, 55
  standard deviation of, 46
  variance of, 46
Random variables
  correlation coefficient of, 46
  covariance of, 46
  definition of, 40
  independent, 41
  joint distribution function of, 41
  joint probability function of, 42

joint probability generating function of, 56
probability function of a function of, 43
Recurrence relation
  definition of, 9
  linear, 9
  linear complete
    general solution of, 11
    particular solution of, 11
  linear homogeneous, 9
    general solution of, 10
  linear with constant coefficients, 9
  solution of, 10
Sample points, 16
Sample space, 16
  partition of, 20
Sampling with replacement and addition, 190
  sequential, 240
Sampling without replacement, 189
  sequential, 239
Schlömilch's formula for Stirling numbers, 115
Sequence of trials, 25
Sequential occupancy model
  classical, 217
  pseudo-contagious, 245
  restricted, 219
Sequential random occupancy model, 216
  extended, 247
  reduced, 242
Set, 1
  cardinal of, 4
  complement of, 4
  countable, 3
  empty, 1
  finite, 3
  infinite, 3
    countably, 3
  null, 1
  uncountable, 3
  universal, 1
Sets
  difference of, 4
  disjoint, 4
  equal, 2
  equivalent, 3
  intersection of, 3
  union of, 3
Shift operator, 12
Societies represented, in central committee
  binomial moments of, 177
  probability function of, 177
Standard deviation, 46
  conditional, 46
Stirling numbers of the first kind
  as moment connection constants, 83
  definition of, 75
  generalized, 139

    as derivatives of generalized factorials, 141
  noncentral
    as derivatives of factorials, 76
    as moment connection constants, 84
    definition of, 75
    generating function for, 79–80
    properties of, 118–119
    recurrence relation for, 88
  recurrence relation for, 115
  Schlömilch's formula for, 115
  signless
    application of, 259, 268, 295
    asymptotic expression of, 117
    definition of, 75
    expressed by multiple sum, 78–79, 86–87
    in enumeration of permutations by cycles, 111
    in records, 116
    recurrence relation for, 90–91
    table of, 90
  signless, generalized, 140, 145
    concavity property of, 153
    expressed by multiple sum, 140
    in non-identical Bernoulli trials, 145
    recurrence relation for, 143
  signless, noncentral, 76, 94
    application of, 279
    definition of, 75
    expressed by central, 77
    expressed by multiple sum, 78
    generating function for, 81
    in enumeration of permutation by cycles, 112
    in non-identical Bernoulli trials, 82
    properties of, 118
    recurrence relation for, 89
q-Stirling numbers of the first kind, 141, 154
  expressed by multiple sum, 155
  noncentral, 141, 154, 156
    explicit expression of, 156
    expressed by central, 154
    expressed by multiple sum, 155
    generating function for, 156
    recurrence relation for, 155
Stirling numbers of the second kind
  application of, 85, 171, 183, 186, 188, 211, 226, 229, 238, 258, 268, 287–288
  as moment connection constants, 84
  asymptotic expression of, 117
  definition of, 75
  expressed by multiple sum, 86–87, 94
  generalized, 139, 145
    as divided differences of powers, 142
    concavity property of, 153
    explicit expression of, 142
    expressed by multiple sum, 144
    generating function for, 144

in non-identical Bernoulli trials, 146
recurrence relation for, 143–144
in Bell numbers expression, 88
in enumeration of partitions by subsets, 110
noncentral, 94
    application of, 171, 173, 213, 229, 246,
      248–249, 279–280, 289–290, 333
    as differences of powers, 76
    as moment connection constants, 84
    definition of, 75
    explicit expression of, 85
    expressed by central, 77
    expressed by multiple sum, 93
    generating function for, 79–80, 93
    in a Markov chain, 96
    in enumeration of partitions by subsets, 112
    properties of, 118–119
    recurrence relation for, 88
recurrence relation for, 92, 115
table of, 90
q-Stirling numbers of the second kind, 141, 155
    noncentral, 141, 155–156
        explicit expression of, 156
        expressed by central, 155
        expressed by multiple sum, 156
        generating function for, 156–157
        recurrence relation for, 155–156
Stirling numbers
    connection of, 115–116
    generalized
        orthogonality relation for, 142
    noncentral
        as coefficients in inverse relations, 83
        orthogonality relation for, 79
    q-Stirling numbers
        orthogonality relation for, 155
Stochastic model, 16
Stopped sum distribution, 281
Subset, 1
Subsets of a finite set, number of, 8
Test
    of empty squares in hemacytometer counts, 177
    of zero frequency
        for continuous population, 183
        for discrete population, 183
Thomas distribution, 332
    binomial moments of, 333
    probability function of, 332
Total probability theorem, 20
    conditional, 62
Trinomial distribution
    generalized, 71
UMVU estimation
    for left truncated
        geometric distribution, 279
        logarithmic distribution, 273, 277

        negative binomial distribution, 278
        Poisson distribution, 274, 278
        power series distribution, 272
    for logarithmic distribution, 268, 277
    for zero truncated
        binomial distribution, 269
        negative binomial distribution, 269, 278
        Poisson distribution, 268, 277
        power series distribution, 268
    of population classes, based on
        fixed size random sample, 186
        sampling with replacement and addition, 190
        sampling without replacement, 189
        sequential sampling, 238
        sequential sampling with replacement and
            addition, 240
        sequential sampling without replacement,
            239
    of population size, based on
        capture-recapture sampling, 186
        multiple capture-recapture sampling, 199
        sequential capture-recapture sampling, 238
Union, 3
Updated prior probabilities, 63
Usable numbers
    in a sample of random numbers
        binomial moments of, 173
        probability function of, 173
    in a sequential sample of random numbers
        complementary waiting-time probability
            function of, 229
        waiting-time probability function of, 229
Vandermonde's formula, 14
q-Vandermonde formula, 154
Variance, 46
    conditional, 46
Waiting-time occupancy distribution
    Bernoulli probabilities
        binomial moments of, 244
        probability function of, 244
    binomial moments of, 221
        under Bose-Einstein model, 246
    classical
        binomial moments of, 218, 226
        expected value of, 227
        probability function of, 217, 226
        structural property of, 228
        variance of, 228
    classical, extended
        binomial moments of, 248
        probability function of, 248
        structural property of, 248
    general probabilities
        binomial moments of, 242
        probability function of, 242
    general probabilities, extended

SUBJECT INDEX   **415**

  binomial moments of, 247
  probability function of, 247
  structural property of, 248
geometric probabilities
  binomial moments of, 243
  probability function of, 243
probability function of, 221
  under Bose-Einstein model, 245
pseudo-contagious
  binomial moments of, 234, 245
  expected value of, 235
  probability function of, 234, 245
  structural property of, 235
randomized

  binomial moments of, 246
  probability function of, 246
  structural property of, 246
restricted
  binomial moments of, 219, 230
  expected value of, 231
  probability function of, 219, 230
  structural property of, 232
restricted, extended
  binomial moments of, 248
  probability function of, 248
  structural property of, 249
structural property of, 223
  under Bose-Einstein model, 246
Weight function in mixing, 286

WILEY SERIES IN PROBABILITY AND STATISTICS
ESTABLISHED BY WALTER A. SHEWHART AND SAMUEL S. WILKS

Editors: *David J. Balding, Noel A. C. Cressie, Nicholas I. Fisher, Iain M. Johnstone, J. B. Kadane, Geert Molenberghs. Louise M. Ryan, David W. Scott, Adrian F. M. Smith, Jozef L. Teugels*
Editors Emeriti: *Vic Barnett, J. Stuart Hunter, David G. Kendall*

The *Wiley Series in Probability and Statistics* is well established and authoritative. It covers many topics of current research interest in both pure and applied statistics and probability theory. Written by leading statisticians and institutions, the titles span both state-of-the-art developments in the field and classical methods.

Reflecting the wide range of current research in statistics, the series encompasses applied, methodological and theoretical statistics, ranging from applications and new techniques made possible by advances in computerized practice to rigorous treatment of theoretical approaches.

This series provides essential and invaluable reading for all statisticians, whether in academia, industry, government, or research.

ABRAHAM and LEDOLTER · Statistical Methods for Forecasting
AGRESTI · Analysis of Ordinal Categorical Data
AGRESTI · An Introduction to Categorical Data Analysis
AGRESTI · Categorical Data Analysis, *Second Edition*
ALTMAN, GILL, and McDONALD · Numerical Issues in Statistical Computing for the Social Scientist
AMARATUNGA and CABRERA · Exploration and Analysis of DNA Microarray and Protein Array Data
ANDĚL · Mathematics of Chance
ANDERSON · An Introduction to Multivariate Statistical Analysis, *Third Edition*
* ANDERSON · The Statistical Analysis of Time Series
ANDERSON, AUQUIER, HAUCK, OAKES, VANDAELE, and WEISBERG · Statistical Methods for Comparative Studies
ANDERSON and LOYNES · The Teaching of Practical Statistics
ARMITAGE and DAVID (editors) · Advances in Biometry
ARNOLD, BALAKRISHNAN, and NAGARAJA · Records
* ARTHANARI and DODGE · Mathematical Programming in Statistics
* BAILEY · The Elements of Stochastic Processes with Applications to the Natural Sciences
BALAKRISHNAN and KOUTRAS · Runs and Scans with Applications
BARNETT · Comparative Statistical Inference, *Third Edition*
BARNETT and LEWIS · Outliers in Statistical Data, *Third Edition*
BARTOSZYNSKI and NIEWIADOMSKA-BUGAJ · Probability and Statistical Inference
BASILEVSKY · Statistical Factor Analysis and Related Methods: Theory and Applications
BASU and RIGDON · Statistical Methods for the Reliability of Repairable Systems
BATES and WATTS · Nonlinear Regression Analysis and Its Applications
BECHHOFER, SANTNER, and GOLDSMAN · Design and Analysis of Experiments for Statistical Selection, Screening, and Multiple Comparisons
BELSLEY · Conditioning Diagnostics: Collinearity and Weak Data in Regression

*Now available in a lower priced paperback edition in the Wiley Classics Library.
†Now available in a lower priced paperback edition in the Wiley–Interscience Paperback Series.

† BELSLEY, KUH, and WELSCH · Regression Diagnostics: Identifying Influential Data and Sources of Collinearity
BENDAT and PIERSOL · Random Data: Analysis and Measurement Procedures, *Third Edition*
BERRY, CHALONER, and GEWEKE · Bayesian Analysis in Statistics and Econometrics: Essays in Honor of Arnold Zellner
BERNARDO and SMITH · Bayesian Theory
BHAT and MILLER · Elements of Applied Stochastic Processes, *Third Edition*
BHATTACHARYA and WAYMIRE · Stochastic Processes with Applications
† BIEMER, GROVES, LYBERG, MATHIOWETZ, and SUDMAN · Measurement Errors in Surveys
BILLINGSLEY · Convergence of Probability Measures, *Second Edition*
BILLINGSLEY · Probability and Measure, *Third Edition*
BIRKES and DODGE · Alternative Methods of Regression
BLISCHKE AND MURTHY (editors) · Case Studies in Reliability and Maintenance
BLISCHKE AND MURTHY · Reliability: Modeling, Prediction, and Optimization
BLOOMFIELD · Fourier Analysis of Time Series: An Introduction, *Second Edition*
BOLLEN · Structural Equations with Latent Variables
BOROVKOV · Ergodicity and Stability of Stochastic Processes
BOULEAU · Numerical Methods for Stochastic Processes
BOX · Bayesian Inference in Statistical Analysis
BOX · R. A. Fisher, the Life of a Scientist
BOX and DRAPER · Empirical Model-Building and Response Surfaces
* BOX and DRAPER · Evolutionary Operation: A Statistical Method for Process Improvement
BOX, HUNTER, and HUNTER · Statistics for Experimenters: An Introduction to Design, Data Analysis, and Model Building
BOX and LUCEÑO · Statistical Control by Monitoring and Feedback Adjustment
BRANDIMARTE · Numerical Methods in Finance: A MATLAB-Based Introduction
BROWN and HOLLANDER · Statistics: A Biomedical Introduction
BRUNNER, DOMHOF, and LANGER · Nonparametric Analysis of Longitudinal Data in Factorial Experiments
BUCKLEW · Large Deviation Techniques in Decision, Simulation, and Estimation
CAIROLI and DALANG · Sequential Stochastic Optimization
CASTILLO, HADI, BALAKRISHNAN, and SARABIA · Extreme Value and Related Models with Applications in Engineering and Science
CHAN · Time Series: Applications to Finance
CHARALAMBIDES · Combinatorial Methods in Discrete Distributions
CHATTERJEE and HADI · Sensitivity Analysis in Linear Regression
CHATTERJEE and PRICE · Regression Analysis by Example, *Third Edition*
CHERNICK · Bootstrap Methods: A Practitioner's Guide
CHERNICK and FRIIS · Introductory Biostatistics for the Health Sciences
CHILÈS and DELFINER · Geostatistics: Modeling Spatial Uncertainty
CHOW and LIU · Design and Analysis of Clinical Trials: Concepts and Methodologies, *Second Edition*
CLARKE and DISNEY · Probability and Random Processes: A First Course with Applications, *Second Edition*
* COCHRAN and COX · Experimental Designs, *Second Edition*
CONGDON · Applied Bayesian Modelling
CONGDON · Bayesian Statistical Modelling
CONOVER · Practical Nonparametric Statistics, *Third Edition*
COOK · Regression Graphics
COOK and WEISBERG · Applied Regression Including Computing and Graphics

*Now available in a lower priced paperback edition in the Wiley Classics Library.
†Now available in a lower priced paperback edition in the Wiley–Interscience Paperback Series.

COOK and WEISBERG · An Introduction to Regression Graphics
CORNELL · Experiments with Mixtures, Designs, Models, and the Analysis of Mixture Data, *Third Edition*
COVER and THOMAS · Elements of Information Theory
COX · A Handbook of Introductory Statistical Methods
* COX · Planning of Experiments
CRESSIE · Statistics for Spatial Data, *Revised Edition*
CSÖRGŐ and HORVÁTH · Limit Theorems in Change Point Analysis
DANIEL · Applications of Statistics to Industrial Experimentation
DANIEL · Biostatistics: A Foundation for Analysis in the Health Sciences, *Eighth Edition*
* DANIEL · Fitting Equations to Data: Computer Analysis of Multifactor Data, *Second Edition*
DASU and JOHNSON · Exploratory Data Mining and Data Cleaning
DAVID and NAGARAJA · Order Statistics, *Third Edition*
* DEGROOT, FIENBERG, and KADANE · Statistics and the Law
DEL CASTILLO · Statistical Process Adjustment for Quality Control
DeMARIS · Regression with Social Data: Modeling Continuous and Limited Response Variables
DEMIDENKO · Mixed Models: Theory and Applications
DENISON, HOLMES, MALLICK and SMITH · Bayesian Methods for Nonlinear Classification and Regression
DETTE and STUDDEN · The Theory of Canonical Moments with Applications in Statistics, Probability, and Analysis
DEY and MUKERJEE · Fractional Factorial Plans
DILLON and GOLDSTEIN · Multivariate Analysis: Methods and Applications
DODGE · Alternative Methods of Regression
* DODGE and ROMIG · Sampling Inspection Tables, *Second Edition*
* DOOB · Stochastic Processes
DOWDY, WEARDEN, and CHILKO · Statistics for Research, *Third Edition*
DRAPER and SMITH · Applied Regression Analysis, *Third Edition*
DRYDEN and MARDIA · Statistical Shape Analysis
DUDEWICZ and MISHRA · Modern Mathematical Statistics
DUNN and CLARK · Basic Statistics: A Primer for the Biomedical Sciences, *Third Edition*
DUPUIS and ELLIS · A Weak Convergence Approach to the Theory of Large Deviations
* ELANDT-JOHNSON and JOHNSON · Survival Models and Data Analysis
ENDERS · Applied Econometric Time Series
ETHIER and KURTZ · Markov Processes: Characterization and Convergence
EVANS, HASTINGS, and PEACOCK · Statistical Distributions, *Third Edition*
FELLER · An Introduction to Probability Theory and Its Applications, Volume I, *Third Edition,* Revised; Volume II, *Second Edition*
FISHER and VAN BELLE · Biostatistics: A Methodology for the Health Sciences
FITZMAURICE, LAIRD, and WARE · Applied Longitudinal Analysis
* FLEISS · The Design and Analysis of Clinical Experiments
FLEISS · Statistical Methods for Rates and Proportions, *Third Edition*
FLEMING and HARRINGTON · Counting Processes and Survival Analysis
FULLER · Introduction to Statistical Time Series, *Second Edition*
FULLER · Measurement Error Models
GALLANT · Nonlinear Statistical Models
GHOSH, MUKHOPADHYAY, and SEN · Sequential Estimation
GIESBRECHT and GUMPERTZ · Planning, Construction, and Statistical Analysis of Comparative Experiments
GIFI · Nonlinear Multivariate Analysis

*Now available in a lower priced paperback edition in the Wiley Classics Library.
†Now available in a lower priced paperback edition in the Wiley–Interscience Paperback Series.

GIVENS and HOETING · Computational Statistics
GLASSERMAN and YAO · Monotone Structure in Discrete-Event Systems
GNANADESIKAN · Methods for Statistical Data Analysis of Multivariate Observations,
    Second Edition
GOLDSTEIN and LEWIS · Assessment: Problems, Development, and Statistical Issues
GREENWOOD and NIKULIN · A Guide to Chi-Squared Testing
GROSS and HARRIS · Fundamentals of Queueing Theory, *Third Edition*
† GROVES · Survey Errors and Survey Costs
* HAHN and SHAPIRO · Statistical Models in Engineering
HAHN and MEEKER · Statistical Intervals: A Guide for Practitioners
HALD · A History of Probability and Statistics and their Applications Before 1750
HALD · A History of Mathematical Statistics from 1750 to 1930
† HAMPEL · Robust Statistics: The Approach Based on Influence Functions
HANNAN and DEISTLER · The Statistical Theory of Linear Systems
HEIBERGER · Computation for the Analysis of Designed Experiments
HEDAYAT and SINHA · Design and Inference in Finite Population Sampling
HELLER · MACSYMA for Statisticians
HINKELMANN and KEMPTHORNE · Design and Analysis of Experiments, Volume 1:
    Introduction to Experimental Design
HINKELMANN and KEMPTHORNE · Design and Analysis of Experiments, Volume 2:
    Advanced Experimental Design
HOAGLIN, MOSTELLER, and TUKEY · Exploratory Approach to Analysis
    of Variance
HOAGLIN, MOSTELLER, and TUKEY · Exploring Data Tables, Trends and Shapes
* HOAGLIN, MOSTELLER, and TUKEY · Understanding Robust and Exploratory
    Data Analysis
HOCHBERG and TAMHANE · Multiple Comparison Procedures
HOCKING · Methods and Applications of Linear Models: Regression and the Analysis
    of Variance, *Second Edition*
HOEL · Introduction to Mathematical Statistics, *Fifth Edition*
HOGG and KLUGMAN · Loss Distributions
HOLLANDER and WOLFE · Nonparametric Statistical Methods, *Second Edition*
HOSMER and LEMESHOW · Applied Logistic Regression, *Second Edition*
HOSMER and LEMESHOW · Applied Survival Analysis: Regression Modeling of
    Time to Event Data
† HUBER · Robust Statistics
HUBERTY · Applied Discriminant Analysis
HUNT and KENNEDY · Financial Derivatives in Theory and Practice
HUSKOVA, BERAN, and DUPAC · Collected Works of Jaroslav Hajek—
    with Commentary
HUZURBAZAR · Flowgraph Models for Multistate Time-to-Event Data
IMAN and CONOVER · A Modern Approach to Statistics
† JACKSON · A User's Guide to Principle Components
JOHN · Statistical Methods in Engineering and Quality Assurance
JOHNSON · Multivariate Statistical Simulation
JOHNSON and BALAKRISHNAN · Advances in the Theory and Practice of Statistics: A
    Volume in Honor of Samuel Kotz
JOHNSON and BHATTACHARYYA · Statistics: Principles and Methods, *Fifth Edition*
JOHNSON and KOTZ · Distributions in Statistics
JOHNSON and KOTZ (editors) · Leading Personalities in Statistical Sciences: From the
    Seventeenth Century to the Present
JOHNSON, KOTZ, and BALAKRISHNAN · Continuous Univariate Distributions,
    Volume 1, *Second Edition*

---

*Now available in a lower priced paperback edition in the Wiley Classics Library.
†Now available in a lower priced paperback edition in the Wiley–Interscience Paperback Series.

JOHNSON, KOTZ, and BALAKRISHNAN · Continuous Univariate Distributions, Volume 2, *Second Edition*
JOHNSON, KOTZ, and BALAKRISHNAN · Discrete Multivariate Distributions
JOHNSON, KOTZ, and KEMP · Univariate Discrete Distributions, *Second Edition*
JUDGE, GRIFFITHS, HILL, LÜTKEPOHL, and LEE · The Theory and Practice of Econometrics, *Second Edition*
JUREČKOVÁ and SEN · Robust Statistical Procedures: Aymptotics and Interrelations
JUREK and MASON · Operator-Limit Distributions in Probability Theory
KADANE · Bayesian Methods and Ethics in a Clinical Trial Design
KADANE AND SCHUM · A Probabilistic Analysis of the Sacco and Vanzetti Evidence
KALBFLEISCH and PRENTICE · The Statistical Analysis of Failure Time Data, *Second Edition*
KASS and VOS · Geometrical Foundations of Asymptotic Inference
† KAUFMAN and ROUSSEEUW · Finding Groups in Data: An Introduction to Cluster Analysis
KEDEM and FOKIANOS · Regression Models for Time Series Analysis
KENDALL, BARDEN, CARNE, and LE · Shape and Shape Theory
KHURI · Advanced Calculus with Applications in Statistics, *Second Edition*
KHURI, MATHEW, and SINHA · Statistical Tests for Mixed Linear Models
* KISH · Statistical Design for Research
KLEIBER and KOTZ · Statistical Size Distributions in Economics and Actuarial Sciences
KLUGMAN, PANJER, and WILLMOT · Loss Models: From Data to Decisions, *Second Edition*
KLUGMAN, PANJER, and WILLMOT · Solutions Manual to Accompany Loss Models: From Data to Decisions, *Second Edition*
KOTZ, BALAKRISHNAN, and JOHNSON · Continuous Multivariate Distributions, Volume 1, *Second Edition*
KOTZ and JOHNSON (editors) · Encyclopedia of Statistical Sciences: Volumes 1 to 9 with Index
KOTZ and JOHNSON (editors) · Encyclopedia of Statistical Sciences: Supplement Volume
KOTZ, READ, and BANKS (editors) · Encyclopedia of Statistical Sciences: Update Volume 1
KOTZ, READ, and BANKS (editors) · Encyclopedia of Statistical Sciences: Update Volume 2
KOVALENKO, KUZNETZOV, and PEGG · Mathematical Theory of Reliability of Time-Dependent Systems with Practical Applications
LACHIN · Biostatistical Methods: The Assessment of Relative Risks
LAD · Operational Subjective Statistical Methods: A Mathematical, Philosophical, and Historical Introduction
LAMPERTI · Probability: A Survey of the Mathematical Theory, *Second Edition*
LANGE, RYAN, BILLARD, BRILLINGER, CONQUEST, and GREENHOUSE · Case Studies in Biometry
LARSON · Introduction to Probability Theory and Statistical Inference, *Third Edition*
LAWLESS · Statistical Models and Methods for Lifetime Data, *Second Edition*
LAWSON · Statistical Methods in Spatial Epidemiology
LE · Applied Categorical Data Analysis
LE · Applied Survival Analysis
LEE and WANG · Statistical Methods for Survival Data Analysis, *Third Edition*
LePAGE and BILLARD · Exploring the Limits of Bootstrap
LEYLAND and GOLDSTEIN (editors) · Multilevel Modelling of Health Statistics
LIAO · Statistical Group Comparison
LINDVALL · Lectures on the Coupling Method

---

*Now available in a lower priced paperback edition in the Wiley Classics Library.
†Now available in a lower priced paperback edition in the Wiley–Interscience Paperback Series.

LINHART and ZUCCHINI · Model Selection
LITTLE and RUBIN · Statistical Analysis with Missing Data, *Second Edition*
LLOYD · The Statistical Analysis of Categorical Data
MAGNUS and NEUDECKER · Matrix Differential Calculus with Applications in Statistics and Econometrics, *Revised Edition*
MALLER and ZHOU · Survival Analysis with Long Term Survivors
MALLOWS · Design, Data, and Analysis by Some Friends of Cuthbert Daniel
MANN, SCHAFER, and SINGPURWALLA · Methods for Statistical Analysis of Reliability and Life Data
MANTON, WOODBURY, and TOLLEY · Statistical Applications Using Fuzzy Sets
MARCHETTE · Random Graphs for Statistical Pattern Recognition
MARDIA and JUPP · Directional Statistics
MASON, GUNST, and HESS · Statistical Design and Analysis of Experiments with Applications to Engineering and Science, *Second Edition*
McCULLOCH and SEARLE · Generalized, Linear, and Mixed Models
McFADDEN · Management of Data in Clinical Trials
* McLACHLAN · Discriminant Analysis and Statistical Pattern Recognition
McLACHLAN, DO, and AMBROISE · Analyzing Microarray Gene Expression Data
McLACHLAN and KRISHNAN · The EM Algorithm and Extensions
McLACHLAN and PEEL · Finite Mixture Models
McNEIL · Epidemiological Research Methods
MEEKER and ESCOBAR · Statistical Methods for Reliability Data
MEERSCHAERT and SCHEFFLER · Limit Distributions for Sums of Independent Random Vectors: Heavy Tails in Theory and Practice
MICKEY, DUNN, and CLARK · Applied Statistics: Analysis of Variance and Regression, *Third Edition*
* MILLER · Survival Analysis, *Second Edition*
MONTGOMERY, PECK, and VINING · Introduction to Linear Regression Analysis, *Third Edition*
MORGENTHALER and TUKEY · Configural Polysampling: A Route to Practical Robustness
MUIRHEAD · Aspects of Multivariate Statistical Theory
MULLER and STOYAN · Comparison Methods for Stochastic Models and Risks
MURRAY · X-STAT 2.0 Statistical Experimentation, Design Data Analysis, and Nonlinear Optimization
MURTHY, XIE, and JIANG · Weibull Models
MYERS and MONTGOMERY · Response Surface Methodology: Process and Product Optimization Using Designed Experiments, *Second Edition*
MYERS, MONTGOMERY, and VINING · Generalized Linear Models. With Applications in Engineering and the Sciences
† NELSON · Accelerated Testing, Statistical Models, Test Plans, and Data Analyses
† NELSON · Applied Life Data Analysis
NEWMAN · Biostatistical Methods in Epidemiology
OCHI · Applied Probability and Stochastic Processes in Engineering and Physical Sciences
OKABE, BOOTS, SUGIHARA, and CHIU · Spatial Tesselations: Concepts and Applications of Voronoi Diagrams, *Second Edition*
OLIVER and SMITH · Influence Diagrams, Belief Nets and Decision Analysis
PALTA · Quantitative Methods in Population Health: Extensions of Ordinary Regressions
PANKRATZ · Forecasting with Dynamic Regression Models
PANKRATZ · Forecasting with Univariate Box-Jenkins Models: Concepts and Cases
* PARZEN · Modern Probability Theory and Its Applications
PEÑA, TIAO, and TSAY · A Course in Time Series Analysis
PIANTADOSI · Clinical Trials: A Methodologic Perspective

*Now available in a lower priced paperback edition in the Wiley Classics Library.
†Now available in a lower priced paperback edition in the Wiley–Interscience Paperback Series.

PORT · Theoretical Probability for Applications
POURAHMADI · Foundations of Time Series Analysis and Prediction Theory
PRESS · Bayesian Statistics: Principles, Models, and Applications
PRESS · Subjective and Objective Bayesian Statistics, *Second Edition*
PRESS and TANUR · The Subjectivity of Scientists and the Bayesian Approach
PUKELSHEIM · Optimal Experimental Design
PURI, VILAPLANA, and WERTZ · New Perspectives in Theoretical and Applied Statistics
† PUTERMAN · Markov Decision Processes: Discrete Stochastic Dynamic Programming
QIU · Image Processing and Jump Regression Analysis
* RAO · Linear Statistical Inference and Its Applications, *Second Edition*
RAUSAND and HØYLAND · System Reliability Theory: Models, Statistical Methods, and Applications, *Second Edition*
RENCHER · Linear Models in Statistics
RENCHER · Methods of Multivariate Analysis, *Second Edition*
RENCHER · Multivariate Statistical Inference with Applications
* RIPLEY · Spatial Statistics
RIPLEY · Stochastic Simulation
ROBINSON · Practical Strategies for Experimenting
ROHATGI and SALEH · An Introduction to Probability and Statistics, *Second Edition*
ROLSKI, SCHMIDLI, SCHMIDT, and TEUGELS · Stochastic Processes for Insurance and Finance
ROSENBERGER and LACHIN · Randomization in Clinical Trials: Theory and Practice
ROSS · Introduction to Probability and Statistics for Engineers and Scientists
† ROUSSEEUW and LEROY · Robust Regression and Outlier Detection
* RUBIN · Multiple Imputation for Nonresponse in Surveys
RUBINSTEIN · Simulation and the Monte Carlo Method
RUBINSTEIN and MELAMED · Modern Simulation and Modeling
RYAN · Modern Regression Methods
RYAN · Statistical Methods for Quality Improvement, *Second Edition*
SALTELLI, CHAN, and SCOTT (editors) · Sensitivity Analysis
* SCHEFFE · The Analysis of Variance
SCHIMEK · Smoothing and Regression: Approaches, Computation, and Application
SCHOTT · Matrix Analysis for Statistics, *Second Edition*
SCHOUTENS · Levy Processes in Finance: Pricing Financial Derivatives
SCHUSS · Theory and Applications of Stochastic Differential Equations
SCOTT · Multivariate Density Estimation: Theory, Practice, and Visualization
* SEARLE · Linear Models
SEARLE · Linear Models for Unbalanced Data
SEARLE · Matrix Algebra Useful for Statistics
SEARLE, CASELLA, and McCULLOCH · Variance Components
SEARLE and WILLETT · Matrix Algebra for Applied Economics
SEBER and LEE · Linear Regression Analysis, *Second Edition*
† SEBER · Multivariate Observations
† SEBER and WILD · Nonlinear Regression
SENNOTT · Stochastic Dynamic Programming and the Control of Queueing Systems
* SERFLING · Approximation Theorems of Mathematical Statistics
SHAFER and VOVK · Probability and Finance: It's Only a Game!
SILVAPULLE and SEN · Constrained Statistical Inference: Inequality, Order, and Shape Restrictions
SMALL and McLEISH · Hilbert Space Methods in Probability and Statistical Inference
SRIVASTAVA · Methods of Multivariate Statistics
STAPLETON · Linear Statistical Models

*Now available in a lower priced paperback edition in the Wiley Classics Library.
†Now available in a lower priced paperback edition in the Wiley–Interscience Paperback Series.

STAUDTE and SHEATHER · Robust Estimation and Testing
STOYAN, KENDALL, and MECKE · Stochastic Geometry and Its Applications, *Second Edition*
STOYAN and STOYAN · Fractals, Random Shapes and Point Fields: Methods of Geometrical Statistics
STYAN · The Collected Papers of T. W. Anderson: 1943–1985
SUTTON, ABRAMS, JONES, SHELDON, and SONG · Methods for Meta-Analysis in Medical Research
TANAKA · Time Series Analysis: Nonstationary and Noninvertible Distribution Theory
THOMPSON · Empirical Model Building
THOMPSON · Sampling, *Second Edition*
THOMPSON · Simulation: A Modeler's Approach
THOMPSON and SEBER · Adaptive Sampling
THOMPSON, WILLIAMS, and FINDLAY · Models for Investors in Real World Markets
TIAO, BISGAARD, HILL, PEÑA, and STIGLER (editors) · Box on Quality and Discovery: with Design, Control, and Robustness
TIERNEY · LISP-STAT: An Object-Oriented Environment for Statistical Computing and Dynamic Graphics
TSAY · Analysis of Financial Time Series
UPTON and FINGLETON · Spatial Data Analysis by Example, Volume II: Categorical and Directional Data
VAN BELLE · Statistical Rules of Thumb
VAN BELLE, FISHER, HEAGERTY, and LUMLEY · Biostatistics: A Methodology for the Health Sciences, *Second Edition*
VESTRUP · The Theory of Measures and Integration
VIDAKOVIC · Statistical Modeling by Wavelets
VINOD and REAGLE · Preparing for the Worst: Incorporating Downside Risk in Stock Market Investments
WALLER and GOTWAY · Applied Spatial Statistics for Public Health Data
WEERAHANDI · Generalized Inference in Repeated Measures: Exact Methods in MANOVA and Mixed Models
WEISBERG · Applied Linear Regression, *Third Edition*
WELSH · Aspects of Statistical Inference
WESTFALL and YOUNG · Resampling-Based Multiple Testing: Examples and Methods for *p*-Value Adjustment
WHITTAKER · Graphical Models in Applied Multivariate Statistics
WINKER · Optimization Heuristics in Economics: Applications of Threshold Accepting
WONNACOTT and WONNACOTT · Econometrics, *Second Edition*
WOODING · Planning Pharmaceutical Clinical Trials: Basic Statistical Principles
WOODWORTH · Biostatistics: A Bayesian Introduction
WOOLSON and CLARKE · Statistical Methods for the Analysis of Biomedical Data, *Second Edition*
WU and HAMADA · Experiments: Planning, Analysis, and Parameter Design Optimization
YANG · The Construction Theory of Denumerable Markov Processes
* ZELLNER · An Introduction to Bayesian Inference in Econometrics
ZHOU, OBUCHOWSKI, and McCLISH · Statistical Methods in Diagnostic Medicine

---

*Now available in a lower priced paperback edition in the Wiley Classics Library.
†Now available in a lower priced paperback edition in the Wiley–Interscience Paperback Series.